THE GLOBAL CASINO

THE GLOBAL CASINO

AN INTRODUCTION TO ENVIRONMENTAL ISSUES

Third Edition

Nick Middleton

A member of the Hodder Headline Group
LONDON
Distributed in the United States of America by
Oxford University Press Inc., New York

First published in Great Britain in 1995
Second Edition 1999
Third Edition 2003 by Hodder Arnold, a member of the Hodder Headline Group,
338 Euston Road, London NW1 3BH

http://www.arnoldpublishers.com

Distributed in the United States of America by
Oxford University Press Inc.,
198 Madison Avenue, New York, NY10016

British Library Cataloguing in Publication Data
A catalogue record for this book is available from the British Library

Library of Congress Cataloging-in-Publication Data
A catalog record for this book is available from the Library of Congress

ISBN 0 340 80949 3 (pb)

1 2 3 4 5 6 7 8 9 10

Typeset in 10.5/13pt Legacy Serif Book by Dorchester Typesetting Group Ltd, Dorset
Printed and bound in Great Britain by J. W. Arrowsmith Ltd, Bristol

CONTENTS

Preface *vii*
Acknowledgements *ix*

1 The Physical Environment *1*
2 The Human Environment *19*
3 Sustainable Development *35*
4 Tropical Deforestation *49*
5 Desertification *68*
6 Oceans *85*
7 Coastal Problems *103*
8 Rivers, Lakes and Wetlands *122*
9 Big Dams *141*
10 Urban Environments *159*
11 Climatic Change *177*
12 Acid Rain *198*
13 Food Production *214*
14 Soil Erosion *232*
15 Biodiversity Loss *248*
16 Transport *267*
17 Waste Management *284*
18 Energy Production *299*
19 Mining *317*
20 War *334*
21 Natural Hazards *351*
22 Conclusions *373*

Glossary *391*
Bibliography *398*
Index *424*

PREFACE

This book is about environmental issues: concerns that have arisen as a result of the human impact on the environment and the ways in which the natural environment affects human society. It deals both with the workings of the physical environment and the political, economic and social frameworks in which the issues occur. Using examples from all over the world, I aim to highlight the underlying causes behind environmental problems, the human actions that have made them issues, and the hopes for solutions.

Eighteen chapters on key issues follow the three initial chapters that outline the background contexts of the physical and human environments and introduce the concept of sustainable development. The conclusion complements the book's thematic approach by looking at the issues and efforts towards sustainable development in a regional context. The organisation of the book allows it to be read in its entirety or dipped into for any particular topic, since each chapter stands on its own. Each chapter sets the issue in a historical context, outlines why the issue has arisen, highlights areas of controversy and uncertainty, and appraises how problems are being, and can be, resolved, both technically and in political and economic frameworks. For this edition, the information in every chapter has been expanded and updated to keep pace with the rapid increase in research and understanding of the issues. The chapters are followed by expanded critical guides to further reading on the subjects, guides to relevant sites on the World Wide Web, and sets of questions that can be used to spark discussion or as essay questions.

I decided on the title *The Global Casino* because there are many parallels between the issues discussed here and the workings of a gambling joint. Money and economics underlie many of the 18 issues covered here, which can be thought of as different games in the global casino, separate yet interrelated. Just like a casino, environmental issues involve winners and losers. The casino's chance element and the players' imperfect knowledge of the outcomes of their actions are relevant in that our understanding of how the Earth works is far from perfect. The casino metaphor also works on a socio-economic level, since some individuals and groups of individuals can afford actively to take part in the games while others are less able. Some groups are more responsible for certain issues than others, yet those who have little influence are still affected by the consequences. Different individuals and groups of people also choose to play the different games in different ways, reflecting their cultural, economic and political backgrounds, and the information available to them.

The stakes are high: some observers believe that the global scale on which many of the issues occur represents humankind gambling with the very future of the planet itself. Everyone who reads this book has some part to play in the 'Global Casino'. I hope that the information presented here will allow those players to participate with a reasonable knowledge of how the games work, the consequences of losing, and the benefits that can be derived from winning.

ACKNOWLEDGEMENTS

I am indebted to many people who have helped in a variety of ways during the research and writing of three editions of this book. I would like to thank the following for their helpful comments on draft chapters. Colleagues at the School of Geography and the Environment in Oxford: Tim Burt (now at Durham), Rachael McDonnell, Alisdair Rogers, Heather Viles and Robert Whittaker, and Andrew Goudie who commented on the entire manuscript; Brenda Boardman, John Boardman and Dave Favis-Mortlock (now at Belfast) at Oxford's Environmental Change Institute; Mark Carwardine; Helen Ellwood; David Thomas of the University of Sheffield Geography Department; Charles Toomer of Willis, and Phillip Crowson, John Hughes, Jim Stevenson and J.H. Schickler of Rio Tinto.

I also thank innumerable undergraduates at the Oxford colleges of St Anne's and Oriel who have been exposed to my efforts at communicating the essence of environmental issues. These many tutorials provided the basis for this book.

My thanks go to the technical staff at the School of Geography and the Environment, University of Oxford, for their patient and efficient help: Martin Barfoot for photographic work and Peter Hayward, Ailsa Allen, David Sansom and Neil McIntosh who drew most of the figures. Paul Cole of the Cartographic Unit, University of Sheffield Geography Department drew the figures in Chapter 5. Thanks also to my editors at Arnold, particularly Laura McKelvie, for her encouragement and for taking me out to lunch a lot, and her successors, Luciana O'Flaherty, Liz Gooster and Lesley Riddle.

I am very grateful to the following for allowing me to reproduce their photographs: Mark Carwardine (Figures 1.11, 6.6, 15.4 and 20.3), Charles Toomer (Figures 1.5 and 9.3), Rio Tinto plc (Figures 19.1, 19.5 and 19.6), National Aeronautics Space Administration (Figures 4.4, 4.5, 7.9 and 9.8), Australian Bureau of Meteorology (Figure 14.5), Dan Downes (Figure 3.5), David Howard (Figure 8.1), Stephen Stokes (Figure 13.5), UNEP (Figure 5.5), UNEP GRID (Figure 13.6), and the US Department of Energy (Figure 18.5). All the other photographs are my own.

Thanks are also due to the following for allowing me to reproduce figures: Academic Press Ltd (Figure 19.3); American Association for the Advancement of Science (Figures 9.7, 15.2 and 20.1); *American Scientist* (Figure 6.4); Arnold Publishers (Figures 1.1, 1.9, 1.10, 2.4, 13.1 and 21.5); Blackwell Publishers (Figure 7.2); Cambridge University Press (Figure 2.1, after Kates, R.W., Turner, B.L. II and Clark, W.C., 1990, The great transformation, in Turner, B.L. II, Clark, W.C., Kates, R.W., Richards, J.F., Mathews, J.T. and Meyer, W.B. (eds) *The earth as transformed by human action*. Cambridge, Cambridge University Press: 1–17; Figures 2.3, 2.6, 2.8 from Lonergan, S.C., 1993, Impoverishment, population, and environmental degradation: the case for equity, *Environmental Conservation* 20: 328–34; Figure 14.4 from Lal, R., 1993, Soil erosion and conservation in West Africa, in Pimental D. (ed.) *World soil erosion and conservation*. Cambridge, Cambridge University Press: 7–25; and Figure 14.7 from Davis, M.B., 1976, Erosion rates and land use history in southern Michigan, *Environmental Conservation* 3: 139–48); Chapman & Hall (Figure 15.1); David Fulton Publishers Ltd (Figure 9.1); Elsevier Science (Figure 9.5, reprinted

from *Engineering Geology* 10, Soboleva, O.V. and Mamadaliev, U.A., The influence of Nurek Reservoir on local earthquake activity: 293–305, copyright 1976, with permission from Elsevier Science); Environment and Policy Institute, Hawaii (Figure 11.6); European Environment Agency (Figures 16.9 and 22.2); Geological Society Publishing House (Figure 17.3); Helen Dwight Reid Educational Foundation (Figure 21.7 from *Weatherwise* 47: 18–22, 1994, reprinted with permission of the Helen Dwight Reid Educational Foundation, published by Heldref Publications, 1319 Eighteenth Street NW, Washington DC 20036-1802, copyright 1994); HMSO (Figure 17.1) – Crown Copyright is reproduced with the permission of the Controller of HMSO; International Tanker Owners Pollution Federation Limited (Figure 6.8); International Soil Reference and Information Centre/ASSOD project (Figure 14.8); IOP Publishing Ltd (Figure 8.4 after Wood, L.B., 1982, *The restoration of the tidal Thames*. London, Hilger); Island Press (Figure 18.2); Jared Schneidman Design (Figure 20.7); John Wiley & Sons (Figure 7.3 from Bird, E.F.C., 1985, *Coastline changes: a global review*; Figure 9.4, after Tolouie, E., West, J.R. and Billam, J., 1993, Sedimentation and desiltation in the Sefid-Rud reservoir, Iran, in McManus, J. and Duck, R.W. (eds) *Geomorphology and sedimentology of lakes and reservoirs*: 125–38; and Figure 9.6 after Chien, N., 1985, Changes in river regime after the construction of upstream reservoirs, *Earth Surface Processes and Landforms* 10: 143–59, all copyright John Wiley & Sons Limited, reproduced with permission); Kluwer Academic Publishers (Figure 16.6 from Noble, A.G., 1980, Noise pollution in selected Chinese and American cities, *GeoJournal* 4: 573–5, with kind permission from Kluwer Academic Publishers); McGill-Queen's University Press (Figures 11.5 and 21.1); Norwegian Polar Institute (Figure 6.9); Oxford University Press (Figure 4.7 from Whitmore, T.C., 1998, *An introduction to tropical rain forests*, 2nd edn; Figure 14.1, after Cooke, R.U. and Doornkamp, J.C., 1990, *Geomorphology in environmental management*, 2nd edn; and Figure 20.2 from Middleton, N.J., 1991, *Desertification*, all by permission of Oxford University Press); Pan American Health Organisation (Figure 10.8 from Briscoe, J., 1987, A role for water supply and sanitation in the child survival revolution, *Bulletin of the Pan American Health Organization* 21: 92–105; to obtain information about PAHO publications, visit its website at http://www.paho.org or write to Pan American Health Organization, Publications Program, 525 Twenty-third Street NW, Washington DC 20037, Fax: (202) 338-0869); Penguin Books Ltd (Figure 15.5 from Heath, J., Pollard, E. and Thomas, J.A., 1984, *Atlas of butterflies in Britain and Ireland*. Harmondsworth, Viking); Routledge (Figure 8.3 from Hinrichsen, D. and Láng, I., 1993, Hungary, in Carter, F.W. and Turnock, D. (eds) *Environmental problems in Eastern Europe*: 89–106, Figure 21.5, after Smith, K., 2001, *Environmental hazards: assessing risk and reducing disaster*, 2nd edn; and Figure 21.9 after Blaikie, P., Cannon, T., Davis, I. and Wisner, B., 1994, *At risk: natural hazards, people's vulnerability and disasters*); Royal Swedish Academy of Sciences (Figure 12.8); Scott W. Nixon (Figure 7.4); Swedish Society for Anthropology and Geography (Figure 14.2); Swiss Re (Figure 21.3 from sigma 1/02); Taylor & Francis Group Ltd (Figure 7.7, after Maragos, J.E., 1993, Impact of coastal construction on coral reefs in the US-affiliated Pacific Islands, *Coastal Management* 21: 235–69); American Geographical Society (Figure 13.7); Swedish NGO Secretariat on Acid Rain (Figures 12.3 and 12.4); Tokyo Metropolitan Government (Figure 10.3); United Nations Environment Programme (Figures 5.1, 5.6, 6.5, 8.6(b), 10.4, 10.6, 11.3, 12.7, 13.6 and 17.2); University of Chicago Press (Figure 4.3 from Kummer, D.M., 1991, *Deforestation in the postwar Philippines*); World Bank (Figure 11.8 from World Bank, 1992, *World development report 1992: development and the environment*. New York, Oxford University Press); World Meteorological Organization – The Intergovernmental Panel on Climate Change – (Figures 11.1 and 11.4).

Every effort has been made to trace the copyright holders of quoted material. If, however, there are inadvertent omissions and/or inappropriate attributions these can be rectified in any future editions.

1

THE PHYSICAL ENVIRONMENT

TOPICS COVERED

Classifying the natural world
Natural cycles
Timescales
Spatial scales
Time and space scales
The state of our knowledge

Key words: biome, productivity, feedback, threshold, timelag, resistance, resilience, natural archive, palaeoenvironmental indicator

The term environment is used in many ways. This book is about issues that arise from the physical environment, which is made up of the living (biotic) and non-living (abiotic) things and conditions that characterise the world around us. While this is the central theme, the main reason for the topicality of the issues covered here is the way in which people interact with the physical environment. Hence, it is pertinent also to refer to the social, economic and political environments to describe those human conditions characteristic of certain places at particular times, and to explain why conflict has arisen between human activity and the natural world. This chapter looks at some of the basic features of the physical environment, and Chapter 2 is concerned with the human factors that affect the ways in which the human race interacts with the physical world.

CLASSIFYING THE NATURAL WORLD

Geography, like other academic disciplines, classifies things in its attempt to understand how they work. The physical environment can be classified in numerous ways, but one of the most commonly used classifications is that which breaks it down into four interrelated spheres: the lithosphere, the atmosphere, the biosphere and the hydrosphere. These four basic elements of the natural world can be further subdivided. The lithosphere, for example, is made up of rocks that are typically classified according to their modes of formation (igneous, metamorphic and sedimentary); these rock types are further subdivided according to the processes that formed them and other factors such as their chemical composition. Similarly, the workings of the atmosphere are manifested at the Earth's surface by a typical distribution of climates; the biosphere is made up of many types of flora and fauna, and the hydrosphere can be subdivided according to its chemical constituents (fresh water and saline, for example), or the condition or phase of the water: solid ice, liquid water or gaseous vapour.

These aspects of the natural world overlap and interact in many different ways. The nature of the soil in a particular place, for example, reflects the underlying rock type, the climatic conditions of the area, the plant and animal matter typical of the region, and the quantity and quality of water available. Suites of characteristics are combined in particular areas called ecosystems. These ecosystems can also be classified in many ways. One approach uses the amount of organic matter or biomass produced per year – the net production – which is simply the solar energy fixed in the

TABLE 1.1 *Annual net primary production of carbon by major world ecosystem types*

Ecosystem type	Mean net primary productivity (g C/m^2/year)	Total net primary production (billion tonnes C/year)
Tropical rain forest	900	15.3
Tropical seasonal forest	675	5.1
Temperate evergreen forest	585	2.9
Temperate deciduous forest	540	3.8
Boreal forest	360	4.3
Woodland and shrubland	270	2.2
Savanna	315	4.7
Temperate grassland	225	2.0
Tundra and alpine	65	0.5
Desert scrub	32	0.6
Rock, ice and sand	1.5	0.04
Agricultural land	290	4.1
Swamp and marsh	1125	2.2
Lake and stream	225	0.6
Total land	324*	48.3
Open ocean	57	18.9
Upwelling zones	225	0.1
Continental shelf	162	4.3
Algal bed and reef	900	0.5
Estuaries	810	1.1
Total oceans	69*	24.9
Total for biosphere	144*	73.2

*The means for land, oceans and biosphere are weighted according to the areas covered by specific ecosystem types
Source: after Whittaker and Likens (1973)

biomass minus the energy used in producing it by respiration (see below). The annual net primary production of carbon, a basic component of all living organisms, by major world ecosystem types is shown in Table 1.1. Clear differences are immediately discernible between highly productive ecosystems such as forests, marshes, estuaries and reefs, and less productive places such as deserts, tundras and the open ocean. All of the data are averaged and variability around the mean is perhaps greatest for agricultural ecosystems which, where intensively managed, can reach productivities as high as any natural ecosystem. One of the main reasons for agriculture's low average is the fact that fields are typically bare of vegetation for significant periods between harvest and sowing.

One of the main factors determining productivity is the availability of nutrients, key

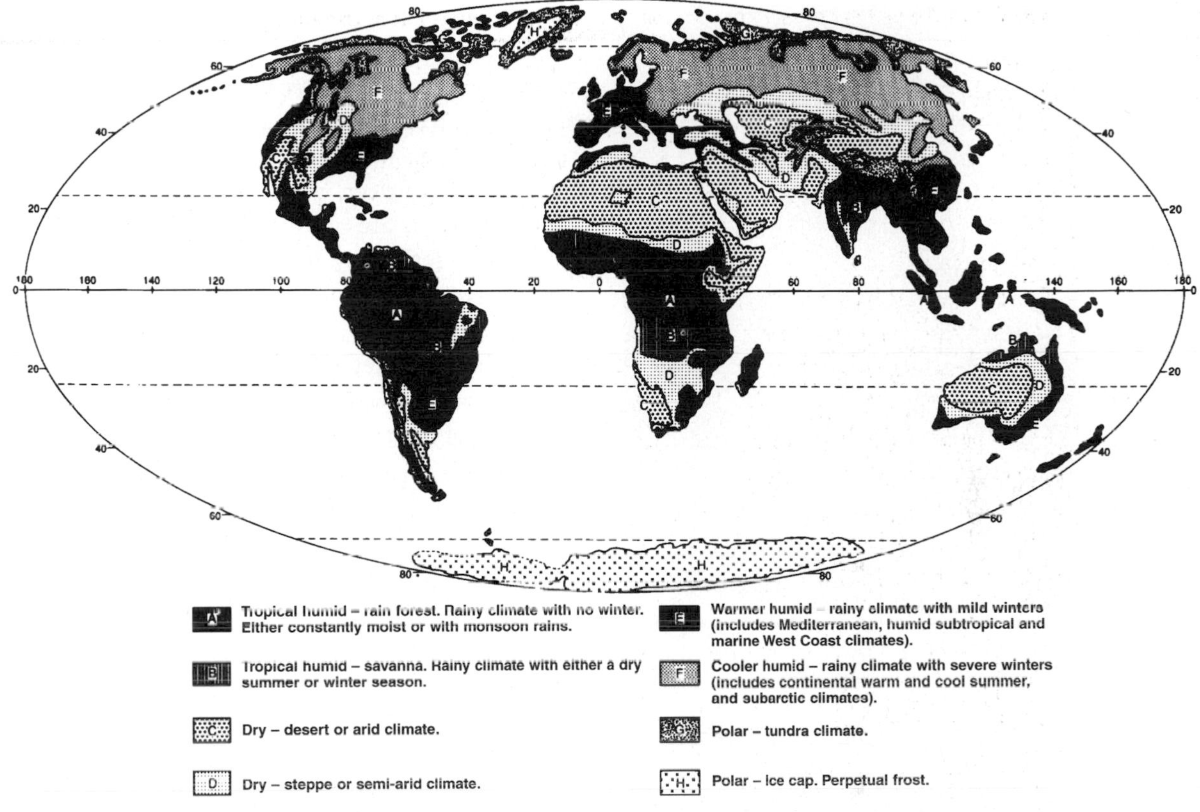

Figure 1.1 Present-day morphoclimatic regions of the world's land surface (Williams *et al.*, 1993).

substances for life on Earth: a lack of nutrients is often put forward to explain the low productivity in the open oceans, for example. Climate is another important factor. Warm, wet climates promote higher productivity than cold, dry ones. Differences in productivity may also go some way towards explaining the general trend of increasing diversity of plant and animal species from the poles to the equatorial regions. Despite many regional exceptions such as mountain tops and deserts, this latitudinal gradient of diversity is a striking characteristic of nature that fossil evidence suggests has been present in all geological epochs. The relationship with productivity is not straightforward, however, and many other hypotheses have been advanced, such as the suggestion that minor disturbances promote diversity by preventing a few species from dominating and excluding others (Connell, 1978).

The relationships between climate and the biosphere are also reflected on the global scale in maps of vegetation and climate, the one reflecting the other. Figure 1.1 shows the world's morphoclimatic regions, which are a combination of both factors. Despite wide internal variations, immense continental areas clearly support distinctive forms of vegetation that are adapted to a broad climatic type. Such great living systems, which also support distinctive animals and to a lesser extent distinctive soils, are called biomes, a concept seldom applied to aquatic zones. Different ecologists produce various lists of biomes, and the following eight-fold classification may be considered conservative (Colinvaux, 1993):

1 tundra
2 coniferous forest (also known as boreal forest or taiga)
3 temperate forest
4 tropical rain forest

Figure 1.2 Coniferous forest in Finland, the eastern end of a broad region of boreal or taiga forest that stretches to the Russian Far East.

5 tropical savanna
6 temperate grassland
7 desert
8 maquis (also known as chaparral).

A notable aspect of the tundra biome is the absence of trees. Vegetation consists largely of grasses and other herbs, mosses, lichens and some small woody plants that are adapted to a short summer growing season. The cold climate ensures that the shallow soils are deeply frozen (permafrost) for all or much of the year, a condition that underlies about 20 per cent of the Earth's land surface. Many animals hibernate or migrate in the colder season, while others, such as lemmings, live beneath the snow. The main tundra region is located in the circumpolar lands north of the Arctic circle, which are bordered to the south by the evergreen, needle-leaved boreal or taiga forests (Fig. 1.2). Here, winters are very cold, as in the tundra, but summers are longer. These forests are subject to periodic fires, and a burn-regeneration cycle is an important characteristic to which populations of deer, bears and insects, as well as the vegetation, are adapted. Much of the boreal forest is underlain by acid soils.

The temperate forests, by contrast, are typically deciduous. They are, however, like the boreal forests in that they are found almost exclusively in the northern hemisphere. This biome is characteristic of northern Europe, eastern China, and eastern and midwest USA, with small stands in the southern hemisphere in South America and New Zealand. Tall broadleaf trees dominate, the climate is seasonal but water is always abundant during the growing season, and this biome is less homogeneous than tundra or boreal forest. Amphibians, such as salamanders and frogs, are present, while they are almost totally absent from the higher-latitude biomes.

The tropical rain forest climate has copious rainfall and warm temperatures in all months of the year. The trees are always green, typically broadleaved, and most are pollinated by animals (trees in temperate and boreal forests, by contrast, are largely pollinated by wind). Many kinds of vines (llianas) and epiphytes, such as ferns and orchids, are characteristic. Most of the nutrients are stored in the biomass and the soils contain little organic matter. Above all, tropical rain forests are characterised by a large number of species of both plants and animals.

Savanna belts flank the tropical rain forests to the north and south in the African and South American tropics. The trees of tropical savannas are stunted and widely spaced, which allows grass to grow between them. Herds of grazing mammals typify the savanna landscape, along with large carnivores such as lions and other big cats, jackals and hyenas. These mammals, in turn, provide a food source for large scavengers such as vultures. The climate is warm all year, but has a dry season several months long when fires are a common feature.

Figure 1.3 Temperate grassland in central Mongolia is still predominantly used for grazing. In many other parts of the world such grasslands have been ploughed up for cultivation.

The greatest expanses of the temperate grassland biome are located in Eurasia (where they are commonly known as steppe – Fig. 1.3), North America (prairie) and South America (pampa), with smaller expanses in South Africa (veldt). The climate is temperate, seasonal and dry. There are certain similarities with savannas in terms of fauna and the occurrence of fire, but unlike savannas, trees are absent. Temperate grassland soils tend to be deep and rich in organic matter.

In many parts of the world, where climates become drier, temperate grasslands fade into the desert biome. Hyper-arid desert supports very little plant life and is characterised by bare rock or sand dunes, but some species of flora and fauna are adapted to the high and variable temperatures and general lack of moisture, since some water is always available (Fig. 1.4). Sporadic rain promotes rapid growth of annual plants, and animals such as locusts, which otherwise lie dormant for several years as seeds or eggs.

Figure 1.4 This strange-looking plant, the welwitschia, is found only in the Namib Desert. Its adaptations to the dry conditions include long roots to take up any moisture in the gravelly soil and the ability to take in moisture from fog through its leaves. The welwitschia's exact position in the plant kingdom is controversial, but it is grouped with the pine trees.

A very distinctive form of vegetation is commonly associated with Mediterranean climates in which summers are hot and dry, and winters are cool and moist. It is found around much of the Mediterranean Basin (where it is known as maquis), in California (chaparral), southern Australia (mallee), Chile (mattöral) and South Africa (fynbos). Low evergreen trees (forming woodland) and shrubs (forming scrub) have thick bark and small, hard leaves. They are frequently exposed to fire during the arid summer period.

All these natural biomes have been affected to a greater or lesser extent by human action. Much of the maquis, for example, may represent a landscape where forests have been degraded by people, through cutting, grazing and the use of fire. The human use of fire may also be an important factor in maintaining, and possibly forming, savannas and temperate grasslands. The temperate forests have been severely altered over long histories with high population densities as people have cleared trees for farming and urban development (Fig. 1.5). Conversely, biomes considered by people to

Figure 1.5 The US city of Boston, part of one of the world's most extensive areas of urban development. Only a few of the original temperate forest trees in the area survive in parks and gardens. Urban areas are now so widespread that they are often treated as a type of physical environment in their own right (Photo: Charles Toomer).

be harsh, such as the tundra and deserts, show less human impact. The anthropogenic influence is but one factor that promotes change in terrestrial as well as oceanic and freshwater ecosystems, because the interactions between the four global spheres have never been static. Better understanding of the dynamism of the natural world can be gained through a complementary way of studying the natural environment. Study of the processes that occur in natural cycles also takes us beyond description, to enable explanation.

NATURAL CYCLES

A good method for understanding the way the natural world works is through the recognition of cycles of matter in which molecules are formed and reformed by chemical and biological reactions, and are manifested as physical changes in the material concerned. The major stores and flows of water in the global hydrological cycle are shown in Fig. 1.6. Most of the Earth's water (about 97 per cent) is stored in liquid form in the oceans. Of the 3 per cent fresh water, most is locked as ice in the ice caps and glaciers, and as a liquid in rocks as groundwater. Only a tiny fraction is present at any time in lakes and rivers. Water is continually exchanged between the Earth's surface and the atmosphere - where it can be present in gaseous, liquid or solid form - by evaporation, transpiration from plants and animals, and precipitation. The largest flows are directly between the ocean and the atmosphere. Smaller amounts are exchanged between the land and the atmosphere, with the difference accounted for by flows in rivers and groundwater to the oceans. Fresh water on the land is most directly useful to human society, since water is an essential prerequisite of life, but the oceans and ice caps play a key role in the workings of climate.

Similar cycles, commonly referred to as biogeochemical cycles, can also be identified for other forms of matter. Nutrients such as nitrogen, phosphorus and sulphur are similarly distributed among the four major environmental spheres and are continually cycled between them. Carbon is another key element for life on Earth, and the stores and flows of the carbon cycle are shown in Fig. 1.7. The major stores of carbon are the oceans and rocks, particularly carbonate sedimentary rocks such as limestones, and the hydrocarbons (coal, oil and natural gas). Much smaller proportions are present in the atmosphere and biosphere. The length of time carbon spends in particular stores also varies widely. Under natural circumstances, fossil carbon locked in rocks remains in these stores for millions of years.

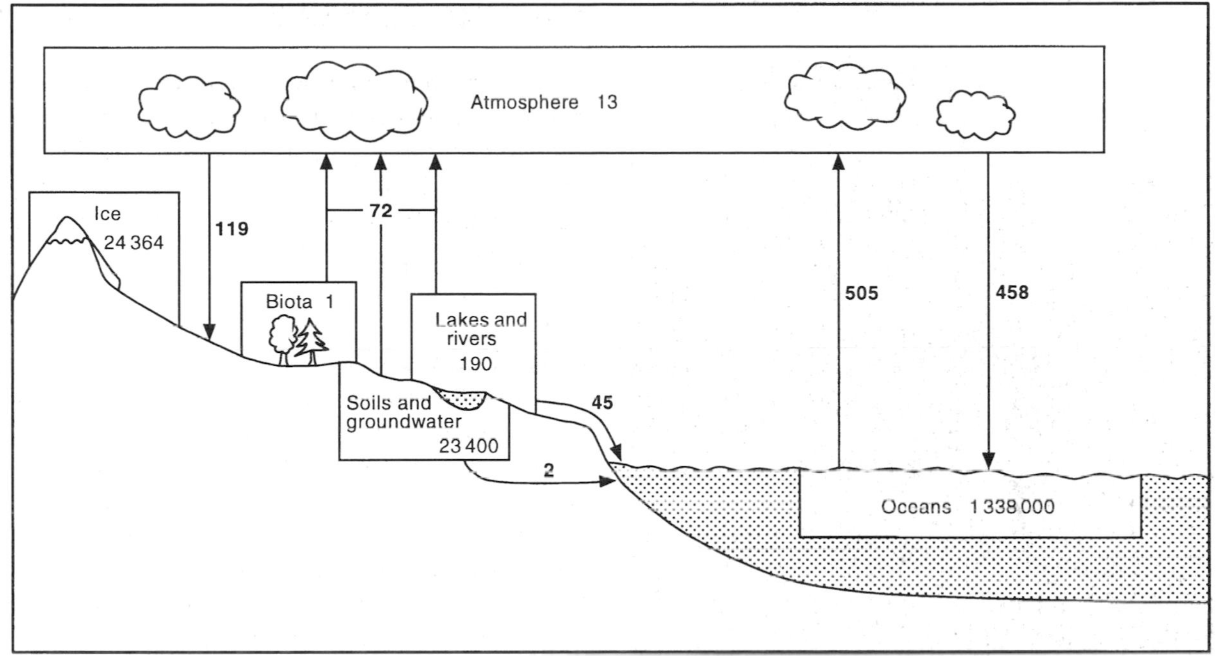

Figure 1.6 Global hydrological cycle showing major stores and flows (data from Shiklomanov, 1993). The values in stores are in thousand km^3, values of flows in thousand km^3 per year.

Carbon reaches these stores by the processes of sedimentation and evaporation, and is released from rocks by weathering, vulcanism and sea-floor spreading. In recent times, however, the rate of flow of carbon from some of the lithospheric stores – the hydrocarbons or fossil fuels – has been greatly increased by human action: the burning of fossil fuels, which liberates carbon by oxidation. Hence, a significant new flow of carbon between the lithosphere and the atmosphere has been introduced by human society, and the natural atmospheric carbon store is being increased as a consequence.

Carbon also reaches the atmosphere by the respiration of plants and animals, which in green plants, blue-green algae and phytoplankton is part of the two-way process of photosynthesis. Photosynthesis is the chemical reaction by which these organisms convert carbon from the atmosphere, with water, to produce complex sugar compounds (which are either stored as organic matter or used by the organism) and oxygen. The reaction is written as follows:

$$6CO_2 + 6H_2O \rightarrow C_6H_{12}O_6 + 6O_2$$

This equation shows that six molecules of carbon dioxide and six molecules of water yield one molecule of organic matter and six molecules of oxygen. The reaction requires an input of energy from the sun, some of which is stored in chemical form in the organic matter formed.

The process of respiration is written as the opposite of the equation for photosynthesis. It is the process by which the chemical energy in organic matter is liberated by combining it with oxygen to produce carbon dioxide and water. All living things respire to produce energy for growth and the other processes of life. The chemical reaction for respiration is, in fact, exactly the same as that for combustion. Humans, for example, derive energy for their life needs from organic matter by eating (just as other animals do) and also by burning plant matter in a number of forms, such as fuelwood and fossil fuels.

The flow of converted solar energy through living organisms can be traced up a hierarchy of

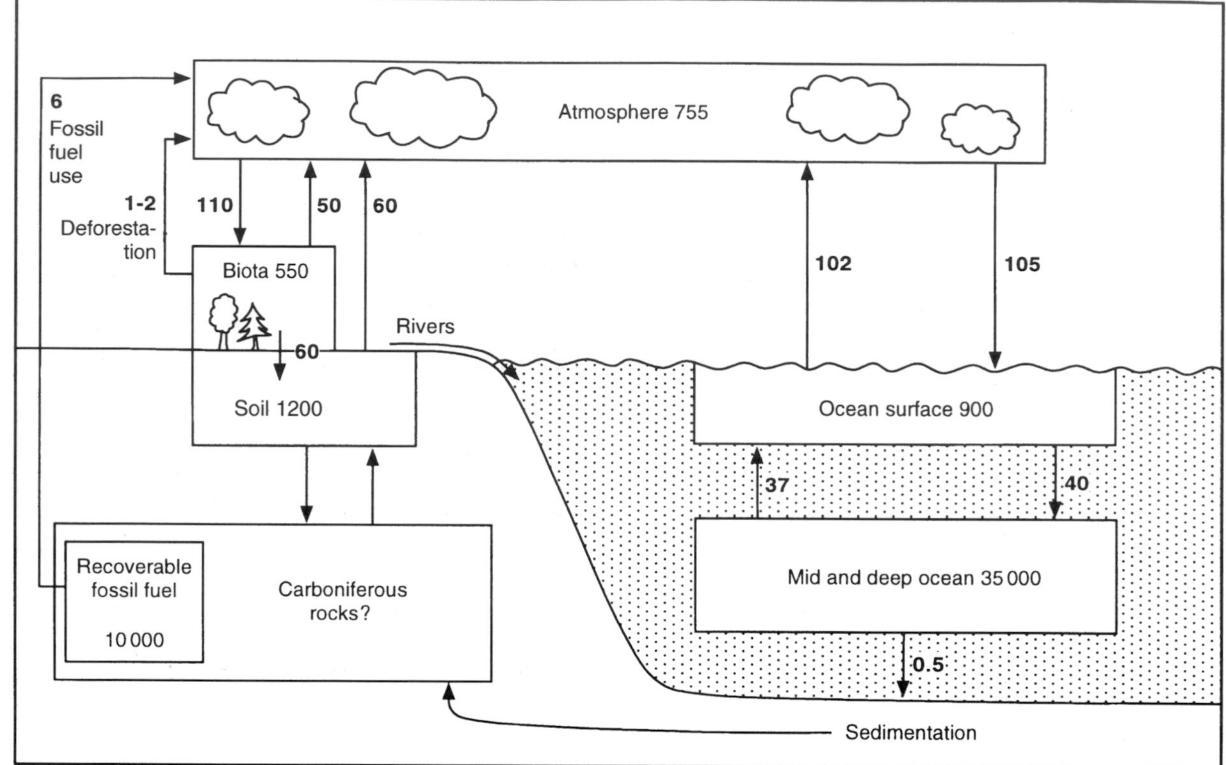

Figure 1.7 Global carbon cycle showing major stores and flows (after Schlesinger, 1991). The values in stores are in units of Pg C, values of flows in Pg C per year. 1Pg C = 10^{15}g C = 1 billion tonnes of carbon as CO_2.

life-forms known as a food chain. Figure 1.8 shows a simple food chain in which solar energy is converted into chemical energy in plants (so-called producers), which are eaten by herbivores (so-called first-order consumers), which, in turn, are eaten by other consumers (primary carnivores), which are themselves eaten by secondary carnivores. An example of such a food chain on land is:

grass → cricket → frog → heron

Each stage in the chain is known as a trophic level. In practice, there are usually many, often interlinked, food chains that together form a food web, but the principles are the same. At each trophic level some energy is lost by respiration, through excretory products and when dead organisms decay, so that available energy declines along the food chain away from the plant. In general terms, animals also tend to be bigger at each sequential trophic level enabling them to eat their prey safely. This model helps us to explain the basic structures of natural communities: with each trophic level, less energy is available to successively larger individuals and thus the number of individuals decreases. Hence, while plants are very numerous because they receive their energy directly from the sun they can only support successively fewer larger animals. With the exception of humans, predators at the top of food chains are therefore always rare.

Food chains, the carbon cycle and the hydrological cycle are all examples of 'systems' in which the individual components are all related to each other. Most of the energy that drives these systems comes from the sun, although energy from the Earth also contributes. All the cycles of energy and matter referred to in this section are affected by human action, deliberately manipulating natural cycles to human advantage. One of the human

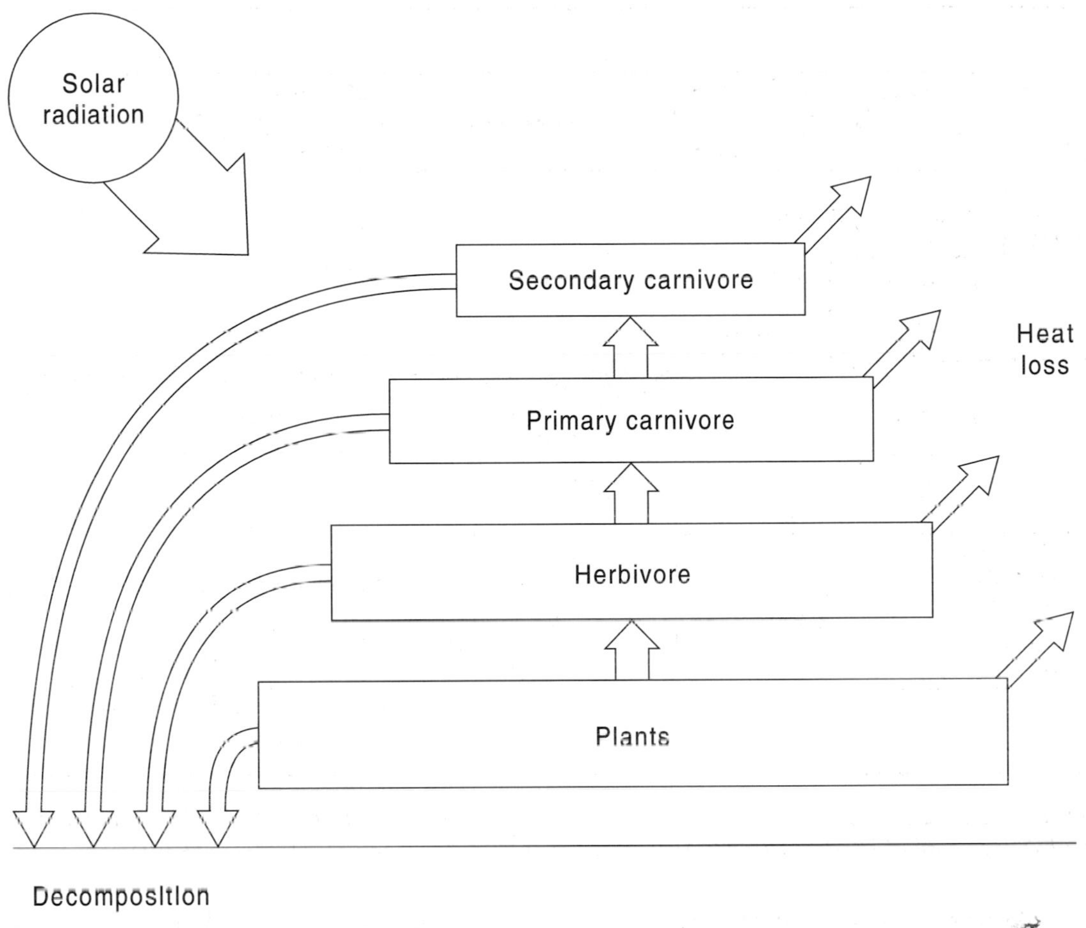

Figure 1.8 Energy flow through a food chain.

impacts on the carbon cycle has been mentioned, but humans also affect other cycles. The cycle of minerals in the rock cycle is affected by the construction industry, for example. Human activity affects the hydrological cycle by diverting natural flows: the damming of rivers or extracting groundwater for human use. The nitrogen cycle is affected by concentrating nitrogen in particular places such as by spreading fertilisers on fields. Food chains are widely affected: human populations manipulate plants and animals to produce food.

However, since all parts of these cycles are interrelated, human intervention in one part of a cycle also affects other parts of the same and other cycles. These knock-on effects are the source of many environmental changes that are undesirable from human society's viewpoint. Our manipulation of the nitrogen cycle by using fertilisers also increases the concentration of nitrogen in rivers and lakes when excess fertiliser is washed away from farmers' fields. This can have deleterious effects on aquatic ecosystems. Excess nitrogen can also enter the atmosphere to become a precursor for 'acid rain'. One of the effects of acid rain is to accelerate the rate of weathering of some building stones. A better appreciation of these types of changes can be gained by looking at the various scales of time and space through which they occur.

TIMESCALES

Changes in the natural environment occur on a wide range of timescales. Geologists believe that

TABLE 1.2 *Geological timescale classifying the history of the Earth*

Era	Period		Start (million years BP)	Important events
		Holocene	0.01	Early civilisations
	Quaternary	Pleistocene	1.8	First humans
		Pliocene	5	First hominids
		Miocene	22.5	
		Oligocene	38	
		Eocene	54	
Cenozoic	Tertiary	Palaeocene	65	Extinction of dinosaurs
	Cretaceous		136	Main fragmentation of Pangaea
	Jurassic		190	
Mesozoic	Triassic		225	First birds
	Permian		280	Formation of Pangaea
		Pennsylvanian	315	
	Carboniferous	Mississippian	345	
	Devonian		395	
	Silurian		440	First land plants and animals
	Ordovician		500	First vertebrates
Palaeozoic	Cambrian		570	
Precambrian			4600	Formation of Earth

Source: after Goudie (1993a); Colinvaux (1993); Williams *et al.* (1993).

the Earth is about 4600 million years old, while fossil evidence suggests that modern humans (*Homo sapiens*) appeared between 100 000 and 200 000 years before present (BP), developing from the hominids whose earliest remains, found in Africa, date to around 3.75 million BP (Table 1.2). The very long timescales over which many changes in the natural world take place may seem at first to have little relevance for today's human society other than to have created the world we know. It is difficult for us to appreciate the age of the Earth and the thought that the present distribution of the continents dates from the break-up of the supercontinent Pangaea, which began during the Cretaceous. Indeed, relative to the forces and changes due to tectonic movements, the human impact on the planet is very minor and short-lived. However, such Earth processes do have relevance on the timescale of a human lifetime. Tectonic movements cause volcanic eruptions that can affect human society as natural disasters at the time of the event. Some volcanic eruptions also affect day-to-day human activities on slightly longer timescales, by injecting dust into the atmosphere, which affects climate, for example, and by providing raw materials from which soils are formed. This example also illustrates the fact that the same event may be interpreted as bad

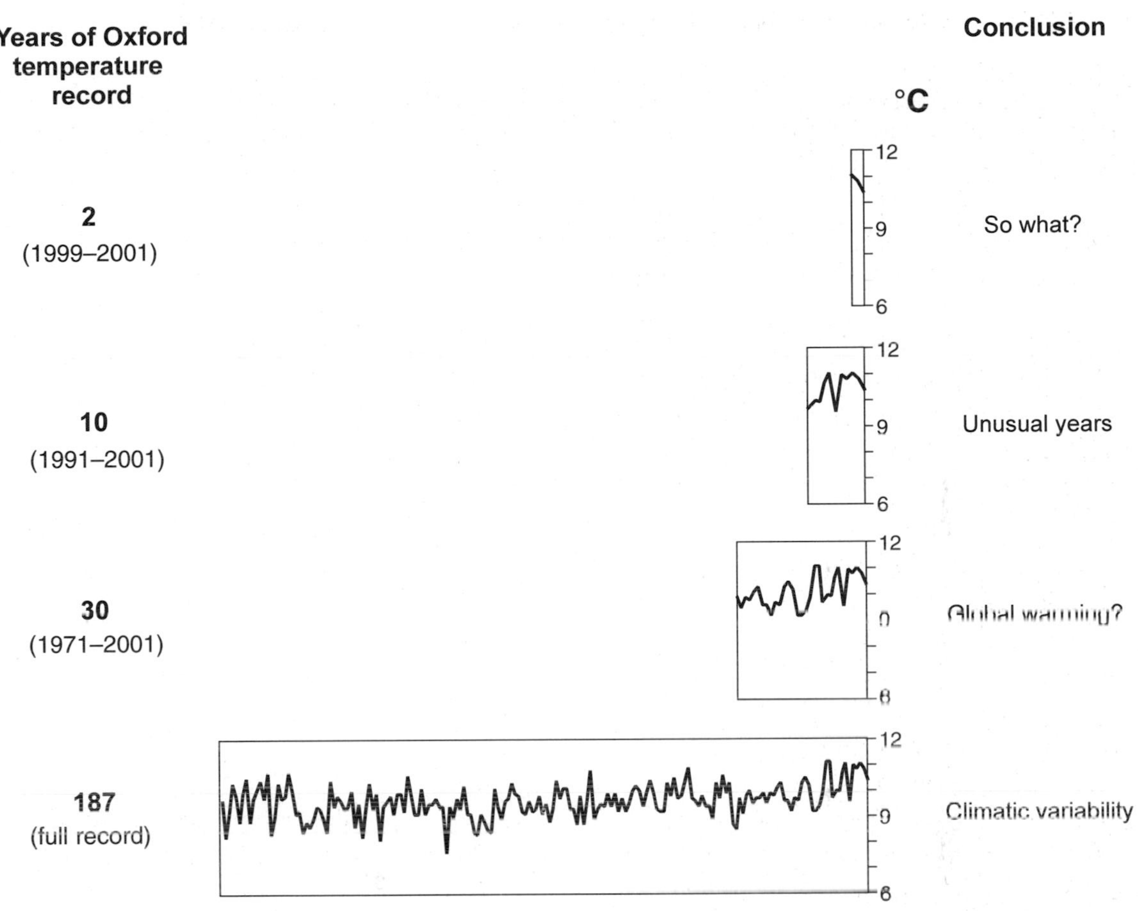

Figure 1.9 Demonstration of how the length of data record (here annual mean air temperature at Oxford, England) can influence conclusions about environmental variability (updated after Burt, 1994).

from a human viewpoint on one timescale (a volcanic disaster) and good on another timescale (fertile volcanic soils).

It is important to realise that the timescale we adopt for the study of natural systems can affect our understanding as well as our perception of them. Many such systems are thought to be in 'dynamic equilibrium', meaning that the input and output of matter and energy is balanced, but recognition of dynamic equilibrium in natural systems depends upon the timescale over which the system is studied. To take the Earth as a whole, for example, the idea of dynamic equilibrium has been proposed to explain why the temperature of the Earth has remained relatively constant for the last 4 billion years, despite the fact that the sun's heat has increased by about 25 per cent over that period. The Gaia hypothesis suggests that life on the planet has played a key role in regulating the Earth's conditions to keep it amenable to life (Lovelock, 1989). The theory is not without its critics, but even if we accept it, the dynamic equilibrium only holds for the few billion years of the Earth's existence. Astronomers predict that eventually the sun will destroy the Earth, so that over a longer timescale, dynamic equilibrium does not apply. This example applies over a very long timescale, and from our perspective the destruction of the Earth by the sun is not imminent, but the principle is relevant to all other natural systems. Adoption of different timescales of analysis dictates which aspects of a system we see and understand because the importance of different factors changes with different timescales. Indeed, even within the lifespan of the Earth, dramatic changes are known to have occurred, such as the progression of glacial and interglacial periods. The longer the timeframe, the coarser the resolution, and vice versa. In a simple example, a human being who contracts a cold might feel miserable for a few days, but in terms of that person's lifetime career the cold is a very minor influence. This principle is depicted for measurements of mean annual air temperature at Oxford in England in Fig. 1.9.

One of the key components of natural cycles and dynamic systems is the operation of feedbacks. Feedbacks may be negative, which tend to dampen down the original effect and thereby maintain dynamic equilibrium, or they may be positive, and hence tend to enhance the original effect. An example of negative feedback can be seen in the operation of the global climate system: more solar energy is received at the tropics than at the poles, but the movement of the atmosphere and oceans continually redistributes heat over the Earth's surface to redress the imbalance. Positive feedbacks can result in a change from one dynamic equilibrium to another: if a forest is cleared by human action, for example, the soil may be eroded to the extent that recolonisation by trees is impossible.

This last example illustrates another important aspect of natural cycles: the existence of thresholds. A change in a system may not occur until a threshold is reached: snow will remain on the ground, for example, until the air temperature rises above a threshold at which the snow melts. Crossing a threshold may be a function of the frequency or intensity of the force for change: a palm tree may be able to withstand, or be 'resilient' to, winds up to a certain speed, but will be blown out of the soil by a hurricane-force wind that is above the tree's threshold of resilience. Conversely, thresholds may be reached by the cumulative impacts of numerous small-scale events: regular rainfall inputs of moisture to a slope may build up to a point at which the slope fails, or in the erosion example, the gradual loss of soil reaches the point at which there is not enough soil left for trees to take root and grow.

These two illustrations of how thresholds can be crossed also embody two important ideas on the way change occurs in the physical environment. The hurricane represents a high-magnitude, low-frequency event, for which some use the term 'catastrophe'. An opposing view ascribes change in the environment to small-scale, commonly occurring processes. Of course, most environments are affected by both catastrophic and gradual changes.

To complicate things further, there may be a lag in time between the onset of the force for change and the change itself: the response time of the system. An animal seldom dies immediately upon contracting a fatal disease, for example; its body ceases to function only after a period of time. Likewise in the erosion example, trees are unable to colonise only after a certain amount of soil has been lost. Consideration of feedbacks, thresholds and lags leads to some other characteristics of natural systems: their sensitivity to forces for change, which dictates their ability to maintain or return to an original condition following a disturbance: their 'stability'. A natural system's ability to maintain its original condition is dictated by its 'resistance' to disturbance, while the ability to return to an original condition is commonly referred to as 'resilience'.

The variability in natural disturbances affecting some environments means that assuming them to be in a more or less stable dynamic equilibrium is not reasonable, however. These are commonly referred to as 'non-equilibrium' environments, and drylands are a good example. Drylands are highly dynamic and are currently thought of as being in a constant state of change, driven by disturbances such as the variability of fire and insect attack and, perhaps most importantly, the variability of moisture from rainfall. Amounts vary widely, from one intense rainy day in a dry month, through seasonal variations to longer periods such as droughts. Many aspects of the physical environment respond accordingly. Perennial plants and small animals respond particularly quickly, so that a different level of their populations can be expected at each particular

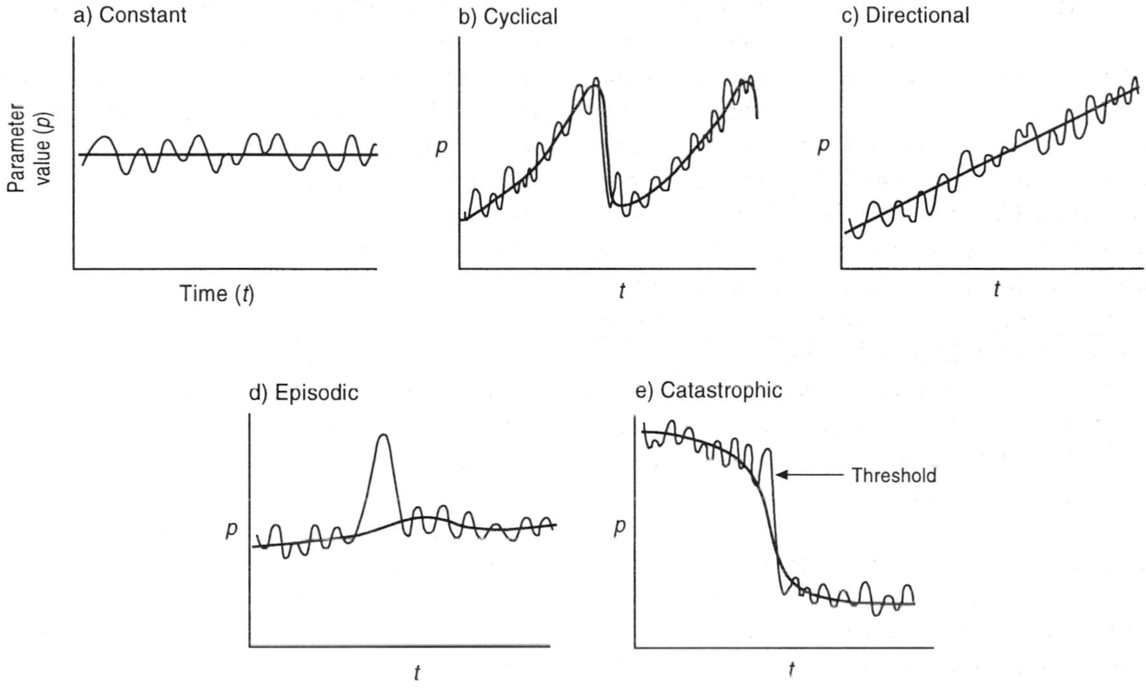

Figure 1.10 Main types of long-term trends in ecosystems, with shorter-term fluctuations superimposed (after Burt, 1994).

time. Larger animals respond to such a dynamic environment by moving, sometimes over very large distances, to take advantage of the spatial changes in water and food availability.

All these considerations on changes in natural ecosystems through time can be assembled into some typical patterns that are illustrated hypothetically in Fig. 1.10. The parameters represented on the y axis of the graphs could be a measurement of any physical thing, such as soil organic matter content, species diversity, carbon dioxide concentration in the atmosphere, or the volume of water flowing along a river channel. Figure 1.10a might represent a mature forest with small variations in biomass with the seasons, and as individual trees grow and die. This constant system could equally be described as stable or as one that is in dynamic equilibrium. It contrasts with the cyclical pattern in Fig. 1.10b, which could represent natural cycles of heather burning and regeneration. Fig. 1.10c could illustrate natural succession of vegetation in an area with a long-term directional trend. The pattern induced by episodes such as drought, which allow recovery in systems with sufficient resilience (Fig. 1.10d) contrasts with a catastrophic disturbance that exceeds resilience, so that the system crosses a threshold resulting in long-term change from one state to another (Fig. 1.10e), such as when soil erosion proceeds to a level where certain types of vegetation can no longer survive in the area.

SPATIAL SCALES

Just as the choice of timescales is important to our understanding of the natural environment, so too is the spatial scale of analysis. Studies can be undertaken at scales that range from the microscopic – the effects of salt weathering on a sand grain, for example – through an erosion plot measured in square metres, to drainage basin studies that can reach subcontinental scales in the largest cases, to the globe itself. As with time, the resolution of analysis becomes coarser with increasing

Figure 1.11 Tundra in Canada's northern Manitoba: part of a biome that may be particularly sensitive to a human-induced warming of global climate and that could create a positive feedback by releasing large amounts of methane, a greenhouse gas (see Chapter 11) (Photo: Mark Carwardine).

spatial scale. We draw a line on a world map to divide one biome from another, but on the ground there is usually no line, more a zone of transition, which may itself vary over different timescales.

Similarly, the types of influence that are important differ according to spatial scale. To use another example involving humans, a landslide that results in the loss of a farmer's field may have a significant impact on the farmer's ability to earn a living, but the same landslide has a minimal impact on national food production.

Thresholds and feedbacks also have relevance on the spatial scale. Certain areas may be more sensitive to change than others and if a threshold is crossed in these more sensitive areas, wider-scale changes may be triggered. Soil particles entrained by wind erosion from one small part of a field, for example, can initiate erosion over the whole field. On a larger scale, sensitive areas such as the Labrador–Ungava plateau of northern Canada appear to have played a key role in triggering global glaciations during the Quaternary because they were particularly susceptible to ice-sheet growth. A contemporary large-scale example can be seen in the tundra biome (Fig. 1.11), which could release large amounts of methane locked in the permafrost if the global climate warms due to human-induced pollution of the atmosphere. Methane is a greenhouse gas so positive feedback could result, enhancing the warming effect worldwide.

TIME AND SPACE SCALES

The key factors influencing natural events also vary at different combined spatial and temporal scales. Individual waves breaking on a beach constantly modify the beach profile, which is also affected by the daily pattern of tides dictating where on the beach the waves break. Individual storms alter the beach too, as do the types of weather associated with the seasons. However, all these influences are superimposed upon the effects of factors that operate over longer timescales and larger spatial scales, such as sediment supply and the sea level itself (Clayton, 1991).

The range of temporal and spatial scales is illustrated for some biological and climatic processes in Fig. 1.12. This emphasises the fact that various processes in the natural environment (e.g. climatic changes, tornadoes) exist at specific scales, as do its elements (e.g. species, biomes). It is also important to note that no part of the physical environment is a closed self-supporting system; all are a part of larger interacting systems.

The environmental issues in this book have arisen as a consequence of human activity conflicting with environmental systems. Resolution of such conflicts can only be based on an understanding of how natural systems work. For issues that stem from human impact upon the physical environment, as most do, we need to be able to rank the temporal and spatial scale of human impact in the natural hierarchy of influences on the natural system in question. Inevitably, we tend to focus on scales directly relevant to people, but we should not forget other scales, which may have less direct but no less significant effects. Indeed, successful management of environmental issues relies on the successful identification of appropriate scales and their linkages.

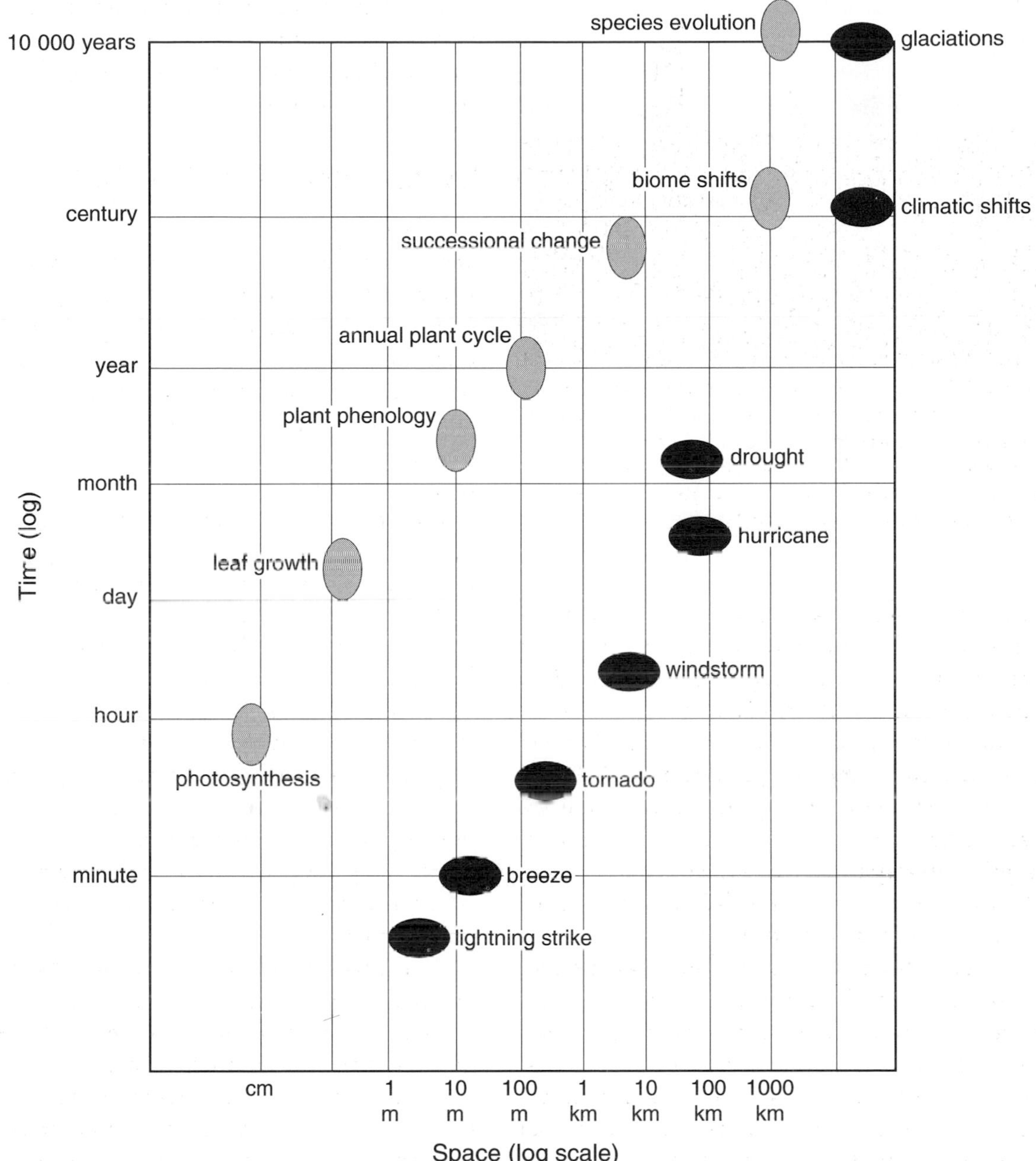

Figure 1.12 The range of temporal and spatial scales at which ecosystem processes exist and operate (after Holling, 1995).

THE STATE OF OUR KNOWLEDGE

We already know a great deal about how the natural world works, but there remains a lot more to learn. We have some good ideas about the sorts of ways natural systems operate, but we remain ignorant of many of the details. Some of the difficulties involved in ascertaining these details include a lack of data and our own short period of residence on the Earth. Direct measurements using instruments are used in the contemporary

TABLE 1.3 *Spatial and temporal availability and limitations of instrumental data and some proxy variables for temperature in the Holocene*

Variable	Spatial extent	Timescale		
		Interannual	Decades to centuries	Centennial and longer
Instrumental data	Europe from early 1700s, most other coastal regions during nineteenth century. Continental interiors by 1920s, Antarctica by late 1950s	Should be 'perfect' if properly maintained – changes assessable on daily, monthly and seasonal timescales	Site moves, observation time changes and urbanisation influences present problems – changing frequency of extremes assessable	As previous, but rates of change to site, instrumentation and urbanisation mean absolute levels increasingly difficult to maintain
Proxy indicators Contemporary written historical records (annals, diaries, etc.)	Europe, China, Japan, Korea, eastern N. America. Some potential in Middle East, Turkey and S. Asia and Latin America (since 1500s)	Depends on function of diary information (freeze dates, harvest dates and amounts, snowlines, etc.). Very difficult to compare with instrumental data	Depends on diary length and observer age. Lower frequencies increasingly likely to be lost due to human lifespan	Only a few indicators are objective and might provide comparable information (e.g. snowlines, rain days)
Tree-ring widths	Trees growing poleward of 30° or at high elevations in regions where cool season suspends growth	Generally dependent upon growing season months. Exact calendar dates determined by cross-dating	Standardisation method potentially compromises interpretation on longer timescales	Highly dependent on standardisation method. Likely to have lost variability, but difficult to assess
Ice-core melt layers	Coastal Greenland and high-latitude and high-altitude ice caps, where temperatures rise above freezing for a few days each summer	Depends on summer warmth. Unable to distinguish cold years that cause no melt. Rarely compared with instrumental records. Dating depends on layer counting – increasingly difficult with depth	May not respond to full range of temperature variability. Whole layer may melt if too warm; no melt layers if too cold	Increasingly depends on any flow model and layer compaction. Veracity can be assessed using other cores
Coral growth and isotopes	Tropics (between 30° N and S) where shallow seas promote coral growth	Response to annual and seasonal water temperature and salinity. Dating depends on counting. Rarely cross-dated	As coral head grows, low-frequency aspects may be affected by amount of sunlight, water depth, nutrient supply, etc.	Only achieved in a few cases. Veracity can be assessed by comparison with other corals

Source: after Jones *et al.* (1998: Tables 1 and 2)

era to monitor environmental processes. Historical archives, sometimes of direct measurements, otherwise of more anecdotal evidence, can extend these data back over decades and centuries. Good records of high and low water levels for the River Nile at Cairo extend from AD 641 to 1451, although they are intermittent thereafter until the nineteenth century, and continuous monthly mean temperature and precipitation records have been kept at several European weather stations since the early eighteenth century. Other types of written historical evidence date back to ancient Chinese and Mesopotamian civilisations as early as 5000 BP. The further back in time we go, however, the patchier the records become, and in some parts of the world historical records begin only in the last century.

These data gaps for historical time, and for longer time periods of thousands, tens of thousands and millions of years, can be filled in using natural archives. The geological timescale given in Table 1.2 is based on fossil evidence. Such 'proxy' methods are based on our knowledge of the current interrelationships between the different environmental spheres. Particular plants and animals thrive in particular climatic zones, for example, so that fossils can indicate former climates. The variability of climate during an organism's lifetime can also be inferred in some cases. Study of the width of the annual growth rings of trees gives an insight into specific ecological events that changed the tree's ability to photosynthesise and fix carbon. Essentially similar methods can be used to infer environmental variability from changes in the rate of growth of coral reefs. Other proxy 'palaeoenvironmental indicators' include pollen types found in cores of sediment taken from lake or ocean beds, and the rate of sediment accumulation in such cores can tell us something about erosion rates on the surrounding land. Landforms, too, become fossilised in landscapes to provide clues about past environmental conditions. Examples include glacial and periglacial forms in central and northern Europe, indicating colder conditions during previous glaciations, and fossilised sand dunes in the Orinoco Basin of South America, also dating from periods of high-latitude glaciation, which indicate an environment much drier than that of today.

As with instrumental data and historical archives, natural archives used as proxy variables are patchy in both their spatial and temporal extent. Coral reefs only grow in tropical waters, ice only accumulates under certain conditions, and not many trees live longer than 1000 years. Even instrumental data may not be perfectly reliable over long periods of time because methods and instrumentation can change, monitoring sites can be moved and external factors may alter the nature of the reading. The availability and limitations through time and space of some of the variables used to indicate temperature, a key palaeoenvironmental variable, in the Holocene are shown in Table 1.3.

It is clear that our understanding of how environments change can only be built up slowly in a patchwork fashion, but the understanding gained from all these lines of evidence can then be used to predict future environmental changes, incorporating any human impact, using models that simulate environmental processes. The human impact may still provide further complications, however, because in many instances through prehistory, history, and indeed in the present era, it can be difficult to distinguish between purely natural events and those that owe something to human activities (temperature readings at a town that becomes a city are an obvious example, because urbanisation affects temperature). It is the interrelationships between human activities and natural functions that form the subject matter of this book.

FURTHER READING

Burt, T.P. 1994 Long-term study of the natural environment – perceptive science or mindless monitoring. *Progress in Physical Geography* 18: 475–96. A cogent argument for long-term study as the basis for environmental science.

Clayton, K. 1991 Scaling environmental problems. *Geography* 76: 2–15. An interesting appraisal of the importance of scale in our analysis of environmental issues.

Colinvaux, P. 1993 *Ecology 2*. New York, Wiley. This comprehensive textbook covers all the basics of ecology, and highlights many areas of ignorance and conflicting interpretations of the natural world.

Goudie, A.S. 2001 *The nature of the environment*, 4th edn. Oxford, Blackwell. A good basic overview of physical geography, including the influence of geology and climatology, profiles of major world environments and the workings of landscapes and ecosystems.

Hannah, L., Lohse, D., Hutchinson, C., Carr, J.L. and Lankerani, A. 1994 A preliminary inventory of human disturbance of world ecosystems. *Ambio* 23: 246–50. A global assessment of the human impact.

Nicholls, K.E. 1997 Planktonic green algae in western Lake Erie: the importance of temporal scale in the interpretation of change. *Freshwater Biology* 38: 419–25. A cautionary illustration of how interpretation of data can vary according to the length of data set.

Websites

http://gcmd.gsfc.nasa.gov/ a wide range of data on the earth sciences is available through the Global Change Master Directory.

http://inqua.nlh.no/ the International Union for Quaternary Research oversees scientific research on environmental change during the Quaternary.

http://lternet.edu/ the Long Term Ecological Research Network supports research on long-term ecological phenomena in the USA and Antarctic.

http://www.fao.org/gtos/ the Global Terrestrial Observing System is a programme for observations, modelling and analysis of terrestrial ecosystems to support sustainable development.

http://www.igbp.kva.se/ the International Geosphere–Biosphere Programme's mission is to deliver scientific knowledge to help human societies develop in harmony with Earth's environment.

POINTS FOR DISCUSSION

- How and why can we recognise large-scale ecosystem types called biomes?
- It is a biological fact that predators at the top of food chains are always rare. The only exception is humans and this is why we face so many environmental issues. How far do you agree?
- Prepare a report for a national agency outlining the arguments for and against the funding of long-term environmental monitoring.
- Explain why studies of the physical environment should be carried out at a range of temporal and spatial scales.

2

THE HUMAN ENVIRONMENT

TOPICS COVERED

Human perspectives on the physical environment
Human forces behind environmental issues
Human-induced imbalances
Time and space scales
Interest in environmental issues

Key words: resource, technology, driving forces, mitigating forces, tragedy of the commons, common property resources, transnational corporations (TNCs), exploitation, dependency, technocentric, ecocentric

Knowledge of the physical environment only illuminates one half of any environmental issue, since an appreciation of factors in the human environment is also required before an issue can be fully understood. Relationships between human activity and the natural world have changed greatly in the relatively short time that people have been present on the Earth. A very large increase in human population, along with widespread urbanisation associated with advances in technology and related developments of economic, political and social structures have all combined to make the interaction between humankind and nature very different from the situation just a few thousand years ago. This chapter is concerned with these aspects of humanity, which together provide the human dimensions of environmental issues.

HUMAN PERSPECTIVES ON THE PHYSICAL ENVIRONMENT

Nobody knows what prehistoric human inhabitants thought about the natural world, but fossil and other archaeological evidence can be marshalled to give us some idea of how they interacted with it. The fact that humans have always interacted with the physical environment is obvious, since all living things do so by definition, but the ability of humans to conceptualise has allowed us to formalise our view of these interactions. One such framework, which must have been around in one form or another for as long as people have inhabited the planet, is the idea of resources. We call anything in the natural environment that may be useful to us a natural resource. Aspects of the human environment, such as people and institutions, can also be thought of as resources, while anything that we perceive as detrimental to us is sometimes called a negative resource. Since resources are simply a cultural appraisal of the material world, individual aspects of the environment can vary at different times and in different places between being resources and negative resources. A tiger, for example, can be viewed as a resource (for its skin) or a negative resource (as a dangerous animal). Similarly, different cultures recognise resources in different ways (the Aztecs, for example, were mystified by the Spaniards' insatiable demand for gold, thinking that perhaps they ate it or used it for medical purposes), and some groups see resources where others do not (a grub is a food resource to an Aboriginal Australian, but not to

TABLE 2.1 *A classification of resources*

PHYSICAL ENVIRONMENT
Continuous resources, which to all intents and purposes will never run out (e.g. solar energy, wind, tidal energy)
Renewable resources, which can naturally regenerate so long as their capacity for so doing is not irreversibly damaged, perhaps by natural catastrophe or human activity (e.g. plants, animals, clean water, soil)
Non-renewable resources, which are available in specific places and only in finite quantities because although they are renewable, the rate at which they are regenerated is extremely slow on the timescale of the human perspective (e.g. fossil fuels and other minerals, some groundwaters)
HUMAN ENVIRONMENT
Extrinsic resources, which include all aspects of the human species, all of which are renewable (e.g. people, their skills, abilities and institutions)

most Europeans). Hence, resources vary in character and some are also more difficult to manage than others. The classification shown in Table 2.1 gives some idea as to their characteristics and manageability.

All environmental issues can be seen as the result of a mismatch between extrinsic resources and natural resources: they stem from people deliberately or inadvertently misusing or abusing the natural environment. The reasons for such inappropriate uses are to be found within the nature of human activity.

HUMAN FORCES BEHIND ENVIRONMENTAL ISSUES

The interactions between humankind and the physical environment result from our attempts to satisfy real and perceived needs and wants. The specific actions that cause environmental issues directly are well known. They include such things as modifying natural distributions of vegetation and animals, overusing soils, polluting water and air, and living in hazardous areas. At a deeper level, the forces that enable, encourage or compel people to act in inappropriate ways can be traced back to humankind's underlying behaviour,

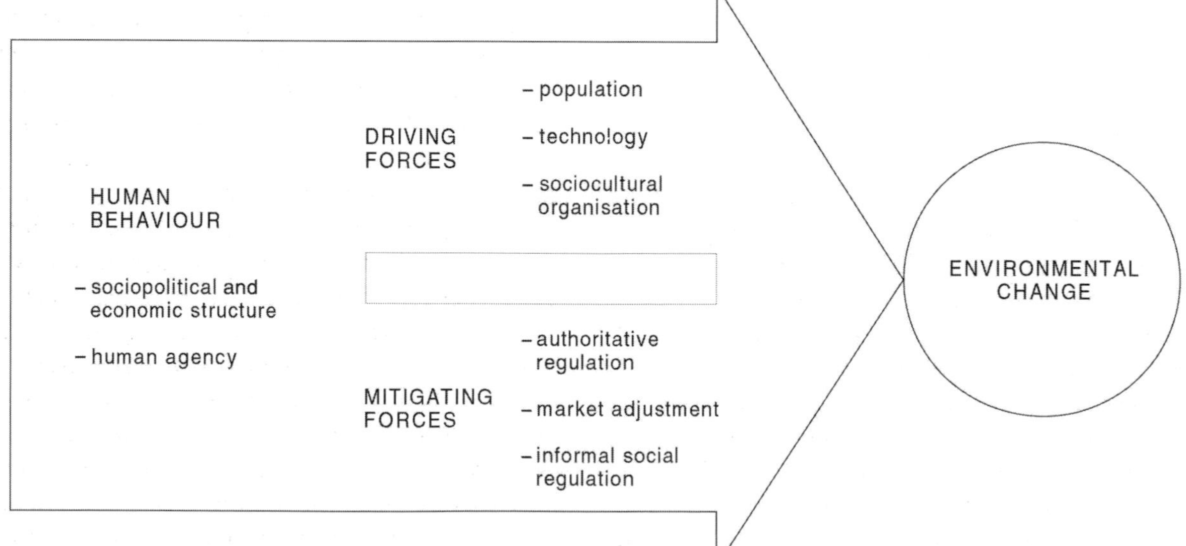

Figure 2.1 Human forces of environmental change (after Kates *et al.*, 1990).

TABLE 2.2 *World human population and growth rates*

Year	Total population (millions)	Annual growth rate (%)	Doubling period (years)
1 million BC	A few thousand	—	—
8000 BC	8	0.0007	100 000
AD 1	300	0.046	1500
1750	800	0.06	1200
1900	1650	0.48	150
1970	3678	1.9	36
2000	6199	1.7	41

Source: Otzen (1993)

which is expressed as a series of driving forces and mitigating forces in Fig. 2.1. To delve as deep as the nature of human behaviour is beyond the scope of this book, but further investigation of the driving forces, which promote human impact on the environment, and the mitigating forces, which act as checks to the driving forces or their impacts, will comprise most of the remainder of this chapter.

Population

Growth in the global population of human beings is widely recognised as one of the most clear-cut driving forces behind increased human impact on the environment. The exponential rise in human population numbers is clearly indicated in Table 2.2. While in 1750 the world population of about 800 million had a doubling period of more than 1000 years, 250 years later in the year 2000, the world population of more than 6 billion had a doubling period of less than 50 years. The logic is simple and undeniable: more people means a greater need for natural resources. As the population grows, more resources are used and more waste is produced.

The relationship between the human population, whose numbers can change, and natural resources, which are essentially fixed, has been an issue for a long time. At the end of the eighteenth century, the Englishman Thomas Malthus suggested that population growth would eventually outstrip food production and lead to famine, conflict and human misery for the poor as a consequence (Malthus, 1798). This Malthusian perspective continues to influence many interpretations of environmental issues. A rapid increase in the population of an area can result in over-exploitation of the area's resources, as numerous examples from refugee camps illustrate (see page 345). On a larger scale, the rapid rise in the population of Inner Mongolia since the 1950s is thought to have made a significant contribution to enhanced land degradation in the area (Takeuchi *et al.*, 1995). However, in many cases the relationship between population numbers and environmental degradation is not straightforward and is complicated by numerous other factors. There are, for example, no clear relationships between population density and soil erosion (Blaikie, 1985). Indeed, rapid population decline can also lead to environmental degradation if some aspects of good management are discarded because there are not enough people to keep them up. This sort of stagnation and decline occurred in many rural agricultural areas of the former Yemen Arab Republic in the 1970s when large numbers of workers emigrated to work in neighbouring oil-rich states (Carapico, 1985).

Furthermore, a high population density in

itself does not necessarily lead to greater degradation of resources. A small number of people living in a particular area can cause greater damage than larger numbers in a similar area, for example. One important factor influencing this equation is the level of technology used. A clear illustration of this comes from a recent study of the Machakos district in Kenya where an increased population density has had positive influences on both environmental conservation and productivity (Tiffen *et al.*, 1994). In the 1920s and 1930s, the landscape of Machakos was badly degraded, with widespread loss of vegetation and severe gullying. Over the period 1930–90, the population of Machakos increased nearly six-fold, but changes in agricultural practices, such as the widespread construction of terraces to control erosion, have helped to transform the condition of the environment.

Political and economic forces also affect the ways in which people use or abuse their resources. The relationship between population and environmental degradation is further complicated by the fact that many human impacts are not direct: people living in a city, for example, have influence on resource use far from their immediate surrounds. A walk down a supermarket aisle in any city will indicate the great distances some products have been transported before they reach the urban consumer, an apt reflection of the globalisation of the economy. Further investigation of the importance of human population numbers is made in Chapter 3.

Technology

Developments in technology have been closely associated with population growth. One view sees technological developments as a spur to growth, so that agricultural innovations provide more food per unit area and enable more people to be supported, for example. An opposite perspective sees technological change as a result of human inventiveness reacting to the needs created by more people (Boserüp, 1965). This latter view can be used to suggest that a growing human population is not necessarily bad from an environmental perspective. Any problems created by increasing population will be countered by new innovations that ease the burden on resources by using them more effectively. One illustration of this line of argument can be drawn from society's use of non-renewable fossil fuel resources. Since fossil fuel supplies will not last forever, and their current use is causing a range of environmental problems, such as global warming and acid rain, energy conservation is being promoted and alternative forms of renewable energy supplies are being developed (see Chapter 18).

However, many arguments can be presented to indicate that technological developments are responsible for much environmental degradation. Technology influences demand for natural resources by changing their accessibility and people's ability to afford them, as well as creating new resources (uranium, for example, was not considered a resource until its energy properties were recognised). The Industrial Revolution has been associated with high population growth and a greatly enhanced level of resource use and misuse; it has promoted urban development, improved transport and resulted in the globalisation of the economy. But although these developments have undoubtedly increased the scale of human impact on the environment, this is not to say that pre-industrial technologies did not also allow severe impacts. Polynesians who reached the planet's last habitable areas, the Pacific islands, within the last 1000 to 4000 years, are thought to have exterminated more than 2000 bird species (some 15 per cent of the world total) with only Stone Age technology (Pimm *et al.*, 1995, see also page 257).

Technology has a direct impact on resources, of course, and virtually every environmental issue can be interpreted as a consequence, either deliberate or inadvertent, of the impact of technology on the natural world. Advances in earth-moving and concrete technology, for example, have precipitated a marked escalation in the rate and scale of construction of big dams on rivers all over the world in recent times, and the adverse environmental impacts associated with these structures have attracted increasing public attention (see Chapter 9). The numerous examples of inadvertent impacts are usually the result of ignorance: a new technology being tried and tested and found to have undesirable consequences. The use of pesticides and fertilisers has helped to improve levels of food production, but they have also had numerous unintended negative effects on other aspects of the world in which we live (Conway and Pretty, 1991). On the other hand, technology is not environmentally damaging by definition, since technological applications can be designed and used as a mitigating force, in many cases in response to a previous undesirable impact. Measures introduced to reduce harmful emissions from motor vehicles, such as the catalytic converter (see page 277), are just one illustration of this point. When and where technology is developed, used or abused is dictated by the nature of society and its organisation.

Sociocultural organisation

The influences of population and technology are intimately linked to the organisation of human society. Economic, political and social values, norms and structures are a diverse set of driving forces that influence environmental change. They also underpin the mitigating forces that have been developed by societies to offset some of the damaging aspects of environmental issues. Dramatic transformations in population and technology have been associated with the two most significant changes in the history of humankind – the Agricultural Revolution of the late Neolithic period and the Industrial Revolution of the eighteenth and nineteenth centuries – but these changes have also been reflected in equally marked modifications to the ways that society is organised.

One of the most striking of these social changes is the rise of the city. Although cities have been a feature of human culture for about 5000 years, virtually the entire human species lived a rural existence just 300 years ago. Today, the proportion of the world's population living in cities is approaching 50 per cent, and individual urban areas have reached unprecedented sizes. In most cases, urban areas have much higher population densities than rural areas, and the consumption of resources per person in urban areas is greater than that of their rural counterparts. The environmental impacts of cities within urban boundaries are obvious, since cities are a clear illustration of the human ability to transform all the natural spheres, but the influence of cities is also felt far beyond their immediate confines. The high level of resources used by city dwellers has been fuelled by extending their resource flows. Economic and political forces have developed to facilitate this extension, so that the world is now an integrated whole.

Although integrated, the world has also become a very unbalanced place in terms of human welfare and environmental quality. Much attention has been focused on how economic and political forces have produced global imbalances in the way human society interacts with the environment, not just between city and countryside, but between groups within society with different levels of access to power and influence (e.g. women and men, different political parties, ethnic groups.) On the global scale, there are clear imbalances between richer nations and poorer nations. In general terms, the wealthiest few are disproportionately responsible for environmental issues, but at the other end of the spectrum the poorest are also accused of a responsibility that is greater than their numbers warrant. The motivations for these disproportionate impacts are very different, however. The impact of the wealthy is driven by their intense resource use, many say their overconsumption of resources. The poor, on the other hand, may degrade the environment because they have no other option (Fig. 2.2). Economic power is seen as a vital determinant: the wealthy have become wealthy because of their high-intensity resource use, and can afford to continue overconsuming and to live away from the problems this creates. The poor cannot afford to do anything other than overuse and misuse the resources that are immediately available to them, and as a consequence they are often the immediate

Figure 2.2 The health of this argun tree in south-west Morocco is not improved by goats browsing in its canopy, but herders often have few options but to overuse the resources that are immediately available to them.

victims of environmental issues. This difference in economic power is also manifested in political power: the wealthy generally have more influence over decisions that affect interaction with the environment than the poor, although when marginalised people are pushed to the edge of environmental destruction they may become active in forcing political changes (Broad, 1994). Some of the causes and manifestations of these global and regional imbalances are investigated further in the following sections.

HUMAN-INDUCED IMBALANCES

Human society has created a set of sociocultural imbalances that is superimposed on geographical patterns of environmental resources on all spatial scales from the global to the individual. These patterns of uneven distribution have been developed by, and are maintained by, the processes and structures of economics, politics and society. These

TABLE 2.3 *Some theories to explain why environmental issues occur*

Theory	Explanation
Neo-Malthusian	Demographic pressure leads to overuse and misuse of resources
Ignorance	Ignorance of the workings of nature means that mistakes are made, leading to unintentional consequences
Tragedy of the commons	Overuse or misuse of certain resources occurs because they are commonly owned
Poor valuation	Overuse or misuse of certain resources occurs because they are not properly valued in economic terms
Dependency	Inappropriate resource use by certain groups is encouraged or compelled by the influence of more powerful groups
Exploitation	Overuse and misuse of resources is pursued deliberately by a culture driven by consumerism
Human domination over nature	Environmental issues result from human misapprehension of being above rather than part of nature

Source: after Barrow (1991)

imbalance theories, combined with the human driving force theories outlined above, together make up a diverse set of explanations that have been put forward to explain the human dimension behind environmental issues (Table 2.3).

Ownership and value

Two interrelated theories that explain the underlying causes of imbalance between human activities and the environment stem from differential ownership of certain resources and the values put on them. Some environmental resources are owned by individuals while others are under common ownership. One theory argues that resources under common ownership are prone to overuse and abuse for this very reason – the so-called 'tragedy of the commons' (Hardin, 1968). The example often given to illustrate this principle is that of grazing lands which are commonly owned in pastoral societies. It is in the interest of an individual to graze as many livestock as possible, but if too many individuals all have the same attitude the grazing lands may be overused and degraded: the rational use of a resource by an individual may not be rational from the viewpoint of a wider society. The principle can also be applied to explain the misuse of other commonly owned resources, such as the pollution of air and water or catching too many fish in the sea.

It is important to note, however, that common ownership does not necessarily lead to over-exploitation of resources. In many areas where resources are commonly owned, strong social and cultural rules have evolved to control the use of resources. In situations like this, resource degradation usually occurs because the traditional rules for the control of resource use break down for some reason. Reasons include migration to a new area, changes in ownership rights, and population growth. In examples like overfishing in the open oceans, by contrast, the tragedy of the commons applies because there is no tradition of rules developed to limit exploitation.

A related concept is the undervaluation of certain resources. Air is a good example. To all intents and purposes, air is a commonly owned

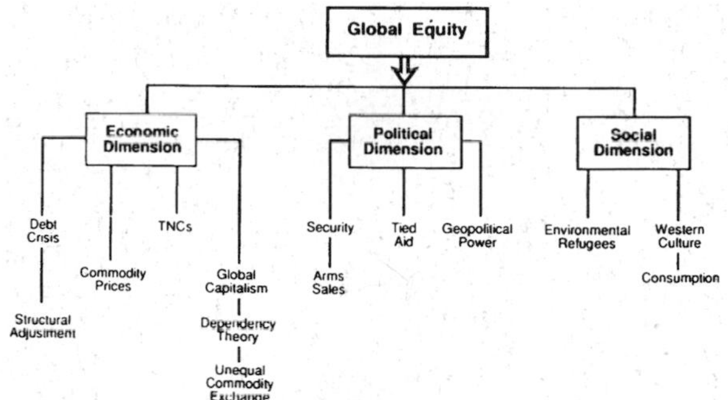

Figure 2.3 Structural inequalities in the global system (Lonergan, 1993).

continuous resource that, in practice, is not given an economic value. The owner of a windmill does not pay for the moving air the windmill harnesses, nor does the owner of a factory who uses the air as a sink for the factory's wastes. Since air has no economic value it is prone to be overused. A simple economic argument suggests that if an appropriate economic value was put on the resource, the workings of the market would ensure that as the resource became scarce so the price would increase. As the value of the resource increased, theory suggests that it would be managed more carefully.

Exploitation and dependency on the global scale

A complex series of economic, political and social processes has resulted in patterns of exploitation and dependency, which are associated with the misuse of resources. Inequalities exist between many different groups of people (see above) and can be identified on several different scales. Three levels are identified by Lonergan (1993): two spatial (international and national) and one temporal (intergenerational).

Some of the main structural inequalities of the global system are shown in Fig. 2.3. They have evolved from colonial times to the point, today, where direct political control of empires has been superseded by more subtle control by wealthier countries over poorer ones. The economic

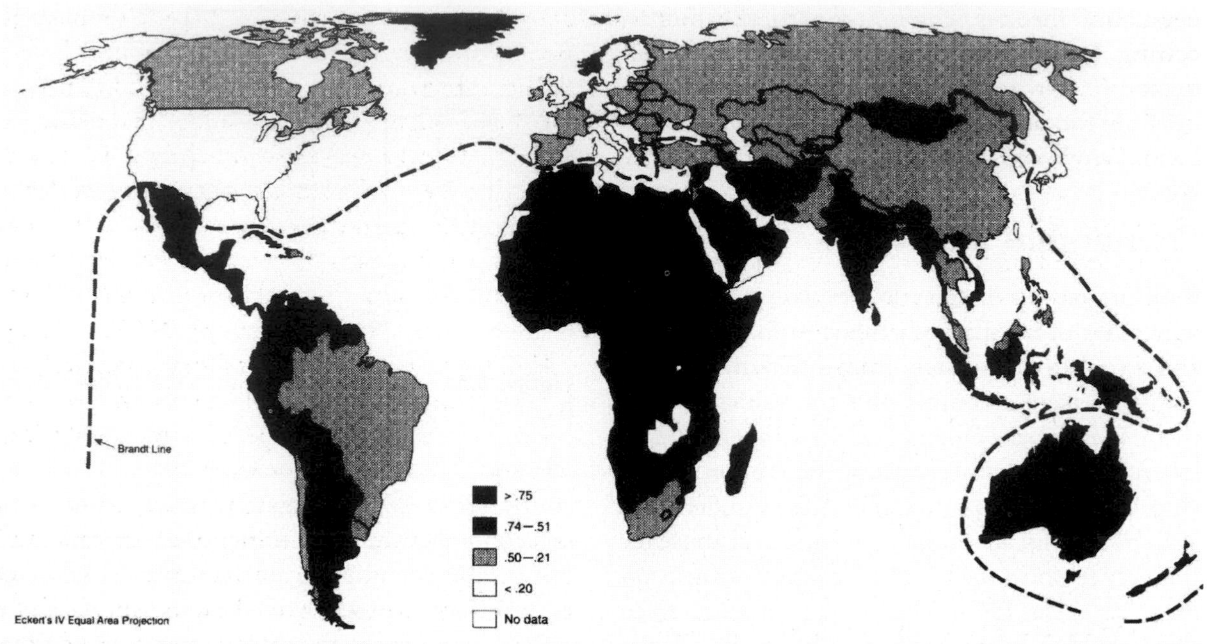

Figure 2.4 Index of commodity concentration of exports (Knox and Agnew, 1994). The 'Brandt line' represents the division between countries of the more affluent 'North' and the poorer countries of the 'South'.

dimension is particularly important, since it influences the rate of exploitation of natural resources in particular countries, the relative levels of economic development apparent in different countries, and the power of certain governments to control their own future.

The realities of global inequality in economic terms are stark. One billion people live in unprecedented luxury, while one billion live in destitution. Children in the USA even have more in pocket money (US$230 a year) than the half-billion poorest people alive (Durning, 1991). The gap between countries has also been widening in recent decades. In 1960, the richest 20 per cent of the world's population, who lived in high-income regions including western Europe, Japan, North America and Australia, absorbed 70 per cent of global income, about 30 times more than the world's poorest 20 per cent, who took just 2.3 per cent of global income (UNDP, 1993). By 2000, the richest 20 per cent of people accounted for 86 per cent of total private consumption, while the poorest 20 per cent accounted for only 1.3 per cent (UNFPA, 2001). On an individual level, a child born at the turn of the century in an industrialised country will add more to consumption and pollution over his or her lifetime than 30 to 50 children born in developing nations.

The structural aspects of this economic dimension have many facets, including the fact that many poorer countries are in debt to the countries and banks of the rich world (the so-called 'debt crisis'), and many less-developed countries rely on a limited number of exports, which are usually primary products such as agricultural goods and minerals. Figure 2.4 illustrates this picture globally with regard to exports. A low index of commodity concentration of exports reflects a diverse export base while a high index reflects a country's reliance on the export of a few agricultural or mineral resources. This pattern is maintained by terms of trade and prices for primary exports ('unequal commodity exchange'), which are set largely by the countries of the North that represent the major markets for these exports. Overexploitation of resources in poorer countries often occurs in response to the falling commodity prices that have been typical of recent decades, and the need to service debts.

Transnational corporations (TNCs), the very large majority of which have headquarters in the advanced capitalist regions, particularly Japan, North America and western Europe, also play an important role in the workings of global economics. Comparison of the annual turnover of TNCs with the Gross National Product (GNP) of entire nation-states gives an indication of this role:

> *By this yardstick, all of the top 50 transnational corporations – including the likes of Exxon, General Motors, Ford Motor Company, Matsushita Electronics, IBM, Unilever, Philips, ICI, Union Carbide, ITT, Siemens and Hitachi – carry more economic clout than many of the world's smaller peripheral nation-states; while the very biggest transnationals are comparable in size with the national economies of semi-peripheral states like Greece, Ireland, Portugal, and New Zealand.*
>
> (Knox and Agnew, 1994: 38–9)

In development terms, investment by TNCs in developing countries has been portrayed as an engine of growth capable of eliminating economic inequality on the one hand, and as a major obstacle to development on the other. Some view such investment as a force capable of dramatically changing productive resources in the economically backward areas of the world, while others see it as a primary cause of underdevelopment because it acts as a major drain of surplus to the advanced capitalist countries (Jenkins, 1987).

The negative view has been illustrated by comparing the development paths of Japan and Java, which had a similar level of development in the 1830s. While Japan has subsequently developed much further, in spite of a poorer endowment of natural resources, Java has remained underdeveloped, despite a position on main trade routes and a good stock of natural resources, because few of the profits from resource use have been reinvested in the country (Geertz, 1963).

One aspect of these relationships, which some consider to be detrimental to the environment in poorer countries, is the movement of heavily polluting industries from richer countries that have developed high pollution-control standards, to poorer countries where such regulations are less stringent. Fear of an increase in this trend was expressed during the negotiations for the Uruguay round of the General Agreement on Tariffs and Trade (GATT) and the establishment of the North American Free Trade Agreement (NAFTA) (Daly, 1993). Daly notes examples of 'maquiladoras', US factories that have located mainly in northern Mexico to take advantage of lower pollution-control standards and labour costs.

Economic power is closely related to political power, which can be seen in the sale of arms and the giving of aid by richer countries to poorer ones. The workings of global capitalism also have social dimensions, including the spread of western cultural norms and practices (Fig. 2.5). The sale of western products can also be seen to reinforce the role of developing nations as suppliers of raw materials, as Grossman (1992) shows for the increasing use of pesticides on Caribbean agricultural plantations where produce is destined for export markets, a trend influenced by foreign aid, among other criteria.

Figure 2.5 Domination of the world's poorer countries by their richer counterparts takes many forms, including what some term 'cultural imperialism', epitomised by this advertisement for Pepsi in Ecuador. Transfer of western technology is often in the form of polluting industries, as shown behind.

Exploitation and dependency on the national scale

Perhaps the most important underlying aspect of the global inequalities is the way in which they combine to maintain a situation where a minority of wealthy nations and most of their inhabitants are able to live an affluent life consuming large quantities of resources at the expense of many more poorer nations and their inhabitants. Nevertheless, some countries do move up the economic development scale, and the terms 'North' and 'South' disguise much internal variability (Fig. 2.4). Structural inequalities also exist at the national level, however, and these inequalities similarly have implications for the ways in which people interact with the environment within countries. Some of the aspects of inequality on the national scale, categorised according to their economic, political and social dimensions, are shown in Fig. 2.6.

As at the global level, the national scene is often characterised by a small élite group that has more economic, political and social power than the majority of people in more marginal groups. In general terms, this pattern is often manifested in a rural/urban divide: cities are centres of power, although they also typically contain a poorer 'underclass' in both rich and poor countries. Outside the city in the South, most rural people do not have access to economic power: in Africa, Asia and Latin America, for example, more than 85 per cent of farmers are estimated to lack access to credit (Holmberg, 1991). Corruption and centralised control are all too frequent characteristics of political and economic management in developing countries, which often means that local communities lack power over their own resources and how they are managed. These aspects are often tied up with social dimensions such as human rights.

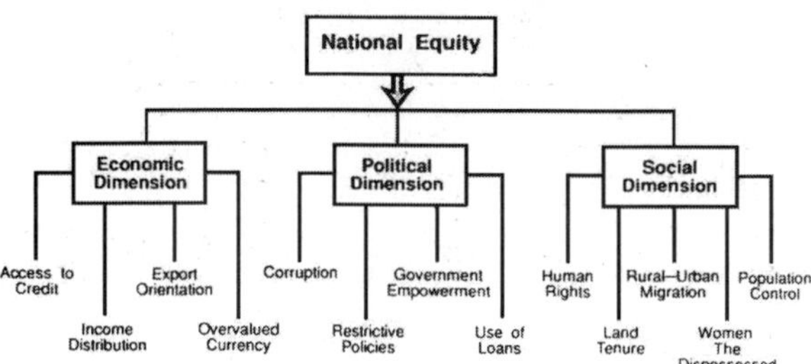

Figure 2.6 Structural inequalities in national systems (Lonergan, 1993).

Another important social dimension is land tenure; poorer, marginalised groups tend to lack ownership of land and its associated resources. One study of rural villages in India found that 14–24 per cent of average household income was derived from natural resources that were open for all to use, so-called 'common property resources' (Jodha, 1992). When people do not own the resources they use, there is a danger of overexploiting them: the tragedy of the commons. There may also be little incentive to invest in the future productivity of an area if a person's future entitlement to that area is in question. Fertilising a field to ensure that it retains its productivity means not using resources for immediate benefit: the animal dung used as fertiliser, for example, could have been used as fuel. Instead, those resources are invested for benefits received some time in the future. But if a farmer is not sure that he will still be able to use the field in future years, he may decide not to invest his animal dung.

The ways in which national inequalities translate into environmental issues often mean that the poor and disadvantaged are both the victims and agents of environmental degradation. The poor, particularly poor women, tend to have access only to the more environmentally sensitive areas and resources. Even where they do have ownership it is often of low-quality resources. They therefore suffer greatly from productivity declines due to degradation such as deforestation or soil erosion. Their poverty also means that there may be little alternative but to extract what they can from the sparse resources available to them, with degradation often being the result. The high fertility rates typical of poor households put further strains on the stock of natural resources.

Of course, not all poor people 'mine the future' in this way. The poor have used natural

Figure 2.7 Refugees in northern Thailand who have fled their native lands in neighbouring Myanmar (Burma). A long-running armed struggle between indigenous peoples and the government has been exacerbated in recent years by government-sponsored clearance of tropical forests in north-east Myanmar.

resources for many generations, and they can live in harmony with the environment using traditional methods. Problems tend to arise when a poor society is subject to some kind of perturbation or shock, whether natural (e.g. floods or droughts) or human-induced (e.g. land tenure changes). Just like in the physical environment, human societies display resistance and resilience to such perturbations, but if a shock exceeds a certain threshold, common responses are either to migrate from the affected area or to stay and try to survive.

The impacts can be seen in numerous examples. Political power in the hands of a minority group can be used to force large concentrations of marginal groups into small parts of the national land area, causing environmental degradation through overpopulation. Examples include the homelands system of apartheid South Africa, and numerous similar inequalities exerted by more powerful ethnic groups in many other countries. The exploitation of resources by élite groups can also result in coerced migration of indigenous inhabitants. Such 'environmental refugees' often take refuge in neighbouring countries (Fig. 2.7). Political, economic and social forces also combine to make marginal groups more vulnerable to natural hazards (Blaikie *et al.*, 1994).

Intergenerational inequalities

The spatial dimensions of exploitation and dependency are also closely related to the temporal dimension. As wealthy, powerful groups exploit available resources on the global and national scales, the inheritance of future generations is compromised.

Marginalised groups who are compelled to overexploit the resources available to them are doing so at the expense of their own futures. Overusing a renewable resource today leaves that resource degraded for future users. Using a non-renewable resource today, means that less of that resource will be available for tomorrow. Mining the future in this way will be offset to some extent, for example by changes in technology which can change our perception of resources, as well as changes in political, economic and

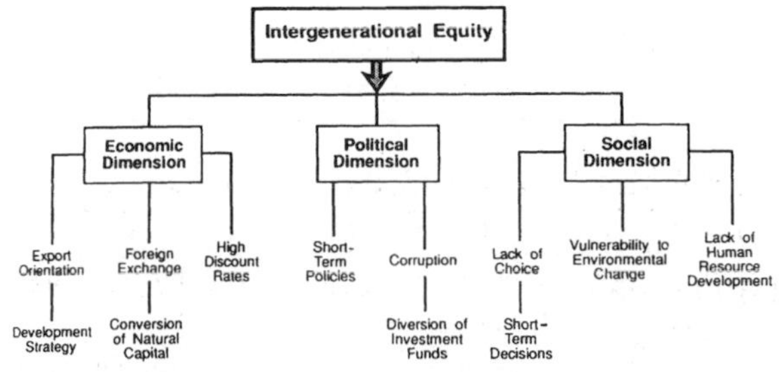

Figure 2.8 Structural inequalities across generations (Lonergan, 1993).

demographic factors. Various aspects of intergenerational inequalities are shown in Fig. 2.8. Resolving these inequalities across generations is the central aim of 'sustainable development', a concept dealt with in more detail in Chapter 3.

TIME AND SPACE SCALES

The above sections on exploitation and dependency illustrate the fact that, just as in the physical environment, key factors influencing events in the human environment vary at different combined spatial and temporal scales. Decisions about our interaction with the physical environment are taken at all levels, from the individual through the local neighbourhood, city, region, country and continent, up to the global level. None of these decision-makers operates in a closed self-supporting system; all are a part of larger interacting systems. Hence, when surveying the economic and social objectives of an area, for example, it might transpire that local community interests differ from interests at the national level.

Similarly, environmental views and decisions are taken with a wide range of time perspectives. A number of time perspectives have also been illustrated above: contrast the poverty-stricken individual who is compelled to overexploit resources today in order survive regardless of the longer-term consequences of his actions, with the time taken to develop effective pesticides designed to enhance food production. Inevitably, we tend to focus on scales directly relevant to people, but we should not forget other scales that may have less direct but no less significant effects. Indeed, successful management of environmental issues relies on the successful identification of appropriate scales and their linkages in the human environment.

INTEREST IN ENVIRONMENTAL ISSUES

People who derive their livelihoods directly from the physical environment have always been interested in environmental issues. When *Homo sapiens* first emerged as a species in his own right, the availability of wild plants and animals was literally a matter of life and death. Farmers and other groups directly involved in extracting physical environmental resources continue to rely on their knowledge and understanding of those resources to make a living or simply to survive. Today, environmental issues are also a prominent concern of government and the general public in most countries of the world (Fig. 2.9), perhaps more widely so than at any time in the past; but worry over environmental matters among those not directly involved in extracting natural resources is also by no means a recent phenomenon. Observers of the natural world in classical Greece, imperial Rome and Mauryan India expressed reservations about the wisdom of human actions that promoted accelerated soil erosion and deforestation. In more recent times, some of the earliest environmental legislation was introduced to protect the quality of natural resources. At the end of the thirteenth century, for example, laws were passed by the Mongolian state to protect trees and animals from overexploitation, and in England atmospheric pollution prompted a royal decree in 1306 forbidding the burning of coal in London. Warning voices were also raised at the damage caused by deforestation and plantation agriculture during the earliest periods of European colonial expansion, on the Canary Islands and Madeira from the early fourteenth century and in the Caribbean after 1560.

Indeed, the origins of today's global concern for the environment can be traced to some of these early European colonial experiences (Grove, 1990). While European colonies were often the sites of environmental degradation, they were also the cradles of modern scientific views on the issues. Colonial administrators and professional scientists working for trading companies in the eighteenth and nineteenth centuries developed western conservationist ethics and environmentally sound management practices. Although such approaches in themselves were not new, since many traditional rural societies had employed similar practices locally for a long time

Figure 2.9 Public hoarding to promote environmental awareness in Mombassa, Kenya. Concern over environmental issues is increasing in most parts of the world.

before, this was the first time that a global perspective could be attained. Hence, international travel and observations by naturalist explorers such as Charles Darwin and Alexander von Humboldt allowed them to develop scientific theories that revolutionised western thinking on the way nature was organised and worked.

The recent rise in interest in the natural environment has thrown up a phenomenal range of different approaches to dealing with the issues. Many of the western schools of thought, with their unfortunately jargonistic labels, are outlined in Table 2.4. This plethora of different movements and interest groups can be divided roughly into two camps, those adopting a 'technocentric' approach and those favouring an 'ecocentric' line (O'Riordan and Turner, 1983). Essentially, technocentrism is based on the belief that the way human society interacts with the natural environment only needs minor modifications to deal successfully with environmental issues. Technology and economics will provide the answers to environmental problems, hence existing structures of political power should be maintained, but political, planning and educational institutions should be more responsive to environmental issues. Ecocentrism is rooted in different beliefs. Environmental issues are more the result of problems with our ethical and moral approaches to nature, rather than simply technical difficulties. Fundamental changes in our worldview are needed to solve the issues successfully, along with a redistribution and decentralisation of power, and more emphasis on informal economic and social transactions. Among the approaches shown in Table 2.4, the cornucopian school of thought is an extreme example of the technocentric approach, while at the other end of the spectrum deep ecology represents an extreme of the ecocentric position.

TABLE 2.4 *A typology of western environmentalist thought in the latter half of the twentieth century*

School of thought	Main tenets
Cornucopian	Any problems easily solved by human ingenuity and the market economy, but no major problems perceived
Market-based approaches	Emphasis on the use of market-based instruments to internalise environmental externalities and tackle sources of market failure
Managerialists and technical fixers	Reliance on organisational or technical solutions
Institutional reform school	Promotion of sustainable development through better integration of environmental policy with economic development. Focus on the north–south dimension and the need for change in global monetary and trade policies
Environmental Keynesianism	Emphasis on the employment creation potential of public expenditure on environmental protection
Post-industrialists and liberal pluralists	Ideology of industrialism seen as underlying problem. Focus on need to change individual attitudes
The Limits to Growth school	Promotion of scientifically based public policy and the need to control both population and economic growth
Sociobiology and authoritarian ecology	Promotion of neo-Malthusian ideas and social Darwinism in public policy
Deep ecology	Preservation of wilderness areas and public policy modelled on intrinsic values of nature
Gaia hypothesis	Promotion of homeostatic and sustainable policies
Utopianism and anarchism	Transformation of society into numerous decentralised and largely self-sufficient communities
Orthodox Marxism	Environmental degradation seen as an inevitable consequence of capitalism
Ecofeminism	Domination and destruction of nature seen as a consequence of women's oppression by men
Post-Marxist structuralists	Environmental problems seen as stemming from structural features in the economy and society identifiable through historical analysis
Environmental planning and ecoarchitecture	Concerned with the design of the human environment and making cities and landscapes beautiful

Source: after Gandy (1996)

Cornucopians believe that physical environmental resources are there to be used, to fuel economic growth, the key to solving environmental problems, and that technological innovations driven by economic forces will provide substitutes when physical resources start causing problems for society. Deep ecologists advocate a minimum use of physical environmental resources, emphasising nature's intrinsic value, regardless of human perceptions, and conferring moral rights

TABLE 2.5 *Timetable of major political acceptance of some environmental issues and the dates when they were first identified as potential problems*

Environmental issue	Potential problem first identified	Decade of major political acceptance
Air and water pollution	Carson (1962)	1960s
Tropical deforestation	Marsh (1874)	
Acid rain	Smith (1852)	
Stratospheric ozone depletion	Molina and Rowland (1974)	1970s
Carbon dioxide-induced climatic change	Wilson (1858)	
Additional greenhouse gases	Wang *et al.* (1976)	1980s

Source: after Döös (1994)

on non-human species. When a cornucopian says 'use it', a deep ecologist says 'preserve it'.

Many of the schools of thought outlined in Table 2.4 tend to rely on a scientific approach to environmental issues, certainly when identifying and assessing the importance of an issue, and to some extent in the approaches taken to deal with it. Some movements, however, are critical of science. From the deep ecology viewpoint, environmental problems are the result of crises in society and in science itself (Capra, 1982). Identification of many of today's global environmental concerns was first made more than 100 years ago but most have only gained major political acceptance in recent decades (Table 2.5). Some observers blame the delays in acceptance on scientists themselves, for being poor communicators to a wider audience and for being overcautious in their conclusions (Döös, 1994), but scientists need to gather data and test theories before they can recommend action. Their endeavours are also passed to wider audiences through a multitude of public and private organisations interested in environmental issues. Such organisations include government ministries and research bodies as well as many non-governmental organisations (NGOs) that operate at international, national and local scales. Some would argue that the lag time between identification and acceptance of issues is reason enough to reject a technocentric approach. Indeed, on some occasions political recognition is forced by civil agitation from environmental activists, but science usually plays a role somewhere in the process. The difficulties scientists face in understanding the natural environment and the human effects on it have been outlined in Chapter 1.

FURTHER READING

Durning, A.B. 1989 *Poverty and the environment: reversing the downward spiral.* Worldwatch Paper 92. Worldwatch Institute, Washington, DC. This pamphlet looks at the many interactions between poverty and the environment, and proposes ways in which situations can be improved.

Naess, A. 1972 The shallow and the deep, long-range ecology movement. *Inquiry* 16: 95–100. A landmark paper in the rise of the deep ecology movement.

Rees, J. 1990 *Natural resources: allocation, economics and policy,* 2nd edn. London, Routledge. This book reviews the spatial distribution of renewable and non-renewable resources, their development and consumption. It also considers the distribution of wealth derived from resources, and the roles of management, economics and politics.

Simmons, I.G. 1996 *Changing the face of the Earth*, 2nd edn. Oxford, Blackwell. The author traces the interactions between people and the natural environment from the earliest times to the present.

Tiffen, M., Mortimore, M. and Gichuki, F. 1994 *More people, less erosion: environmental recovery in Kenya*. Chichester, Wiley. This study offers an alternative to the Malthusian model of how population growth affects environmental degradation.

Wildes, F.T. 1995 Recent themes in conservation philosophy and policy in the United States. *Environmental Conservation* 22: 143–50. A review of the various schools of thought behind nature conservation in the USA.

Websites

http://home.coqui.net/rosah/ site of the Comunidades Unidas contra la Contaminación, a grassroots environmental organisation fighting to stop pollution in an area of Puerto Rico.

http://www.worldbank.org/poverty a World Bank site providing resources and support for people working to alleviate poverty.

http://www.greenpeace.org/ one of the largest international NGOs devoted to environmental issues has a wide range of information on campaigns and data.

http://www.panda.org/ another of the largest international environmental NGOs with information and data on a wide range of issues.

http://www.unfpa.org/ the UN Population Fund site includes annual *State of the World Population* reports.

http://www.zpg.org/ Zero Population Growth, a US NGO.

POINTS FOR DISCUSSION

- Human society needs physical resources and environmental issues are an inevitable outcome of using them. Do you agree?
- Discuss the premise that all environmental issues are political issues.
- Explain the ways in which the ownership of physical resources is important in determining how they are used.
- Do you agree that the eradication of poverty is necessary before human society can live within the limits imposed by the physical environment?

3

SUSTAINABLE DEVELOPMENT

TOPICS COVERED
The precautionary principle
Valuing environmental resources
Growth and development
Population and technology

Key words: environmental services, maximum sustainable yield, lifecycle analysis, green GNP, *I = PAT*, carrying capacity

All the environmental issues covered in this book are, by definition, made up of physical and human elements. The approach throughout is to analyse both the physical problems and the human responses to understand why the issues are issues, and how solutions can be found. Although particular issues are apparent at particular spatial scales, all have some manifestation on the global scale. In the case of climatic change brought about by atmospheric pollution, the issue affects the entire planet because climate is a global phenomenon. In other cases, more localised issues have become so commonplace that, cumulatively, they occur on a worldwide scale. Environmental problems associated with agriculture, urban areas, industrial pollution, waste management, warfare, deforestation and soil erosion are just a few examples of such cumulative global issues. Since so many environmental issues now have a global nature, many believe that the time has come for a complete rethink of the way we view these issues, for a change in the philosophy that lies behind the ways in which people interact with the environment, and for a change in the methodology with which the interactions occur. The approach most widely advocated by this rethink is sustainable development, and this chapter looks in more detail at the need for sustainable development and some of the ways in which it might work.

THE NEED FOR CHANGE

The socio-economic system is just one part of the natural ecosystem in which materials are transformed and energy converted to heat (Fig. 3.1). The operation of the socio-economic system is dependent upon the ecosystem as a provider of energy and natural resources, and as a sink for wastes. The ecosystem also provides numerous 'environmental services' by virtue of its processes. These include the operation of climate and the hydrological cycle, recycling of nutrients, the generation of soils, pollination of crops, and so on. This ecological, economic perspective emphasises the fact that resource use and waste disposal take place in the same environment, and that both activities affect the life-support functions of the environment. Hence, the socio-economic system cannot expand indefinitely since it is limited by the finite global biosphere. Until recently, for example, the number of fish that could be sold at market was limited by the number of boats at sea; now it is limited by the number of fish in the sea.

Through much of the history of human occupation of the planet, the socio-economic system has been small relative to the biosphere, so that resources were plentiful, the environment's capacity for assimilating wastes was large, and biospheric functions were relatively little affected by

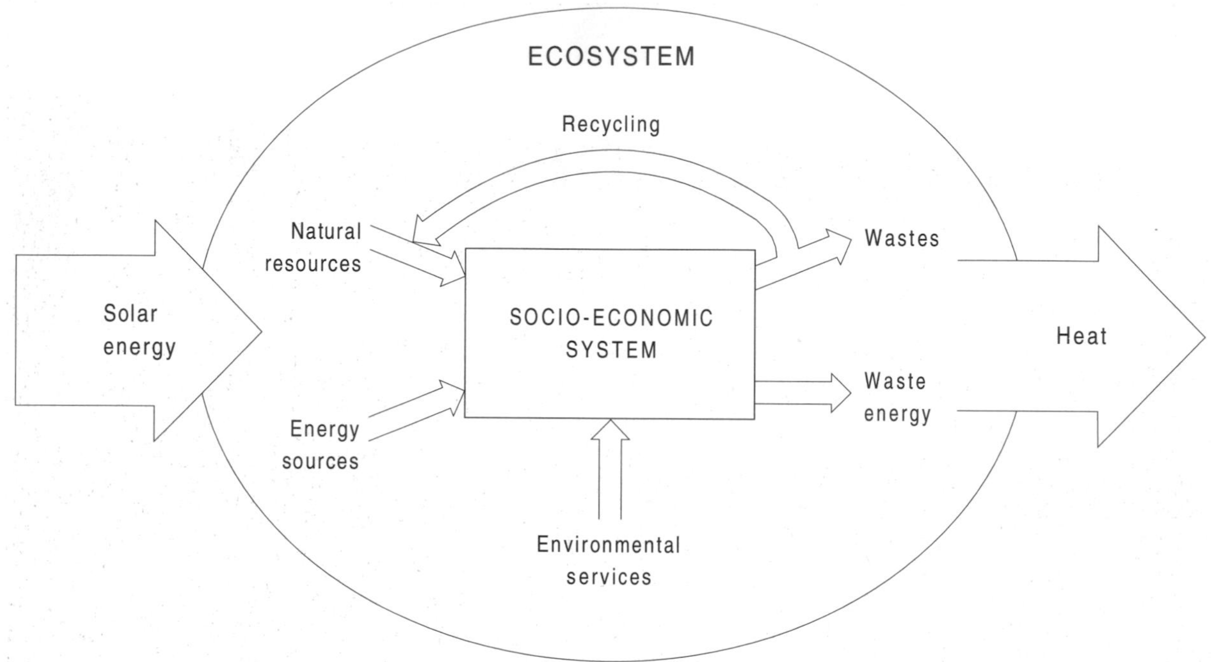

Figure 3.1 The socio-economic system as part of the global ecosystem (after Folke and Jansson, 1992; Daly, 1993).

human impact. Today the situation is different. As Goodland *et al.* (1993a: 298) put it: 'now the world is no longer "empty" of people and their artefacts, the economic subsystem having become large relative to the biosphere'. As the socio-economic system has grown larger, fuelled by increased 'throughput' of energy and resources, its capacity for disturbing the environment has increased. Since this disruption ultimately feeds back on the operation of society itself, society's behaviour must conform more closely with that of the total ecosystem because otherwise it may destroy itself (Folke and Jansson, 1992).

The potential for socio-economic degeneration and eventual collapse to occur when the human system operates too far out of harmony with the environment can be illustrated from both contemporary and historical examples, since although human activity has until recently been within the capacity of the environment on the global scale, breakdown has occurred on more local scales. Folke and Jansson (1992) note the recent rapid self-generated collapse of coastal shrimp industries in Taiwan and Thailand, within a decade of their beginnings, due to clearance of mangroves, which act as nursery and feeding areas for shrimps, and the deterioration of water quality through eutrophication and disease. A lack of harmony between the socio-economic system and the natural ecosystem has also been suggested as a cause of collapse for several ancient cultures. The decline of the Mayan civilisation in Central America that began around the year 900 may have been due to excessive use of soils and an over-reliance on maize, which failed due to a virus (Fig. 3.2). Similarly, the civilisation that flourished in the tenth to twelfth centuries around Angkor Wat in present-day Cambodia was based on a sophisticated irrigation system, but forest clearance for cropland resulted in high silt loads, which clogged the canals in rainy-season floods. Irrigation channels were abandoned to become stagnant swamps where mosquitoes bred prolifically, and malaria epidemics swept through the city. This weakened Angkor Wat's capacity to adapt to change and the city was abandoned (McNeely, 1994).

Figure 3.2 The ruins at Palenque in southern Mexico, a reminder of the collapse of the Mayan civilisation. No one is sure of the true reasons for the downfall of the Mayas, but one theory implicates human-induced environmental degradation as a key factor.

A schematic representation of three ways in which human society interacts with the environment is shown in Fig. 3.3. The model can be applied on various scales, from the global to the local, and across differing timescales. In cycle A, which is typical of the global economy in historical times, wealth is accumulated largely by degrading the environment. This wealth has

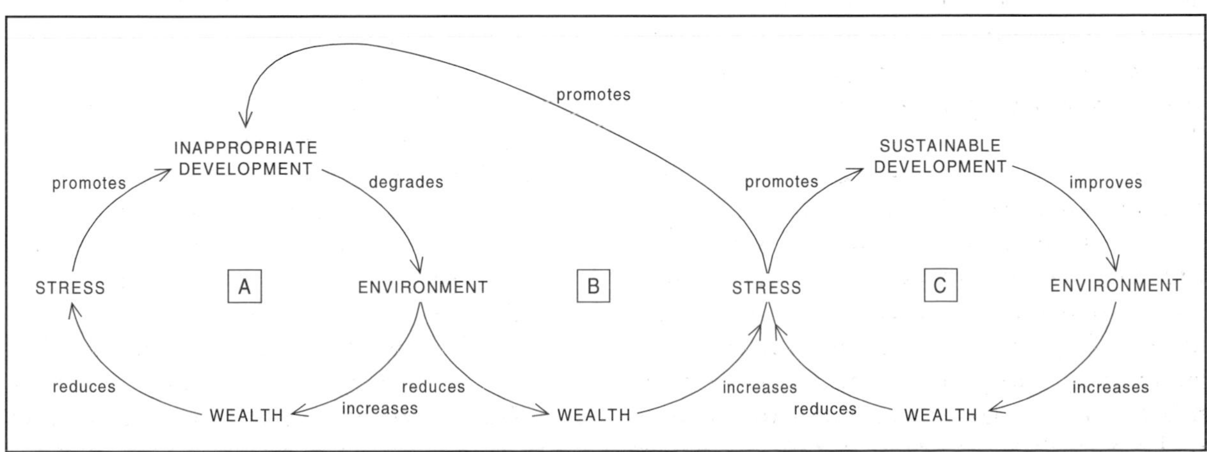

Figure 3.3 Three cycles showing the relationship between modes of development and the environment.

brought numerous advances to most parts of the world, which reduce 'stress', a term used here in a wide sense to reflect the general well-being of society. Such advances include sanitation and other facilities, improved health and higher living standards. These improvements have promoted further inappropriate development to continue on cycle A.

At some point in time, however, cycle A crosses an environmental threshold and the society may enter cycle B. The degraded environment begins to feed back on socio-economic wealth, and stress is increased. Increased stress in cycle B promotes further inappropriate development, particularly when the society in question has limited options as examples of poor, vulnerable groups quoted in this book show (e.g. accelerated soil erosion caused by farmers in Haiti and Ethiopia is discussed in Chapter 14), and on the national scale where rapid deforestation in many tropical countries has been pursued in response to the burden of debt and declining commodity prices (see Chapter 4). Continuing on cycle B can ultimately lead to socio-economic collapse, as in the examples of modern shrimp industries and ancient civilisations cited above.

Many people believe that, globally, we have now entered cycle B due to the sheer numbers of people on the Earth and the resulting scale of human impact on natural systems. There are numerous examples throughout this book which indicate that the productivity of many of the world's renewable resource-producing systems has reached its peak and in some cases is in decline. Currently, as much as 20–40 per cent of the potential global net primary productivity in terrestrial ecosystems is estimated to be diverted to human activities (Wright, 1990); about 20 per cent of atmospheric carbon dioxide results from human action; about 20 per cent of bird species on Earth have become extinct in the last two millennia, almost all due to human activities; more than 40 per cent of the planet's land surface has been transformed by human society; we use about 50 per cent of all accessible surface fresh water; 60 per cent of fixed nitrogen released each year owes its origin to human activity, and 66 per cent of marine fisheries are fully exploited, overexploited or depleted (Vitousek *et al.*, 1997b). Increased stress is manifested by, amongst other things, increasing levels of world poverty, fears over the effects of human-induced global warming and the depletion of the ozone layer.

The only appropriate way to exit cycle B is to enter cycle C. To use a biological metaphor, the socio-economic system needs to be adjusted from its present dominantly parasitic relationship with the environment to a more symbiotic one. The global community does have this option; by adopting sustainable development we can live within the confines of the biosphere, enhance environmental quality and achieve greater wealth by changing the pattern of resource consumption. Adopting cycle C must be a permanent long-term strategy, however, because there is still a danger that reduced stress encourages a reversion to the old ways of cycles A and B.

Examples of societies that have managed to attain a sustainable use of resources can be quoted from both historical and recent times, although the view that all traditionalists were conservationists is somewhat romantic, as evidence of pre-Hispanic accelerated soil erosion in Mexico (see Chapter 14) and the widespread species extinctions that occurred in Hawaii prior to the arrival of European settlers (see Chapter 15) indicate. Contemporary examples of commercial resource exploitation, in which local communities have had to develop social methods to sustain the resources on which they depend, can also be cited (e.g. Acheson, 1975; Berkes, 1989). The experience of these societies can be put to good use by other groups seeking to adopt sustainable development.

SUSTAINABLE DEVELOPMENT

Despite the fact that the term sustainable development has become common currency among many groups, it is a confused and sometimes contradictory idea, and there is no widespread agreement as to how it should work in practice (Dovers and Handmer, 1993). The concept has developed in the international forum from a document called the 'World Conservation Strategy' (IUCN/UNEP/WWF, 1980), which argued that three priorities should be incorporated into all development programmes:

1 maintenance of ecological processes
2 sustainable use of resources
3 maintenance of genetic diversity.

Sustainable development gained further credence thanks to the World Commission on Environment and Development (also known as the Brundtland Commission after its chair Gro Harlem Brundtland of Norway), which was formed by the UN in 1983 and reported four years later (WCED, 1987). The Commission emphasised that the integration of economic and ecological systems is all-important if sustainable development is to be achieved, and coined a broad definition for sustainable development that is often quoted: 'development which meets the needs of the present without compromising the ability of future generations to meet their own needs' (WCED, 1987: 43). Although this definition is concise, it is none the less open to varying interpretations. What exactly is a *need*, for example, and how can it be defined? Something that is considered a need by one person, or cultural group, may not necessarily be thought of as such by another person or cultural group. Needs may also vary through time, as does the *ability* of people to meet their needs. Resources are simply a cultural appraisal of the world (see page 19), and our ability to identify and use them is a function of technology amongst other things, and technology changes. Likewise, the meaning of *development* can be interpreted in many different ways.

Despite the difficulties in pinning down sustainable development and understanding how it should be applied, similar calls for its adoption have been made subsequently. These have come from various international groupings (e.g. IUCN/UNEP/WWF, 1991), notably at the UN Conference on Environment and Development (UNCED), otherwise known as the Earth Summit, in Rio de Janeiro in 1992. But although use of the term sustainable development has become increasingly widespread, the fact remains that it is still an ambiguous concept. Perhaps this should not be surprising, since the word 'sustainable' itself is used with different connotations. When we 'sustain' something, we might be supporting a desired state of some kind or,

TABLE 3.1 *Extremes in the range of views on the guiding principles of sustainability*

Dominant economic worldview	Deep ecology worldview
Dominance over nature	Natural harmony: symbiosis
Natural environment is a resource for human use	All nature has intrinsic worth: biospecies equality
Material and economic growth for increasing human populations	Simple material needs: goal of self-realisation
Belief in ample reserves of resources	Earth's resources are limited
High technological progress and solutions	Appropriate technology and non-dominating science
Consumerism and growth in consumption	Make do with enough: recycling
National/centralised community	Minority traditions and religious community knowledge

Sources: after Gibbon *et al.* (1995); Colby (1989)

TABLE 3.2 *Three key areas for integrated action in the attempt to build a sustainable way of life discussed at the World Summit on Sustainable Development in Johannesburg in 2002*

Area	Needs
Economic growth and equity	Today's interlinked, global economic systems demand an integrated approach in order to foster responsible long-term growth while ensuring that no nation or community is left behind
Conserving natural resources and the environment	To conserve our environmental heritage and natural resources for future generations economically viable solutions must be developed to reduce resource consumption, stop pollution and conserve natural habitats
Social development	Throughout the world, people require jobs, food, education, energy, healthcare, water and sanitation. While addressing these needs, the world community must also ensure that the rich fabric of cultural and social diversity, and the rights of workers, are respected, and that all members of society are empowered to play a role in determining their futures

Source: United Nations (2001)

conversely, we might be enduring an undesired state (de Vries, 1989). These different meanings have allowed the concept to be used in varying, often contradictory, ways. The range of views on the matter is illustrated in Table 3.1 where the guiding principles of sustainability as perceived by the dominant economic worldview are contrasted with those put forward by deep ecology.

Further confusion over the meaning of the term sustainable stems from its use in a number of different contexts, such as ecological sustainability and economic sustainability. A central tenet of ecological sustainability is that human interaction with the natural world should not impair the functioning of natural biological processes. Hence concepts such as 'maximum sustainable yield' have been developed to indicate the quantity of a renewable resource that can be extracted from nature without impairing nature's ability to produce a similar yield at a later date. Economic sustainability tends to give a lower priority to ecosystem functions and resource depletion. It can result in different approaches to the environment depending on the economic objectives of the environmental manager concerned.

One strength of the sustainability idea is that it draws together environmental, economic and social concerns. In practice, most would agree on a number of common guiding principles for sustainable development:

- continued support of human life
- continued maintenance of environmental quality and the long-term stock of biological resources
- the right of future generations to resources that are of equal worth as those used today.

Deciding in greater detail just how these principles should be applied was the subject of the World Summit on Sustainable Development in Johannesburg in 2002, a follow-up to the Earth Summit. The preparations for Johannesburg indicated that efforts to build a truly sustainable way of life require the integration of action in three key areas, as shown in Table 3.2.

THE PRECAUTIONARY PRINCIPLE

As Chapters 1 and 2 have shown, our knowledge and understanding of both the physical and human environments are not complete and one of the greatest challenges of sustainability is society's need to formulate policy in the face of this uncertainty. We are uncertain of many aspects of environmental issues, from the local and discrete to the global and complex. These aspects include the shape of future societies, the nature and causes of environmental change and the severity of human impacts. One approach that has been widely adopted to help make decisions about environmental issues in the light of uncertainty is the precautionary principle.

The precautionary principle, which evolved in Europe, advocates that human responses to environmental issues should err on the side of caution, a sort of 'better safe that sorry' approach to interacting with the environment. It appears in the 1992 Rio Declaration and has been incorporated into European Union environmental policy. Like sustainable development itself, the precautionary principle is interpreted in different ways by different people, but Dovers and Handmer (1995) identify some common themes:

- uncertainty is unavoidable in sustainability issues
- uncertainty over the severity of environmental impacts resulting from a development decision or a current human activity should not be an excuse to avoid or delay environmental protection measures
- the principle recommends we should anticipate and prevent environmental damage rather than simply react after it has occurred
- the burden of proof should shift from the victim to the developer, so that those proposing an action must show that it will not harm the environment or that whatever practical measures available for preventing damage will be taken.

The principle can help society in making decisions about how it interacts sustainably with the physical environment, although it should be remembered that the answers to questions about sustainable development will continue to be informed as much by moral and political issues as by scientific understanding.

VALUING ENVIRONMENTAL RESOURCES

Much research and thinking about sustainable development has focused on modifying economics to better integrate its operation with the workings and capacity of the environment, to use natural resources more efficiently, and to reduce flows of waste and pollution. The full cost of a product, from raw material extraction to eventual disposal as waste, should be reflected in its market price (so-called 'lifecycle analysis'), although in practice such a 'cradle to grave' approach may prove troublesome for materials such as minerals (see Chapter 19). Many economists contend that a root cause of environmental degradation is the simple fact that many aspects of the environment are not properly valued in economic terms. Commonly owned resources – such as the air, oceans and fisheries – are particularly vulnerable to overexploitation for this reason, but when such things have proper price tags, it is argued that decisions can be made using cost–benefit analysis. Putting a price on environmental assets and services is one of the central aims of the discipline of environmental economics. This can be done by finding out how much people are willing to pay for an aspect of the environment or how much people would accept in compensation for the loss of an environmental asset. One of the justifications of environmental pricing is the fact that money is the language of government treasuries and big business, and thus it is appropriate to address environmental issues in terms that such influential bodies understand.

There are problems with the approach, however. People's willingness to pay is heavily dependent on their awareness and knowledge of the resource and of the consequences of losing it. Information, when available, is open to

TABLE 3.3 *Illustrative values of biological resources*

Resource	Location	Value
DIRECT VALUES		
Consumptive use		
Firewood, dung	Nepal, Tanzania, Malawi	90% of total national primary energy
Wild animals as food	Ghana	Main protein source for 75% of population
	Nigeria	20% of animal protein in rural areas
Wild pigs	Sarawak	US$100 million/year
Productive use		
Wild plants and animals	USA	4.5% of GDP (US$87 billion/year 1967–80)
Timber (exports)	Asia, Africa, South America	US$8.1 billion/year (1981–83)
Plant-based drugs	OECD countries	US$43 billion (1985) but rising to US$200 billion to 1.8 trillion if benefits from better health, etc., costed in
INDIRECT VALUES		
Non-consumptive use		
Each lion	Kenya, Amboseli National Park	US$27 000/year
Each elephant herd	Kenya, Amboseli National Park	US$610 000/year
National parks (via tourism)	Kenya	US$40/ha/year
Wetlands area as flood protection	Boston, USA	US$17 million
Option		
Wild relatives of crop species with future potential (e.g. teosinté maize)	Mexico	Incalculable by definition until used, but e.g. US$6.8 billion/year contribution to perennial maize hybrid
Existence		
Aesthetic value of existence of species	Worldwide	US$100 million/year taken by WWF

Source: McNeely (1988)

manipulation by the media and other interest groups. In instances where the resource is unique in world terms – such as an endangered species or a feature such as the Grand Canyon – who should be asked about willingness to pay? Should it be local people, national groups or an international audience? Our ignorance of how the environment works, and the nature of the consequences of environmental change and degradation also present difficulties. In the case of climatic change due to human-induced atmospheric pollution, for example, all we know for sure is that the atmospheric concentrations of greenhouse gases have been rising and that human activity is most likely to be responsible. However, we do not know exactly how the climate will change nor what effects any changes will have upon human society. We can only guess at the consequences, so we can only guess at the costs. Economists are undeterred by these types of problem: 'Valuation may be imperfect but, invariably, some valuation is better than none' (Pearce, 1993: 5).

McNeely (1988) has classified such economic values into two groups:

1. direct values, which can be subdivided into values for consumptive use and values for productive use, and
2. indirect values, which are made up of non-consumptive value use, option value and existence value.

Table 3.3 shows some examples of values that, where they have been calculated, appear to provide compelling arguments for the conservation and proper management of natural systems. For example, the value to the national economy of wilderness areas in Kenya, when conserved as National Parks, is estimated to be at least an order of magnitude greater per hectare than the land would yield if put to pastoral use, for example. However, differing values potentially derived from particular areas can be of benefit to different sectors of society. In the Kenyan example, management of the Masai Mara Reserve as a game park principally benefits the operators of tourist lodges and tours, while the pastoral Masai derive only 1 per cent or less of the accruing revenue. For the Masai, a more financially worthwhile use of the reserve would be to open it up to pastoralism and/or for poaching ivory and other wildlife products (Holdgate, 1991).

Other difficulties stem from the different ways groups perceive environmental value. Some groups may consider that certain parts of the environment lie outside the economists' realm, because economists are not actually dealing with values at all, but with prices. Adams (1993) gives the example of a mining company that wanted to exploit a site in Australia's Kakadu National Park, an area sacred in Aboriginal culture. In this case, the application of cost–benefit analysis amounts to immorality: 'The aborigines say "It is sacred." The economist replies "How much?"' (Adams, 1993: 258). It is for these sorts of reasons that some people reject the economic approach to valuing and hence preserving the environment and biological diversity. They argue that economic criteria of value can change and are opportunistic in their practical application. The greed that underlies such systems will ultimately destroy them (e.g. Ehrenfeld, 1988).

GROWTH AND DEVELOPMENT

A key issue in the sustainable development debate is the relative roles of economic growth (the quantitative expansion of economies) and development (the qualitative improvement of society). In its first report, the World Commission on Environment and Development (WCED) suggested that sustainability could only be achieved with a five- to ten-fold increase in world economic activity in 50 years. This growth would be necessary to meet the basic needs and aspirations of a larger future global population (WCED, 1987). Subsequently, however, the WCED played down the importance of growth (WCED, 1992). This makes sense, because many believe that it has been the pursuit of economic growth, and neglect of its ecological consequences, that created most of the environmental problems in the first place.

The change of thinking on economic growth has been reflected in the two types of reaction to calls for sustainability that have been made to date. On the one hand, to concentrate on growth as usual, though at a slower rate, and on the other hand, to define sustainable development as 'development without growth in throughput beyond environmental capacity' (Goodland *et al.*, 1993a: 297). The idea of controlling throughput, the flow of environmental matter and energy through the socio-economic system (see Fig. 3.1), can be referred back to the ways in which human society interacts with the environment, as shown in Fig. 3.3: cycles A and B are based on increased throughput, while in cycle C throughput is controlled to a level within the environment's capacity to support it. Sustainable development – cycle C – means that the level of throughput must not exceed the ability of the environment to replace resources and withstand the impacts of wastes (see Table 3.4, which is effectively a set of rules to promote intergenerational equity). This does not necessarily mean that further economic growth is impossible, but it does mean that growth should be achieved by better use of resources and improved environmental management (Daly, 1987) rather than by the traditional method of increased throughput.

One indication of the degree of change necessary to make this possible is in the ways we measure progress and living standards at the national level. Measurements such as Gross National Product (GNP) and Gross Domestic Product (GDP) are the principal means by which economic progress is judged, and thus form key elements in government policies. But GNP is essentially a measure of throughput, and it has severe limitations with respect to considerations of environmental and natural resources. The calculation of GNP does not take into account any depletion of natural resources or adverse effects of economic activity on the environment, which have feedback costs on such things as health and welfare. Indeed, conventional calculations of GNP frequently regard the degradation of resources as contributing to wealth, so that the destruction of an area of forest, for example, is recorded as an increase in GDP. The need to introduce environmental parameters into national accounting systems is now widely recognised, and adjusted measures of 'green GNP' now being worked upon could provide a good measure of national sustainability (Pearce and Mäler, 1991).

TABLE 3.4 *Key rules for the operation of environmental sustainability*

OUTPUT RULE
Waste emissions should be within the assimilative capacity of the environment to absorb, without unacceptable degradation of its future waste-absorptive capacity or other important services

INPUT RULE
Renewable resources: Harvest rates of renewables used as inputs should be within the capacity of the environment to regenerate equal replacements
Non-renewable resources: Depletion rates of non-renewables used as inputs should be equal to the rate at which sustained income or renewable substitutes are developed by human invention and investment. Part of the proceeds from using non-renewable resources should be allocated to research in pursuit of sustainable substitutes

Source: after Goodland *et al.* (1993b)

POPULATION AND TECHNOLOGY

Further investigation of the limiting throughput approach to sustainable development can be conducted by representing human society's impact on the environment by the equation

$$I = P \times A \times T$$

where I is impact, P is population, A is affluence (measured as the consumption of environmental resources per person) and T is technology. These three factors can provide the keys to keeping throughput within environmental capacity. This can be done by one of three means (Goodland *et al.* 1993a):

1 limit population
2 limit affluence
3 improve technology.

This approach to achieving sustainability is attractive if only because of its simplicity. It certainly provides researchers with variables that can

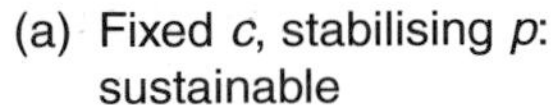

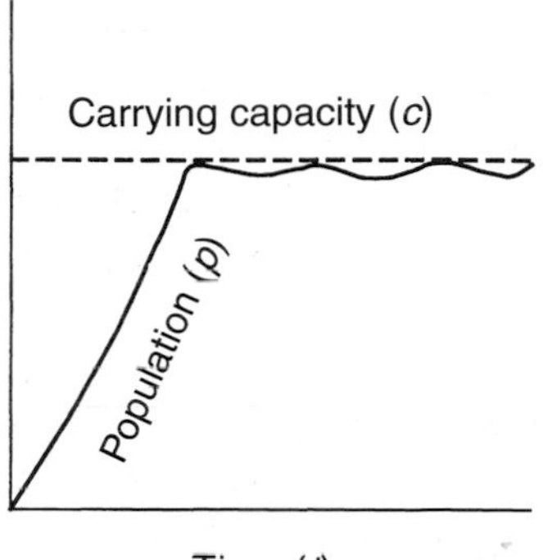

(b) Declining c, falling p: unsustainable

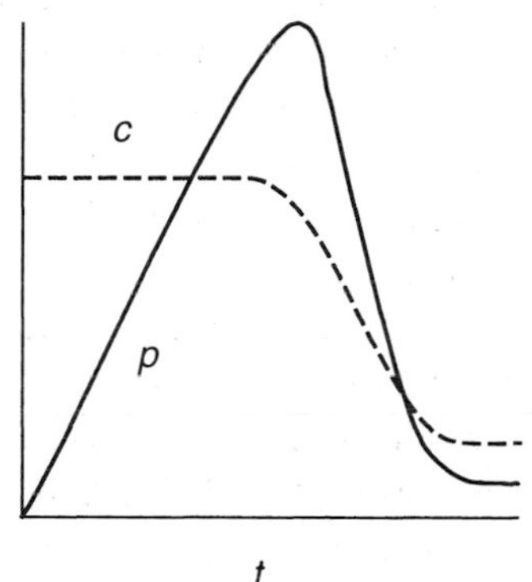

(c) Variable c, increasing p: sustainable

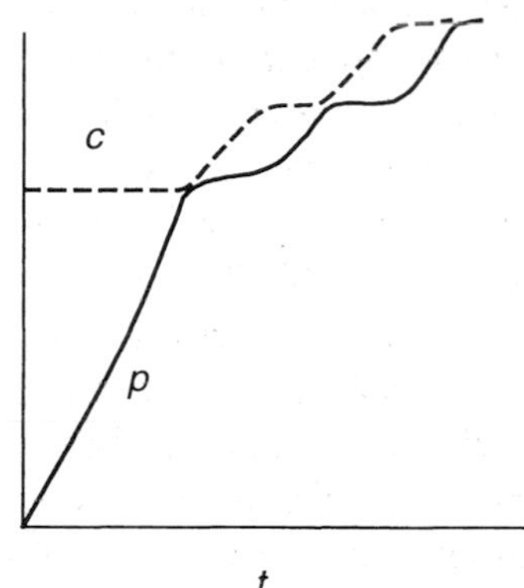

Figure 3.4 Three theoretical variations in carrying capacity and population totals.

be measured in an attempt to understand complex issues. While the equation can have applications, there are some flaws in it and some assumptions behind it that not all agree with. Some researchers argue that technological changes are far more significant than population or affluence in determining human-induced environmental change (e.g. Raskin, 1995). Indeed, absolute population numbers, density and growth rates are not always clearly related to degradation (see page 21). Although greater affluence, as defined above, can equate to greater environmental impact this is not necessarily so, and there is also a threshold below which decreasing affluence leads to greater degradation, as comments on the relationships between poverty and impact indicate (see page 23). Indeed, poverty is a key issue in the sustainability debate, an outcome of inequalities at the global and national scales. As a former executive director of the UN Environment Programme put it: 'The harsh fact remains: conservation is incompatible with absolute poverty' (Tolba, 1990: 10).

The $I = PAT$ equation also gives little recognition to such factors as beliefs, attitudes or politics, all important influences on a society's environmental impact. Another concern surrounds the geographical scale at which the equation can be applied. While the formula can be used to sum the human pressures on resources in a closed system, it does not indicate where within the system those pressures will be felt. Particular regions may be subject to environmental impact that is driven by factors in a different region (Turner *et al.*, 1995).

Carrying capacity

Another facet of the $I = PAT$ formula is its implicit acceptance of the concept of a finite carrying capacity. Carrying capacity is a term developed in ecology that formally means the point when the rate of growth of a population becomes zero (Caughley, 1979). However, the term has subsequently been widely used in many different areas of resource management, often meaning different things to different users. One of the most commonly used meanings is to indicate the point at which human use of an ecosystem can reach a maximum without causing degradation. In other words, carrying capacity is a threshold point of stability. If human activity occurs in a particular area at or below its carrying capacity, such activity can proceed at equilibrium with the environment, but if the carrying capacity of a particular area is exceeded, then the area's resources will be degraded. The important point is that the carrying capacity of an area is defined by the ecology of the area, not by people.

Three simple theoretical views of the relationship between carrying capacity and population

Figure 3.5 Ancient statues on the deforested Easter Island in the Pacific Ocean. Many believe that widespread clearance of trees exceeded the island's carrying capacity, eventually resulting in a collapse of the island's human population (Photo: Dan Downes).

are shown in Fig. 3.4. Figure 3.4a illustrates a sustainable situation in which population numbers remain more or less stable and within the limits imposed by a fixed environmental carrying capacity. Figure 3.4b, by contrast, illustrates how carrying capacity declines when resources are degraded and human population declines as a consequence. The decline and fall of the Mayan civilisation has already been mentioned as an example. Another often-quoted case is thought to have occurred on Easter Island where widespread deforestation probably took place principally for rollers to move the enigmatic statues found all over the island (Fig. 3.5). Deforestation resulted in accelerated soil erosion and falling crop yields leading to food shortages. The inability to erect more statues undermined belief systems and social organisation, resulting in armed conflicts over remaining resources. This combination of effects resulted in a severe decline in human population (Bahn and Flenly, 1992). Easter Island is an extremely remote Pacific landmass, the nearest inhabited land being more than 2000 km distant, so effectively the island was a closed system. A parallel can be drawn with planet Earth isolated in an essentially uninhabitable universe.

This idea of a biologically defined carrying capacity underpins many notions of environmental degradation. It is inherent in Malthusian ideas of overpopulation and the tragedy of the commons (Fig. 3.4b), and is directly related to the idea of maximum sustainable yield (Fig. 3.4a). This idea that human society is subject to natural biological laws strikes right at the heart of the sustainable development debate. Some believe that human ingenuity puts us above the laws of nature. The situation in Fig. 3.4c is also sustainable, but human population grows because carrying capacity is not fixed; it is actually increased because of new technologies and new resource perceptions (see also below). Meanwhile, some ecologists suggest that the carrying capacities of some natural environments are naturally variable, because the environments are not in equilibrium. Such environments are constantly dynamic, varying in time and space, rendering the idea of carrying capacity meaningless (see page 12).

Population

Population is regarded by many as the key to the whole sustainable development issue. In biological terms, several authors have invoked the Malthusian argument in recent times to suggest that essentially there is a finite limit to the size of human population on this planet (e.g. Ehrlich, 1968) and that this limit is imposed by the planet's finite resources (e.g. Meadows *et al.*, 1972). Needless to say, not everyone agrees with these scenarios. The opposing school of thought points out that population alone is not a reliable predictor of resource use, because people's use of resources is influenced by economic, environmental and cultural factors. They see the world's human population not as the passive victims of immutable biological laws of carrying capacity, but as active creators of new resources thanks to human ingenuity developing technological innovations (Boserüp, 1965 – see page 22). This view is epitomised by the idea that humans are in fact the 'ultimate resource' (Simon, 1981). While population growth may give rise to problems in the short term, these problems become the driving forces behind innovations and the creation of new resources. Central to this perspective is the idea that physical resources should not be seen as fixed because they are culturally defined. Hence, the fixed carrying capacity argument does not apply.

The following chapters on individual environmental issues put the theory and concepts outlined in these first three chapters into context by showing how the issues have arisen and how human societies have acted in their attempts to resolve the problems that have been created.

FURTHER READING

Ehrlich, P.R. and Ehrlich, A.H. 2002 Population, development, and human natures. *Environment and Development Economics* 7: 158–70. An argument for constraining global population growth, limiting consumption and for changing human culture, followed by responses from other authors.

Myers, N. and Simon, J.L. 1994 *Scarcity or abundance: a debate on the environment.* New York, WW Norton. Two extremes in the debate over population, resource use and environmental issues.

Odum, H.T. 1989 *Ecology and our endangered life-support systems.* Sunderland, MA, Sinauer Associates. An ecological view of sustainable development.

Pearce, D.W. and Turner, R.K. 1990 The historical development of environmental economics, in Pearce, D.W. and Turner, R.K. (eds) *Economics of natural resources and the environment.* New York, Harvester Wheatsheaf: 3–28. A comprehensive review of the different approaches to environmental economics and how they have evolved.

Redclift, M. 1993 Sustainable development: needs, values, rights. *Environmental Values* 2: 3–20. This in-depth paper explores the rise of sustainable development as an idea, arguing against the dominance of science and technology in favour of greater emphasis on cultural diversity.

WCED (World Commission on Environment and Development) 1987 *Our common future.* Oxford, Oxford University Press. This is a key report on the environmental and development problems faced by the world's human population, with proposals for their solution.

Websites

http://www.iisd.ca/ a multimedia resource for environment and development run by the International Institute for Sustainable Development.

http://www.fao.org/waicent/faoinfo/sustdev/ information and links on Sustainable Development Dimensions.

http://www.un.org/esa/sustdev/ the UN's sustainable development site with information on global and regional agreements, national and international indicators.

http://www.earthsummit2002.org/ site for the World Summit on Sustainable Development in Johannesburg.

http://sdgateway.net/ the sustainable development gateway includes over 1200 documents on a range of topics, a job bank, the Sustainability Web Ring, and sustainable development news sites.

http://www.wbcsd.org/ the World Business Council for Sustainable Development site includes case studies of improving environmental performance and online sustainable development courses

http://www.sei.se/ the Stockholm Environment Institute is an independent, international research institute specialising in sustainable development and environment issues working at local, national, regional and global levels.

POINTS FOR DISCUSSION

- Consider what you can do to make your life more sustainable.
- Can environmental economics solve all environmental problems?
- Design a study to test the $I = PAT$ formula in your local area.
- Are people the ultimate resource?

4

TROPICAL DEFORESTATION

TOPICS COVERED
Deforestation rates
Causes of deforestation
Consequences of deforestation
Tropical forest management

Key words: tropical moist forest, primary forest, fragmentation, frontier zone, low-intensity harvesting

Forests have been cleared by people for many centuries, to use both the trees themselves and the land on which they stand. Throughout the history of most cultures, deforestation has been one of the first steps away from a hunter-gathering and herding way of life towards sedentary farming and other types of economy. Forests were being cleared in Europe in Mesolithic and Neolithic times, but in central and western parts of the continent an intense phase occurred in the period 1050–1250. Later, when European settlers arrived in North America, for example, deforestation took place over similarly large areas but at a much faster rate: more woodland was cleared in North America in 200 years than in Europe in over 2000 years. Estimates suggest that in pre-agricultural times the world's forest cover was about 5 billion hectares (Mather, 1990). In 2000, the Food and Agriculture Organization (FAO) concluded that natural and plantation forests covered 3.9 billion hectares (FAO, 2001). Although most of this loss has taken place in the temperate latitudes of the northern hemisphere, in recent decades the loss in this zone has been largely halted, and in many countries reversed by planting programmes. Meanwhile, in the tropics, rapidly increasing human populations and improved access to forests have combined in recent times to create an accelerating pace of deforestation, which has become a source of considerable concern both at national and international levels. This chapter concentrates on the humid tropics, while information on forest clearance in the drier parts of these latitudes can be found in Chapter 5.

DEFORESTATION RATES

Despite the high level of interest in deforestation, our knowledge of the rates at which it is currently occurring is far from satisfactory. In part, this is because of the lack of standard definitions of just what a forest is and what deforestation means. A distinction is often made between 'open' and 'closed' forest or woodland, which is sometimes defined according to the percentage of the land area covered by tree crowns. A 20 per cent crown cover is sometimes taken to be the cut-off point for closed forest, for example, while open woodland is defined as land with a crown cover of 5–20 per cent. Measurements of deforestation rates by the FAO define deforestation as the clearing of forestlands for all forms of agriculture and for other land uses such as settlements, other infrastructure and mining. In tropical forests this entails clearing to reduce tree crown cover to less than 10 per cent, and thus does not include some damaging activities such as selective logging, which can seriously affect soils, wildlife and its

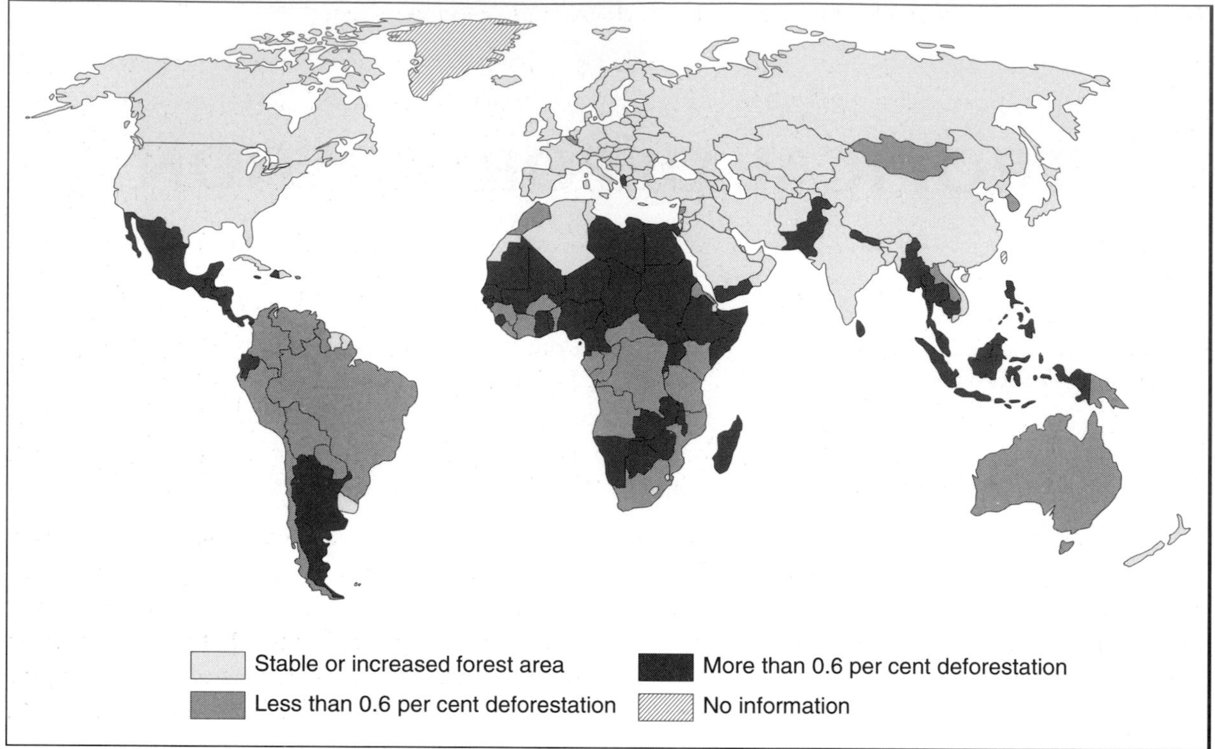

Figure 4.1 Estimated annual rates of change in forest cover by country, 1990–2000 (from data in FAO, 2001).

habitat. In this case, the change in the status of the forest is referred to as degradation rather than deforestation.

The collection of data for all countries of the world and their compilation by different experts inevitably causes problems, since different countries and individuals often define the terms in different ways, and the quality and availability of data varies between countries. One tool that is being used increasingly to avoid some of the problems of definition and poor data is satellite imagery, which can be evaluated using one set of criteria. But even these data can vary, depending on the satellite sensors used, differing methods of interpretation, and other difficulties such as that of distinguishing between primary and secondary forests. Estimates, from numerous sources, of the annual rate of deforestation of closed forests in the humid tropics (tropical moist forest, which includes two main types: tropical rain forest and tropical monsoon/seasonal forest) have varied from 11 to 15 million ha for the early 1970s, 6.1 to 7.5m ha for late 1970s, and 12.2 to 14.2m ha for the 1980s (Grainger, 1993b). This author suggests that the uncertainties are primarily the result of lack of attention to remote sensing measurements, and overconfidence in the use of expert judgement. One of the most recent assessments of the rates of change of forest cover by country, which has been prepared by the FAO, is shown in Fig. 4.1. This map shows clearly that while forest cover in the high latitudes is generally stable or increasing, virtually all tropical countries are experiencing a loss. This survey estimated the deforestation rate in all tropical forests for the period 1990–2000 to be 14.2m ha a year (FAO, 2001). The overall annual global loss of all tropical moist forests over the period 1981–90 was estimated at 13.1m ha (FAO, 1995).

CAUSES OF DEFORESTATION

A concise summary of the causes of deforestation in the humid tropics is no easy task, since cutting

TABLE 4.1 *Important factors influencing deforestation in the tropics, by major world region*

Region	Main factors
Latin America	Cattle ranching, resettlement and spontaneous migration, agricultural expansion, road networks, population pressure, inequitable social structures
Africa	Fuelwood collection, logging, agricultural expansion, population pressure
South Asia	Population pressure, agricultural expansion, corruption, fodder collection, fuelwood collection
South-East Asia	Corruption, agricultural expansion, logging, population pressure

Source: after Kummer (1991)

down trees is the end result of a series of motivations and driving forces that are interlinked in numerous ways. On a worldwide basis, the people who actually cut down the trees can generally be agreed upon. They are:

- agriculturalists
- ranchers
- loggers.

However, a proper understanding of the deforestation process requires a deeper investigation of the driving forces behind these agents. Access to forests is an important aspect, for example, and this is usually provided by road networks. Hence, construction of a new road, whether it be by a logging company or as part of a national development scheme, is an integral part of deforestation. Another part of the equation is the role played by the socio-economic factors that drive people to the forest frontier: poverty, low agricultural productivity and an unequal distribution of land are often important, while the rapid population growth rates that characterise many countries in the tropics also play a part. The role of national government is another factor in encouraging certain groups to use the forest resource, through tax incentives to loggers and ranchers, for example, or through large-scale resettlement schemes. On the global scale, international markets for some forest products, such as lumber and the produce from agricultural plantations, must also be considered.

The importance of these factors varies from country to country and from region to region, and may change over time. Table 4.1 is an attempt to identify the prime factors that lie behind deforestation in the major forest regions of the tropics, based on a review of the large literature on the subject.

The scale of clearance that has occurred in some countries, and the influence of external factors, can be illustrated by the experience of Viet Nam, a country that in pre-agricultural times was almost entirely covered in forests but that has lost more than 80 per cent of its original forest area, much of it during the second half of the last century (Fig. 4.2). Clearance of the coastal plains and valleys for agriculture took place over the last few centuries, and during the French colonial period when large areas in the south were cleared for banana, coffee and rubber plantations, but 45 per cent of the country was still forested in the 1940s. That proportion had fallen to about 17 per cent in the late 1980s as extensive zones were destroyed during the Viet Nam War (see Chapter 20) and still greater areas have since been destroyed by a rapidly growing population rebuilding after the war.

The forces of international economics that are integral to many areas of deforestation have played a central role in West Africa, where in most countries there is hardly any stretch of natural, unmodified vegetation left. A classic example is

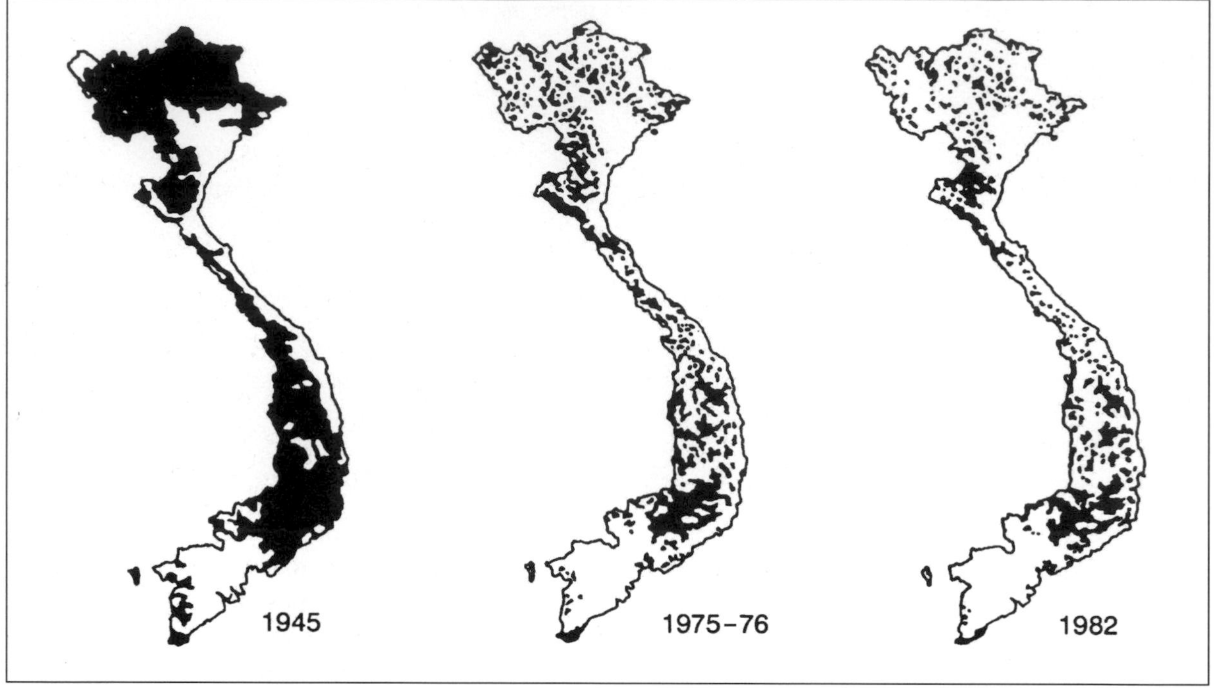

Figure 4.2 Decline of forest cover in Viet Nam, 1945–82 (Viet Nam, 1985).

Côte d'Ivoire. Virtually every study of deforestation in the tropics has concluded that Côte d'Ivoire has experienced the most rapid forest clearance rates in the world, at 2800–3500 km^2 per year for the past 40 years.

Like most African governments, Côte d'Ivoire has viewed the nation's forests as a source of revenue and foreign exchange. However, given the country's high external debt, and declining international prices received for agricultural export commodities, the country has been faced with little alternative but to exploit its forests heavily. In 1973, logs and wood product exports provided 35 per cent of export earnings, but this figure had fallen to 11 per cent by the end of that decade due to the rapidly declining resource base. In the late 1970s, about 5.5 million m^3 of industrial roundwood was extracted annually but this production had fallen below 3 million m^3 by 1991. In 1997 Côte d'Ivoire introduced restrictions on the export of logs with the aim of increasing its income by processing them at home. This has resulted in increased log milling and increased manufacture of wood products, but also in lower log prices for forest owners, devaluation of the forest resource and negative impacts on forest management (FAO, 2000). Deforestation has also been fuelled by a population that grew from 5 million in 1970 to nearly 15 million in 2000, uncontrolled settlement by farmers, and clearance for new coffee and cacao plantations encouraged by government incentives.

Logging and agriculture have also been the primary agents of deforestation in the Philippines since the 1940s, and the series of factors that lie behind the loggers and agriculturalists, as interpreted by Kummer (1991), is shown graphically in Fig. 4.3. Forest cover in the Philippines declined from 70 per cent of the national land area to 50 per cent between 1900 and 1950, and declined further to below 25 per cent in the early 1990s, by which time lowland forest had virtually disappeared. Natural forests were estimated to cover just 17 per cent of the national land area in 2000 (FAO, 2001).

The most common pattern of events has been the conversion of primary forest to secondary forest by logging, followed by clearance of the

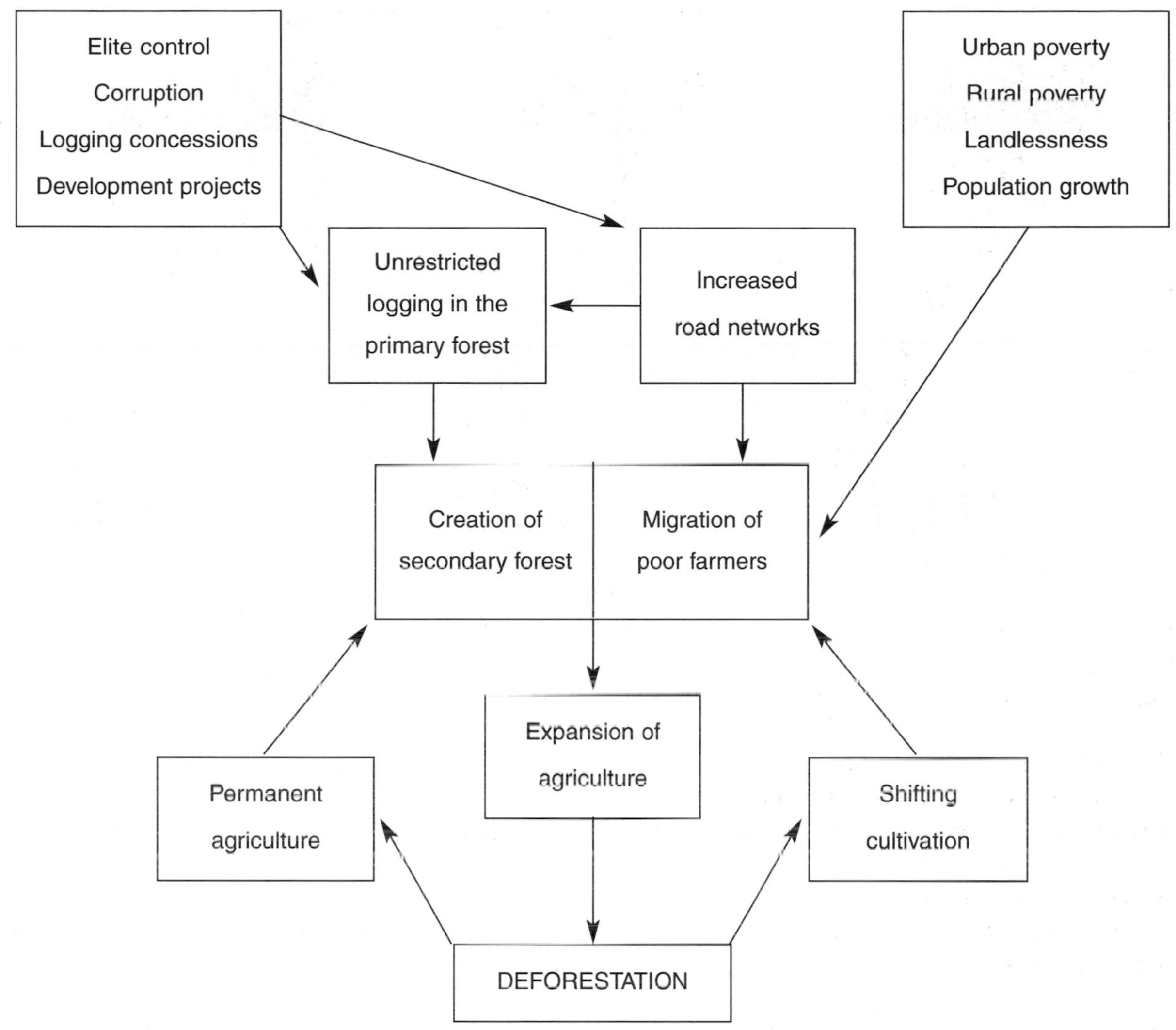

Figure 4.3 Factors affecting deforestation in the Philippines (Kummer, 1991).

secondary forest for the expansion of agriculture. Both activities are preceded by the construction of roads, built by provincial and national governments for general development purposes, or by loggers to access their concessions. Kummer suggests that the granting of logging concessions has occurred to foster development but also as political favours to Filipino élites and foreign-based transnationals. Much of the financial gain has also flowed to a small number of well-connected individuals. This concentration of the benefits from the Philippines' forest resource has been partly responsible for widespread and increasing poverty in the country, with no improvement in the living standards of the bottom 50–75 per cent of the population over the past 40 years. Most of the agriculturalists who clear the secondary forests are subsistence farmers, spurred on by poverty, a rising population and a lack of land.

The role of government policy and practice has also been central to the deforestation of the Amazon Basin in Brazil, which began on a large scale in the mid-1970s with a concerted effort to develop the country's tropical frontier. Agricultural expansion was the most important factor responsible for forest clearance at this time,

Figure 4.4 The distinctive herringbone pattern of tropical forest clearance by slash-and-burn agriculturists along transport routes in Rondônia, Brazil. Dark areas in this space shuttle photograph are remaining forest. Roads are 4 to 5 km apart (Photo: NASA).

both by smallholders and large-scale commercial agriculturalists, including ranchers producing beef for the domestic market. The Brazilian government's view of the Amazon as an empty land rich in resources spurred programmes of resettlement, particularly in the states of Rondônia and Pará under the slogan 'people without land in a land without people' (Moran, 1981), along with agricultural expansion programmes and plans to exploit mineral, biotic and hydroelectric resources.

The various agents of deforestation in the Brazilian Amazon have been summarised by Fearnside (1990). Cattle ranching has been encouraged by government subsidies and has been additionally attractive as a means of storing wealth, both in land and cattle, during periods of high inflation. Slash-and-burn agriculture practised by pioneer farmers arriving from outside the Amazon, as opposed to the shifting cultivation long practised by the indigenous peoples of the region, is another serious cause of deforestation, as it has long been in the Amazonian parts of Peru and Ecuador. Pioneers' slash-and-burn occurs at too great a human density of population, often leaves insufficient fallow periods and/or follows an initial crop with pasture planting, making it unsustainable. The construction of hydroelectric dams, with the inundation of reservoirs, has the potential for further major impacts to follow on from some of those existing, such as the Tucurúi and Balbina dams. The destruction caused by mining is also important, although usually on a local scale. In the case of the Grande Carajás

Programme in the state of Pará, however, the proposed development of iron ore, bauxite, copper and manganese involves local smelting, a new railway and highway network, and a very large agricultural scheme. Lumbering is rapidly increasing in importance; agribusiness, until now a minor cause of deforestation in Brazil, could expand significantly, and the military bases being constructed along the Brazilian Amazon's northern frontier under the Calha Norte Programme pose another serious threat.

The broader set of dynamic circumstances underlying this policy of frontier development in Brazil has been highlighted by Turner *et al.* (1993) who trace them back to the OPEC (Organization of Petroleum Exporting Countries) oil crisis of the 1970s, which resulted in a large transfer of economic wealth from industrial countries to oil producers who, in turn, deposited these revenues in US and European banks. Large-scale lending of these monies to Brazil, amongst other developing countries, allowed development programmes to be financed. In Brazil, agricultural modernisation took much of this finance and was channelled into export crops such as wheat, soybeans and coffee. Soybean production was boosted particularly, because relative to coffee the international market for soybeans was much more dependable, and this expansion was concentrated in two states: Rio Grande do Sul and Paraná. However, widespread replacement of coffee cultivation, a labour-intensive crop, with soybean, which is more capital and energy intensive, has resulted in large-scale emigration, particularly from Paraná. Large numbers of migrants went to new opportunities on the forest frontier in Rondônia, practising small-scale slash-and-burn agriculture along the World Bank-financed Highway BR-364, as it was extended into the state (Fig. 4.4).

CONSEQUENCES OF DEFORESTATION

Much of the concern over the issue of tropical deforestation stems from the perturbation it represents to the forests' role as part of human life-support systems, by regulating local climates, water flow and nutrient cycles, as well as their role as reservoirs of biodiversity and habitats for species. The numerous deleterious environmental consequences resulting from the loss of tropical forests range in their scale of impact from the local to the global. Human clearance is by no means the only form of disturbance that forests experience, however, since the forest landscape is in a state of dynamic equilibrium. Gaps are continually formed in the forest canopy as older trees die, are struck by lightning or are blown over. Regrowth occurs continually in such gaps so that small areas are in a state of perpetual flux while the landscape as a whole does not appear to change, prompting the description of tropical rain forests as a shifting mosaic steady state (Bormann and Likens, 1979). Natural perturbations also occur on much larger scales. A tropical cyclone, for example, can cause tremendous damage over large areas, while volcanic eruptions, earthquakes and associated landslides are frequent in some tectonically active areas such as Papua New Guinea. Fire has been confirmed as a natural disturbance in the Amazon Basin (Sanford *et al.*, 1985), while on the upper tributaries of the Amazon in Peru violent annual flooding causes frequent changes in the rivers' courses, eroding banks and depositing fresh sediment on which primary succession occurs (Salo *et al.*, 1986). These disturbances result in a markedly striped pattern in the forest canopy representing different species composition at various stages of succession. These continual disturbances, evident across about one-eighth of the Peruvian Amazon forests, mean that climax is never reached. Through geological time, the global extent of tropical forests has also varied with climate and is thought to have been much smaller than its current extent during periods of glacial maxima (e.g. Endler, 1982).

The human impact on tropical forests has a long history too. In practice it can be difficult to distinguish in the field between so-called 'primary forest' and 'secondary forest' that is the result of human impact, sometimes dating back over long

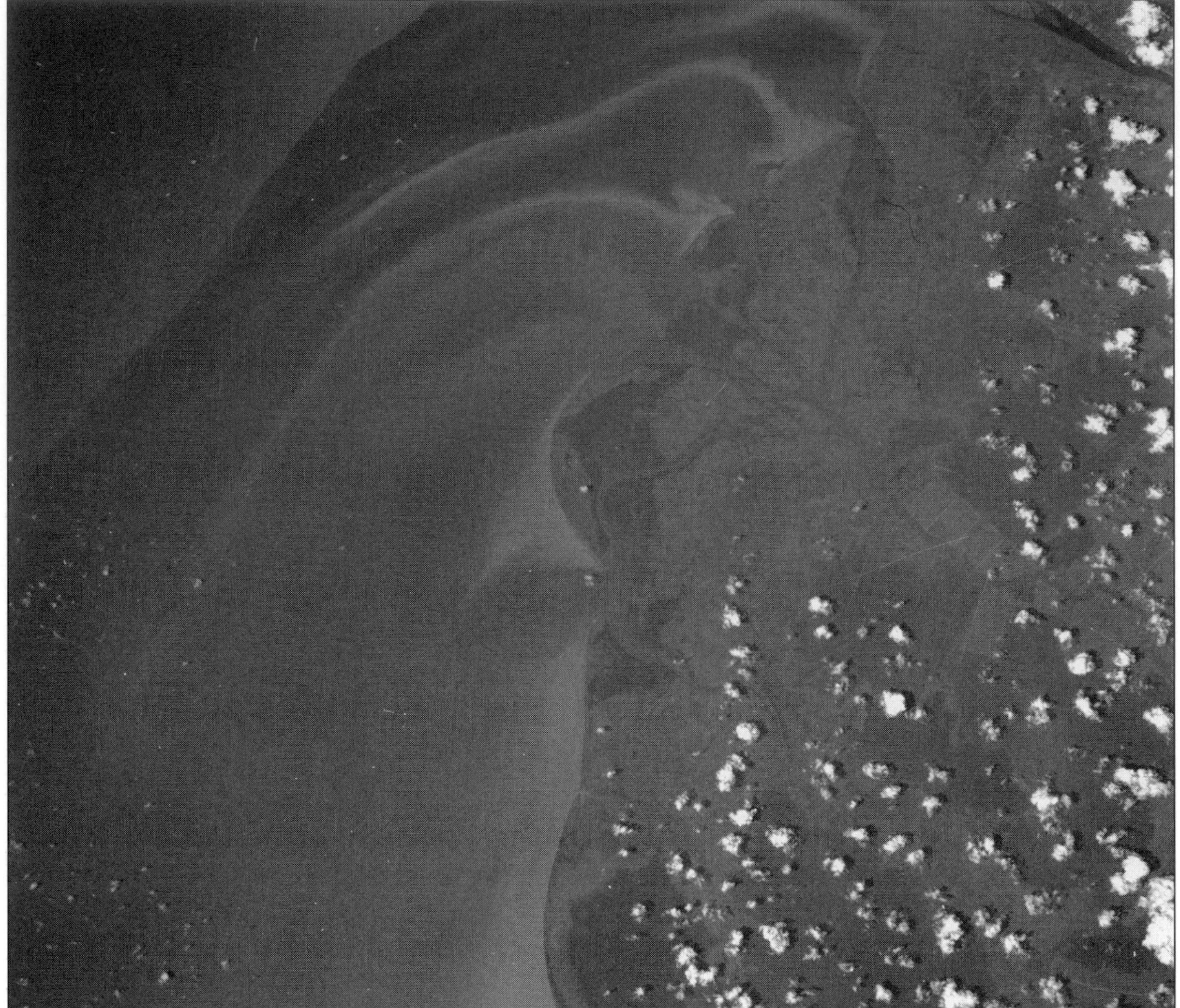

Figure 4.5 Plumes of sediment and vegetable matter in the rivers and coastal seas of Kalimantan Barat province in Borneo, Indonesia. Deforestation in this area has been caused by settlers encouraged to emigrate from the main Indonesian island of Java by the government. The lighter, deforested coastline is distinguishable from the darker, undisturbed patches of dense rain forest (Photo: NASA).

periods. For instance, large areas of tropical forest in the Yucatan Peninsula in Mexico, long believed to be primary forest, are now considered to be secondary forest managed by the Mayan people more than 1000 years ago (Gómez-Pompa *et al.*, 1987). To the casual observer, secondary forests more than 60–80 years old are often indistinguishable from undisturbed primary forests and are, in fact, treated as primary forests in the FAO assessments of tropical forests. Secondary forests less than 60–80 years of age are extensive in the tropics, making up about 40 per cent of the total forest area (Brown and Lugo, 1990). However, although humans have registered an impact on tropical forests over many centuries, which cumulatively has affected great areas, the intensity, speed and relative permanence of clearance by human populations in the modern era puts this form of disturbance into a different category.

The effects of widespread deforestation on hydrology have been widely discussed and an example of the multiple impacts upon aquatic

TABLE 4.2 *Some of the major constraints to agricultural development of two dominant Amazon Basin soils*

Constraint	Oxisols	Ultisols
Low plant nutrient levels	×	×
Nitrogen losses through leaching	×	×
High levels of exchangeable aluminium	×	×
Low cation exchange capacity	×	
Weak retention of bases	×	
Strong fixation or deficiency of phosphate	×	
Soil acidity	×	
Very low calcium content	×	
Essential to continue to maintain soil organic matter levels		×
Weak structure in surface layers		×

Source: after Nortcliff and Gregory (1992)

habitats in the rivers of Madagascar is described in Chapter 8 (see Fig. 8.2). A particular concern stems from recognition of the forests' role in regulating the flows of rivers and streams from upland catchments. The feared type and scale of regional impact on natural systems consequent upon deforestation is well illustrated in the Himalayas, where the theory that widespread deforestation since 1950 has been responsible for increased flooding of the Ganges and Brahmaputra plains has been widely accepted. The theory has not been without its critics, however, who, upon looking at the evidence, have shown that there is, in fact, little reliable data with which to support the proposed link between deforestation and increased flooding (Ives and Messerli, 1989). Although there is clearly a need to conserve forest cover on catchments, the role of deforestation in flooding appears to have been exaggerated in many cases.

A less controversial effect is that on sediment loads in rivers due to enhanced runoff and erosion (Fig. 4.5). High sediment loads have been the cause of problems, particularly in reservoirs – shortening their useful lives as suppliers of energy and irrigation water, for example, and requiring costly remediation (see Chapter 9). Another facet of hydrology likely to be affected by deforestation is local water tables. When deep-rooted trees are replaced by shallow-rooted crops or grasses the increased recharge of aquifers can lead to a rise in the water table. This effect has been reported from the uplands of north-eastern Thailand where local forest clearance for kenaf and cassava cultivation has resulted in salinisation problems on lower slopes and valley floors due to rising saline groundwater (Löffler and Kubinok 1988).

Soil degradation – including erosion, land slides, compaction and laterisation – is a common problem associated with deforestation in any biome, but is often particularly severe in the tropics. Tropical rainfall is characterised by large annual totals, it is continuous throughout the year and highly erosive. Hence, clearance of protective vegetation cover quickly leads to accelerated soil loss, although the rate will vary according to what land use replaces the forest cover (see Chapter 14). Soil erosion can also be exacerbated by compaction caused by heavy machinery, trampling from cattle where pasture replaces forest, and exposure to the sun and rain. A compacted soil surface yields greater runoff and often more soil loss as a consequence. Laterisation, the formation of a hard, impermeable surface, was once feared to be a very widespread consequence of deforestation in the tropics, but is probably a danger on just 2 per cent of the humid tropical land area (Grainger, 1993a).

The poor nutrient content of many tropical forest soils is one factor that tends to limit the potential of many of the land uses for which forest is cleared. In parts of the Amazon Basin, for example, the available phosphorus and other nutrients in the soil initially peaks after the burning of cleared vegetation, but declines thereafter due to leaching, and is insufficient to maintain pasture growth after ten years. Soil nutrient depletion, compaction and invasion by inedible weeds quickly depletes the land's usefulness as pasture. Similarly, unsustainable shifting cultivation quickly depletes soil nutrients and, in both cases, rapidly diminishing yields from inappropriate land uses encourage settlers to move on and clear further forest areas in a process of positive feedback. Some of the other major constraints for two of the most dominant soil orders found in the Amazon Basin are shown in Table 4.2.

The possible climatic impacts of large-scale forest clearance, resulting from effects on the hydrological and carbon cycles, may occur on regional and even global scales. In a large region like Amazonia, where about half of the rainfall is returned to the atmosphere via evapotranspiration, widespread removal of forest could have a serious impact on the hydrological cycle, with consequent negative effects on rainfall and, hence, continued forest survival and agriculture. Such theories have been tested by simulating future conditions with general circulation models, and the results to date generally agree on a reduction in precipitation and a small change in the surface temperature (McGuffie *et al.*, 1998). The changes brought about by tropical deforestation to surface properties such as albedo and roughness could also have knock-on effects in middle and high latitudes via the Hadley and Walker circulations and Rossby wave propagation. Further global effects of large-scale deforestation have been predicted via the carbon cycle. Deforestation releases carbon dioxide into the atmosphere, either by burning or decomposition, possibly contributing to enhanced global warming through the greenhouse effect. Estimates of just how much additional carbon dioxide reaches the atmosphere due to tropical deforestation are uncertain, due both to our imprecise knowledge of deforestation rates and of the amount of carbon locked into a unit area of forest. Although deforestation's contribution to rising atmospheric carbon dioxide concentrations is less than that of fossil fuels, it could contribute up to a third as much again. The possible consequences of climatic change due to the resulting greenhouse warming are discussed in Chapter 11.

One of the most widely discussed consequences of deforestation in the tropics is the loss of plant and animal species. The diversity of life in tropical rain forests relative to other biomes is legendary. Study of a single tree in Peru, for example, yielded no less than 43 ant species belonging to 26 genera, roughly equivalent to the ant diversity of the entire British Isles (Wilson, 1989). Examples of this level of diversity are numerous: on just 1 ha of forest in Amazonian Ecuador, researchers have recorded no less than 473 species, 187 genera and 54 families of tree (Valencia *et al.*, 1994), the highest number of tree species found on a tropical rain forest plot of this size, which is more than half of the total number of tree species native to North America. Although our knowledge of the total number of species on the planet is scant (see Chapter 15), most estimates agree that more than half of them live in the tropical moist forests that cover only 6 per cent of the world's land surface. Many species are highly localised in their distribution and are characterised by intimate links or narrow ecological specialisation, resulting in a delicate interdependence between one species and another. Hence they can be destroyed without clearing a very large area. Overall, there is widespread agreement in the scientific community that deforestation in the tropics is a severe threat to the species and genetic diversity of the planet.

Not all authorities are so pessimistic, however (see, e.g., Lugo, 1988), and several studies have shown that populations of larger faunal species, such as birds and mammals, are surprisingly resilient in the face of single logging operations. Although the relative abundance of species can change markedly, total species numbers often show little change in response to such disturbances. Individuals retreat to untouched forest pockets during logging operations and return after logging has been completed, as Johns and Johns (1995) show for primates. Despite this minor note of optimism, less versatile and adaptable species, particularly insects, which tend to be very specialised feeders, are less likely to survive such disturbances. From a purely practical viewpoint, the loss of species, most of which have not been documented, has been likened to destroying a library of potentially useful information having only flicked through the odd book. The types of goods that could be derived from tropical forests include new pharmaceutical products, food and industrial crops, while a diversity of genetic material is of importance to improve existing crops as they become prone to new strains of diseases and pests.

Forest structure and diversity are not just affected by clear-cutting, but also by fragmentation (see Chapter 15) and lesser forms of disturbance. Work in central Amazonia suggests that forest fragmentation is having a disproportionately severe effect on large trees, the loss of which will have major impacts on the rain forest ecosystem (Laurance *et al.*, 2000). Large trees are unusually vulnerable in fragmented rain forests for several reasons. They may be especially prone to uprooting and breakage near forest edges, where wind turbulence is increased, because of their tall stature and relatively thick, inflexible trunks. Large, old trees are also particularly susceptible to infestation by lianas (parasitic woody vines that reduce tree survival), which increase markedly near edges. Further, because their crowns are exposed to intense sunlight and evaporation, large tropical trees are sensitive to droughts and so may be vulnerable to increased desiccation near edges.

The loss of large tropical forest trees is alarming because they are crucial sources of fruits, flowers and shelter for animal populations. Their loss is also likely to diminish forest volume and structural complexity, promote the proliferation of short-lived pioneer species, and alter biogeochemical cycles affecting evapotranspiration, carbon cycling and greenhouse gas emissions. Equally worrying is the possibility that, because tree mortality is increased in forest fragments and large trees range in age from a century to well over a thousand years old, their populations in fragmented landscapes may never recover.

Fire is another threat to the health of the forest ecosystem affected by logging. The 1982/83 drought fires that seriously affected 950 000 hectares of dryland forest in Sabah and 2.7 million

Figure 4.6 The outsiders' perception of indigenous tropical rain forest inhabitants as savages, as this engraving from a seventeenth century book of travels in Brazil aptly illustrates, still tends to prevail. The all too common consequences for indigenous inhabitants include loss of cultural identity, and suffering due to the introduction of new diseases.

hectares of swamp and dryland forest in East Kalimantan, for example, raged mainly in selectively logged areas where dry combustible material from tree extraction littered the forest floor (Wirawan, 1993). Forest degradation and land clearance were also the root causes of the 1997/98 fire disaster that blanketed nearly 20 million people across South-East Asia in smoke for months, with disastrous consequences for local health (Siegert *et al.*, 2001). While the small-scale, low-intensity, low-frequency fires commonly associated with traditional shifting cultivation systems are considered to be a moderate-level disturbance that is likely to stimulate the maintenance of high biodiversity in tropical rain forests, such large-scale fires represent a much greater disturbance from which it will take a very long time to recover.

Another all too common casualty of deforestation in the tropics is indigenous peoples, for whom death or loss of cultural identity usually follows the arrival of external cultures and the frontier of forest clearance (Fig. 4.6). Whether it is deliberate annexation of their territory by the dominant society, or a less formalised erosion of their way of life, the result is often the same. The indigenous Indian population of the Amazon Basin, for example, is thought to have numbered 10 million when Europeans first arrived in South America, but has been reduced to about 250 000 today (Layrisse, 1992). To extend the library metaphor with regard to the potential wider uses of biodiversity, the impacts of deforestation on tribal peoples is equivalent to killing off the librarians who know a great deal about what the library contains. At a more fundamental level, the plight of indigenous peoples is a simple question of human rights.

TROPICAL FOREST MANAGEMENT

The answers to the problems caused by exploitation of tropical forests are by no means straightforward to formulate or easy to implement. One difficulty that should be appreciated is that most remaining tropical forest areas are remote from national seats of government and that the rule of law in these frontier zones may be less than total. Therefore passing legislation that supposedly limits the agents of deforestation is often meaningless in practice. The World Bank (2000) estimates that in many countries illegal logging in natural forests accounts for at least half of the total timber supply. A case in point are the lowland forests of the Sunda Shelf in Indonesia where mafia-like logging gangs, backed by army and rebel groups, operate within forestlands that overlap official park boundaries, ignoring the Indonesian State Forestry Department's sustainable management policy (Jepson *et al.*, 2001). Government officials who attempt to stop illegal logging face serious intimidation, which includes arson and even murder, whilst many of the local communities who live on the land have signed away their rights to it to the loggers, both for immediate cash benefit and to avoid retribution from logging gangs.

Preventing this type of systematic tropical forest abuse, a symptom of the underlying causes of deforestation noted above, means that it is these causes that must be addressed if development is to be directed towards a more sensible use of the forest resource. Many of the structural changes needed to strike at the root causes of deforestation will require years of endeavour. Nevertheless, as Fearnside (1990) suggests for the Brazilian Amazon, a start can be made now.

Among Fearnside's proposals for Brazil are some deep-seated national economic policies that need to be addressed, such as slowing population growth and improving employment opportunities both in labour-intensive agriculture and in urban areas outside the Amazon. Among his suggestions for the Brazilian government's immediate action are:

- stop highway construction in Amazonia
- abolish subsidies for cattle pastures
- end energy subsidies in Amazonia
- ensure proper protection for natural forest reserves and stop reneging on previous commitments to reserves whenever the land is wanted for another purpose.

While these proposals might seem sensible from one perspective, the needs of national economic development cannot be denied. For Brazil to cease development of its largest sovereign natural resource is clearly neither desirable nor feasible. In practical terms, too, the fact that such suggestions often come from developed countries that have previously all but destroyed their own forests in the course of their development can quickly be interpreted as 'ecoimperialism'. There are many parallels between today's developing countries and the past experience of those that have developed further. In Britain, for example, almost the entire land area was covered by forest about 5000 years ago, about the time that human activity began to have an impact on the landscape. Today, the UK is one of the least-wooded countries in Europe. Woodland covers just 8 per cent of England and Wales, but most of this is recent plantings. The area covered by ancient woodland is about 2.6 per cent (Spencer and Kirby, 1992); most of the lost forests have been converted into farmland, which occupies some 72 per cent of the total land area today, and many wildlife species have become extinct in Britain as a result of this destruction of habitat, which made them more susceptible to hunters (see Table 15.2).

Outsiders can make a positive contribution to controlling the poor use of forest resources, although direct pressure on governments from well-intentioned individuals abroad can backfire. Improvements in the environmental performance of such players as commercial banks and other investors can be called for, and moves made to tighten their financial approach to reduce the temptation of borrowing countries to run up heavy debts and promote short-sighted land use policies to pay them back, as in the case of investment in soybean production in southern Brazil. Criticism of the World Bank for its association with wasteful and illegal deforestation, such as the aforementioned Highway BR-364 in Brazil's Rondônia state, led to the Bank changing its approach towards forests in 1991. The World Bank stopped financing major infrastructure investments that were likely to harm forests and introduced a ban on the financing of commercial logging (World Bank, 2000). Debt-for-nature swaps have also proved to be a successful approach in several countries, making a positive contribution to conservation and developing countries' debt burdens (see Chapter 22).

A pragmatic approach to the problems of deforestation cannot seriously obstruct all extension of cultivated lands or all clearance of forest to provide new industrial or urban sites. But this process requires surveys, inventories and evaluation to ensure that each hectare of land is used in the optimum way, to meet both the needs of today and those of future generations. A series of questions that should be answered in deciding whether or not to alter a tropical forest has been proposed by Poore and Sayer (1987):

- What are the dominant ecological and social constraints in the tropical forest area in question? Soil analysis is essential, many tropical forest lands (especially on poor, wet soils) can sustain nothing but an intact tropical forest cover.
- Is there degraded, altered – or at least non-primary tropical forest available that could also sustain the required enterprise? Such land usually supports fewer species, provides fewer ecological benefits, is nearer to settlements and infrastructure, and can be rehabilitated by development. A notable example is timber production from old secondary tropical forest, which often yields a much higher proportion of desirable species than primary tropical forest.
- Are there ways in which the enterprise could be sustained within the existing tropical forest? In other words, can this be done without altering the forest structure?
- If the tropical forest were to be cleared and a new use established, what would be the net change in costs and benefits (including all tropical forest values, such as watershed protection)?

These authors suggest that the full range of ecological processes and species diversity of tropical

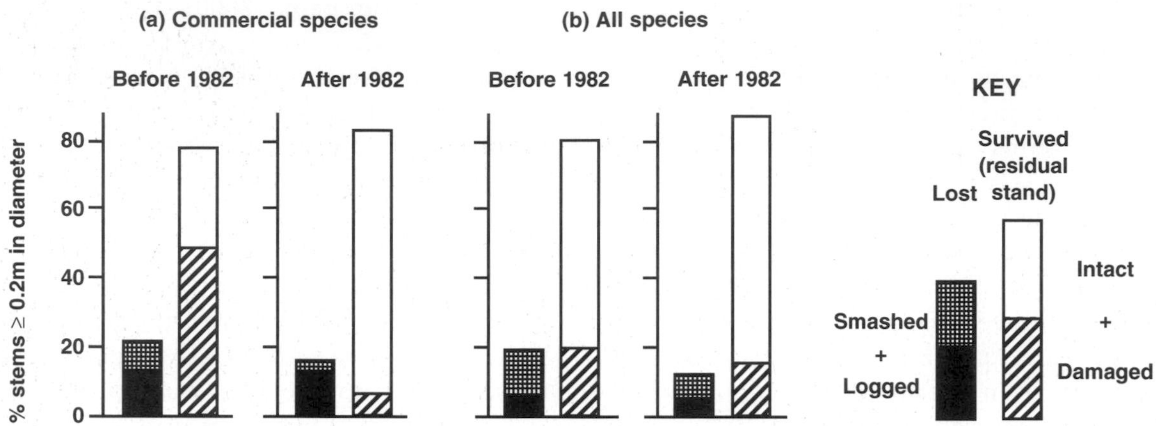

Figure 4.7 Effects of selective logging in Queensland, Australia, before and after the introduction of strict silvicultural regulations in 1982 (Whitmore, 1998).

forests can only be maintained if large undisturbed areas are established in perpetuity for conservation. Yet very little of the remaining tropical forest area is currently legally protected as national parks or other reserves, and even if existing proposals are implemented, the proportion will not be greater than 7 per cent of remaining tropical forests. In practice, the amount that is properly protected from intrusion by hunters and settlers, or excluded from oil and mineral prospecting, hydroschemes and highway development is very much less. Conservationists agree that a particular effort is needed to increase protected areas and to strengthen effective protection. However, totally protected areas can never be sufficiently extensive to provide for the conservation of all ecological processes and all species. In this case, the obvious avenues to explore are ways in which a natural forest cover can be used sustainably, to harvest products without degrading the forest. In this case, sustainable forestry means to harvest forest in a way that provides a regular yield of forest produce without destroying or radically altering the composition and structure of the forest as a whole. In other words, sustainable operations need to mimic natural ecosystem processes.

The obvious product in this respect is timber, but attitudes towards the logging industry have become polarised, with some people suggesting that it is essentially a primary cause of tropical deforestation and that the international trade in tropical timber should therefore be banned. In practice, however, logging in the tropics almost always involves the removal of selected trees rather than the clear-felling that is much more common in temperate forestry operations. Some damage to other trees also occurs even during selective logging, but less damaging techniques are available and if well enforced can greatly reduce smashing and other forms of degradation. The introduction of strict silvicultural rules in Queensland, Australia, in 1982 proved this point (Fig. 4.7).

However, although logging is rarely a direct cause of total forest clearance, it is often responsible for opening the way to other agents of deforestation, such as agriculturalists. Hence, others argue that despite the poor track record of loggers in the past, they should be encouraged to move away from a mining approach towards timber harvesting through sustainable management. Although, to date, there are few examples of a sustainable approach to logging actually being carried out, many foresters believe that most tropical moist forests could be managed as a truly renewable resource if human intervention operated within the inherent limits of the natural cycle of

Figure 4.8 Processing of tropical rain forest logs here at Samreboi in Ghana, where machinery is outdated and inefficient, means wastage rates of 65 per cent for plywood and 70 per cent for planks.

growth and decay found in all forests (Johns, 1992). Minimising damage to the forest, and particularly the forest floor, during timber extraction is important. Directional felling reduces damage to the remaining canopy and facilitates log removal, while pulling out logs by cable reduces damage to the forest floor. Careful planning of extraction tracks can reduce the severity of soil erosion, as can uphill logging, as opposed to downhill logging. Although these low-impact techniques are gaining in popularity, following intense pressure on logging companies from environmentalists, most timber operations are still controlled by entrepreneurs, to whom short-term profits are of prime importance, rather than by foresters, whose duty is to the long-term maintenance of the resources. Ironically, a considerable portion of tropical forest trees felled for timber in many areas is wasted due to inefficient processing using outdated equipment (Fig. 4.8).

One key policy change that could encourage the wider adoption of low-impact techniques would be to lengthen the periods over which concessions were granted. Typically, concessions are granted for periods of between 5 and 20 years, but if they were increased to at least 60 years, it would be in the interest of the logging companies to protect their forest and produce timber sustainably on perhaps 30- to 70-year rotations. One problem here, however, is that no one can guarantee that there will be a market for tropical timber so far into the future.

Two international initiatives have been established with the aim of managing timber production in a more sustainable way. One is the International Tropical Timber Agreement (ITTA), administered by the International Tropical Timber Organization (ITTO), which came into being in 1983. Although it is primarily a commodity trade agreement, the ITTA also has a component that is rare for such agreements which states that one of its objectives is to encourage national policies aimed at maintaining the ecological balance in timber-producing regions. The ITTO has established some important projects on multiple-use forestry practices, natural forest management and plantation development, which aim to demonstrate appropriate management of tropical forests, but the organisation also needs to encourage policy changes at national level, more realistic pricing and the international cooperation needed to encourage these changes. With such moves in mind, the ITTO set a target that all tropical forests should be managed sustainably by the year 2000 for its producer country members, which encompass more than 90 per cent of tropical rain forests; that target is still some way off being achieved.

By far the largest initiative to have focused on the conservation and sustainable management of tropical rain forests is the Tropical Forestry Action Plan (TFAP), set up in 1985 by several international bodies under the philosophy that since poverty is the root cause of deforestation, richer countries should fund projects to alleviate poverty in poorer tropical countries and thus reduce the problem. The basic objective of the TFAP is to restore, conserve and manage forests sustainably to benefit rural people, agriculture and the economy, in general, of the countries concerned. It is offered to developing countries rather than forced on them and, although there is a good deal of scepticism over whether development agencies can effectively discourage deforestation, since their previous track records have not been good, more than 70 countries have joined. To date, however, progress in implementing projects has been slow and a lack of appropriate funding has been a serious problem.

Although logging has received particular attention for sustainable forestry practices, the range of uses for tropical rain forest plants is very wide, as Table 4.3 shows from studies in Mexico. The top position of medicinal plants indicated in this table has been highlighted by some campaigning groups in recent years, claiming that rain forests contain untold numbers of 'new' drugs just waiting to be developed. However, such claims are debatable (Swanson, 1995). The collection and screening of plants, followed by purification and testing of extracts, is an expensive procedure that takes many years and it may not

TABLE 4.3 *Useful plants from the tropical rain forests of Mexico*

Use	Number of plants
Medicinal	780
Edible	360
Construction	175
Timber	102
Fuel	93
Honey	84
Forage	73
Domestic use	69
Crafts	59
Poison	52
Ornamental	51
Tools	51
Ceremonial use	50
Fibres	38
Dyes	34
Shade plants	31
Flavour enhancer	24
Gums	20
Stimulants	16
Fertiliser	15
Hedge plants	15
Tannins	12
Insecticides	11
Perfume	11
Nurse crop	10
Latex	10
Soap	10

Source: Toledo *et al.* (1995)

TABLE 4.4 *Economic returns from different uses of one hectare of lowland tropical rain forest in eastern Peru*

Use	Annual income (after labour and transport)*	Net present value (20-year discounted)*	Comments
EXTRACTIVE USES			
Latex harvest	$22	$440	Does not include other forest products or tourism
Edible fruits harvest	$400	$6000	
Total (harvesting)	**$422**	**$6440**	
Sustainable selective logging	$15	$490	
Total (harvesting + sustainable logging)	**$437**	**$6930**	
CONVENTIONAL 'DEVELOPMENT' USES			
One-time removal of marketable timber		$1000 (no 20-year discount)	Cutting destroys extractive resources
Reforestation with *Gmelina arborea*	$159	$3184	Not sustainable
Total (forestry)		**$4184**	
Intensive cattle ranching on ideal pasture	$148	<$2960	Not sustainable
Total (ranching)		**<$2960**	

*US$
Source: after Peters *et al.* (1989)

be as economical as the computer modelling of molecules and their synthesis. Some programmes to collect samples from rain forests and to evaluate their medicinal potential are being undertaken (e.g. Blum, 1993), but if and when a marketable new drug emerges from such procedures, the issue of who should share the resulting profits, and in what proportion, is an important one. Efforts are being made to protect the property rights of local communities and to share profits with them, but the western legal principles upon which such agreements are usually based may undermine patterns of community ownership.

Other forest products have also been the subject of some research. Low-intensity harvesting of fruits, nuts, rubber and other produce should result in a minimum of genetic erosion and other forms of disturbance, and a maximum of conservation. One economic study of the Peruvian Amazon (Table 4.4) has shown that this form of sustainable resource use is the most economically profitable in the medium to long term (Peters *et al.*, 1989). The results from this study in Peru may not be as widely applicable as hoped, however. A review of 24 such economic analyses found that the average annual value per hectare of sustainably produced non-timber forest products was US$50 a year (Godoy *et al.*, 1993), about half the annual average value of timber if harvested sustainably.

Adequate attention also needs to be given to appropriate markets and sufficient transport and distribution networks for non-timber products. If the products concerned do not have an international market, such an approach is likely to be less attractive than timber for national governments, because of the hard currency revenues an internationally tradeable commodity can command. Conversely, some believe that since the root causes of overexploitation of tropical forests are often traced to national and international élites, the highest priority should be given to increasing the control of small-scale forest dwellers over their resources (Dove, 1993). Where existing small-scale harvesting from tropical forests is already carried out for local markets, the diversity of forest products represents a great strength for the producers, allowing them to respond well to the changing needs and tastes of the market and furnishing them with a good annual income (Anderson and Ioris, 1992).

Some observers are confident that countries can develop and still maintain a high forest cover given sufficient investment in agriculture, and government dedication to sustainable forest management and to a strong national network of protected areas (Grainger, 1993a). Inevitably, some further tropical forest areas will still be lost due to economic problems, international disputes, organisational shortcomings and as policies evolve to make balanced decisions on various alternative land uses. There is a great deal of will, both inside and outside tropical forest countries, to use forests in a sensible manner, but a sustainable future for the world's tropical forests will also depend upon a greater level of international cooperation and equity than we have seen to date.

FURTHER READING

Blum, E. 1993 Making biodiversity conservation profitable: a case study of the Merck/INBio agreement. *Environment* 35(4): 16–45. Appraisal of an agreement between a multinational pharmaceutical company and the Costa Rican Biodiversity Institute to collect plant, insect and soil samples for new drug development.

Furley, P.A. (ed.) 1994 *The forest frontier: settlement and change in Brazilian Roraima.* London, Routledge. A collection of papers on both physical and human aspects of forest use in one of Brazil's states.

Grainger, A. 1993 *Controlling tropical deforestation.* London, Earthscan. This book looks at the causes of tropical rain forest destruction, and presents an integrated strategy for controlling the losses and using forests sustainably.

Mather, A.S. 1990 *Global forest resources.* London, Belhaven. An assessment of global forest resources, their use through history and in the present day, and policies and management techniques.

Morley, R.J. 2000 *Origin and evolution of tropical rain forests.* Chichester, Wiley. A comprehensive overview of tropical rain forest ecology.

Poore, D. 1989 *No timber without trees: sustainability in the tropical forest.* London, Earthscan. This book focuses on the prospects for timber production in the wider context of tropical rain forest conservation. It is based on a study for the International Tropical Timber Organization.

Skole, D. and Tucker, C. 1993 Tropical deforestation and habitat fragmentation in the Amazon: satellite data from 1978 to 1988. *Science* 260: 1905–10. A detailed case study of the Brazilian Amazon showing how remote sensing can be used to measure deforestation processes and rates.

Whitmore, T.C. 1998 *An introduction to tropical rain forests*, 2nd edn. Oxford, Oxford University Press. An excellent introductory text covering flora, fauna and forest dynamics, as well as human impact and management.

Websites

http://www.fao.org/forestry/ the FAO site has an extensive forestry section including data and up-to-date reports.

http://www.itto.or.jp/ the International Tropical Timber Organization site includes details on the organisation and its activities.

http://www.tropenbos.nl/ the Tropenbos Foundation is dedicated to the conservation and wise use of tropical rain forests.

http://www.cifor.org/ the Center for International Forestry Research is a global organisation committed to enhancing the benefits of forests for people in the tropics.

http://www.inpe.br/ the Brazilian Institute for Space Research site includes current deforestation rates in the Brazilian Amazon (in Portuguese).

http://www.lei.or.id/ the Indonesian Ecolabeling Institute aims to ensure sustainable forest resource management by means of a credible ecolabelling certification system.

http://www.globalforestwatch.org/ Global Forest Watch is an international data and mapping network covering forests inside and outside the tropics.

POINTS FOR DISCUSSION

- Should the international trade in tropical timber be banned?
- Deforestation in the tropics is proceeding at a rapid rate. So what? (Answer the question from the viewpoint of (a) a landless Brazilian peasant, (b) a biotechnology company, and (c) a deep ecologist.)
- To what extent do the arguments in favour of conserving biodiversity also apply to forests in the humid tropics?
- Should the future of tropical rain forests in any particular country be solely a matter for the country concerned?

5

DESERTIFICATION

TOPICS COVERED

Definition of desertification
Areas affected by desertification
Causes of desertification
Understanding desertification

Key words: drought, desiccation, monoculture, rangeland, piosphere, sacrifice zone, secondary salinisation, saline seep

Human societies have occupied deserts and their margins for thousands of years, and the history of human use of these drylands is punctuated with many examples of productive land being lost to the desert. In some cases the cause has been human, through overuse and mismanagement of dryland resources, while in other cases natural changes in the environment have reduced the suitability of such areas for human habitation. Often a combination of human and natural factors is at work. Salinisation and siltation of irrigation schemes in Lower Mesopotamia are thought to have played a significant role in the eventual collapse of the Sumerian civilisation 4000 years ago, and similar mismanagement of irrigation water has been implicated in the abandonment of the Khorezm oasis settlements of Uzbekistan dating from the first century AD. The decline of Nabatean towns in the Negev Desert from about AD 500 has been attributed to Muslim–Arab invasions, but a gradual change in climate towards a more arid regime probably played an important role. The belief that climates in post-glacial times were becoming gradually drier, so-called 'progressive desiccation', generated widespread concern for the future human use of drylands in Central Asia, southern and West Africa during the late nineteenth and early twentieth centuries, but the concept was subsequently disproved. Invasion and settlement of areas by outsiders has often been highlighted as a reason for the onset of dryland environmental degradation in historical times. The Spanish conquest is cited in South America and a more diverse mix of European settlers was responsible for the dramatic years of the Dust Bowl in the US Great Plains in the 1930s. Desertification became a major global environmental issue in the 1970s after international concern over famine in West Africa focused attention on both drought and desertification as insidious causes. A UN Conference on Desertification (UNCOD) was convened in 1977, at which desertification was seen as a global threat affecting drylands all over the world, in rich and poor countries alike. Since that time, however, desertification has become surrounded in controversy, over its nature, extent, causes and effects, and even its very definition.

DEFINITION OF DESERTIFICATION

The word 'desertification' was first coined by a French forester to describe the change towards more desert-like conditions in humid areas adjacent to the Sahara Desert in West Africa as forests in the region were gradually cleared (Aubreville, 1949). Literally, desertification means the making

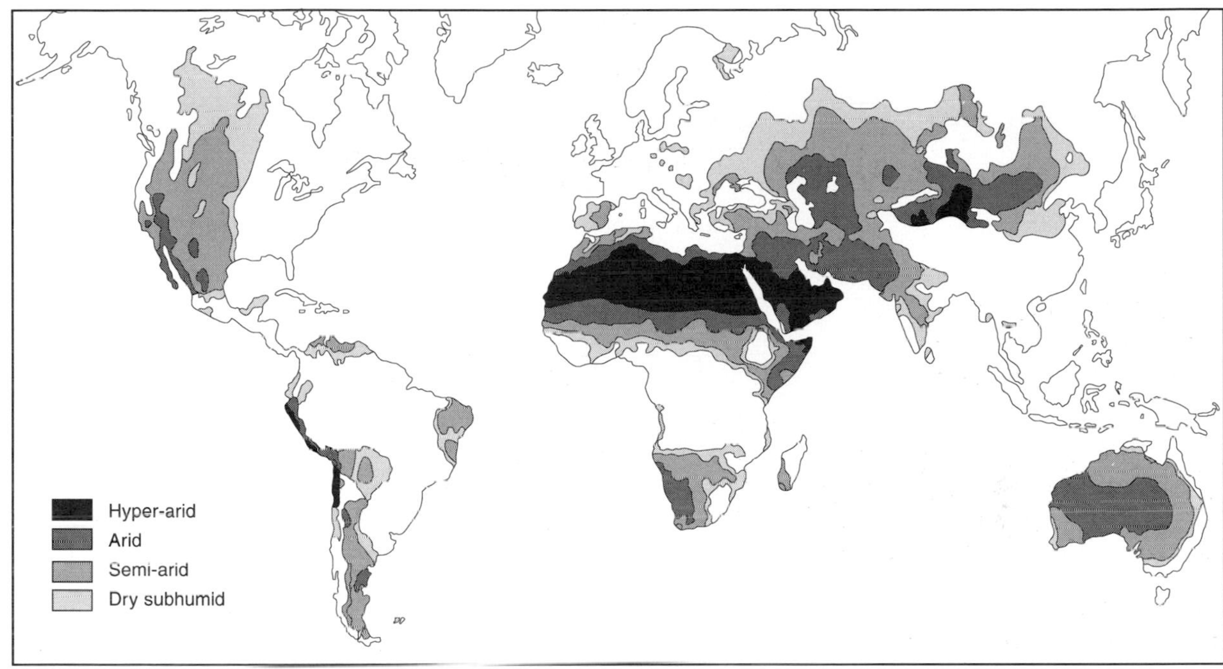

Figure 5.1 The world's drylands (after UNEP, 1992a).

of a desert, but although the word has been in use for more than 50 years, a universally agreed definition of the term has only recently been arrived at. A survey of the literature on the subject has revealed more than a hundred different definitions (Glantz and Orlovsky, 1983) and most of these suggest that the extent of deserts is increasing, usually into desert-marginal lands. The idea of a loss of an area's resource potential, through the depletion of soil cover, vegetation cover or certain useful plant species, for example, is included in many definitions. Some suggest that such losses are irreversible over the timescale of a human lifetime.

Most, though not all, authorities agree that desertification occurs in drylands, which can be defined using the boundaries of climatic classifications (Fig. 5.1) and make up about one-third of the world's land surface. Arid, semi-arid and dry subhumid zones can be taken together as comprising those drylands susceptible to desertification, excluding hyper-arid areas because they offer so few resources to human populations that they are seldom used and can hardly become more desert-like. Recent thinking on desertification refers to it as a process of land degradation in drylands. The concept of degradation is linked to using land resources – which include soil, vegetation and local water resources – in a sustainable way. Land that is being used in an unsustainable manner is being degraded, which implies a reduction of resource potential caused by one or a series of processes acting on the land. The official definition adopted by the UN Convention to Combat Desertification, which came into force in 1996, is 'land degradation in arid, semi-arid and dry subhumid areas resulting from various factors, including climatic variations and human activities'. The definition illustrates one of the difficulties of the desertification issue: the term itself covers many different forms of land degradation.

The relative roles of human and natural factors, such as drought, have long been a subject of debate. Drylands are, by definition, areas that experience a deficiency in water availability on an annual basis, but precipitation in drylands is characterised by high variability in both time and space. Rainfall typically falls in just a few rainfall

TABLE 5.1 *UNEP estimates of types of drylands deemed susceptible to desertification, proportion affected and actual extent*

	1977	1984	1992
Climatic zones susceptible to desertification	Arid, semi-arid and subhumid	Arid, semi-arid and subhumid	Arid, semi-arid and dry subhumid
Total dryland area susceptible to desertification (million ha)	5281	4409	5172
Proportion of susceptible drylands affected by desertification (%)	75	79	70
Total of susceptible drylands affected by desertification (million ha)	3970	3475	3592

Source: Thomas and Middleton (1994)

events and its spatial pattern is often highly localised since much comes from convective cells. In addition to this variability on an annual basis, longer-term variations such as droughts occur over periods of years and decades. Dryland ecology is attuned to this variability in available moisture and is therefore highly dynamic. On the large scale, satellite imagery has been used to observe changes in the vegetational boundary of the Sahara of up to 200 km from one dry year to a following wet one (Tucker *et al.*, 1991).

In practice, it can be difficult to distinguish in the field between the adverse effects of human action and the response of drylands to natural variations in the availability of moisture. A good example is the dramatic rise in the amount of soil lost to wind erosion in parts of the Sahel region of Africa during the 1970s and 1980s as recorded by the annual number of dust storm days. At Nouakchott, the capital of Mauritania, dust storms blew on average less than 10 days a year during the 1960s, but in the mid-1980s the average had increased to around 80 days per year (Middleton, 1985). This increase in soil loss can be explained both by drought, which characterised the area during the 1970s and 1980s, and by human action, but the relative roles of these factors are difficult to quantify.

AREAS AFFECTED BY DESERTIFICATION

In preparation for UNCOD, UNEP (the UN Environment Programme) produced an estimate of the global area at risk from desertification. The estimate suggested that at least 35 per cent of the Earth's land surface was threatened, an area inhabited by 20 per cent of the world's population. This estimate was based upon what limited data were available at the time and a lot of expert opinion and guesswork. It was a fair first attempt to assess a very difficult problem, but subsequent efforts to assess the global proportions of desertification have not greatly reduced the reliance upon estimates. Two subsequent assessments have been made (Table 5.1) and UNEP has been accused of treating these estimates as if they were based upon scientific data (Thomas and Middleton, 1994). It has even been suggested that since UNCOD, desertification has become an 'institutional fact', one that UNEP wanted to believe for its own purposes, for the continued existence of its desertification control unit, rather than because the evidence was there to support it (Warren and Agnew, 1988). Indeed, the lack of good scientific evidence to support the concept has led some people to question the very existence

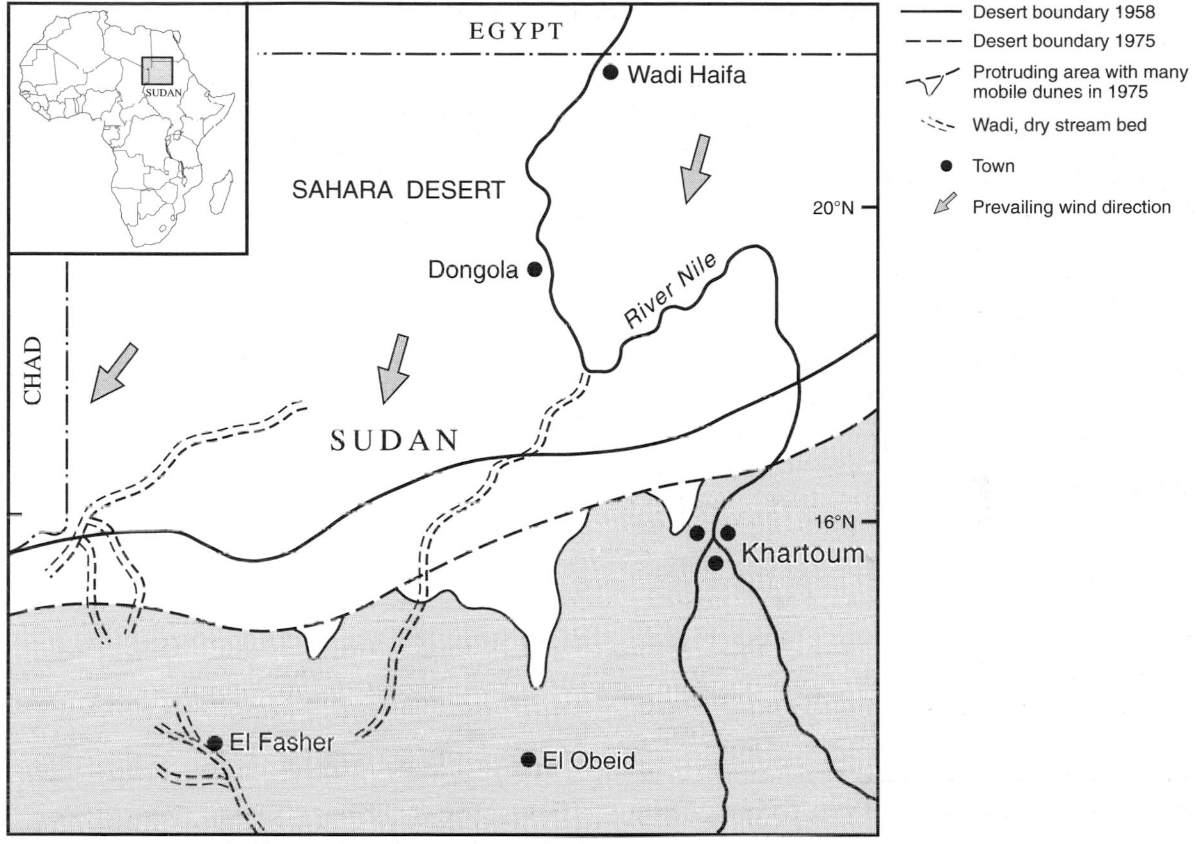

Figure 5.2 Lamprey's 'advancing Sahara', a study that purported to show that the desert was advancing at 5.5 km a year. See text for criticism of these findings.

of desertification. As Hellden (1991: 372) puts it: 'the word "myth" has been mentioned'.

Some people question whether a global assessment of something as poorly understood as desertification is possible, or even useful, given that solutions to the problem should be locally oriented. However, a case can be made for a global assessment if only to put the issue into perspective, to identify specific problem areas, and to generate the political and economic will to do something about it.

'Spreading deserts'

Another area in which many authorities have helped to 'propagate the myth' has been in the numerous statements concerning the rate at which desertification is happening. These statements have been made in the academic literature and in popular and political circles for publicity and fund-raising purposes. Warren and Agnew (1988) quote several examples, including one made in 1986 in which US Vice-President Bush was being urged to give aid to the Sudan because 'desertification was advancing at 9 km per annum' (Warren and Agnew, 1988: 2). Such pronouncements have become so commonly used that their validity has seldom been questioned, but in reality they are rarely supported by any scientific evidence. They also reinforce an image of sand dunes advancing to engulf agricultural land. Although undoubtedly a powerful vision for publicity purposes, and one that is appropriate in some places, the idea is oversimplified and not a generally accurate image of the way desertification works. It is a much more complex set of processes that is more commonly manifested as patches of degradation in the desert fringe.

One well-known study that did set out to measure the rate of desert advance was made in northern Sudan (Lamprey, 1975). The vegetational edge of the desert was mapped by reconnaissance in 1975 and then compared to the boundary drawn from another survey carried out in 1958. The two boundaries indicated that the desert's southern margin had advanced by 90–100 km between the two dates (Fig. 5.2), at an average rate of advance of 5.5 km per year over 17 years.

An immediate criticism can be made of the methodology used for the study in that given the inherent variability of drylands, two 'snapshot' surveys can only be of very limited use in determining significant changes in vegetation, particularly since the 1970s was a period of drought in the Sudan while the 1950s was a decade characterised by above-average rainfall. Reliable conclusions can only be drawn after long-term monitoring that takes into account seasonal and year-to-year fluctuations in vegetation cover. When Lamprey's conclusions were scrutinised by Swedish geographers they were found to be inaccurate. It transpired that the 1958 desert vegetation line was not based on a ground survey but followed the 75 mm isohyet, and satellite imagery and ground survey data compiled over several years did not confirm Lamprey's findings (Hellden, 1988).

CAUSES OF DESERTIFICATION

The methods of land use that are applied too intensively and, hence, contribute to desertification are widely quoted and apparently well known. They can be classified under the headings of intensive grazing, overcultivation and overexploitation of vegetation. Salinisation of irrigated cropland is often viewed as a separate category from these. Although the way in which these inappropriate land uses lead to desertification is also well known in theory, in practice, particular areas deemed to be desertified due to a particular cause have often been so described according to subjective assessments rather than after long-term scientific monitoring. Furthermore, although specific land uses have been the subject of most interest from desertification researchers, it can be argued that permanent solutions to specific problem areas can only be found when the deeper reasons for people misusing resources are identified. Such reasons, many of which are rooted in social, economic and political systems (Table 5.2), enable, encourage or force inappropriate practices to be used. Although an understanding of the physical processes involved in overcultivation, for example, is important, there has generally been too much emphasis on this physical side of the equation, resulting in an over-reliance upon technical solutions. It is just as important to understand the underlying driving forces, the amelioration of which is of equal significance in the quest to find long-lasting solutions to desertification problems.

Intensive grazing

The overuse of pastures, caused by allowing too many animals or inappropriate types of animals to graze, has been the cause of degradation over the largest areas of desertified land on the global scale according to the estimates produced by UNEP. Of the 3592 million hectares estimated in 1992 to be suffering from desertification, no less than 2576m ha, just under 72 per cent, were considered to be experiencing degradation of vegetation. Intensive grazing can result in both the actual removal of biomass by grazing animals and other effects of livestock such as trampling and consequent compaction. A common consequence of heavy grazing pressure is a decrease in the vegetative cover, leading to increased erosion by water or wind.

Another widespread effect of intensive grazing is the encroachment of unpalatable or noxious shrubs into grazing lands. Long-term grazing of semi-arid grasslands can typically lead to an increase in the spatial and temporal heterogeneity of water, nitrogen and other soil resources, which promotes invasion by desert shrubs, in turn

TABLE 5.2 *Suggested root causes of land degradation*

Natural disasters	Degradation due to biogeophysical causes or 'acts of God'
Population change	Degradation occurs when population growth exceeds environmental thresholds (neo-Malthusian) or decline causes collapse of adequate management
Underdevelopment	Resources exploited to benefit world economy or developed countries, leaving little profit to manage or restore degraded environments
Internationalism	Taxation and other forces interfere with the market, triggering overexploitation
Colonial legacies	Trade links, communications, rural–urban linkages, cash crops and other 'hangovers' from the past promote poor management of resource exploitation
Inappropriate technology and advice	Promotion of wrong strategies and techniques, which result in land degradation
Ignorance	Linked to inappropriate technology and advice: a lack of knowledge leading to degradation
Attitudes	People's or institutions' attitudes blamed for degradation
War and civil unrest	Overuse of resources in national emergencies and concentrations of refugees leading to high population pressure in safe locations

Source: Thomas and Middleton (1994)

leading to a further localisation of soil resources under shrub canopies in a process of positive feedback. In the barren area between shrubs, soil fertility is decreased by erosion and gaseous emissions (Schlesinger *et al.*, 1990). Increased runoff and erosion result in stripping of the soil surface layer, the formation of desert pavement in inter-shrub areas, and the development of rills. This degradation of the plant resource, in turn, reduces the number of livestock that are able to graze the area.

Traditional herders, for whom larger numbers of livestock represent greater personal wealth and social standing, have often been seen as the culprits in this process in the developing world. Such arguments have been used as a reason for encouraging settlement of nomadic pastoralists in many dryland countries of Africa, the Middle East and Asia. Intensive grazing problems have also been identified on commercial ranches in North America and Australia.

In practice, the reasons for increased grazing pressure on pastures are numerous and complex. They include competition for land as cultivated areas increase, pushing herders into more marginal pastures. In many parts of the Sahel, expanding areas of sorghum and millet cultivation have been primary factors responsible for a major decrease in the availability of range, as Ringrose and Matheson (1992) have identified. The position of herders at the margins of society in the eyes of many central governments has often

Figure 5.3 Mixed herds are traditionally kept by herders in drylands, as here in Mongolia's Gobi Desert. Different animals eat different types of vegetation and have different susceptibilities to moisture availability.

meant that they are situated at the end of a chain of events that sees the expansion of irrigated land for cash cropping displacing rainfed subsistence cultivators who encroach into traditional grazing grounds, forcing herders into smaller ranges. This situation is described by Janzen (1994) in southern Somalia where expansion of irrigated agriculture on the Jubba and Shabelle rivers has forced small farmers to clear large areas of bushland for cultivation. For nomadic herders, this savanna zone and the river valleys themselves were important grazing lands during the dry season. Resultant intensification of grazing pressure into smaller areas has also been driven by the sedentarisation of some nomadic groups, a trend encouraged by government policy in the 1970s, and accelerated more recently by drought. The issue of government-sponsored sedentarisation continues in many other countries in Africa and elsewhere.

Sinking boreholes to provide new and more reliable water supplies has also led to increased grazing pressures across much of the Sahel and in the Kalahari of Botswana where, between 1965 and 1976, livestock numbers and the accessible grazing resources were increased by about two and a half times (Sandford, 1977). In some parts of the world, degradation has been seen as the result of a change in the approach to pasture use: from a flexible strategy typical of traditional pastoralists in which the natural dynamism of dryland vegetation is used by regular movement of herds and maintaining several different animal species (Fig. 5.3), to a less flexible westernised approach in which the pasture may be fenced off and only cattle are grazed.

Although these ideas on intensive grazing problems have been followed by many scientists and policy-makers, recent thinking on the nature of dryland vegetation and ecology has questioned much of this conventional view. Distinguishing between degradation of vegetation and other forms of vegetation change is very difficult, and the importance of intensive grazing as a cause of desertification has probably been exaggerated in many areas.

One commonly quoted situation considered to be typical of desertified areas is the loss of vegetation around wells or boreholes, bare areas that supposedly grow and coalesce. In fact, several studies have failed to show that these 'piospheres' expand over time (e.g. Hanan *et al.*, 1991). Halos of bare, compacted soil 50–100 m in circumference caused by grazing and trampling are undoubtedly characteristic of many rangeland watering holes (Fig. 5.4). But such areas are also typified by higher levels of soil nutrients than surrounding areas thanks to regular inputs from dung and urine (Barker *et al.*, 1990), which may balance out any negative effects of possible soil loss by erosion. Piospheres are perhaps best interpreted as 'sacrifice zones' in which the loss of the vegetation resource is outweighed by the advantages of a predictable water supply (Perkins and Thomas, 1993). However, despite this reassessment of piospheres, the areas beyond the sacrifice zone around these types of focal point in the landscape can become desertified by bush encroachment. High cattle densities are thought to encourage invasion by thorny bushes as described above. Although cows will browse from bushes as well as graze on grasses, they tend to avoid some species because of the thorns, and these species become more abundant as a result. A thick cover of thorny bushes deters grass growth and prevents cows from entering thickets. This type of bush encroachment has led to a significant reduction in the extent of high-quality rangeland in Botswana (Moleele *et al.*, 2002).

Other ideas about the effects of grazing on rangeland vegetation are also being substantially revised. Formerly, most claims that a particular area was being 'overgrazed' were based on the idea that the area had a fixed 'carrying capacity', or theoretical number of livestock a unit area of pasture could support without being degraded. This idea was in turn based on Clements' (1916) model of vegetation succession and ecological stability. But current thinking depicts semi-arid ecosystems as seldom, if ever, reaching equilibrium. Rather they are in a state of more or less constant flux, driven by disturbances such as drought, fire and insect attack. Hence, while calculation of a carrying capacity is possible in relatively unchanging environments, it is difficult to apply to non-equilibrium environments like dryland pastures because the number of animals an area can support varies on several timescales: before and after a rainy day, between a wet season and a dry season, and between drought years and wet years. Grazing the number of livestock appropriate for a wet year on the same pasture during a drought could result in degradation, but several studies cited by Warren and Khogali (1992) show that such disastrous concentrations of livestock are rarely reached because, under conditions of environmental stress, the animals are often the first component of the system to fail, by dying. Traditional pastoralists have also developed social and economic mechanisms for coping with such occurrences, as McCabe (1990) has documented for the Turkana in East Africa.

In the light of these changes in our understanding of rangeland ecology, the relative importance of intense grazing and natural stresses is being reassessed. One study of open woodlands in north-west Namibia, an area long thought to have been desertified by local herders overgrazing, found very little evidence of degradation (Sullivan, 1999). This study concluded that

Figure 5.4 Cattle around a water hole in Niger. Such 'piospheres' are typically devoid of vegetation but the soil has a high nutrient content.

previous assessments of the area's ecological health have been based more on perceptions than scientific assessment. These perceptions have been clouded by an adherence to ideas of a fixed carrying capacity based on rangelands as equilibrium systems, and remnants of a colonial ideology that views traditional communal herding as environmentally degrading.

Many authorities now believe that rainfall variability is a more important determinant of the health of rangeland and its soils than overgrazing (Behnke *et al.*, 1993). This change in understanding of dryland ecology is an important development because if natural environment factors are the major determinant of available rangeland resources, this has major implications for pastoral management (Scoones and Graham, 1994).

Overcultivation

Several facets of overcultivation are commonly quoted. Some stem from intensification of farming, which can result in shorter fallow periods, leading to nutrient depletion and eventually reduced crop yields. Soil erosion by wind and water is another result of the overintensive use of soil, resulting from a weakened soil structure and reduced vegetation cover. Monocultures can lead to all these forms of soil degradation as data from 27 years of monitoring on cropland in the

TABLE 5.3 *Examples of overcultivation*

Area (date)	Root causes	Land use	Environmental response
US Great Plains (1930s)	Pioneer spirit	Deep ploughing for cereals	Wind erosion
Former USSR's Virgin Lands (1950s)	Political ideology, food security	Deep ploughing for cereals	Wind erosion
Niger (1950s and 1960s)	Colonial influence, cash crops	Extensification of groundnut cultivation	Nutrient depletion

semi-arid pampa of Argentina show (Buschiazzo *et al.*, 1999). Long-term cultivation of millet was demonstrated to have the most deleterious effects on both the physical and chemical properties of soils, leading to decreases in dry aggregate stability (by 10 per cent), soil organic matter (30 per cent), and available nutrients such as phosphorus (44 per cent), iron (20 per cent) and zinc (90 per cent). The depletion of nutrients means that greater amounts of fertilisers have to be applied to maintain crop yields, while the declines in organic matter and soil stability have meant a greater susceptibility to erosion.

In many cases, erosion has also been caused by the introduction of mechanised agriculture with its large fields and deep ploughing, which further disturbs soil structure and increases its susceptibility to erosive forces. Similar outcomes have also been noted in areas where cultivation has expanded into new zones that are marginal for agricultural use because they are more prone to drought, or are made up of steeper slopes that are more prone to erosion. Some classic examples of overcultivation, their root causes, inappropriate land uses and resulting degradation, are shown in Table 5.3. The wind erosion resulting in the Great Plains and the Virgin Lands is discussed on page 242.

Overexploitation of vegetation

Clearance of forested land is undertaken for a number of reasons, but most commonly to make way for the expansion of grazing or cultivation and/or to provide fuelwood. Such action is a degradation of vegetation resources and also reduces the protection offered to soil by tree cover. Accelerated erosion may result, and over the longer term soil is deprived of inputs of nutrients and organic matter from decomposing leaf litter. Both of these processes can lead to the degradation of soil structure and fertility. Water tables may also be affected.

In many areas, vegetation clearance has a long history. In northern Argentina, the semi-arid Chacoan forests have been exploited for more than 100 years, initially because the wood made excellent railway sleepers, but more recent clearance has been for agricultural expansion, prompted by high grain prices and positive rainfall balances since the 1970s (Margarita and Loyarte, 1996). In African drylands, more or less complete removal of the vegetation cover is occurring in many parts of the Sahel region. The expansion of agriculture is a prime cause of deforestation in Burkina Faso, where an estimated 50 000 ha of woodland was being cleared every year in the early 1980s. Similarly, widespread replacement of tree savanna by cultivation has been reported over the period 1957–87 from the Nara area on the Mali/Mauritania border, where the agricultural area virtually doubled over the period (IGN, 1992).

Overexploitation for domestic purposes (particularly fuelwood and charcoal making) does not usually lead to the complete removal of all

TABLE 5.4 *Loss of closed forest cover around selected dryland urban centres in India*

Urban centre (state)	Closed forest cover (km^2)		Loss (%) 1972–82
	1972–75	1980–82	
Ajmer (Rajasthan)	259	124	52
Amritsar (Punjab)	208	111	47
Bhavnagar (Gujarat)	112	9	92
Bhopal (Madhya Pradesh)	3031	1417	53
Gwalior (Madhya Pradesh)	1353	515	62
Hyderabad (Andhra Pradesh)	40	26	35
Indore (Madhya Pradesh)	3770	1070	71
Jaipur (Rajasthan)	1534	786	49

Source: after Bowonder *et al.* (1988)

vegetation, but it represents the use of vegetation to a degree that is beyond its natural capability to renew itself and thus results in a degraded vegetation cover. As trees are removed, villagers may supplement wood for fires with dried animal dung that would otherwise be left on the soil to provide valuable nutrients. One study in Ethiopia estimated that the economic value of manure diverted from this fertilising role to the cooking stove was US$123 million annually, which could increase grain harvests by 1–1.5 million tonnes a year (Newcombe, 1984). An expanding human population is the ultimate driving force behind this form of desertification.

In Pakistan, the few remaining stands of tropical thorn forests that once covered the Punjab Plains, largely cleared for irrigated agriculture, are under continuing pressure for their long-standing use as a source of fuelwood (Khan, 1994). Fuelwood collection around many urban centres in India has significantly reduced forested areas in their hinterlands in recent decades, as the urban poor have been forced to turn increasingly to fuelwood by rising prices of kerosene, coal and charcoal. One study of major Indian cities, using satellite imagery (Bowonder *et al.*, 1988), has shown that more than half of the closed forest cover within a 100-km radius around many dryland cities was lost in the ten years to 1982 (Table 5.4).

In the 1970s and 1980s, the environmental problems caused by fuelwood collection were thought to be so acute in Sahelian Africa that a 'fuelwood crisis' was feared. The World Bank (1985) estimated that sustainably harvested fuelwood could support just two-thirds of the human population in the Sahel. However, this regional environmental disaster has not materialised and in the late 1990s the fuelwood situation in the Sahel was being reassessed. The methods used, comparing current woodfuel consumption with current stocks, annual tree growth and annual population growth, has been criticised by some (e.g. Leach and Mearns, 1988). Poor survey data, the importance of replacement fuels such as crop residues and dung, and the effect of increasing woodfuel prices as local supplies dwindle are highlighted as making the situation more complex than was previously proposed. Overall, it seems that the surveys conducted in the 1970s and 1980s underestimated the amount of fuelwood available in the Sahel (World Bank, 1996).

Although the predicted fuelwood crisis in the Sahel has not taken place on the scale once feared, environmental degradation has occurred in some areas. Much of this form of desertification is

TABLE 5.5 *Major causes and effects of secondary salinisation*

Cause	Effect
Direct effects of poor irrigation techniques	Four principal causes: water leakage from supply canals; overapplication of water; poor drainage, and insufficient application of water to leach salts away; all contribute to the build-up of salts in the root zone and/or to the creation of perched saline water tables under irrigated fields
Indirect effects of irrigation schemes	River barrages and dams can result in soil salinisation on valley terraces above the dam should a reservoir raise local water tables; irrigation schemes can also increase the salinity of soils downstream due to an absolute reduction in downstream discharge and/or the discharge of saline irrigation wastewater
Effects of vegetation changes	Replacement of native shrubs, trees and grasses with annual grain crops or grassland for grazing reduces evapotranspiration so that rainfall percolates through subsurface sediments to impermeable horizons where it is conducted laterally to lower slope positions, producing extensive areas of salty waterlogged soil; these so-called 'saline seeps' are most common in non-irrigated dryland farming regions
Sea water incursion	Excessive pumping of coastal freshwater aquifers for irrigation or other uses lowers the water table allowing sea water to seep into the aquifer, often resulting in soil salinisation; a similar effect can also result from reduced river flows, due to excessive irrigation offtakes or dam building (sometimes exacerbated by drought), allowing enhanced up-channel movement of ocean water
Disposal of saline wastes	Dumping of saline wastes from industrial and municipal sources (e.g. petroleum wells, coal mines, desalination plants) directly on to soils

Source: Middleton and van Lynden (2000)

concentrated around urban areas that have expanded very rapidly since the late 1960s as rural migrants fled the effects of drought. Denuded areas susceptible to enhanced erosion have been reported from numerous Sahelian cities, including Khartoum, Dakar, Ouagadougou and Niamey. However, the complexity of the fuelwood issue can be illustrated by the case of Kano in northern Nigeria where Nichol (1989) found that tree density had increased in recent times in the immediate vicinity of the city as the transport of fuelwood by donkey had been displaced by the use of long-distance trucks. A decline in tree density was recorded in the zone 70–250 km from Kano.

Salinisation

Salinisation is a form of soil degradation most commonly encountered in dry climatic regions, although it also occurs in more humid environments. Natural, or 'primary', salt-affected soils are widespread in drylands because the potential evaporation rate of water from the soil exceeds the input of water as rainfall, allowing salts to accumulate near the surface as the soil dries. While these primary salt-affected soils are found extensively under natural conditions, salinity problems of particular concern to agriculturalists arise when previously productive soil becomes salinised as a result of poor land management, so-called 'secondary' salinisation. Secondary, or human-induced, salt-affected soils cover a smaller area than primary salt-affected soils (Oldeman *et al.*, 1990) but secondary salinisation represents a more serious problem for human societies because it mainly affects cropland. Major agricultural crops have a low salt tolerance compared to wild salt-tolerant plants (halophytes), so salinisation rapidly leads to declines in yields. Consequently arable land, a scarce and valuable resource in dryland regions, is frequently abandoned when it becomes salinised, due to the very high cost of remediation.

The human activities that lead to desertification through the build-up of salts in soils are well documented and are summarised into five groups in Table 5.5; secondary salinisation occurs under a number of circumstances but is most commonly associated with poorly managed irrigation schemes. One estimate suggests that nearly 50 per cent of all the irrigated land in arid and semi-arid regions is affected to some extent by secondary salinisation (Abrol *et al.*, 1988) and the process is widely regarded as irrigated agriculture's most significant environmental problem (Ghassemi *et al.*, 1995).

The impact of salinisation on crop yields acts indirectly via effects on the soil and through direct effects on the plants themselves (Rhoades, 1990). Salt accumulation reduces soil pore space and the ability to hold soil air, moisture and nutrients, resulting in a deterioration of soil structure and a reduction in the soil's suitability as a growing medium. Plant growth is also directly impaired by salinisation firstly because salts are toxic to plants, particularly when they are seedlings, and secondly through its effects on osmotic pressures. Contact between a solution containing large amounts of dissolved salts and a plant cell causes the cell's lining to shrink due to the osmotic movement of water from the cell to the more concentrated soil solution. As a result the cell collapses and the plant succumbs.

Salinisation also leads to a range of off-site hazards. Drainage from salinised areas often increases the concentration of salts in streams, rivers and wetlands, adversely affecting freshwater biota and in some cases resulting in a loss of biodiversity (see the example of the Aral Sea tragedy on page 130). Groundwater that has become salinised may no longer be suitable for other human uses of this resource (e.g. for drinking), while the capillary rise of salts from such 'aggressive' groundwater into buildings and other structures can cause severe damage to building materials by salt weathering (Goudie and Viles, 1997).

Table 5.6 shows the serious scale of the salinisation problem on irrigation schemes in some

TABLE 5.6 *Salinisation of irrigated cropland in selected countries*

Country	Irrigated land as a proportion of all cropland (%)	Proportion of irrigated land affected by salinisation (%)
Australia	4	9
China	46	15
Egypt	100	33
Iraq	47	50
Kazakhstan	6	60–70
Pakistan	78	26
Spain	17	10–15
Syria	15	30–35
Turkmenistan	91	87
USA	10	23
Uzbekistan	92	>60

Source: Gleick (1993), Ghassemi *et al.* (1995), Glazovsky (1995) and WRI (1996)

Figure 5.5 Fuelwood harvest from a formerly salinised area on the southern margins of the Taklimakan Desert rehabilitated with *Tamarix* (Photo: UNEP).

individual countries. In Pakistan, where irrigated land supplies more than 90 per cent of agricultural production, about a quarter of the irrigated area suffers from problems of salinisation. In total, some 9.3 million ha are affected, about 11 per cent of the country's land area (Middleton and van Lynden, 2000). Much of the affected area is part of the Indus irrigation system, the largest single irrigation system in the world. Estimates of the costs of salinisation in Pakistan indicate that it depletes the country's potential production of cotton and rice by about 25 per cent. In economic terms, this figure equates to US$2.5 billion a year (Ahmad and Kutcher, 1992).

Financial constraints, maintenance problems, inexperienced management and farmers are factors common to many such irrigation problems and failures. A lack of appropriate advice and training, and credit and marketing facilities for farmers are again commonly encountered, as the experience of Iraq's Greater Mussayeb Project illustrated in the 1950s (Government of Iraq, 1980).

Although expensive, salinisation on irrigation schemes can be improved. In Pakistan's Punjab, Salinity Control and Reclamation Projects (SCARPs) were first introduced in the 1950s to address the long-standing problems of waterlogging and salinisation. The first SCARP extended over an area of nearly 0.5m ha in the Rechna Doab area near Lahore. It involved sinking more than 2000 tube-wells to lower the water table and to provide more water for agriculture; reducing the salinity hazard by leaching and adding amendments, and helping farmers with technical information and economic support. About 73 per cent of the area, some 360 000 ha, had water tables shallower than 3 m in 1959, but this area had been reduced to about 38 per cent or approximately 160 000 ha by 1977–78.

Poorly managed irrigation schemes are not the only human-induced cause of salinity problems in dryland environments. Rising water tables can also result when natural vegetation is cleared and replaced with pasture or a crop that requires less moisture for growth. Less evapotranspiration from crops or pastures than from native deeper-rooted vegetation increases the amount of water percolating through the soil to recharge aquifers. The clearance of native eucalyptus from large areas of south-west Western Australia has caused saline groundwater to rise in this way, leading to the problem of 'saline seep' (Peck and Williamson, 1987). Such salinity problems can also be reversed. Planting of deep-rooted *Tamarix* on the southern margins of the Taklimakan Desert in China has reduced waterlogging and salinity as well as providing a new source of fuelwood for local villagers (Middleton and Thomas, 1997 – see Fig. 5.5).

FOOD PRODUCTION

One of the most important reasons for desertification's position as a major global environmental issue is the link between resource degradation and food production. Desertification reduces the productivity of the land and, hence, less food is available for local populations. Malnutrition, starvation and ultimately famine may result. This link was the central reason behind the calling of the UN Conference on Desertification in 1977 and was a strong theme in the negotiations leading to the UN Convention to Combat Desertification.

However, the links between soil degradation and soil productivity, and consequently crop yields, are seldom straightforward or clear-cut (Stocking, 1987). Productivity is affected by many different factors, such as the weather, disease, pests, the farming methods used and economic forces. Distinguishing the effects of these various factors is, in practice, very difficult, quite apart from the fact that data on crop productivity are often unreliable and many measurements of soil erosion are also open to question (see Chapter 14). While the theory linking desertification to food production is basically sound, it is often oversimplified and deserves further careful research.

The link between desertification and famine is even more difficult to make. This is not the case in theory, since food shortages stem from reduced harvests and thinner or even dead animals, although such shortages are also caused by factors of the natural environment, particularly drought. In practice, however, famines typically occur in areas that are characterised by a range of other factors such as poverty, civil unrest and war. Several authors have emphasised the social causes of severe food shortages (e.g. Sen, 1981; Wijkman and Timberlake, 1984). In the African context, Curtis *et al.* (1988: 3) suggest that 'social, not natural or technological, obstacles stand in the way of modern famine prevention'.

Mass starvation is probably more a function of poor distribution of food, and people's inability to buy what is available, than desertification, although desertification, drought and many other natural hazards may act as a trigger. This tentative conclusion is supported by one of the few studies that specifically aimed to assess the relative importance of these factors, in the Sudanese famine of 1984/85 (Olsson, 1993). Millet and sorghum yields in the provinces of Darfur and Kordofan, the worst affected areas, were reduced to about 20 per cent of pre-drought levels, largely due to climatic factors – the amount and timing of the rains. Desertification, by contrast, was calculated to account for just 10–15 per cent of the variation in crop yields. While drought was undoubtedly a trigger for ensuing famine, the profound causes were different. As fears of crop failures mounted, the price of food in Darfur and Kordofan soared to levels that most rural people were unable to afford. While food was available at the national level, the distribution network was inadequate and famine followed.

UNDERSTANDING DESERTIFICATION

Appropriate solutions to desertification problem areas can only be found after more long-term studies of specific areas. Monitoring is necessary to identify where desertification is happening, how it works and why it is occurring. It must also attempt to distinguish between natural environmental fluctuations and impacts that can be attributed to human action, since it is the human aspects we have the ability to alter. Monitoring programmes also need to combine scientific measurements with an understanding of how social systems are related to the environment, an aspect of the desertification issue that has been relatively neglected until recently. Although, to date, there have been few such long-term investigations, work from two areas illustrates the approach.

The US Great Plains

The Dust Bowl of the 1930s on the North American Great Plains is one of the best-known dryland environmental disasters in history. Over a period of 50 years, grasslands were turned into wheat fields by a culture set on dominating and exploiting natural resources using ploughs and other machinery developed in western Europe (Worster, 1979). When, in the 1930s, drought hit the area, as it periodically does, wind erosion ensued on a huge scale. By 1937, the US Soil Conservation Service estimated that 43 per cent of a 6.5 million-hectare area at the heart of the Dust Bowl had been seriously damaged by wind erosion. The large-scale environmental degradation, combined with the effects of the Great Depression, ruined the livelihoods of hundreds of thousands of American families.

However, although the lessons learnt from the Dust Bowl years inspired major advances in soil conservation techniques, the environmental health of the area has continued to decline. Continued degradation is believed to account for a significant part of declining yields of sorghum and kafir over 30–40 years in the Texas Panhandle (Fryrear, 1981), and soil erosion during the drought in the 1970s was on a comparable scale to that in the 1930s (Lockeretz, 1978). Investigation of the soil losses of the 1970s illustrates some important underlying causes, indicating that political and economic considerations overshadowed the need for sustainable land management. Large tracts of marginal land had been ploughed for wheat cultivation in the early 1970s, driven by high levels of exports, particularly to the then USSR. To encourage farmers to produce, a Federal programme was set up that guaranteed them payment according to the area sown, so that the disincentive to plough marginal areas was removed: the farmers were paid whether these areas yielded a crop or not. At the same time, new centre-pivot irrigation technology was widely adopted to water the new cropland, but the use of the rotating irrigation booms required that wind breaks, planted to protect the soil, be removed (McCauley *et al.*, 1981).

The increasing use of irrigation has also issued in a new form of desertification on the Great Plains: groundwater depletion. Water pumped from the Ogallala aquifer is essentially a non-renewable resource, since recharge of the aquifer is negligible. However, in the 1960s when water levels in irrigation wells began to decline, farmers did not act to conserve dwindling supplies, but actually increased pumping because those who reduced their consumption still experienced declining water tables as others continued to irrigate, a clear example of the tragedy of the commons. Projections indicate that, over the next 30 years, up to half the land irrigated today may have to be withdrawn due to water depletion (Opie, 1993). Despite the booms and busts Great Plains farmers have experienced throughout the last century, they are still not farming in a sustainable manner.

The Sahel

The important role of external factors in the degradation equation can also be illustrated from work on the Sahel catastrophe of the 1970s and 1980s, events that sparked worldwide interest in desertification. Rainfall varies greatly across the Sahel, being patchy in space and fluctuating from year to year and from decade to decade. Sahelian societies have adapted to this variability in

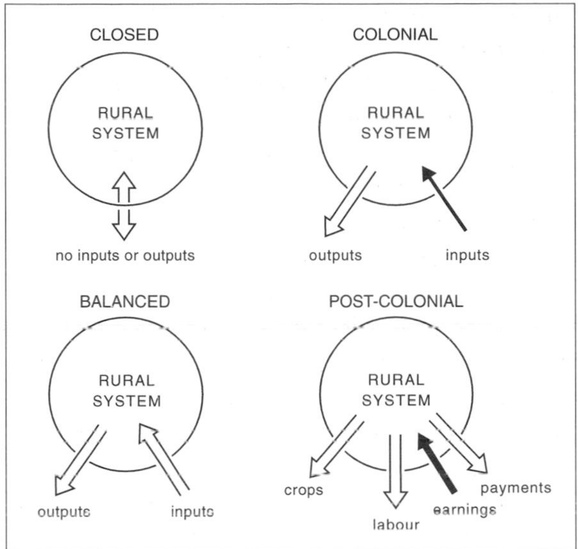

Figure 5.6 Model for the incorporation of rural systems into the market (after Mortimore, 1987).

numerous ways to insure themselves during times of hardship caused by drought, which can be applied equally to desertification, as Mortimore (1987) has shown in northern Nigeria. However, these adaptive systems were put under a greater burden than they could cope with by two main driving forces: the increasingly dry episode that started in the Sahel in the late 1960s and the changing political economy of the region.

The recent desiccation of the Sahelian climate provides the clearest and most dramatic example of climate variability that has been measured anywhere in the world to date (Hulme, 2001): averaged over 30-year periods, annual rainfall in the region declined by between 20 and 30 per cent between the decades (1930s to 1950s) leading to political independence for Sahelian countries and the decades since (post-1960s). The resultant impact on traditional adaptations was combined with stresses from deteriorating regional political and global economic conditions, which illustrate the changing scene faced by inhabitants of many dryland communities in the developing world.

The possibilities for grain storage have been diminished by a number of circumstances, including a growing population and land scarcity. The imposition of taxes payable in currency, the monetisation of the economy and poor terms of trade, changes that began during colonial times, have also encouraged people to produce cash crops in place of subsistence crops.

The generation of savings from activities away from the fields also suffered under changing political policies. Agriculturalists who had moved to urban areas to work in the informal sector were forcibly sent back to the land by the government, while cash savings were further eroded by inflation and sudden devaluations in the national currency.

The desiccation in climate, combined with continued population growth, reduced the effectiveness of social insurance, as people who could previously have been relied upon to help out their poorer relations did not have enough to feed themselves. The traditional advantages of ecological insurance have also been eroded as colonial and subsequent governments have encouraged moves towards specialisation in crop production and discouraged traditional pastoralism, which is inconsistent with modern political policies and boundaries.

A loss of autonomy over their affairs lies at the heart of the consequent problems faced by farming and pastoral communities in developing-country drylands. The absorption of such communities into centralised economies has undermined the traditional ways in which they have coped with environmental stress, increasing their vulnerability (Fig. 5.6). While government welfare or private economic insurance is available to inhabitants of richer countries, such safety nets are seldom available in the so-called 'Third World', and it is in these regions where the need to more fully comprehend the nature and extent of desertification is the most pressing.

FURTHER READING

Batterbury, S. and Warren, A. 2001 The African Sahel 25 years after the great drought: assessing progress and moving towards new agendas and approaches. *Global Environmental*

Change 11: 1–8. Overview of several papers on the Sahel in this volume.

Beaumont, P. 1989 *Drylands: environmental management and development*. London, Routledge. This book gives an overview of the nature of drylands and the history of their use by human society. It uses numerous detailed case studies to illustrate different management approaches.

Hulme, M. and Kelly, M. 1993 Exploring the links between desertification and climate change. *Environment* 35(6): 4–11, 39–46. An interesting look at the connections between two of the world's major environmental issues.

Mortimore, M. 1989 *Adapting to drought: farmers, famines and desertification in West Africa*. Cambridge, Cambridge University Press. A detailed case study of northern Nigeria focusing on the physical effects of drought and desertification, and the social responses.

Scoones, I. 1995 Policies for pastoralists: new directions for pastoral development in Africa. In Binns, T. (ed.) *People and environment in Africa*. Chichester, Wiley: 23–30. A good overview of the recent changes in our understanding of rangelands and the implications for herders.

Stiles, D. (ed.) 1995 *Social aspects of sustainable dryland management*. Chichester, Wiley. A book that continues to redress the imbalance in desertification studies historically marked by concentration on physical and technical aspects to the neglect of the social side.

Thomas, D.S.G. and Middleton, N.J. 1994 *Desertification: exploding the myth*. Chichester, Wiley. This book focuses on the political aspects of the desertification issue and highlights many of the difficulties of identifying desertification in the field.

Websites

http://ag.arizona.edu/OALS/IALC the International Arid Lands Consortium has information and news on dryland research.

http://www.unccd.int/ the UN Convention to Combat Desertification site.

http://www.wcmc.org.uk/dynamic/desert/ the Desertification Information Network provides data and information services with an emphasis on Africa and the Mediterranean basin.

http://www.din.net.cn/ China's Desertification Information Network

http://www.dwe.csiro.au/sites/cazr/ CSIRO Wildlife & Ecology – Centre for Arid Zone Research in Australia.

http://www.badia.gov.jo the Badia Research and Development Programme that conducts research into sustainable development in an area of Jordan.

http://www.serg.sdsu.edu/SERG/ the Soil Ecology and Restoration Group conducts research into the ecosystem dynamics of drylands primarily in south-western USA and Mexico.

http://www.ussl.ars.usda.gov/ research and information on salt-affected soils.

POINTS FOR DISCUSSION

- Solutions to the world's desertification problems lie not in more technical fixes, but in a greater appreciation of the social side of the issue. How far do you agree?
- Outline the ways in which salinisation occurs in drylands and the problems it poses to human society.
- In what ways might global warming affect the desertification issue?
- Is desertification a myth?

6

OCEANS

TOPICS COVERED

Fisheries
Whaling
Pollution

Key words: pelagic, demersal, Exclusive Economic Zone, El Niño/Southern Oscillation (ENSO), teleconnections, Commission for the Conservation of Antarctic Marine Living Resources (CCAMLR), tributyltin (TBT), persistent organic pollutants

Oceans cover nearly 71 per cent of the Earth's surface and contain 97 per cent of the planet's water. They play a fundamental role in the Earth's climatic system, and marine ecosystems have long provided people with fish and other resources, as well as a receptacle for wastes. The perception of oceans as being infinitely vast has tended to undermine concern over their increasing exploitation by human society, despite their economic and ecological importance. Although public outcry focuses on some marine issues, such as the killing of whales and pollution from major oil spills, more pervasive threats – from other pollutants and overfishing, for example – are seldom given such attention. However, consideration of our relatively poor understanding of how oceans work, of the population dynamics of many marine biota and of the effects of certain pollutants, suggests that the arguments for caution in our use and abuse of oceans are strong. In this respect, ownership is an important aspect of many marine environmental issues. Most of the open oceans are common property resources, and the waters that come under national jurisdictions near coasts are subject to influences from other nation-states.

FISHERIES

Global marine fish production is humankind's largest source of either wild or domestic animal protein, and is particularly important in the developing countries. The 1990s saw the end of a 40-year fishing boom. The worldwide catch increased more than four times between 1950 and 1989; but has remained at around the same level since then. In 1997, 86 million tonnes of fish were caught at sea (GESAMP, 2001).

As with terrestrial food crops, we are heavily reliant upon a very limited number of the 13 000 or so marine fish species. Herrings, sardines and anchovies made up 25 per cent of landings in 1989. Continuous growth in global catches throughout the 1980s was largely based on increased landings of these shoaling pelagic (free-swimming) species because most demersal (bottom-dwelling) stocks were, and still are, fully fished. This trend is worrying because shoaling pelagic species tend to be at lower trophic levels than demersal species and although fishing down food webs in this way leads at first to higher catches, the increase eventually levels out and is then likely to decline. Hence the globally declining trend in the average trophic level of fish caught, identifiable over the period 1950–94, is

Figure 6.1 European fishermen, like these in Portugal, have become subject to increasing restrictions on their activities in an effort to reduce pressure on overexploited fish stocks, but to date such restrictions have not been sufficient to prevent the decline in many stocks.

unlikely to be a sustainable means of exploiting marine fish stocks (Pauly *et al.*, 1998).

Many marine biologists also suspect that the currently exploited fishery resources around the world are now close to their maximum sustainable catch limits of about 85 million tonnes, and many regional stocks show signs of biological degradation. While in the early 1950s, 55 per cent of the world's fish stocks were underexploited, by the mid-1990s catches from about 35 per cent of the world's stocks were decreasing, and those of another 25 per cent had stagnated at a high level of exploitation. Only the remaining 40 per cent were continuing to yield more fish (GESAMP, 2001).

The Joint Group of Experts on the Scientific Aspects of Marine Environmental Protection (GESAMP), an international group sponsored by the UN, has identified three main management failings as driving the fisheries crisis (GESAMP, 2001).

1. Many of the world's fisheries, and particularly those on the high seas, are still a free-for-all. Free and open access encourages overfishing, as each boat, and each country, tends to catch what it can, leading to a tragedy of the commons situation. Fisheries bodies and agreements are not particularly effective, largely because their members are only weakly committed to cooperating on conserving stocks.
2. Many nations heavily subsidise their fishing fleets, encouraging unprofitable and unsustainable fishing, making overfishing even worse. Reducing or removing fleets would

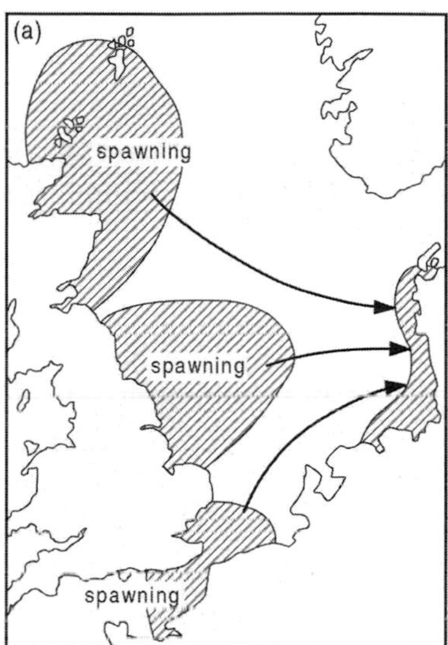

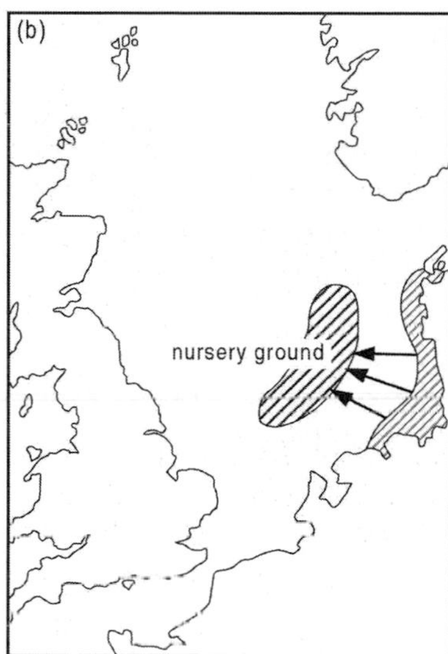

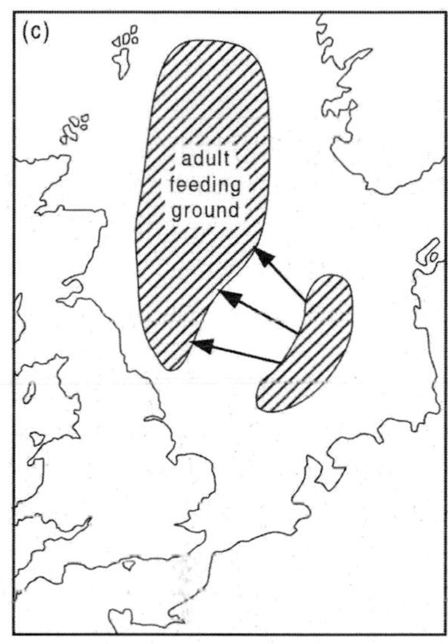

Figure 6.2 Movement of North Sea herring: (a) after hatching from eggs planktonic larvae drift from three main spawning grounds to shallow waters off Germany and Denmark; (b) young herring swim offshore to the nursery ground; (c) young herring join the adult population in their feeding ground (after Meadows and Campbell, 1988).

have short- to medium-term economic and social consequences that governments are reluctant to accept (Fig. 6.1).

3 Some attempts to conserve fisheries (e.g. introducing closed seasons, or setting limits on the total catch but not on the amount that can be caught by each boat) may unintentionally allow fishing fleets to grow too much. If a fishery is profitable enough, owners will continue to build and operate boats even if they have to be kept in port for part of the year.

Overfishing of demersal species in the North Atlantic, for example, is thought to have first occurred 100 years ago in the case of North Sea plaice. The 1980s saw a significant increase in the levels of national and international controls designed to ensure the conservation and sustainable exploitation of fish stocks. The establishment of state jurisdiction zones up to 200 miles (322 km) from national coastlines, so-called Exclusive Economic Zones, has been a critically important development in facilitating this move, since 99 per cent of marine fisheries are currently taken from within this limit. But international cooperation is still needed in many cases simply because fish are no respecters of national zoning. North Atlantic herring, for example, spawn in British and French coastal waters; their planktonic larvae drift to shallow nursery grounds off the Netherlands, Germany and Denmark, and the young herring join the adult population in feeding grounds in British and Norwegian controlled waters (Fig. 6.2). Overfishing of the North Atlantic herring in the North Sea in the 1960s and 1970s resulted in its virtual extinction as a commercial fishery, and a ban on catches between 1977 and 1981. The ban allowed a modest recovery in its stock, but the inability of member states of the EU to control and reduce fleet capacity caused another downturn in North Atlantic herring stocks in the early 1990s (FAO, 1997a). In an assessment of fish stocks in the whole north-east Atlantic at the end of the twentieth century, the North Sea was deemed the most critical region with all nine commercial fish stocks deemed to be outside safe biological limits (OSPAR Commission, 2000).

Identifying the impacts of overfishing is, however, a complex task for fisheries managers. The location of the world's marine fisheries is governed principally by the distribution of floating plants on which they depend for food, and the most important factor determining phytoplankton production is the supply of nutrients, which is greatest in areas of upwelling. Climatic circulations can greatly alter the pattern of ocean circulation, however, and hence the pattern of fisheries. It is often difficult to distinguish between the impacts of overexploitation and the natural variability of stocks, dependent upon species biology, migratory habits, food availability and natural hydrographic factors. Even with a relatively stable and resilient stock like the North Sea herring, a fishery that has been studied as much as any other comparable marine resource, the relative role played by environmental factors in the 1970s' collapse is still not clear (Bailey and Steele, 1992). The difficulties are greater still with some other shoaling pelagic species such as anchoveta and sardine, the populations of which are notoriously difficult to assess and manage, making their fishing a high-risk business (Csirke, 1988).

One of best-known examples of dramatically fluctuating population numbers is the Peruvian anchoveta of the south-eastern Pacific. Disruption to population biomass occurs in some years due to the El Niño phenomenon, which intermittently leads to the near total collapse of coastal fisheries. El Niño refers to a distinct warming of the surface sea water in the south-eastern tropical Pacific that occurs around Christmas time along with a rapid change in atmospheric pressure over coastal South America and Indonesia referred to as the Southern Oscillation (the intimate link between El Niño and Southern Oscillation has generated the widely used acronym ENSO). These warmer surface waters occur when the South Pacific trade winds weaken, decreasing the amount of upwelling of water from depth along the coast of South America. This upwelling is very rich in nutrients, hence the abundant fish stocks, so when it cuts out during an El Niño the number of fish available declines sharply.

The impact of this natural variability on the Peruvian anchoveta is also thought to have been exacerbated by overfishing. The Peruvian anchoveta was the most important single fishery in the world in the 1960s, providing up to 10 million tonnes of animal protein a year. Annual catches rose from a minor 59 000 t in 1955 to peak in 1970 at just over 13 million tonnes (Fig. 6.3). Just prior to 1972, the fishery was so intense that few of the fish caught were more than two years old, and many biologists expressed concern about the dangers of overfishing. The 1972/73 El Niño event caused a reduction in recruitment and subsequent intense exploitation of the remaining stock. Continued heavy fishing after 1972 further reduced the stock. By the late 1970s, annual catches had fallen to less than a million tonnes and directed commercial fishing was abandoned. In 1984, after another intense El Niño event in 1982/83, the landed catch was just 94 000 t. Stocks appear to have collapsed due to the combination of short-term fluctuations caused by El Niño events combined with intense fishing pressures (Caviedes and Fik, 1992). The fortuitous coincidence of favourable environmental conditions and controlled fishing has allowed the stock to recover since the historic lows of the early 1980s, and landed tonnage was back at nearly 12 million t in 1994. However, the high natural variability of the Peruvian anchoveta was apparent again late in the 1990s when the very intense El Niño event of 1997/98 resulted in a steep decline in total fish caught for 1998. Industrial fishing firms closed fishmeal plants and canneries, and kept many vessels idle, while small-scale artisanal fishermen shifted their efforts to catch tropical species only available during El Niño conditions. Nevertheless, their revenues were limited by drops in market prices and climate-related interruptions to transport on land. By September 1998, unemployment in Chimbote, the largest fishing port in Peru, was so severe that food had to be distributed to thousands of fishermen and their families (Pfaff *et al.*, 1999).

Similarly enormous differences between maximum and minimum catches occur for other shoaling pelagic species: the Japanese sardine, for example, appears to be affected by changes in the Kuroshio current that last for periods of one or

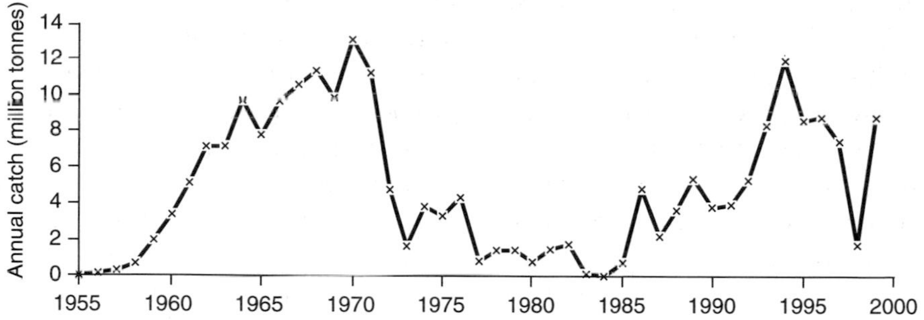

Figure 6.3 Catch history of the Peruvian anchoveta, 1955–99 (from data in FAO *Yearbooks of Fisheries Statistics, Catchings and Landings*).

more decades. Natural variability may also have contributed to the dramatic collapse of the California sardine fishery, which boomed in the 1930s and 1940s but had almost completely disappeared by 1950. Once thought to be a classic example of the effects of overfishing, work by Soutar (1967) has thrown up an alternative explanation. The 2000-year history of species abundance reconstructed from fish scales found in sea-bed sediments suggests that irregularities in the abundance of sardine in this area were common over the period, with decades of almost total absence not unusual.

The spasmodic population dynamics of the Peruvian anchoveta, Japanese sardine and California sardine exhibit one of four types of fish population dynamics described by Caddy and Gulland (1983):

- steady-state
- cyclical
- irregular
- spasmodic.

While the human impact upon steady-state species is relatively easy to detect and management of their exploitation fairly straightforward, irregular and spasmodic stocks are particularly difficult to manage and do not lend themselves to current ideas of sustainable yields on a year-to-year basis. However, as more long-term data become available, a remarkable pattern is emerging of synchronised variability in the growth and abrupt decline in many of the world's largest fish stocks. The phenomenon was first recognised for the simultaneous rise and fall of sardine populations in widely separate areas of the Pacific Ocean (Kawasaki, 1983). Other examples have now been recognised between oceans: dramatic rises in stocks between the late 1950s and early 1970s similar to that of Peruvian anchoveta shown in Fig. 6.3 were recorded for the South African pilchard and the Japanese flying squid. While these stocks were peaking in abundance, the large Pacific pilchard stocks essentially disappeared. The period from the 1970s to the mid-1980s, when catches of the Peruvian anchoveta in the Pacific and the South African pilchard in the Atlantic collapsed, saw huge increases in population sizes of Pacific Ocean sardine stocks.

It should be noted that the data illustrating these variabilities are for catch landings, and that these are probably only rough indicators of actual population size, but similar patterns are evident where actual biomass assessments have been made. It is possible, of course, that these synchronous patterns occur by chance, but it is also possible that they are linked by a climatic mechanism, in the same way that ENSO events in the South Pacific appear to affect weather patterns far removed from the area by so-called 'teleconnections'. It is unlikely that sea temperatures alone could explain the global synchrony in fish population variations, but if a feasible mechanism can be identified it would have important implications for the management and perhaps prediction of global fish supplies (FAO, 1997a).

Southern Ocean

Ignorance of the workings of the marine ecosystem makes management of ocean resources difficult, and no more so is this the case than in the Southern Ocean surrounding Antarctica. The Southern Ocean has been the site of several phases of species depletion caused by overexploitation. In the nineteenth century, seals were pushed to the edge of extinction for their fur and oil. They were followed by the great whales in the first two-thirds of the last century (see below), and in the 1970s and 1980s Southern Ocean finfish stocks were heavily exploited with the marbled Antarctic rock cod and mackerel icefish showing particular signs of decline. Most recently, attention has turned to Antarctic krill, the most abundant living resource of the Southern Ocean. Commercial fishing of the small shrimp-like crustacean began in the early 1970s, but annual catches have not risen above the peak of over half a million tonnes in 1980–81 (Fig. 6.4).

The reason for the relatively modest krill catches is the establishment of a limit on harvesting by the Commission for the Conservation of Antarctic Marine Living Resources (CCAMLR), an international body set up in 1981. The job of the CCAMLR is to manage the sustainable use of Antarctic marine living resources, an attempt to prevent the dangers of overexploitation that came with the common ownership of the Southern Ocean. The devastation of seal and whale populations, and the depletion of some finfish, was a clear example of the tragedy of the commons. Since krill are at the centre of the Antarctic ecosystem – providing the major food source for many larger animals such as seals, fish, squid, seabirds and baleen whales – there was a fear that overharvesting of krill might lead to the collapse of the entire ecosystem.

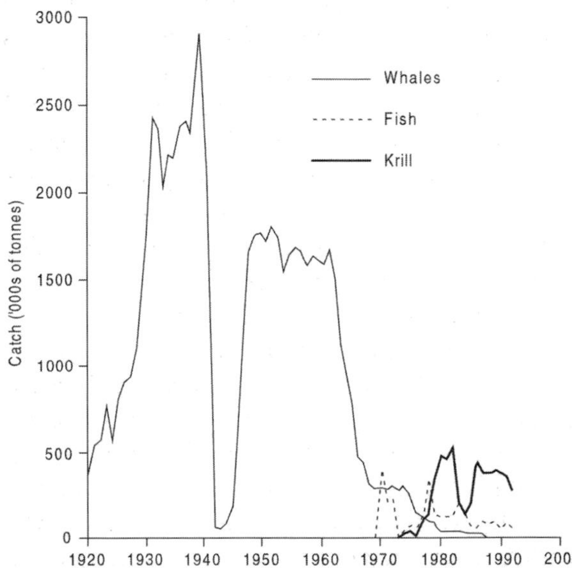

Figure 6.4 Annual catch of whales, finfish and krill in the Southern Ocean, 1920–92 (after Nicol and de la Mare, 1993).

The CCAMLR's task is severely hampered by a basic lack of knowledge and information on the workings of the Antarctic ecosystem in general, and of krill in particular. While most estimates of a sustainable krill harvest are well in excess of the current estimated annual sustainable catch of all other species from the world's oceans (100 million tonnes), making it a potentially huge resource, these estimates are very rough and are based on some rather major assumptions. As Nicol and de la Mare (1993: 38) put it: 'Biologists do not have any reliable estimates of how many krill there are, nor a sense of exactly where they live and how their local distribution is governed. We do not know where they spend the winter, nor the age structure of the population.'

The abundance of krill appears to fluctuate greatly from one year to the next, but overall their numbers seem to have been declining in recent decades as temperatures in the Antarctic have risen since the 1940s and, consequently, the frequency of extensive winter sea-ice has declined. This lower abundance of krill may have significant effects on the marine food web in the Southern Ocean (Loeb *et al.*, 1997). The problems of marine resource managers are further exacerbated by a recent explosion in the population of Antarctic fur seals. Evidence gleaned from seal hairs in lake sediment cores taken from one Antarctic island suggests that the current seal population explosion exceeds the range of natural variability (Hodgson and

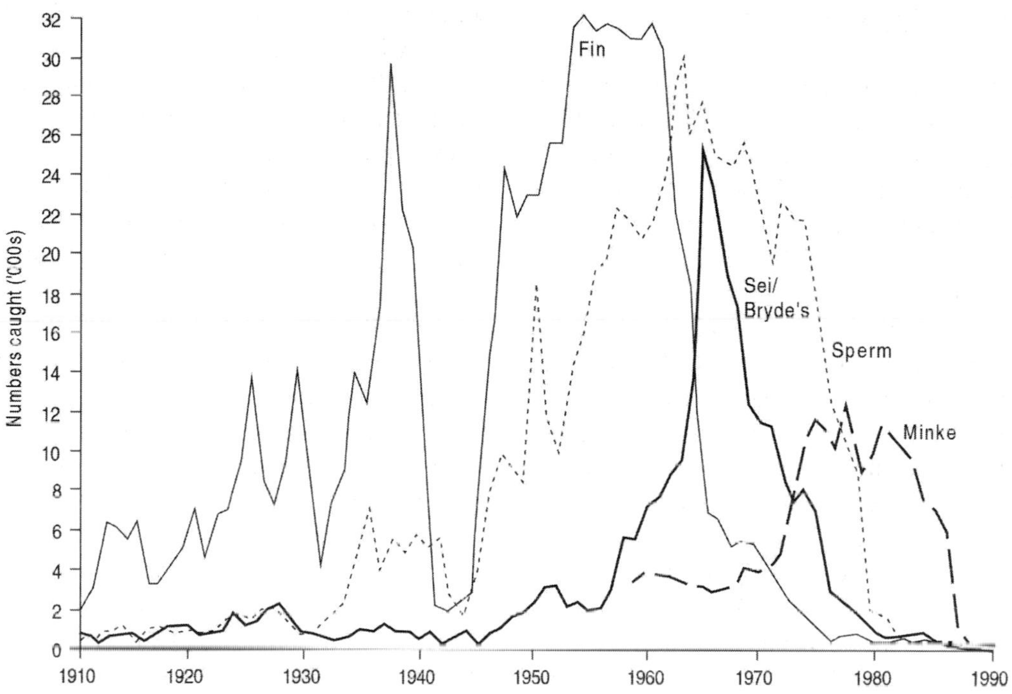

Figure 6.5 World annual whale catch, 1910–91 (from data in UNEP, 1987 and 1993).

Johnston, 1997), implying that the ecosystem is still responding to the dramatic human impacts of the last few centuries. These two recent developments simply underline the importance of understanding the biology and ecology of the entire marine ecosystem before informed decisions can be made about sustainable management.

WHALING

Although whales have been hunted for thousands of years, modern commercial whaling for oil, and in time also for meat, is traced back to the eleventh or twelfth centuries when groups of Basque villagers began to hunt right whales from small rowing boats in the Bay of Biscay. Whaling had become big business by the eighteenth and nineteenth centuries, and large whaling fleets from many nations, including England, North America, Japan and the Netherlands, made enormous profits from their sorties.

Commercial whaling has focused on the largest, most profitable species. These are the so-called 'great whales', of which there are ten species: the blue, fin, humpback, right, bowhead (or Greenland), Bryde's, sei, minke, grey and sperm whales. The history of commercial whaling has seen intensive catches of a species until it has been driven to the brink of extinction, at which point the whalers shifted their attention to a different species in a new area of ocean. The populations of most of the great whale species are now so low as to be effectively extinct from a commercial viewpoint; about 2 million whales were killed in the twentieth century alone, and several species could ultimately disappear altogether.

The Greenland bowhead whale, for example, was first exploited in 1610 and reduced to the brink of extinction by the mid-nineteenth century. The Arctic bowhead whales, discovered in the mid-nineteenth century and quickly exploited, were almost totally depleted by 1914. As many as 30 000 sperm whales were killed in the Atlantic during peak years in the eighteenth and nineteenth centuries, and by the 1920s the Atlantic sperm whale was practically extinct.

It was early in the twentieth century that whalers began to exploit the previously untouched waters of the Southern Ocean where several species go to feed on krill. The catch rates were further increased with the introduction of factory ships in the 1920s. The humpback was the first species targeted in the Antarctic waters, but as its numbers declined, the blue whale became the main quarry. An estimated 250 000 blue whales were quickly killed, leaving around 1000 today. In the second half of the twentieth century it was the turn of the fin and sperm whales. Annual catches of fin whales regularly topped 30 000 in the 1950s (Fig. 6.5). Since 1990, the numbers of whales caught for scientific research purposes (see below) has been around the 1000 mark. These are minke, sperm and Bryde's whales.

Protecting whales

There have been a number of attempts to control the whaling industry over the years. The earliest effort was the Convention for the Regulation of Whaling, drawn up in 1931. But the convention had little beneficial effect. One of its regulatory instruments, the so-called blue whale unit, effectively lumped all great whales together, so that one unit was made up of successively greater numbers of smaller species: two fin, three humpback, five sei, and so on. Countries were spurred by the profit motive to fulfil their quotas with a few large species rather than more smaller ones, resulting in a rush on the blue whales. In 1946, this convention was replaced by the Whaling Convention, which established the International Whaling Commission (IWC) to discuss and adopt regulations. The IWC has had a chequered history, overseeing as it did the all-time peak in whaling, in the 1960–61 season, when 64 000 whales were caught, but after increasing pressure from outside, the IWC declared a moratorium on all commercial whaling, which entered into force in 1986 and remains in place today.

The ban has not been a complete success, since several countries have not abided by it, claiming that whaling should be continued for scientific purposes. Japan, Norway and Iceland have been some of the most reluctant nations to comply with the ban. Conservationists have argued that effective non-lethal techniques for studying whales have been developed, which do away with the need to butcher the object of the scientists' enquiries. It may come as no surprise to learn that the meat and oil of the whales killed for scientific research is still processed and sold in the normal way. At the time of writing, Norway and Japan continue to defy the ban. Conservationists have also voiced fears for the future viability of the moratorium based upon Japan's powerful position as an aid donor. It is not inconceivable that Japan could use aid as a lever to persuade developing-country IWC members to back its position.

Various attempts have been made outside the IWC to dissuade the countries who continue to kill whales. Amendments in US fisheries law designed to bolster IWC regulations have been used by conservation organisations. The amendments enable the US Government to take action against countries that diminish the effectiveness of the IWC, by blocking imports of their fisheries products or by denying them access to fish within US waters. In the mid-1980s, a group of conservation bodies took legal action against the Commerce and State departments because they feared that sanctions would not be imposed against Japan, which was planning a large sperm whale hunt after the species had received official protection. The case eventually reached the US Supreme Court in 1986. The conservationists lost, and in the meantime Japan had caught its 400 sperm whales.

Other actions by conservation organisations have proved more successful. Greenpeace estimated that a boycott of products from Iceland's fishing industry in 1988–89 cost that country £30 million in lost orders. This action contributed to Iceland's abandonment of scientific whaling in late 1989. In 1993, Greenpeace conducted a similar boycott campaign against Norwegian products. The conduct of this campaign saw a new development in such action when Greenpeace

Figure 6.6 A boatload of whale-watchers following two humpback whales in Massachusetts Bay off the north-east coast of the USA. Whale-watching is being promoted as a viable alternative to whaling (Photo: Mark Carwardine).

asked its UK members to shop at Safeway stores on a particular weekend since Safeway had come out strongly against Norway's resumption of commercial whaling.

Among the most recent approaches to persuading the few nations who still catch whales to stop their whaling is the establishment of the Southern Ocean Whale Sanctuary in 1994, in which all whaling is banned, although Japan has a standing objection to this decision. Organised whale-watching tours have also been vigorously promoted in recent years as a viable and sensible alternative way of 'using' these marine mammals. Whale watching is clearly a more sustainable method of exploiting whales than simply killing them, and the income generated from whale-watching 'ecotourists' can make the organisation of these tours very profitable in economic terms (Fig. 6.6).

Commercial whale-watching began in the mid-1950s in southern California where migrating grey whales were the attraction. Since then it has become a major tourist industry. In 1998, about 5 million people went on whale-watching trips in 87 countries. These tourists spent more than US$1 billion on food, travel, accommodation and souvenirs (Hoyt, 2000).

In the areas of the world where whale-watching is well organised, ships carrying eco-tourists abide by strict regulations designed to cause minimal disturbance to the subjects of their searches. Engines must be cut within a certain distance and no vessel is allowed to approach too close. Many of the trips are made on research

TABLE 6.1 *Primary causes and effects of marine pollution*

Type	Primary source/cause	Effect
Nutrients	Runoff approximately half sewage, half from upland forestry, farming, other land uses; also nitrogen oxides from power plants, cars, and so on	Feed algal blooms in coastal waters. Decomposing algae deplete water of oxygen, killing other marine life. Can spur toxic algal blooms (red tides), releasing toxicants into the water that can kill fish and poison people
Sediments	Runoff from mining, forestry, farming, other land uses; coastal mining and dredging	Cloud water. Impede photosynthesis below surface waters. Clog gills of fish. Smother and bury coastal ecosystems. Carry toxicants and excess nutrients
Pathogens	Sewage; livestock	Contaminate coastal swimming areas and seafood, spreading cholera, typhoid and other diseases
Persistent toxicants (e.g. PCBs, DDT, heavy metals)	Industrial discharge; wastewater from cities; pesticides from farms, forests, home use, and so on; seepage from landfills	Poison or cause disease in coastal marine life. Contaminate seafood. Fat-soluble toxicants that bioaccumulate in predators
Oil	46%, runoff from cars, heavy machinery, industry, other land-based sources; 32%, oil tanker operations and other shipping; 13%, accidents at sea; also offshore oil drilling and natural seepage	Low-level contamination can kill larvae and cause disease in marine life. Oil slicks kill marine life, especially in coastal habitats. Tar balls from coagulated oil litter beaches and coastal habitat
Introduced species	Several thousand species in transit every day in ballast water; also from canals linking bodies of water and fishery enhancement projects	Outcompete native species and reduce marine biological diversity. Introduce new marine diseases. Associated with increased incidence of red tides and other algal blooms
Plastics	Fishing nets; cargo and cruise ships; beach litter; wastes from plastics industry and landfills	Discarded fishing gear continues to catch fish. Other plastic debris entangles marine life or is mistaken for food. Litters beaches and coasts. May persist for 200–400 years

Source: Weber (1994: 47, Table 3.3)

ships (non-lethal research, that is), with the income from carrying tourists helping to finance the research itself. In some parts of the world, old whaling vessels themselves are used, a development that offsets one of the arguments proposed in favour of continuing the killing of whales: its importance for employment.

Whale-watching tours have even started in Japan, the first setting out from Tokyo for the Ogasawara Islands in April 1988. Despite stern

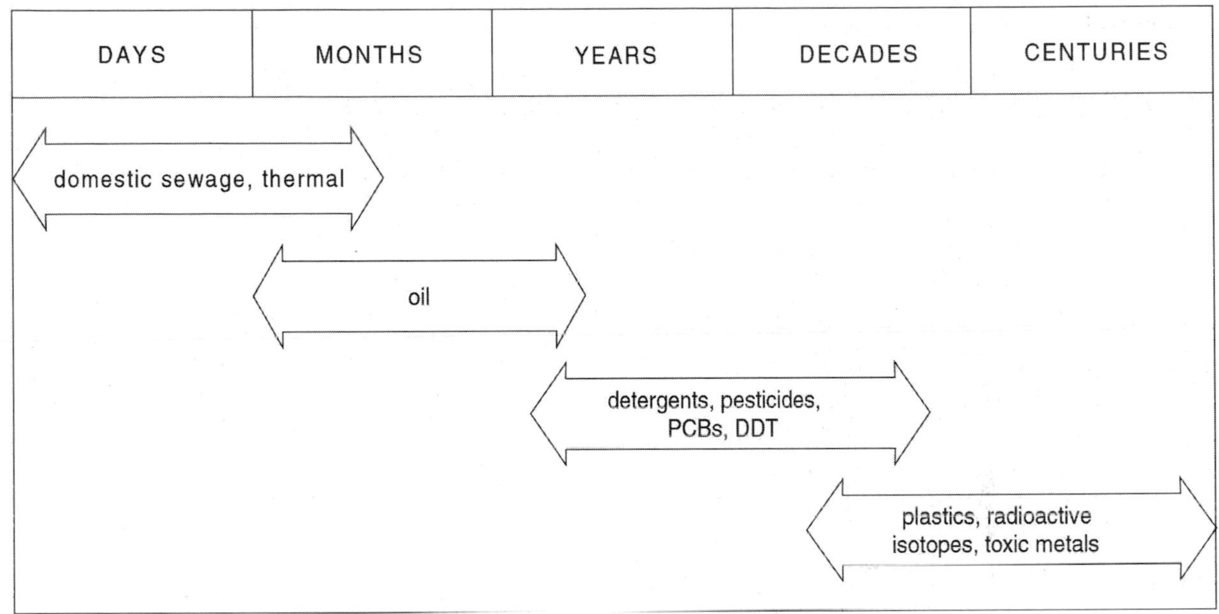

Figure 6.7 Persistence of pollutants in the marine environment (after Meadows and Campbell, 1988).

opposition from both government officials and whalers, Japan now has one of the world's fastest-growing whale-watching industries. In 1992, nearly 20 000 people paid over £0.5 million to join these Japanese tours. If the trend continues, it will not be long before live whales in Japanese waters are worth more in economic terms than dead ones. It is a trend the IWC hopes to encourage with its backing of whale-watching as a sensible, more humane way of treating these creatures.

POLLUTION

The wide variety of sources, causes and effects of human-induced pollution of the marine environment is indicated in Table 6.1, while an indication of the varying persistence of marine pollutants is given in Fig. 6.7. Since land-based sources tend to be dominant, the worst effects of these pollutants are concentrated in coastal waters adjacent to areas with large population densities, and in similarly placed regional seas where the mixing of waters is limited. The open ocean, by contrast, is little affected to date. In this chapter, the focus is on pollution and its effects outside the coastal zone, while coastal pollution is covered in Chapter 7, although in practice the distinction is somewhat arbitrary since coastal pollutants also reach the wider context of the high seas and oceans.

GESAMP (Joint Group of Experts on the Scientific Aspects of Marine Environmental Protection) noted the planet-wide distribution of some pollutants, but gave the open oceans a clean bill of health in one of its reports:

> *The open ocean is still relatively clean. Low levels of lead, synthetic organic compounds and artificial radionuclides, though widely detectable, are biologically insignificant. Oil slicks and litter are common along sea lanes, but are, at present, of minor consequence to communities of organisms living in open-ocean waters.*
>
> *(GESAMP, 1990: 1)*

Others, however, warn that the lack of ecological studies in the open ocean cloud our understanding of the effects of pollutants there. Weber (1994) suggests that there is reason to believe that contaminants may have an effect disproportionate to their concentration, since chemical pollutants tend to amass in surface waters where

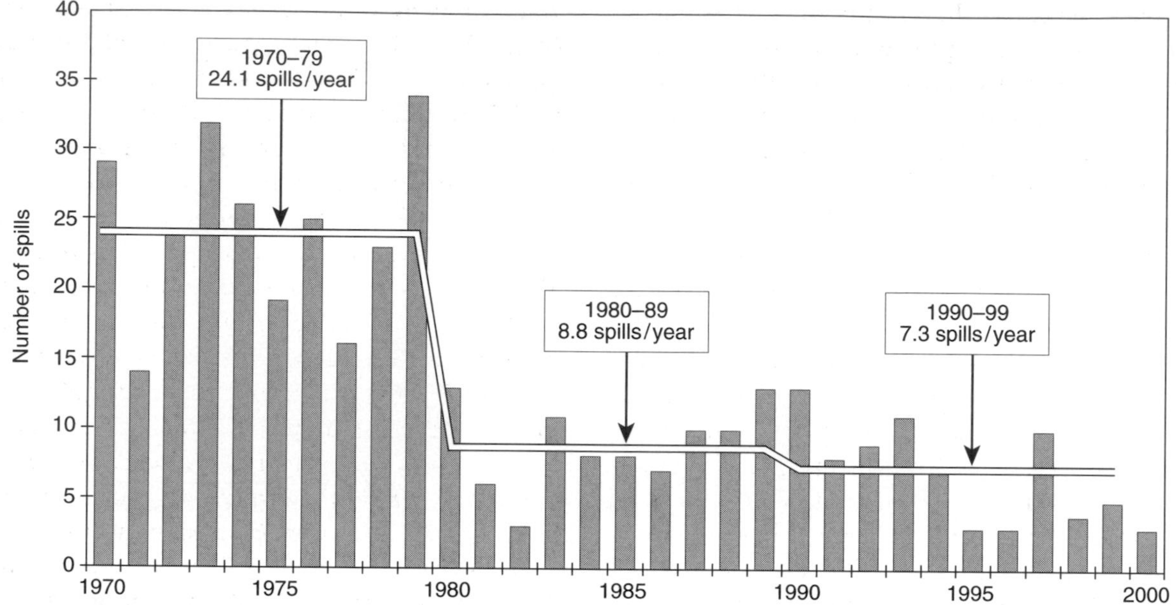

Figure 6.8 Global numbers of oil spills over 700 tonnes, 1970–2000 (http://www.itopf.com/stats.html, accessed September 2002).

larvae, eggs and micro-organisms concentrate: 'heavy metals can be 10 to 100 or more times more concentrated near the surface than in the waters below, and pesticide residues can be millions of times stronger' (Weber, 1994: 50).

Shipping and atmospheric transport are the main sources of pollution to the open oceans. Inputs from shipping arise mainly from operational activities and from deliberate discharges. Oil is the most obvious pollutant involved, but biocides from anti-fouling paints leach into ocean water along shipping routes and, recently, the accumulation of plastic debris from ships has been on the increase. Dumping and accidents also contribute.

Oil affects marine life both because of its physical nature (i.e. by physical contamination and smothering), and because of its chemical components (e.g. toxic effects). Plants and animals may also be affected by clean-up operations or indirectly through damage to their habitats. The animals and plants most at risk from physical smothering are those that come into contact with a contaminated sea surface, including marine mammals and reptiles, birds that feed by diving or form flocks on the sea surface, marine life on shorelines, and animals and plants in mariculture facilities. Lethal concentrations of toxic components leading to large-scale mortality are relatively rare, localised and short-lived because the most toxic components in oil tend to be those lost rapidly through evaporation when oil is spilt. However, lower concentrations of oil can impair the ability of individual marine organisms to reproduce, grow, feed or perform other functions. Some animals that filter large volumes of sea water to extract food, such as oysters, mussels and clams, are particularly likely to accumulate oil components.

Most oil spills from shipping worldwide result from routine operations such as loading, discharging and bunkering – activities that normally take place in ports or at oil terminals. Most of these spills are small, while accidents tend to give rise to much larger spills, with a fifth involving quantities in excess of 700 tonnes. International maritime regulations governing the operational inputs of oil to the oceans have been responsible, in part, for the decline in the number of major spills (>700 t) during the last 30 years. The number of major spills over the period 1970–79 averaged 24 a year, but this average fell to seven a year in the period 1990–99 (Fig. 6.8). Likewise, the total amounts of oil spilled from tankers in the world's oceans has tended to be lower in recent years, although huge individual accidents continue to have significant local impacts, such as the spillage of 260 000 t of oil from the tanker ABT *Summer* off the coast of Angola in 1991. Tankers are not the only source of such large spillages. The blow-out at the Ixtoc 1 drilling platform off the Mexican coast in 1979, for example, continued for ten months and released more than 400 000 t of oil, producing a slick that spread across the Gulf of Mexico and contaminated beaches in Texas.

Another contaminant found in many oceans and seas is tributyltin (TBT), which has been widely used in anti-fouling paints applied to ships' hulls. TBT has been linked to problems of reproduction in aquatic creatures. Exposure to TBT has been found to interfere with the hormonal systems of whelks, producing 'gender bending' effects whereby individuals develop the sexual characteristics of the other sex. The effect has been documented in seas and oceans. The OSPAR Commission (2000) noted impacts on dogwhelks and common whelks in harbours in British and Irish waters, in northern Portugal, north-west Spain, Iceland, Norway, Svalbard and Kattegat.

The application of TBT on small vessels (less than 25 m in length) has been prohibited since 1990 and some biological recovery in the north-east Atlantic has been observed in areas of small-boat use, but the gender-bending effects of TBT occur even at very low concentrations. The OSPAR Commission (2000) found a significant correlation between shipping intensity and TBT levels in marine sediments and the occurrence of gender-bending effects, suggesting that larger ships using TBT anti-fouling paints represent the main source of TBT in the marine environment. The International Maritime Organization has decided to ban the use of TBT and other similar compounds in anti-fouling treatments on ships longer than 25 m. The target is to prohibit their application from 2003 and to require the removal of TBT from ships' hulls by 2008. However, many areas will continue to show the legacy of historic TBT inputs for some years after that date.

The dumping of wastes into the oceans is controlled by the London Dumping Convention adopted in 1972, which has largely been responsible for the gradual decrease in the amounts of industrial wastes and sewage sludge dumped at sea since it came into force. A similar agreement for the north-east Atlantic, including the North Sea and the Baltic, is the Oslo Convention, adopted in 1974 and subsequently modernised and amalgamated into the Convention for the Protection of the Marine Environment of the North-East Atlantic (OSPAR Convention), which came into force in 1998. Dumping of industrial and hazardous wastes in the North Sea has been banned since the end of 1989, and was scheduled to cease in other marine waters by the end of 1995 under the London Dumping Convention. In the North Sea, under the OSPAR Convention, disposal of sewage sludge had stopped completely by 1998.

The Oslo Convention members also agreed, in 1988, to phase out the incineration of wastes at sea by 1994, and the London Dumping Convention contracting parties have agreed, in principle, to discourage this activity. The Oslo decision was not based upon observed harmful effects but on the conviction that adequate means would be available for land disposal. GESAMP suggested that on the available evidence:

> *The environmental implications of incineration at sea cannot be regarded as more significant than those of incineration on land. There are limitations and risks associated with either choice. In practice, the choice will depend on a careful analysis of technical, social and political as well as environmental factors.*
>
> *(GESAMP, 1990: 18)*

The oceans are an attractive place to dispose of wastes due to their vast diluting capacity, but the long residence times of compounds within them does imply that any contaminants will be removed only very slowly (Jickells *et al.*, 1990). Indeed, the dilute-and-disperse philosophy becomes questionable for persistent pollutants that take very long periods to become harmless (e.g. radionuclides, heavy metals, certain pesticide residues) and that can be accumulated up the food chain. The dumping of high- and medium-level radioactive wastes at sea does not occur and the dumping of low-level packaged radioactive waste at sea was voluntarily banned in 1982. During the period 1949–82, the UK and the USA made most use of the 26 sites used in the Atlantic and 21 sites in the Pacific. The total amount dumped was less than 0.1 per cent of the amount that reached the oceans due to atmospheric nuclear weapons testing between 1954 and 1962. This, in turn, is only 1 per cent of the amount found naturally in the oceans. However, the mix of radionuclides involved, their rates of decay and the effects upon biological life are different in each case, and GESAMP (1990: 50) warns that 'dumping cannot be considered safe just because releases of radionuclides are small compared with the natural incidence of radionuclides in the environment'.

A similar statement might be applied to the disposal of low-level liquid waste into the oceans, from nuclear reactors and fuel reprocessing plants, particularly since the discharges are made from point sources. In some parts of the world, these discharges have been made continuously over several decades; the nuclear facilities at Sellafield in north-western England, for example, have been discharging low-level liquid radioactive waste into the Irish Sea since the early 1950s. Such discharges are carefully monitored and limits are set with the aim of maintaining bioaccumulation in organisms well below international standards for various radionuclides. Nevertheless, discharges of nuclear waste to marine and other environments continues to be a source of considerable controversy (see Chapter 18). It is also worth noting that nuclear material for disposal is in many cases derived from military activities, and information on this source of radioactive waste to the oceans, as with other disposal options, is on the whole not available.

Pollutants that reach the marine environment from terrestrial sources via atmospheric pathways are more difficult to quantify, not least because they can be transported many thousands of kilometres from source areas before being deposited. The wide dissemination of persistent organic pollutants, or POPs, is generating increasing international concern because these classes of chemical are long-lived and tend to accumulate up the food chain. POPs include PCBs, polychlorinated dioxins and pesticides such as DDT, aldrin and dieldrin. POPs from industrialised parts of Europe and North America have contaminated marine environments all over the northern hemisphere, and the concentrations of PCB and DDT found in a number of Arctic marine mammals (Fig. 6.9) clearly show the way in which they bioaccumulate up the food chain. Our knowledge of the effects of POPs on the Arctic environment is limited, but the very high levels of PCBs in the polar bear, the top predator, are thought to reduce the level of thyroid hormones and may have a negative effect on the bear's ability to reproduce.

Black Sea

The ecosystem of the Black Sea and the adjoining Sea of Azov has been deteriorating since the late 1960s as a result of pollution, principally from land-based sources. The result has been eutrophication and contamination by pathogenic microbes and toxic chemicals, which has, in turn, had severe repercussions for fisheries and the sea's amenity value (Mee, 1992).

A major part of the Black Sea is now critically eutrophic following a large increase in nutrient loads entering the sea over the last 30 years. Much of the increase has come from increased loads in rivers, reflecting the widespread use of phosphate detergents and agricultural intensification (see Fig. 8.3). The River Danube alone delivered 60 000 t of total phosphorus and 340 000 t of total inorganic nitrogen per year in the early 1990s. In recent years, the volumes of nutrients

Figure 6.9 Bioaccumulation of POPs in Arctic marine mammals with levels in Norwegian women for comparison (after Hansen *et al.*, 1996).

from coastal communities' sewage has probably become comparable to the riverine loads. The north-western portions of the sea are particularly affected, and have been changed from a diverse ecosystem supporting highly productive fisheries to a eutrophic plankton culture leading to a marine environment that is unsuitable for most higher organisms. The effects of nutrient inflows have also been traced to formerly pristine sites on the central and eastern coasts where macrobenthic species have declined significantly since the 1960s (Fig. 6.10). Increased flows of phosphorus and nitrogen via the Danube do not account for the whole story however. Following the construction in the early 1970s of the 'Iron Gates' Dam on the Yugoslavia/Romania border, the dissolved silicate load of the river has been reduced by about two-thirds and is thought to be responsible for a dramatic shift in the composition of phytoplankton species in the Black Sea (Humborg *et al.*, 1997).

The Danube is not the only river in the Black Sea watershed that has been significantly affected by dam construction and consequent water withdrawals. Substantial net decreases in runoff (and probably also silicates) reaching the sea from several other rivers are also evident (Table 6.2). The resulting biogeochemical changes, including increased salinity and shifts in pH, have had a serious effect on anadromous fish catches in the Sea of Azov, which fell by 95 per cent between 1950 and 1975. Data on chemical and microbiological pollutants are scarce, but heavy metals, pesticides, radionuclides and oil are all thought to

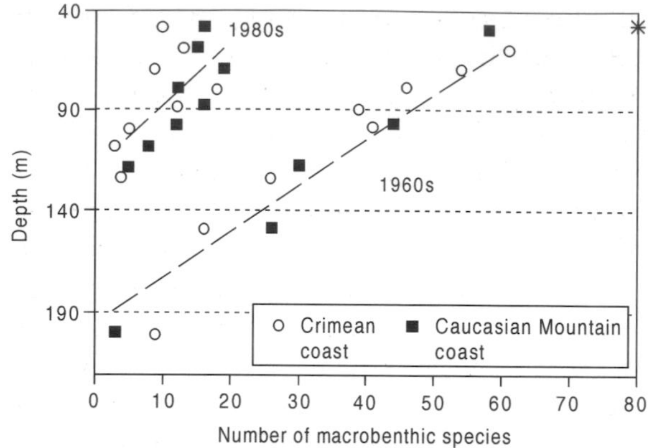

Figure 6.10 Change in the number of macrobenthic species as a function of depth in the Black Sea (after Zaika, 1990).

have suppressed biota. Oil spills, increased trawl fishing and invasion by opportunistic settler species have also contributed to the worsening situation. From the socio-economic perspective, the Black Sea fishing industry, which supported some 2 million fishermen and their dependants, has suffered an almost total collapse and the attractions of the sea's beaches have been severely depleted.

Steps towards dealing with the ecological problems have been made with the signing of a legal Convention for the Protection of the Black Sea and the adoption of a plan of action in 1992. There is still a need for basic data for input into designing the very large investments that will be needed to control pollution, and there is no doubt that rehabilitation will take many years.

Other regional seas

Like the Black Sea, many other regional seas are showing signs of degradation, in many cases with land-based sources being most prevalent. In the Mediterranean, for example, at least half of the nutrient, mercury, lead, chromium, zinc, pesticide, and radioactive pollutants enter from rivers draining areas outside the immediate coastal zone (Grenon and Batisse, 1989). Projected increases in the Mediterranean's population indicate that the pollution loads entering the sea will continue to increase, with the greatest rises in population likely to occur in the poorer North African and Middle Eastern countries that border the sea.

The difficulties of managing environmental problems in regional seas are similar to those posed by other global 'commons', they are multi-disciplinary and they require international cooperation. In contrast to the relatively well-developed international agreements governing marine sources of pollution, such as the London Dumping Convention and the Oslo Convention, agreements to limit problems that emanate from coastal landmasses tend to be conspicuous by their absence. Some progress has been made in

TABLE 6.2 *Reduction in discharge of some of the rivers flowing into the Black Sea*

River	Natural flow (km^3/year)	Reduction (%)		
		1971–75	1981–85	1991–2000*
Don	27.9	19	27	43
Kuban	13.4	39	49	65
Dnieper	53.5	24	52	71
Dniester	9.3	20	40	62

*Estimated
Source: Tolmazin (1985)

TABLE 6.3 *UNEP Regional Seas Programme*

Regional sea area	Action plan adopted
Mediterranean	1975
Kuwait	1978
West and Central Africa	1981
Caribbean	1981
East Asia	1981
South-east Pacific	1981
Red Sea and Gulf of Aden	1982
South Pacific	1982
Eastern Africa	1985
Black Sea	1993
North-west Pacific	1994
South Asia	1995
North-east Pacific	2001

this direction, however, in the form of a series of action plans for regional seas, coordinated by UNEP (Table 6.3).

The Mediterranean was the first regional sea for which such a plan was formulated. The future well-being of the Mediterranean depends principally on land use planning, urban management and pollution prevention, and these are central aspects of the Action Plan for the Protection of the Mediterranean. The plan has succeeded in adopting a number of conventions and protocols to combat pollution and protect the marine environment (e.g. to prevent the dumping of pollutants at sea from ships and aircraft; to compel governments to cooperate in emergencies; to control pollution from land-based sources). A pollution monitoring and research programme (MED POL) has also been established, as well as the so-called Blue Plan, which aims to reconcile the varying needs of environment and development. But the difficulties of organising successful cooperation between the nation-states involved – 21 have coastlines bordering the Mediterranean – have been serious (Skjaerseth, 1993). Amongst many issues raised by the problems of the Mediterranean is that of liability. Most of the pollution entering the sea comes from the northern industrialised countries but the effects are felt, to varying degrees, by all. Should the 'polluter pays' principle be applied? Or should countries pay according to the impact they experience? Other political difficulties arose in the 1990s following the break-up of the former Yugoslavia and continued civil unrest in some of the new Balkan states. Projections for the countries bordering the Mediterranean indicate that the potential threats to the marine environment will continue to increase as one of the main causative factors, the size of the human population, continues to rise. There is no doubt, therefore, that continued successful international cooperation is vital for the management of the Mediterranean, as for other marine environments, and thus such initiatives need to succeed.

FURTHER READING

Clark, R.B. 1997 *Marine pollution*, 4th edn. Oxford, Oxford University Press. A comprehensive study of the causes and consequences of marine pollution by pollution type.

Cuyvers, L. 1984 *Ocean uses and their regulation*. New York, Wiley. This book covers food, minerals, waste disposal, transport, energy and the law of the sea.

GESAMP (Joint Group of Experts on the Scientific Aspects of Marine Environmental Protection) 2001 *A sea of troubles*. GESAMP Report Study 70. UNEP Nairobi. A comprehensive assessment of the current state of the oceans available at http://gesamp.imo.org/publicat.htm.

Gulland, J.A. (ed.) 1988 *Fish population dynamics*, 2nd edn. Chichester, Wiley. A collection of papers on our developing knowledge of natural variations in the oceans' fish stocks.

Nicol, S. and de la Mare, W. 1993 Ecosystem management and the Antarctic krill. *American Scientist* 81: 36–47. An overview of the CCAMLR's attempts to manage the Southern Ocean's most abundant living resource.

Safina, C. 1995 The world's imperilled fish. *Scientific American* 273(5): 30–7. An overview of threats to the world's major fisheries and the economic considerations behind their effective management.

Watson, R. and Pauly, D. 2001 Systematic distortions in world fisheries catch trends. *Nature* 414: 534–6. A cautionary examination of the methods used to monitor fish catches.

Websites

http://www.fao.org/ the FAO site has an extensive fisheries section with data and many up-to-date reports.

http://www.imo.org/ site of the International Maritime Organization, concerned with maritime safety and preventing pollution from ships. Information, reports and links to many other marine sites.

http://www.bsein.mhi.iuf.net/ a regional environmental data centre for the Black Sea with information on research, a bibliography and satellite images.

http://pices.ios.bc.ca/ site of the North Pacific Marine Science Organization (PICES) with datasets, publications and links to allied sites.

http://www.ospar.org/ the OSPAR Convention site has data, publications and details of action plans for the north-east Atlantic.

http://www.unep.ch/seas/ UNEP's Regional Seas Programme site with information on major threats and actions.

http://www.iwcoffice.org/ the International Whaling Commission.

http://www.itopf.com/ the International Tanker Owners Pollution Federation site with information, data and country profiles on marine oil spills.

POINTS FOR DISCUSSION

- How can we manage the oceans' biological resources when we know so little about them?
- How can we justify efforts to save whales when so many people in the world live in destitution?
- The environmental problems of regional seas like the Mediterranean and the Black Sea are good examples of the tragedy of the commons. How far do you agree?
- For an ocean of your choice, assess which is the more damaging to its ecology: pollution or overfishing.

7

COASTAL PROBLEMS

TOPICS COVERED

Physical changes
Historical overfishing
Pollution
Habitat destruction
Management and conservation

Key words: erosion, sedimentation, sea level rise, conservative and non-conservative pollutants, phytoplankton bloom, coral reef, mangrove

The coastal zone, which can be defined as the region between the seaward margin of the continental shelf and the inland limit of the coastal plain, is among the Earth's most biologically productive regions and the zone with the greatest human population. It also embraces a wide variety of landscape and ecosystem types, including barrier islands, beaches, deltas, estuaries, mangroves, rocky coasts, salt marshes and seagrass beds. Coral reefs and coastal wetlands are some of the most diverse and productive ecosystems, and 90 per cent of the world's marine fish catch, measured by weight, reproduces in coastal areas (FAO, 1991). In 1995, over 2.2 billion people (39 per cent of the world's population) lived within 100 km of a coast, an increase from 2 billion people in 1990 (Burke *et al.*, 2000). This coastal area accounts for just 20 per cent of all land area.

This large and increasing density of human occupation inevitably puts great pressures on coastal resources, since as Walker (1990: 276) put it:

> *Virtually every aspect of the coastal zone, whether it be space, climate, beauty, animal or plant life, minerals, or water, must be considered a resource. Thus everything humans do in the coastal zone (and many of the things they do outside it) alters the resource base.*

PHYSICAL CHANGES

The natural operation of the processes of erosion and sedimentation affects human use of coasts by constantly changing the configuration of coastlines, effects that are superimposed upon the larger-scale changes in sea level relative to land, due to crustal movements and changes in the overall volume of marine waters. Sedimentation by the Büyük Menderes River into the Gulf of Miletus on Turkey's Aegean coast over the last 2400 years, for example, has today left the ancient port of Haracleia 30 km from the coast (Bird, 1985). Conversely, erosion of Britain's Humberside coastline between Bridlington and Spurn Head has removed a 3-km strip of coastline, with the loss of 30 villages, since Roman times (Fig. 7.1). These natural processes, which affect the physical make-up of coastlines, have also been modified in numerous ways by human populations. In Japan, for example, more than 40 per cent of the country's 32 000-km coastline is influenced by human action, through land reclamation, urban and port facilities, or by shoreline protection associated with lines of communication (Koike, 1985).

Deliberate efforts to control the sediment budget on certain portions of coasts are often

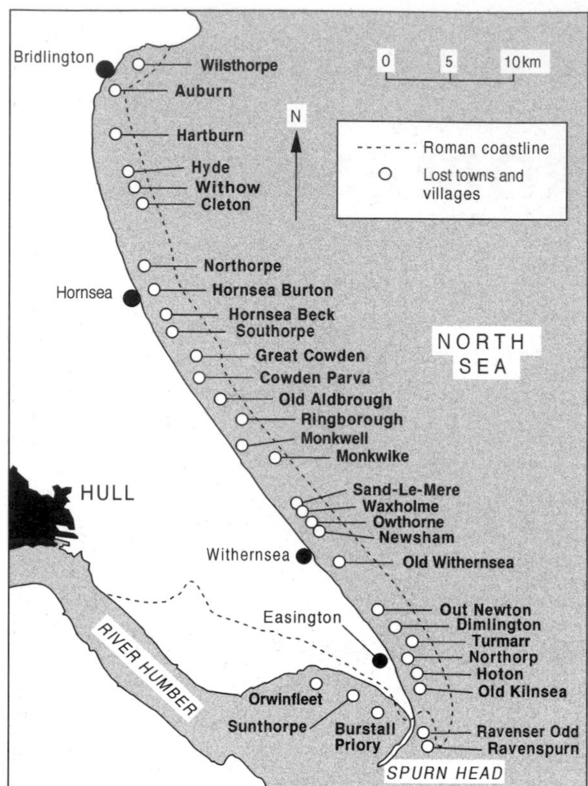

Figure 7.1 Extent of coastal erosion since Roman times in the Humberside region of north-eastern England (after Willson, 1902; Murray, 1994).

carried out in an attempt to maintain a beach that is useful for recreation and/or coastal protection. However, interference with any portion of a coast, whether deliberate or inadvertent, will inevitably have impacts elsewhere through its effect on the coastal processes of erosion and deposition of unconsolidated sediments such as beach sand. Hence, the effects of groynes built for beach protection are often similar to those caused by other structures, such as breakwaters and jetties, built for alternative purposes. The frequent outcome mirrors the effects noted after the completion in 1907 of two moles to protect the West African harbour of Lagos from sedimentation. The result has been a mean erosion rate at Victoria beach to the west of 6 m/year and a mean sedimentation rate at Lighthouse beach to the east of 8 m/year (UNEP/UNESCO/UN-DIESA, 1985). This effect is illustrated in Fig. 7.2, which shows significant modification to Kuta beach and erosion damage to a resort area on the Indonesian island of Bali following the construction of the Denpasar airfield, which extends into the ocean. Similar effects can be caused by the construction of a sea-wall to protect a cliff, which may result in depletion of a beach elsewhere that was fed by the eroding cliff.

Other activities in the coastal zone can have equally significant effects on beaches, not least the removal of coastal materials for construction purposes. Elsewhere off Bali, coral reefs were broken up and brought ashore to produce lime for cement used in the hotel construction boom during the late 1970s and 1980s. At one stage in the 1980s, there were more than 400 coral-burning kilns on the eastern and southern coasts of the island and the volume of coral mined off the south coast has been estimated at up to 150 000 m^3. Without the protection to the coastline provided by the reefs, significant erosion of the beachfront has resulted. The Balinese government outlawed coral mining in 1985, but the initial response to the ban was slow. Mining had stopped in most areas by 1990, but millions of dollars have had to be spent on artificial measures to protect the eroding beaches, such as sea-walls and breakwaters (Bentley, 1998). In other areas, similar effects have been documented due to the direct removal of beach sand for construction. In the northern region of Grand Comore in the Indian Ocean, whole beaches are disappearing due to the collection of sand for cement making (UNEP, 1984).

The risk from marine floods and other natural hazards has inspired many modifications to coastlines around the world. In western Europe, the storm surge hazards posed to the low-lying coastlines of the North Sea have been combated by a number of engineering solutions. In the Netherlands, the Zuider Zee scheme (1927–32) is an early example, which involved a 32 km-long, 19 m-high dam, closing off the vast estuary of the Zuider Zee. In addition to flood protection, closure of the estuary also facilitated land reclamation and improvements to freshwater resource management.

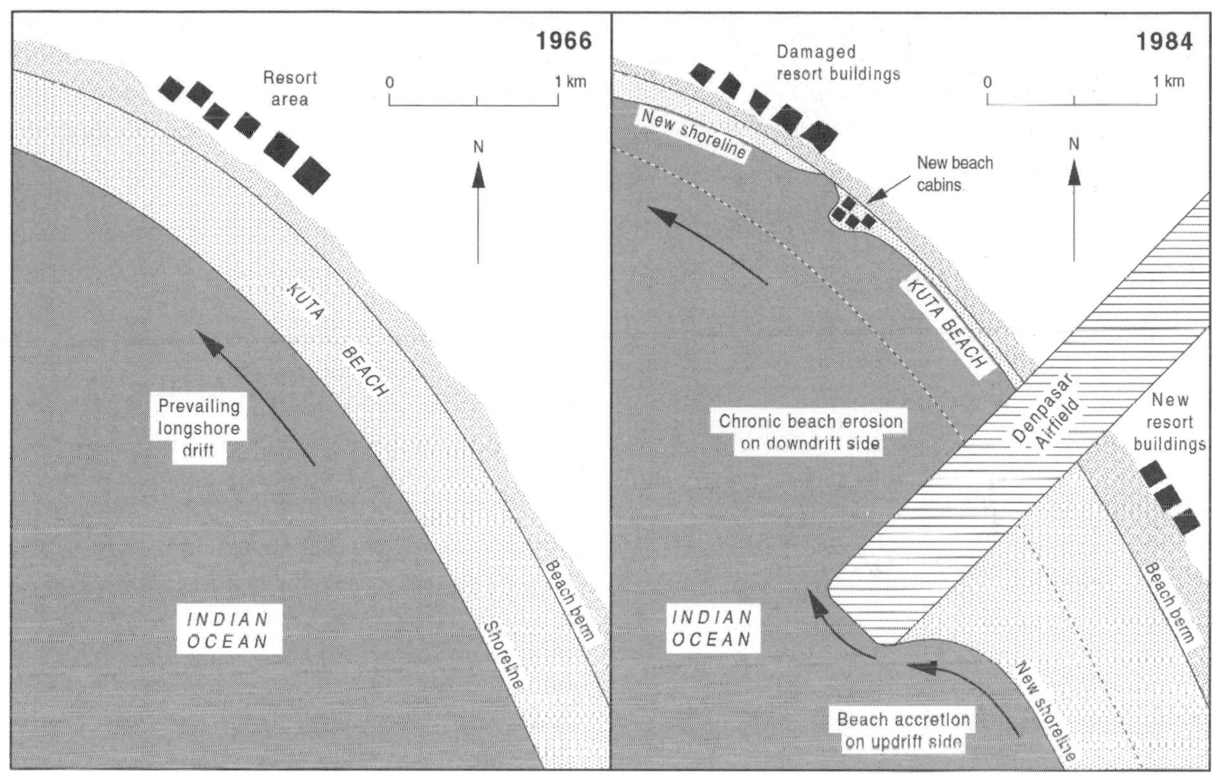

Figure 7.2 Changes to Kuta Beach on the Indonesian Island of Bali caused by the construction of an airport runway out to sea (after Nunn, 1994).

In 1953, when North Sea surge floods drowned 1853 people and seriously damaged 50 000 buildings in the Netherlands, the Delta Project was set in motion to protect the country's south-western coastal area. It provides for the closing off of some of the estuaries discharging into the North Sea – the Rhine, Maas and Scheldt – and three closures have been built: the Veerse Meer, Haringvliet and Grevelingen. The project has effectively shortened the Dutch coastline by 700 km.

A similar threat of marine flooding to London in south-east England, where the dangers from high tides and storm surges are exacerbated by land subsidence, led to the construction of the Thames Barrier, completed in 1983. Such schemes are not without their problems, however. The 25-km barrier dike built across the Gulf of Finland to protect St Petersburg from storm-induced tidal surges, which is currently just over half built, has hampered water circulation in the diked-off portion, turning it into a stagnant settling basin for effluent. The resulting algal blooms and noxious odours have a deleterious effect on fishing and coastal recreation, while siltation behind the dike and ice build-up along it also threaten the port's maritime economy. Worries over similar effects behind tidal power barrages have also been expressed. A further question mark over the efficiency of these and other coastal flood protection schemes is currently posed by the potential for rapid global sea-level rise associated with global warming (see Chapter 11).

It is not just activities in the coastal zone itself that have an impact there. River outlets also supply sediment to the coasts, so that human modification of river regimes can have a downstream impact. Changes in land use, such as cultivation, can increase sediment loads, as in the case of many of Taiwan's rivers and the Huang Ho River

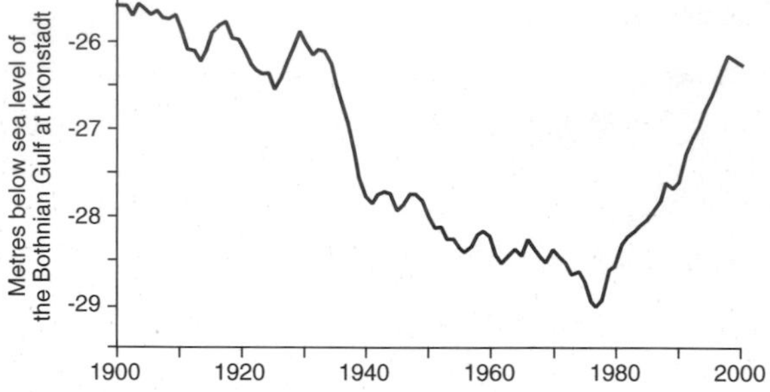

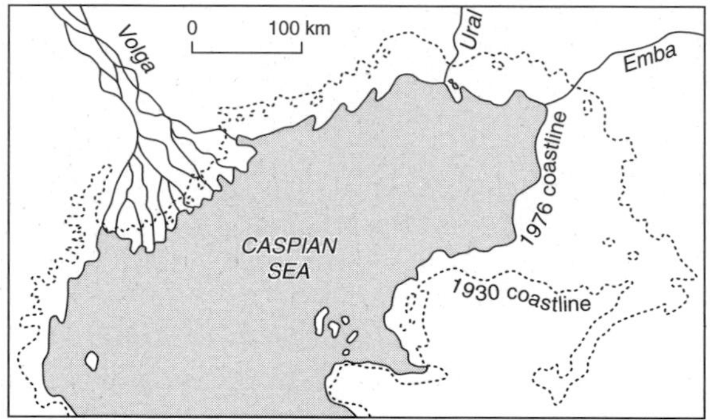

Figure 7.3 Variations in sea level and northern coastline changes of the Caspian Sea (after Kroonenberg *et al.*, 2000; Bird, 1985).

in China, which transports an order of magnitude more sediment than it did prior to widespread cultivation of the loess plateau (Milliman, 1990). Conversely, numerous major rivers have had their sediment budgets severely depleted, commonly by the construction of dams. The Colorado River and the Nile no longer deliver any sediment to the coast; the effects in the latter case are illustrated in Fig. 9.7. The absolute reduction in discharge due to dam construction can also have severe repercussions for coastlines in inland seas. The completion of a dam to facilitate agricultural, industrial and municipal withdrawals across the mouth of the Kora-Bogaz-Gol embayment, an area of high evaporation on the Turkmenistan coast of the Caspian Sea, is thought to have contributed to a decline in sea level in the Caspian of more than 3 m between 1929 and 1977, exposing more than 50 000 km^2 of former sea bed (Fig. 7.3). However, a subsequent rise of nearly 3 m by 1995 suggests that climatic variations influencing fluvial discharge into the Caspian were more important. In fact, large fluctuations in sea level have been identified in the Caspian throughout the Quaternary and although the cause of these changes is much debated, it is most probably related to the fact that the Volga River, which supplies 80 per cent of the inflow of the Caspian Sea, has experienced strong variations in annual discharge due to climatic variability (Kroonenberg *et al.*, 2000).

HISTORICAL OVERFISHING

Some scientists have traced the degradation of many coastlines back over considerable periods of time and concluded that ecological extinctions caused by overfishing precede all other pervasive human impacts to coastal ecosystems (Jackson *et al.*, 2001). Using a range of proxy indicators and historical evidence, the first major human disturbance to all coastal zones studied was found to be overfishing of large marine fauna and shellfish. The effective loss of entire trophic levels has made coastal ecosystems more vulnerable to other human and natural disturbances such as pollution, disease, storms and climate change.

The study examined records from marine sediments dating from about 125 000 years ago; archaeological records from human coastal settlements occupied after about 10 000 years ago; historical records from documents of the first European trade-based colonial expansion in the Americas and South Pacific in the fifteenth century to the present; and ecological studies from the past century to help calibrate the older records. Everywhere, the magnitude of losses in terms of abundance and biomass of large animals such as whales, manatees, dugongs, monk seals, sea turtles, swordfish, sharks, giant codfish and rays, was found to be enormous. These ecological extinctions caused by overfishing are thought to have led to serious changes in coastal marine ecosystems over many centuries as overfished populations no longer interacted with other species.

Such changes were not always immediate. Timelags of decades to centuries were identified

TABLE 7.1 *The decline of kelp forests off southern California due to human impact*

Date	Human impact	Ecology
Pre-nineteenth century	Little	Grazing of kelp by sea urchins kept in check mainly by sea otter predation
Early 1800s	Sea otters effectively eliminated for fur trade	Sheephead fish and spiny lobster replace sea otter as main sea urchin predators; abalone compete with sea urchins
Post-1950s	Sheephead, spiny lobster and abalone effectively eliminated	Widespread loss of kelp due to overgrazing by sea urchins

Source: after Jackson *et al.* (2001)

between the onset of overfishing and consequent ecological changes because other species of similar trophic levels took over the ecological roles of the overfished species until they too were overfished or perished as a result of disease epidemics related to overcrowding. The plight of the southern California kelp forests, which began to disappear on a large scale after the 1950s, is shown in Table 7.1. Kelp forests are usually found in shallow, rocky areas from warm temperate to subarctic regions worldwide, but those off the southern Californian coast began a long process of ecological change dating from the early 1800s when sea otters were hunted to virtual extinction for their fur. The diverse food web of the kelp forest enabled the ecosystem to remain for more than a century as other species took over the sea otters' role in keeping sea urchins, which graze on the kelp, in check. However, widespread loss of the kelp occurred after the 1950s when these other species also became targets for intense human exploitation, enabling a population explosion of sea urchins.

POLLUTION

Pollution of the coastal zone comes from activities on the coast, in coastal waters and from inland; in the latter case, usually arriving at the coastal zone via rivers. Pollutants can be divided into non-conservative types that are eventually assimilated into the biota by processes such as biodegradation and dissipation, and conservative pollutants that are non-biodegradable and are not readily dissipated, and thus tend to accumulate in marine biological systems (Table 7.2, see also Fig. 6.7).

The wide range of human activities that contributes pollutants to coastlines, and the pathways by which they reach the coast (see also Table 6.1), are illustrated in an issue that is causing considerable concern: the increasing discharges of nutrients to coastal environments – a problem that is envisaged to constitute a worldwide issue in the next few decades. Human activities have increased nutrient flows to the coast in numerous ways:

- forest clearing
- destruction of riverine swamps and wetlands
- application of large amounts of synthetic fertilisers
- industrial use of large amounts of nitrogen and phosphorus
- addition of phosphates to detergents
- production of large livestock populations

TABLE 7.2 *Classification of marine pollutants*

NON-CONSERVATIVE

- Degradable wastes (including sewage, slurry, wastes from food processing, brewing, distilling, pulp and paper, and chemical industries, oil spillages)
- Fertilisers (from agriculture)
- Dissipating wastes (principally heat from power station and industrial cooling discharges, but also acids, alkalis and cyanide from industries, which can be important very locally)

CONSERVATIVE

- Particulates (including mining wastes and inert plastics)
- Persistent wastes (heavy metals such as mercury, lead, copper and zinc; halogenated hydrocarbons such as DDT and other chlorinated hydrocarbon pesticides, and polychlorinated biphenyls; radionuclides)

Source: Clark (1992)

- high-temperature fossil fuel combustion, adding large amounts of nitrogen to acid rain, and
- expansion of human populations in coastal areas.

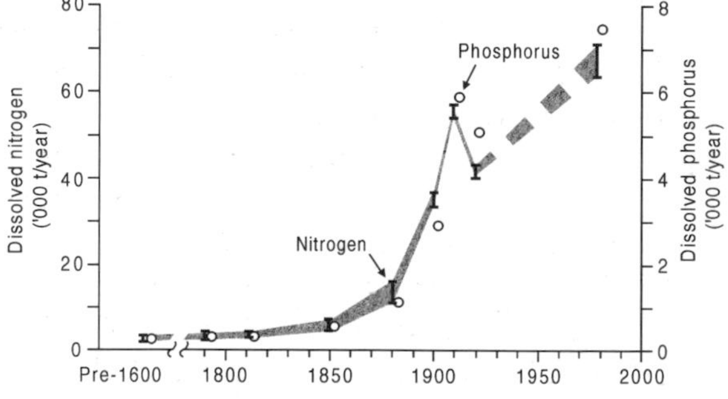

Figure 7.4 Inorganic nitrogen and phosphorus input to Narranganset and Mount Hope bays, north-east USA, pre-seventeenth to late twentieth centuries (after Nixon, 1993).

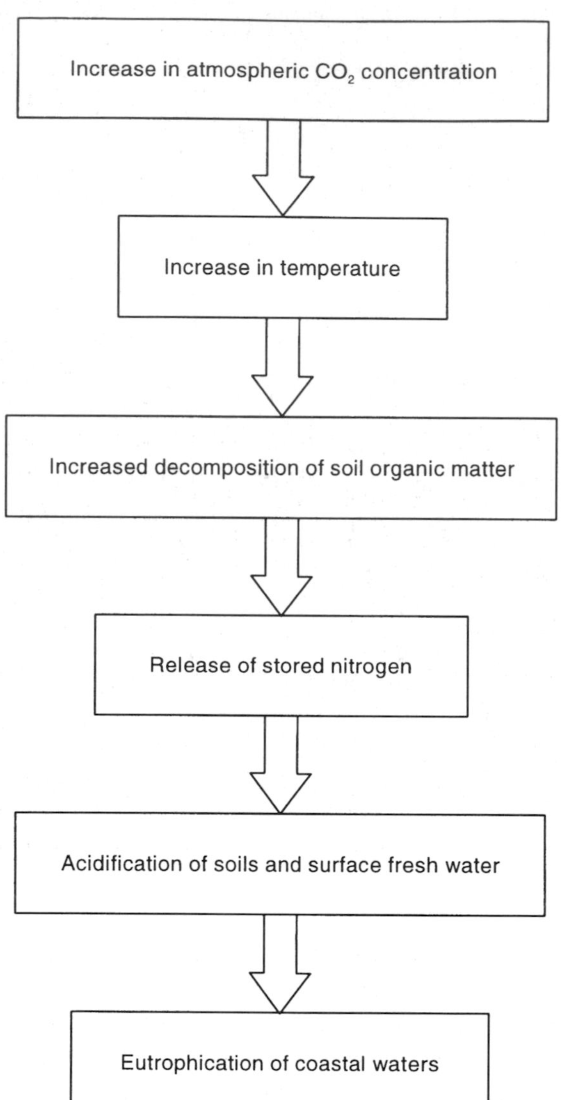

Figure 7.5 Possible effect of global warming on coastal eutrophication through the release of terrestrial nitrates (after Wright and Hauhs, 1991).

The extent of the increase is indicated for a part of the North American seaboard by Nixon (1993), who used a complex mix of historical data to estimate the changes in nutrient inputs to the Narragansett and Mount Hope bays (Fig. 7.4). Up to around 1800, the inflow of nutrients from the ocean was five to ten times larger than inputs from the land and the atmosphere, but today

Figure 7.6 An unusual form of coastal pollution in Pakistan where a 10-km stretch of sands at Gadani is used for ship-breaking

about five times more inorganic nitrogen enters Narragansett Bay from the land and atmosphere than from the coastal ocean. The particularly dramatic rise that occurred over a 30-year period from 1880 to 1910 was a response to the proliferation of public water and sewage systems through much of the urban area of the north-eastern USA at this time.

Excessive nutrient flows to coastal environments cause oxygen depletion, reduced diversity, increased toxic and other undesirable phytoplankton blooms, and generally negative changes in population structure and dynamics. Hence, algal blooms have been associated with catastrophic mass deaths of marine mammals, fish and invertebrates, and the emergence of new disease pathogens, such as a new form of cholera present in several Asian nations, which threatens to become the agent of an eighth cholera pandemic (Levins *et al.*, 1994). Outbreaks of shellfish poisoning affected more than twice as many areas globally in 1990 as in 1970 (Anderson, 1994), and in coming decades this trend may be recognised in part as an early sign of the effects of climatic change. Some scientists fear that the frequency and extent of coastal eutrophication and associated algal blooms may increase with global warming because nitrogen and sulphur stored in soil organic matter will be released into rivers if temperatures rise as predicted. Increased nitrogen losses are particularly likely (Fig. 7.5).

All kinds of urban and industrial activities in the coastal zone contribute pollution (Fig. 7.6), the effects of which can become particularly critical in areas where the mixing of coastal water with the open ocean is relatively slow, as in bays, estuaries and enclosed seas (e.g. see the account of the Black Sea in Chapter 6). For example, in dryland coastal cities where fresh water is produced by desalination plants, the discharge of chlorine back into coastal waters may represent a severe pollution problem, as in Kuwait where about 17 t of chlorine are dumped each day (Khordagui, 1992).

Petroleum hydrocarbons are a particular hazard to coastal ecology, chiefly at oil terminals but also from accidental spillages. Habitats such as mangroves and marshes are especially vulnerable to oil spills, although wildlife is often the most conspicuous victim. Generally, younger individuals are more sensitive than adults, and crustaceans are more sensitive than fish. The release of 36 000 t of oil into Prince William Sound in Alaska from the grounded supertanker *Exxon Valdez* in March 1989 resulted in the deaths of 36 000 seabirds, 1000 sea otters and 153 bald eagles in the following six months (Maki, 1991). The loss of these animals from intermediate and top trophic levels has had many indirect effects at lower levels, and recovery of most biotic communities to their pre-spill levels could take up to 25 years (Shaw, 1992).

Heavy metals and organochlorine compounds have long been recognised as among the most deleterious contaminants to biota in coastal and estuarine waters, and a large proportion of these pollutants arrive at coasts in rivers draining wide areas outside the coastal zone. A study of the Mediterranean found that more than half the mercury, chromium, lead, zinc and pesticides (as well as a similar level of nutrients and radioactive materials) came from such rivers (UNEP, 1989a).

Since heavy metals and organochlorine compounds cause problems even in very low concentrations, and the measurement of such small quantities is technically difficult, many studies of these coastal pollutants focus on concentrations in marine organisms. Mussels and oysters have been widely used because as filter feeders they bioaccumulate contaminants. Several investigations of long-term trends in pollutants have measured concentrations in the feathers of seabirds since old specimens can often be found in museums. Mercury contamination measured in herring gulls and common terns from the German North Sea coast, for example, indicates that the rivers Elbe and Rhine are the major sources of the pollutant. Mercury flows peaked at three times 1880 levels after the Second World War when heavy metals from ignition devices in munitions were released in large quantities (Thompson *et al.*, 1993). Significant declines in the last two decades reflect measures to reduce river-borne pollution, however, and the trend will probably continue.

HABITAT DESTRUCTION

Coastal habitats, particularly mangroves, salt marshes, seagrass beds and coral reefs, are subject to many types of pressure from human activities. Direct destruction of these habitats for urban, industrial and recreational growth, as well as for aquaculture, combined with overexploitation of their resources and the indirect effects of other activities, such as pollution and sedimentation or erosion, are causing rapid losses all over the world. Rates of habitat loss are difficult to assess, but the Joint Group of Experts on the Scientific Aspects of Marine Environmental Protection reported: 'If unchecked, this trend will lead to global deterioration in the quality and productivity of the marine environment' (GESAMP, 1990: 1). The sections below detail the situation for two coastal habitats in the tropics that have come under substantial pressure in recent times.

Coral reefs

Coral reefs, tropical marine ecosystems found in shallow waters (generally 30 m) and largely restricted to the areas between 30°N and 30°S, rank among the most biologically productive and diverse of all natural ecosystems. They cover less than 0.2 per cent of the ocean floor, yet support 25 per cent of marine species. Their high productivity is a function of their efficient biological recycling, high retention of nutrients and their structure, which provides a habitat for a vast array of organisms. From the economic viewpoint, coral reefs

TABLE 7.3 *Human-induced threats to coral reefs, with selected examples from the Pacific Ocean*

Threat	Example
OVERCOLLECTING	
Fish	Futuna Island, France
Giant clams	Kadavu and islands, Fiji
Pearl oysters	Suwarrow Atoll, Cook Islands
Coral	Vanuatu
FISHING METHODS	
Dynamiting	Belau, USA
Breakage	Vava'u Group, Tonga ("tu'afeo")
Poison	Uvea Island, France
RECREATIONAL USE	
Tourism	Heron Island, Great Barrier Reef, Australia
Scuba diving	Hong Kong
Anchor damage	Molokini Islat, Hawaii, USA
SILTATION DUE TO EROSION FOLLOWING LAND CLEARANCE	
Fuelwood collection	Upolu Island, Western Samoa
Deforestation	Ishigakishima, Yaeyama-retto, Japan
COASTAL DEVELOPMENT	
Causeway construction	Canton Atoll, Kiribati
Sand mining	Moorea, French Polynesia
Roads and housing	Kenting National Park, Taiwan
Dredging	Johnston Island, Hawaii, USA
POLLUTION	
Oil spillage	Easter Island, Chile (1983)
Pesticide spillage	Nukunonu Atoll, New Zealand (1969)
Urban/industrial	Hong Kong
Thermal	North-western Guam, USA
Sewage	Micronesia
MILITARY	
Nuclear testing	Bikini Atoll, Marshall Islands (1946–58)
Conventional bombing	Kwajalein Atoll, Marshall Islands (1944)

Source: compiled from sources quoted in Nunn (1990); IUCN (1988)

protect the coastline from waves and storm surges, prevent erosion, and contribute to the formation of sandy beaches and sheltered harbours. They play a crucial role in fisheries, by providing nutrients and breeding grounds, and provide a source of many raw materials such as coral for jewellery and building materials. Most recently, their potential for tourism has been realised – one that plays an important role in the economies of many Caribbean island states particularly.

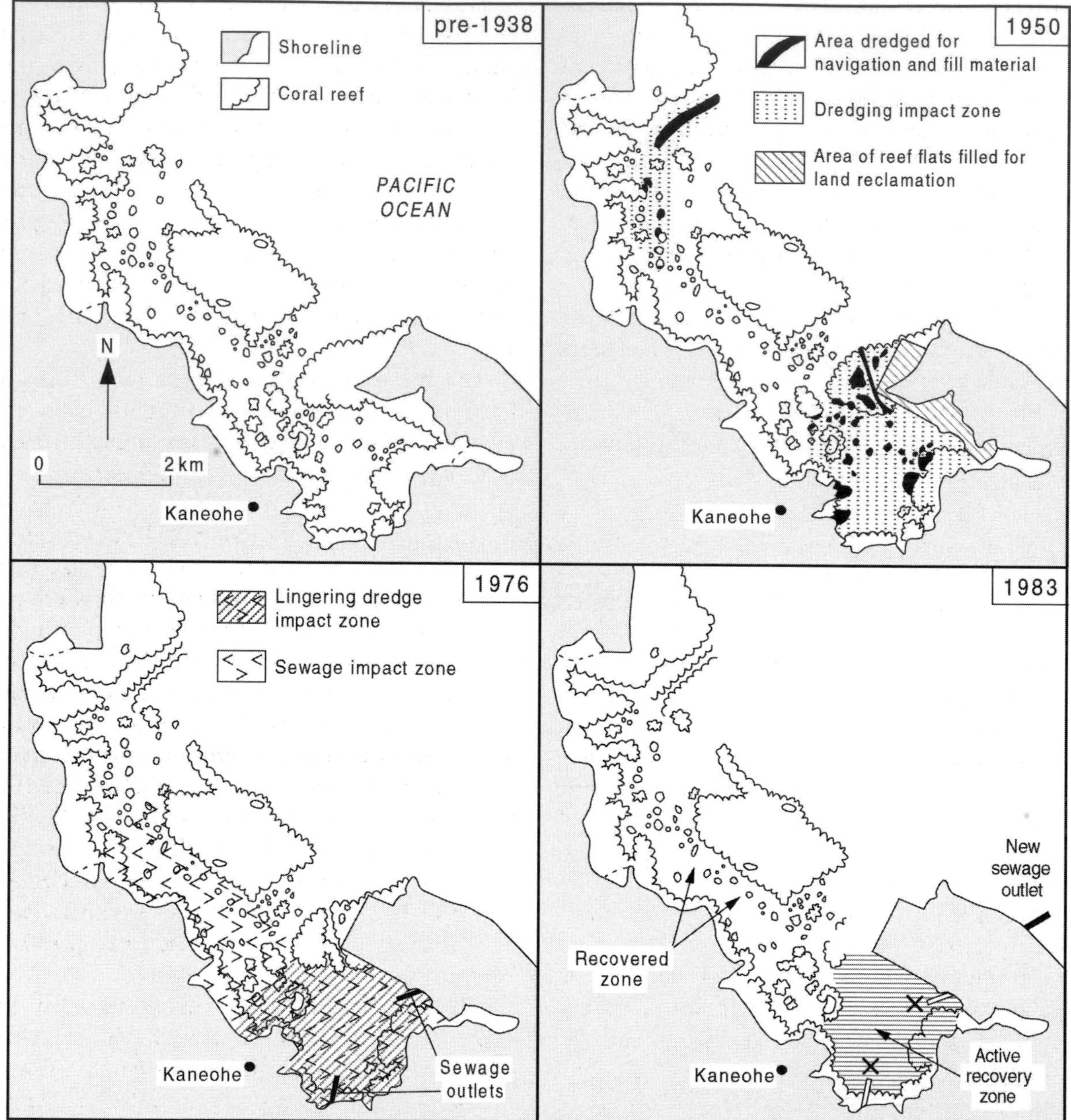

Figure 7.7 Effects of dredging, filling and sewage discharge in Kaneche Bay, Oahu, Hawaii, USA (after Maragos, 1993).

The past 20 years or so have been marked by a growing concern for the future of the world's coral reefs. Reefs have always been subject to natural disturbances such as hurricanes, storms, predators, diseases and fluctuations in sea level, but in recent times they have also been subject to increasing pressures from human action. Island and coastal societies have long used coral reefs for subsistence purposes, as sources of food and craft materials, but the development of commercial fisheries, rapidly increasing populations still dependent upon a subsistence lifestyle, the growth of coastal ports and urban areas, increases in soil loss due to deforestation and poor land use, and most recently the exponential growth of coastal tourism, have combined to put unprecedented pressures on reefs from both direct and indirect impacts (IUCN, 1988).

The wide range of human impacts upon coral reefs is indicated in Table 7.3 with examples from the Pacific. They can be divided roughly into direct impacts caused by overexploitation of reefs (overcollecting, damaging fishing techniques, and careless or poorly supervised recreational use), and indirect impacts resulting from other activities (siltation following land clearance, damage caused by coastal developments and pollution, and that caused by military activities).

In a global assessment carried out in 1997 (Bryant *et al.*, 1998), it was estimated that nearly 0.5 billion people live within 100 km of a coral reef. Mapping of potential threats indicated that 58 per cent of the world's reefs are at risk from human activities. Those most threatened are in South-East Asia – which also happen to be the most species-rich on Earth – where more than 80 per cent are at risk, and in the Caribbean where nearly two-thirds of reefs are in jeopardy. In many cases reefs are subject to a combination of impacts. In the Philippines, where only 30 per cent of remaining reefs are considered to be in good or excellent condition, siltation due to forest denudation, blast fishing and the use of sodium cyanide by tropical fish collectors, are the main causes of coral decline. Chronic overfishing has also been a constant hazard to reefs in Jamaica since the 1960s, and this stress has combined with natural hurricane damage, disease and algal blooms to reduce coral cover around the island's coastline from about 50 per cent in the late 1970s to less than 5 per cent in the mid-1990s (Hughes, 1994).

The changing nature of human threats to the coral reefs of Kaneohe Bay, on the east coast of the Hawaiian island of Oahu, is shown in Fig. 7.7. Dredging and filling by the military between 1938 and 1950 significantly increased problems of turbidity and sedimentation in most of the bay, and caused direct damage in southern parts where coral was used as fill to create new land. In 1970, while northern reefs were recovering from this impact and colonising some of the dredged zones, two new sewage outlets caused further reef decline in the southern portion. Closure of these sewage outlets in 1977–78 allowed a rapid recovery in the central part of the bay and a more prolonged recovery in the south. The new sewage outlet outside the bay has not damaged coral, due to the high level of mixing and flushing at the site.

Given the importance of coral reefs to many local human communities, the economic knock-on effects of damage can also be significant. Reefs in Rosario Coral Reef National Park in Colombia were severely affected after the Canal del Dique, leading to the coast, was redredged and reopened in 1986 (Le Ble and Cuignon, 1988). The resulting cover of sediment, which deprived symbiotic algae in the coral of the sunlight needed to photosynthesise, led to a mortality rate of 55 per cent in 1989. Loss of elkhorn and staghorn corals, in particular, has left the coastline susceptible to wave attack, and the resultant erosion now threatens homes and port facilities. As the reef has declined, the local fish catch has dropped due to the disruption of the food chain, and the diving industry for tourists has suffered as sedimentation spreads. High nutrient levels in the canal water have also led to eutrophication and the spread of brown algae across areas of the park.

The effects of indirect human activities on the health of coral reefs is illustrated by the situation on the east coast of Australia where a significant increase in pollution discharged to the Great Barrier Reef lagoon over the last 150 years (Table 7.4) is having an adverse effect on coral and seagrass ecosystems (Great Barrier Reef Marine Park Authority, 2001). Sediment smothers the living coral and seagrasses, while nutrients inhibit both their growth and reproduction. Herbicides from agricultural runoff also damage seagrasses and other reef organisms by affecting their ability to manufacture food.

Land use in most of the river catchments

TABLE 7.4 *Pollution increases to the Great Barrier Reef since c.1850*

Pollutant	Increase (%)
Sediment loads	300 to 900
Phosphate discharges	300 to 1500
Total nitrogen load	200 to 400
Pesticide residues	now being detected in tidal sediments

Source: Great Barrier Reef Marine Park Authority (2001)

adjacent to the reef has changed significantly in the past 150 years. Grazing has become widespread and large areas of natural vegetation have been cleared to make way for crops. The area used for sugar cultivation in the Great Barrier Reef catchments rose from about 100 000 ha in 1930 to nearly 400 000 ha in the mid-1990s. Pollutants from these catchments, transported to the reef mostly during flood events, are continuing to increase and the decline in water quality has a real potential to adversely affect industries such as tourism, and recreational and commercial fishing. Recognition of these dangers led to the recommendation in 2001 of specific end-of-river pollution targets for all 26 river catchments adjacent to the Great Barrier Reef. Catchments identified as high risk include those of the Johnstone, Herbert and Proserpine rivers. The targets for 2011 represent a first step designed to halt the increase in water pollution. More ambitious targets will be required in the longer term to reverse the overall decline in water quality over the past 150 years.

Although the anthropogenic impact is unmistakable in the above examples of coral reef stress, two other recent issues surrounding reef degradation are not so clear-cut. These are the population explosion of the predatory crown-of-thorns starfish (*Acanthaster planci*) since the 1960s, and the coral-bleaching phenomenon since the 1980s (Salvat, 1992).

Crown-of-thorns populations exploded at the end of the 1950s and throughout the 1960s in many parts of the Indian and western Pacific oceans, causing widespread damage to reefs. No less than 90 per cent of the fringing reefs of Guam and 14 per cent of the Great Barrier Reef were destroyed, the dead corals being replaced by an algae- and sea urchin-dominated community. Initially the finger of suspicion for the outbreaks was pointed at human action, through the destruction of starfish predators and the effects of pollution on larvae. Subsequently, however, researchers have tended to view the phenomenon as part of essentially natural cyclical events.

Similar uncertainty surrounds the widely reported phenomenon of coral bleaching, which has resulted in the widespread mortality of reefs, particularly in the Caribbean. While bleaching commonly occurs naturally when sea surface temperatures rise during events such as El Niño, a worldwide overview concluded that bleaching events were much more usual in the 1980s (Williams and Bunkley-Williams, 1990), leading to suggestions of links with global warming.

The uncertainty surrounding these two issues of coral reef vitality highlights the general lack of a historical perspective on the environmental parameters affecting reefs and their population dynamics. A heightened realisation of the important role of corals in the marine ecosystem, their high biodiversity and their importance to local communities should help to turn concern at their degradation into action towards managing these key habitats more sustainably. Such management is needed most urgently for those atoll island states whose future is threatened by the altered environmental dynamics of a warmer world (see Chapter 11). In response to global sea-level rise, some reefs appear to 'keep up' with rising water levels, others 'catch up' after a lag period by rapid vertical accretion of coral, while others are terminally affected by initial drowning and effectively 'give up' (Neumann and Macintyre, 1985).

Mangroves

Mangrove forests are made up of salt-adapted evergreen trees, and are found in intertidal zones

TABLE 7.5 *Extent of mangroves and major causes of destruction in selected Asian and African countries, late 1980s*

Country	Extent (km^2)	Pre-agricultural extent (km^2)	Major causes of destruction
India	3100	12600	Agriculture, urban development
Philippines	1000	4500	Charcoal production and fuelwood, fish ponds
Singapore	<5	75	Urban/industrial development
Viet Nam	1600	3800	Herbicide spraying during Viet Nam War, agriculture and fish ponds
Côte d'ivoire	29	1600	Fuelwood
Ghana	3	2100	Charcoal production and fuelwood, salt extraction, wood for construction
Guinea	3000	4000	Agriculture, fuelwood
Guinea-Bissau	2400	3100	Agriculture
Sierra Leone	1000	6800	Agriculture, fuelwood, wood for construction

Sources. WRI (1986, 1990); World Bank (1989); Collins *et al.* (1991); Sayer *et al.* (1992); Jagtap *et al.* (1993)

of tropical and subtropical latitudes. These diverse and productive wetland ecosystems are under threat from human action in many parts of the world, despite their importance as breeding and feeding grounds for many species of fish and crustaceans, and their role in protecting coasts from erosion, among the many other functions of wetlands in general (see Table 8.5). Although estimates of mangrove area in the literature vary considerably, due to differing survey methods and definitions, Table 7.5 shows a number of countries in Asia and Africa where mangroves have declined significantly from their pre-agricultural extent. Increasing population numbers, and a general lack of protection and management appear to be the common denominators behind this destruction. Conversion to agriculture, fish ponds, urban/industrial uses, and destruction for fuelwood or charcoal production are the main causes.

Some of the largest losses along the African coastline have been in the west of the continent. The most serious threat to mangroves in Guinea, Guinea-Bissau and Sierra Leone, formerly some of the most extensive areas of this habitat in West Africa, is clearance and conversion of the land to rice farming, although exploitation for fuelwood also plays an important part. Estimates for Guinea suggest that about 0.25 million tonnes of

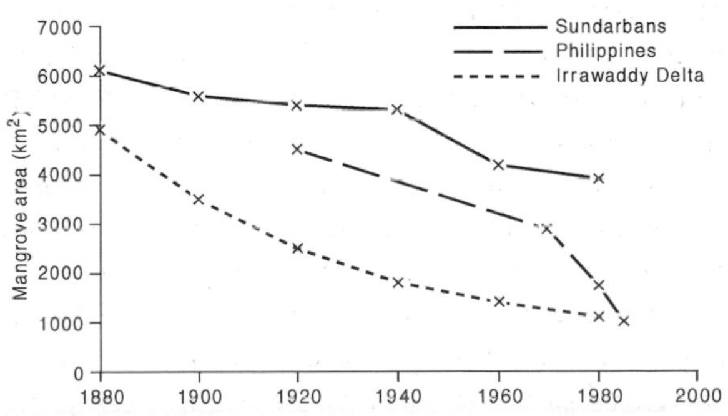

Figure 7.8 Decline of mangroves in the Sundarbans (Bangladesh and India), the Philippines and the Irrawaddy Delta (Myanmar) (from data in Richards, 1990; World Bank, 1989; Collins *et al.*, 1991).

TABLE 7.6 *Use of mangrove fuelwood in Guinea*

Use	Amount (t/year)
Urban	54 000
Rural	52 000
Fish smoking	58 000
Salt making	93 000
Total	257 000

Source: Bertrand (1993)

fuelwood per year are taken from the mangrove forests (Table 7.6). Clearance for fuelwood has been the major factor behind the almost total disappearance of mangroves in Ghana and Côte d'Ivoire (Sayer *et al.*, 1992).

In Asia, as Fig. 7.8 indicates, the clearance of mangrove forests for agriculture has been a progressive process over the last century or so in the delta regions of the Irrawaddy (Fig. 7.9) and Ganges rivers, while in the Philippines, where more than 75 per cent of the mangrove cover has been lost in less than 70 years, cutting for fuelwood and charcoal production has been

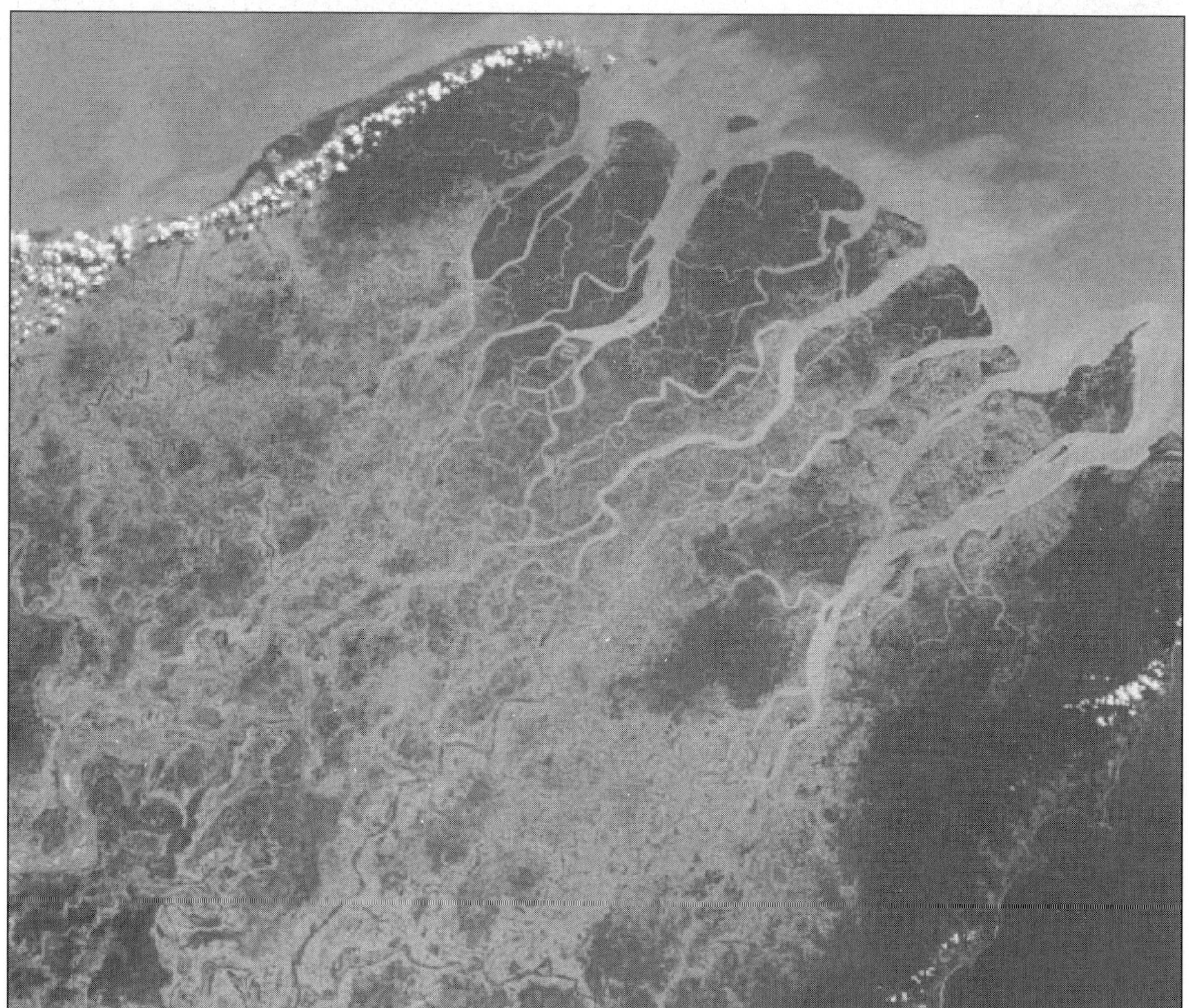

Figure 7.9 The extent of mangroves on the delta of the Irrawaddy River in Myanmar has progressively declined throughout the last century as the area devoted to rice cultivation has expanded. Remaining mangroves are the dark areas at the seaward end of the multi-channelled delta (Photo: NASA).

Figure 7.10 Captive fishing in the mangroves of the Gulf of Guayaquil, Ecuador. The area has also become the centre of a shrimp-farming industry, one of the country's major sources of foreign exchange. Since the mid-1960s, about 20 per cent of mangroves in the gulf have been cleared for shrimp production.

superseded as the most serious cause by clearing for the establishment of fish ponds (see page 229). Fish ponds grew to cover an area of 2050 km^2 in the Philippines in 1987 from 890 km^2 in 1952 (World Bank, 1989). Destruction of Indonesia's estimated 44 000 km^2 of mangroves began more recently, being little affected before 1975. Since then, however, conversion to fish ponds has been widely practised in Sulawesi, Java and Sumatra, and large areas have been cleared on many islands to produce wood chips, which are exported to Japan for cellulose and paper making.

The mangrove forests of the Ganges and Brahmaputra deltas, known as the Sundarbans, remain the largest mangrove ecosystem in the world, and parts of this area have been protected as a Reserved Forest since 1879. It is a key wildlife habitat for a large number of bird, mammal, reptile and amphibian species, and provides the last stronghold of the royal Bengal tiger. About half a million people also depend directly on the Sundarbans for their livelihood. New areas have been planted on offshore islands as part of the management plan, both for fuelwood harvesting and coastal protection against cyclonic storm surges, but the area is under increasing pressure from Bangladesh's growing population. A mass replanting programme has also been undertaken in post-war Viet Nam, successfully re-establishing many thousands of hectares destroyed by spraying with the herbicide Agent Orange; but large areas still remain damaged and there has been virtually no natural regrowth since the war.

Local populations use mangrove forests for a wide variety of their resources, including wood (for fuel, construction materials and stakes), leaves for fodder, fruits for human consumption, the rich mud that is used as manure, tannin extracted from barks, leaf extracts for medicinal purposes, and captive fisheries of prawns, shrimps and fish (Fig. 7.10). The questionable wisdom of destroying this diverse ecosystem is illustrated by data presented for Thailand by Christensen (1983), which show that when converted to rice paddies, a hectare of former mangrove yields US$168 a year, but an undamaged hectare farmed traditionally for shrimp produces US$206 a year and up to ten times that amount if carefully managed. This is quite apart from the mangrove's role in coastal protection, a particularly important one in the storm-prone, low-lying coastal zones where they are found. Realisation of the long-term economic benefits of careful management noted above has led to calls for the Thai government to devote as much attention to sustainability issues as it has to promoting intensive shrimp aquaculture, which has grown tremendously since the mid-1980s (see Fig. 13.9). Mangrove losses attributable to shrimp aquaculture are estimated to be between 16 and 32 per cent of the total mangrove area destroyed in Thailand between 1979 and 1993 (Dierber and Kiattisimkul, 1996).

A similar cost-benefit analysis has been made for the 8700 ha Sarawak Mangroves Forest Reserve in Malaysian Borneo by Bennett and Reynolds (1993), who calculate that the area is worth about US$25 million a year in forestry, marine fisheries and tourism revenues, as well as providing up to 3000 jobs in fisheries alone. Most of this value would be lost with conversion to an alternative form of land use, which would greatly reduce levels of revenue. Conversion to aquaculture ponds or oil palm plantations, the greatest threats, would also necessitate highly expensive engineering works to prevent coastal erosion, flooding and other damage, as well as causing immeasurable loss in terms of the reserve's wildlife importance.

Mangrove loss has also been widespread in Latin America in recent times, and as in some Asian countries much clearance has been to make way for aquacultural production of shrimps for export. The factors affecting this process are often complex, however, as a study of the Gulf of Fonseca on the Pacific coast of Honduras clearly shows (Stonich and DeWalt, 1996). The boom in shrimp aquaculture has been vigorously promoted since the early 1980s by the Honduran government, national and international private investors and several bilateral aid agencies. From the aid agencies' viewpoint, the shrimp export industry was a good way to tackle the country's economic crisis and a strong emphasis was put on integrating poorer households into the process mainly through the formation of cooperatives. Unfortunately, small-scale producers have not been profitable and attention has since shifted to larger-scale operations. Nearly 5000 ha of mature mangrove forests were lost along the Gulf of Fonseca coast in the period 1982–92 as the area of shrimp farms increased tenfold. Only about one-fifth of this mangrove loss was attributed to clearance for aquaculture, however, because among other things, mature mangrove stands are not the most appropriate areas for shrimp farming in this region.

However, a significant portion of the mature mangroves lost was cleared by peasant farmers, and the acid soils and difficulties of managing the shrimp ponds in these areas mean that the probability of failure is high. Further disadvantages for the area's poorer inhabitants have stemmed from the establishment of the larger shrimp operations in the more suitable mudflats. These areas were formerly used by artisanal fishermen when high tides left seasonal ponds rich in fish and shrimp. The transformation of these common property resources into private property, often controlled by foreign investors and national élites, has sparked violent confrontations and grassroots attempts to slow the rapid rise of shrimp aquaculture in the region.

MANAGEMENT AND CONSERVATION

Management of the coastal zone is necessary both to limit damaging activities and to protect coastal resources, as well as to restrict development in areas prone to natural hazards such as hurricanes, tsunamis, subsidence and inundation. On the national scale, Costa Rica has one of the world's most comprehensive and ambitious shoreland restriction programmes (Sorensen and McCreary, 1990) which covers a jurisdictional area 200 m wide in the marine and terrestrial zone. The law divides this zone into two components: the *zona pública* and the *zona restringida* (restricted zone). The *zona pública* extends inland 50 m from the mean high tide level or the inland limit of wetlands and the upstream limit of estuaries as defined by salt or tidal influence, and the *zona restringida* covers the remaining 150 m inland. The *zona pública* is devoted to public use and access, and commercial development is generally prohibited unless it is for enterprises that need to be in a coastal location, such as port installations. In the *zona restringida*, development is controlled by a permit and concessions system based on a detailed regulation plan formulated by local government. A concession is a development right on a specific area of land for a particular use over a fixed time period.

The importance of adequately managing coastal areas from ecological and economic viewpoints, which are often intimately related, can be illustrated at a regional level by Mauritania's coastal Parc National du Banc d'Arguin. The park covers an area of more than 1 million hectares and includes extensive tidal mudflats and extremely rich offshore fishing grounds. The site is internationally significant as the most important coastal wetland in Africa through its role as a breeding and wintering ground for millions of waterfowl: it hosts the largest concentration of wintering waders in the world. It is also a vital spawning ground and nursery for the fish that are Mauritania's main foreign currency earner. The area became a National Park in 1976, and has subsequently been listed under both the Ramsar and World Heritage Conventions. National and international conservation efforts focus on maintaining and strengthening the wise use of the mullet fishery by the Imraguen fishermen who reside in the park and use traditional harvesting methods in which they collaborate symbiotically with wild dolphins to catch grey mullet. The number of Imraguen boats has been limited to 100 in order to maintain a sustainable catch, but despite the Imraguen's exclusive fishing rights, the park's waters are under increasing pressure from pirate commercial fishing vessels. These are both small-scale fishers from Mauritania and Senegal, and industrial vessels, often from Europe. In response to these threats, the park and its partners have raised funds for three rapid patrol boats. Meanwhile, investigation is also being made into the possibilities of developing ecotourism to increase the use of the park's resources.

Sustainable use of resources is well established in the protected Bangladesh portion of the Sundarbans mangrove forest (Halls, 1997). Wood from the forest reserve provides raw material for the country's only newsprint paper mill, as well as a number of match and board mills and more local uses. The simple management regime involves a 20-year cycle of exploitation. The forest is divided into 20 compartments with each compartment harvested in turn and the cycle repeated at year 21. During the harvest, all trees above a certain diameter are removed as long as such removals do not create any permanent gaps in the forest canopy. In a subsequent operation, all dead and deformed trees are removed. Similar regimes control the use of other mangrove resources (Table 7.7). More than a century of sustainable management has caused very little change to the composition, quality and extent of the Sundarbans forest in Bangladesh.

The number of people actually living in the Sundarbans is relatively small and access by outsiders is strictly controlled, but in other coastal areas management problems are made more complex by large permanent human population densities and the consequent heavy pressures put on resources. The links between coastal problems and pollution sources outside the coastal zone also make integrated management over the entire watershed necessary, given the importance of rivers in contributing to pollution. The initiatives taken to improve the quality of Chesapeake Bay,

TABLE 7.7 *Sustainable resource management in the Sundarbans forest, Bangladesh*

Resource	Uses	Management
Timber	Fuelwood, poles, industrial wood, newsprint	20-year harvesting cycle
Nypa leaf	Roof thatch	3-year harvesting cycle
Honey	Food	Annual collection quotas
Fish	Food	Annual and seasonal catch quotas

Source: after Halls (1997)

the largest estuary in North America, are widely considered to be a good model for dealing with such complex pollution problems of coastal zones. The Chesapeake Bay Program, adopted by the six states and the District of Columbia which together have jurisdiction over all of the watershed, aims to reduce nutrient flows into the bay largely through subsidies for improved agricultural practices, reduction of urban runoff and treatment of sewage (USEPA, 1983). Although these measures have had some success in reducing pollutants, further efforts are deemed necessary, and the complex political situation highlights the possibility of conflict between polluter and polluted: 'it is not yet certain whether upstream landowners and governments are willing to restore the bay by restricting their own activities when downstream neighbors pay for the impacts of those activities' (WRI, 1992: 189).

Many of the management decisions made at any level need to balance conflicting interests and must do so with the cost-effectiveness of particular strategies in mind. These issues are illustrated in the small area of Britain shown in Fig. 7.1, where erosion continues to affect existing economic activity. Many farmers along the North Sea coast of Humberside are losing their fields at a rate of up to 2 m/year. The local council is sympathetic to calls for coastal protection schemes, but such projects are costly in financial terms and the priority given to individual farms is likely to be lower than that given to the gas terminals at Easington, which handle 25 per cent of Britain's gas supplies. An added complicating factor is the worry that sediment eroded from the North Sea coast is transported up the Humber estuary and deposited on the river banks. Halting the erosion could therefore increase the flood risk at Hull. The need for an integrated management plan for the coastal zone, which includes a thorough appreciation of marine and terrestrial sediment dynamics, is therefore paramount.

FURTHER READING

Ambio 1995 Research and capacity building for sustainable coastal management. Special issue 24(7–8). A collection of articles on coastal management issues, particularly in Sri Lanka and East Africa.

Anderson, D.M. 1994 Red tides. *Scientific American* 271(2): 52–8. Algal blooms, their effects on marine ecology and possible links to pollution.

Bird, E.F.C. 1985 *Coastline changes: a global review*. Chichester, Wiley. Many case studies of physical changes to coastlines on a country-by-country basis.

Burke, L., Selig, L. and Spalding, M. 2002 *Reefs at risk in Southeast Asia*. Washington, World Resources Institute. A regional assessment of the most biologically diverse coral reefs on the planet.

Hinrichsen, D. 1998 *Coastal waters of the world: trends, threats, and strategies*. Washington, Island Press. A very readable overview of the variety and value of the world's coastlines, and efforts to manage them.

Hutchings, P., Payri, C. and Bagrie, C. 1994 The current status of coral reef management in

French Polynesia. *Marine Pollution Bulletin* 29: 26–33. Case study of management efforts in a society heavily reliant upon its coral reef resources.

Jackson, J., Kirby, M., Berger, W., Bjorndal, K., Dotsford, L., Bourque, B., Brabury, R., Cooke, R., Erlandson, J., Estes, J., Hughes, T., Kidwell, S., Lange, C., Lenihan, H., Pandolfi, J., Peterson, C., Steneck, R., Tigner M. and Warner, R. 2001 Historical overfishing and the recent collapse of coastal ecosystems. *Science* 293: 629–38. A summary of evidence suggesting that historical overfishing lies behind the current problems on coastlines.

Viles, H.A. and Spencer, T. 1995 *Coastal problems*. London, Edward Arnold. A comprehensive review of all types of coastal problems, their origins and management.

Websites

http://www.coastalmanagement.com/ links to websites covering sustainable management, use and research on the world's coasts.

http://www.eucc.nl/ the European Union for Coastal Conservation is dedicated to the integrity and natural diversity of the coastal heritage, and to ecologically sustainable development.

http://coral.aoml.noaa.gov/ The US National Oceanic and Atmospheric Administration's coral health monitoring programme, providing a directory of coral reef scientists and links to other coral reef sites.

http://www.iclarm.org/ the World Fish Center is a food and environment research organisation that joins forces with farmers, scientists and policy-makers worldwide to help the rural poor increase their income, preserve their environment, and improve their lives.

http://www.earthisland.org/map/ official site of the Mangrove Action Project, dedicated to reversing the degradation of mangrove forests worldwide.

http://www.reefbase.org/ a global database on coral reefs and their resources.

http://www.wcmc.org.uk/marine/data/ the World Conservation Monitoring Centre's marine information site.

http://www.gbrmpa.gov.au/ the Great Barrier Reef Marine Park Authority site.

POINTS FOR DISCUSSION

- Do you agree that environmental problems in most coastal zones stem largely from the fact that there are simply too many people living near them?
- Explain why overfishing is thought to be the root cause of ecological problems on many coasts.
- Why might sea levels rise in coming decades, and what problems might this present to coastal environments?
- Why should we conserve (a) coral reefs and (b) mangroves?

8

RIVERS, LAKES AND WETLANDS

TOPICS COVERED
Water pollution
Rivers
Lakes
Wetlands

Key words: hydraulic civilisation, eutrophication, catchment management plan, Ramsar Convention

Fresh water plays an integral role in the functioning of all environments and societies. Since its earliest inception, human society has seen freshwater bodies as a vital resource, and entire ancient 'hydraulic civilisations' developed on certain rivers, notably those of the Tigris–Euphrates, the Nile and the Indus. Society is no less reliant on this fundamental natural resource today. Fresh water from lakes, rivers and wetlands is utilised for municipal, agricultural, industrial, recreational and power generation uses; it forms a convenient medium for transport and a sink for wastes.

Society's use of fresh water, its availability and quality, have thrown up numerous issues of controversy, and these issues will become more acute as global water withdrawals continue to increase. Estimates of the gross global withdrawals for the year 2000 are 5200 km^3, a nine-fold increase over the 1900 level, and this increase continues to accelerate (Shiklomanov, 1993). Humanity currently uses just over half of all reasonably accessible global runoff and about a quarter of total terrestrial evapotranspiration (Postel *et al.*, 1996). Global data mask regional differences, however, which are essentially functions of climatic influences. All countries suffer from periodic excesses of fresh water in the form of floods (see pages 366–71), while others also experience perennial shortages, in some cases due to a complete absence of permanent rivers (e.g. Malta and Saudi Arabia).

An idea of the wide variety of ways in which human activity can adversely affect freshwater ecosystems is given by noting the effects upon freshwater fish, a biological indicator of ecosystem health (Table 8.1). This chapter deals with surface freshwater ecosystems: rivers, lakes and freshwater wetlands, although the specific issue of big dams is covered in Chapter 9, and some of the effects of particular activities such as deforestation, urbanisation and mining on the quality and quantity of freshwater bodies are covered elsewhere (see Chapters 4, 10 and 19). Wetlands found in coastal regions, such as mangroves and estuaries, are covered in Chapter 7, and issues surrounding underground sources of water are covered in Chapters 10 and 13.

WATER POLLUTION

Sources of water pollution can be traced to all sorts of human activity, including agriculture, irrigation, industry, urbanisation and mining. Most forms of water pollution can be classified into three categories:

- excess nutrients from sewage and soil erosion
- pathogens from sewage

TABLE 8.1 *Summary of main pressures facing freshwater fish and their habitats in temperate areas*

Danger	Effects
Industrial and domestic effluents	Pollution, elimination of stocks, blocking of migratory species
Acid deposition	Elimination of stocks in poorly buffered areas
Land use (farming and forestry)	Eutrophication, acidification, sedimentation
Eutrophication	Algal blooms, deoxygenation, changes in species
Industrial development (including roads)	Sedimentation, obstructions, transfer of species
Warm water discharge	Deoxygenation, temperature gradients
River obstruction (dams)	Blocking of migratory species
Fluctuating water levels (reservoirs)	Loss of habitat, spawning and food supply
Infilling, drainage and canalisation	Loss of habitat, shelter, food supply
Water transfer	Transfer of species and disease
Water abstraction	Loss of habitat and spawning grounds, transfer of species
Fish farming	Eutrophication, introductions, diseases, genetic changes
Angling and fishery management	Elimination by piscicides, introductions
Commercial fishing	Overfishing, genetic changes
Introduction of new species	Elimination of native species, diseases, parasites
Water recreation	Disturbance, habitat loss

Source: after Maitland (1991)

- heavy metals and synthetic organic compounds from industry, mining and agriculture.

Three other forms should also be mentioned:

- thermal pollution from power generation and industrial plants
- radioactive substances
- turbidity problems caused by increased sediment loads or decreased water flow.

Organic liquid wastes can be broken down by bacteria and other micro-organisms in the presence of oxygen, and the burden of organics to be decomposed is measured by the biochemical oxygen demand (BOD). Liquid organic wastes include sewage, many industrial wastes (particularly from industries processing agricultural products) and runoff, which picks up organic wastes from land. As a river's dissolved oxygen decreases, with increasing loads of organic wastes, so fish and aquatic plant life suffer and may eventually die. Heavy volumes of organic wastes can overload a riverine system to the point at which all dissolved oxygen is exhausted.

Pollution of water due to temperature increase, so-called 'thermal pollution', also reduces its dissolved oxygen content. This occurs in two ways. An increase in water temperature decreases the solubility of oxygen on the one hand, and increases the rate of oxidation, thereby imposing a faster oxygen demand on a smaller content, on the other. Thermal pollution also has a number of more direct adverse effects on river ecology, including a general increase in undesirable forms of algae, and reduced reproduction and growth of some species of fish (Langford, 1990).

Figure 8.1 A channelised urban watercourse, polluted by all forms of human waste, in a low-income area of Santo Domingo, Dominican Republic. Such open sewers, common in many developing countries, represent a significant health hazard (Photo: David Howard).

Inorganic liquid wastes become dangerous when not adequately diluted. Even in very small concentrations, however, some heavy metals (such as cadmium, lead and mercury) are particularly dangerous and can bioaccumulate up the food chain, ultimately damaging human health.

Pathogens from human waste spread disease and represent the most widespread contamination of water (Fig. 8.1). Water-related diseases can be classified into those that are water-borne (e.g. diarrhoea, cholera and polio), those that are related to a lack of personal cleanliness (e.g. trachoma and typhoid) and those that are related to water as a habitat for certain disease vectors (e.g. schistosomiasis, malaria and onchocerciasis).

Data for freshwater quality are generally sparse, cover limited periods, and are subject to changes in analytical methods and the movement of gauging stations, but a coordinated effort at worldwide monitoring has been made within the framework of UNEP's Global Environmental Monitoring System, or GEMS (UNEP/WHO, 1988). The network, launched in 1977, comprises 240 river stations, 43 lake stations and 61 groundwater stations.

Meybeck *et al.* (1989) indicate how particular sources of water pollution have ebbed and flowed with time in the highly industrialised countries. The growth of urban areas during the Industrial Revolution created the first wave of serious water pollution, from domestic sources, around the turn of the last century. These have been superseded by industrial pollutants, and during the second half of the twentieth century by nutrient pollution (particularly from agricultural sources) and micro-organisms towards the end of the 1990s. A similar pattern of peaks in serious pollution problems is being experienced in the rapidly industrialising countries, but compressed into the post-war decades, and it is here that some of the most serious water pollution problems are being faced today. The other regions where water pollution has reached crisis proportions are in the countries of the former Soviet bloc. As with many other pollution issues, it is often the poorer sectors of society that suffer most from the detrimental effects of poor water quality.

As Meybeck *et al.*'s (1989) analysis suggests, most sources of water pollution are well known, and methods for reducing them or their detrimental effects have been devised. Most of today's continuing water pollution problem areas are due to a lack of political will, poor coordination between governments and/or a lack of funds.

RIVERS

Multiple impacts

In many rivers ecological degradation is caused by multiple human impacts. The extreme vulnerability of freshwater fish in Madagascar, where four

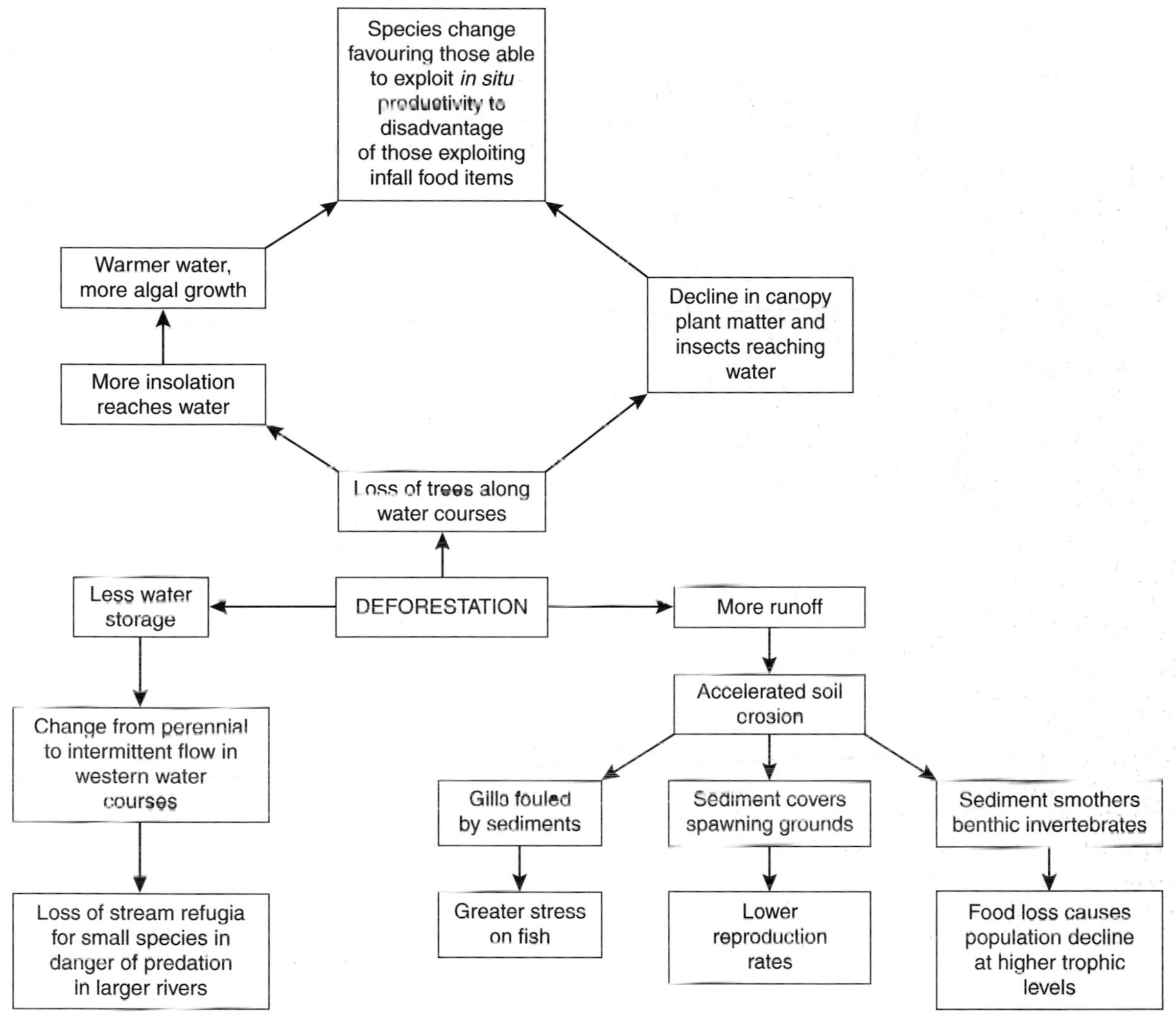

Figure 8.2 Degradation of aquatic habitats due to deforestation in Madagascar (after Benstead *et al.*, 2000).

of the 64 endemic freshwater species are feared extinct and another 38 endangered, have been brought to this point by three main factors: habitat degradation due to deforestation, overfishing, and interactions with exotic species (Benstead *et al.*, 2000).

The numerous ways in which Madagascar's widespread deforestation has contributed to the degradation of aquatic habitats are shown in Fig. 8.2. The effects consequent upon the decline of falling food items (plant matter and insects) and the enhanced growth of algae due to warmer waters are frequently dramatic in the short term as fish that feed on falling food are edged out by those able to feed on algae. But these effects need not be irreversible as other tree species colonise river banks to replace primary forest, providing a new source of insects. The changes caused by a more regional decline in water storage and greater runoff are less readily reversed because the forest is usually cleared for crops that have a dampened effect on hydrology.

Problems caused by overfishing and exotic species are also more intractable. Given the rising demand for fish from a rapidly increasing human population and the great logistical difficulties

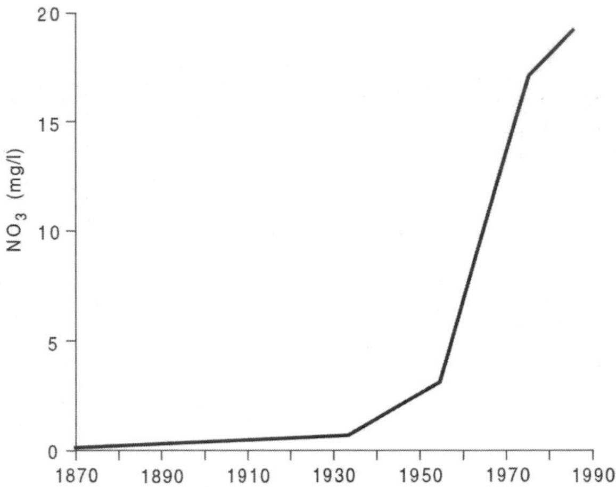

Figure 8.3 Nitrate concentration in the River Danube in Hungary, 1870–1985 (after Hinrichsen and Láng, 1993).

faced in enforcing any sort of environmental regulations, overfishing is likely to remain an issue in Madagascar for a long time to come. Exotics introduced to the island include both aquacultural and ornamental species, and their impact on aquatic ecosystems has been profound. Naturalised exotics have completely replaced native fish in the central highlands of Madagascar and they are widespread in other parts of the island (Stiassny and Raminosoa, 1994).

River pollution

The World Bank (1992a) found from a sample of GEMS monitoring sites, classified according to the development status of countries, that through the 1980s, high-income countries had seen some overall improvement in river water quality, middle-income countries had on average shown no change, and low-income countries experienced continued deterioration.

The major sources of river pollution in the USA are monitored by the Environmental Protection Agency (EPA). In 1988, based on a 30 per cent sample of US river miles, no less than 55 per cent of total river length was deemed to be impaired by pollution from agricultural sources. Municipal pollution impaired 16 per cent of river length, mining 13 per cent and habitat modification 13 per cent (USEPA, 1990). Other sources of pollution, all of which impaired less than 10 per cent of total river length, were storm sewers/runoff, silviculture, industrial, construction, land disposal and combined sewers. As far as causes of pollution are concerned, siltation, nutrients, pathogens, organic enrichment, metals and pesticides were the major culprits.

Agriculture is the leading non-point source of pollutants such as sediments, pesticides and nutrients, particularly nitrogen and phosphorus. The growing use of chemical fertilisers, fuelled by increasing demands for food, has led to a steady rise in the concentration of nitrates in many world rivers and groundwaters in recent decades. It has been estimated that less than 10 per cent of rivers in Europe, for example, are pristine with regard to nitrate concentrations (Meybeck, 1982). While nitrate pollution of waters, linked to cancers of the digestive tract and so-called 'blue baby syndrome', do not currently pose a serious threat to public health, the continued rapid rise in concentrations may be a cause for concern in the future. The startling rise in nitrate concentration in the Hungarian Danube indicated in Fig. 8.3, which mirrors curves in fertiliser usage, is fairly representative of many European rivers. The guideline for drinking water quality in Hungary is 20 mg NO_3/l. Data on pesticide contamination of rivers are not sufficient to give a clear picture of the situation even in some of the better-monitored parts of the world, but rising levels of usage give reasonable cause for concern, particularly in the developing countries where many pesticides now banned in the developed world are still used (see page 223).

Riverine pollution from sewage is certainly an acute problem facing many of the world's poorer countries where sewage treatment is inadequate or non-existent, but even in some of the richer nations the level of treatment is still far from satisfactory. In 1990, the proportion of national population served by primary, secondary or tertiary facilities was just 21 per cent in Portugal and 42 per cent in Japan, for example (OECD, 1993).

TABLE 8.2 *Water quality in the Mississippi*

Dissolved concentration (μg/l)	1970	1975	1980	1985	1991
Lead	18.00	1.50	0.42	4.90	1.00
Cadmium	18.00	0.36	1.40	1.00	1.00
Chromium	18.00	0.36	2.50	1.00	1.80
Copper	98.00	4.10	4.10	5.70	3.80

Source: data from OECD (1993)

The combined effects of various pollutants have had serious impacts in many world rivers. In Metro Manila, capital of the Philippines, for example, just one in ten households was served by a wastewater disposal system in the mid-1980s and more than half of the country's industry is located in the city. All Metro Manila's rivers are biologically dead and virtually unnavigable due to pollution and heavy siltation (World Bank, 1989).

Similarly dramatic pollution black-spots have been recorded in many of the countries of eastern Europe. In Poland, where, biologically, 75 per cent of the rivers were dead in 1988 (Carter, 1993b), the River Vistula, once known as the Queen of Rivers, is now a potent symbol of environmental degradation. Most of Poland's large towns and much of its industry are located along the banks of the Vistula, and its waters provide the country's most important source of drinking and industrial water, and a sink for domestic and industrial wastes. Concentrations of pollutants in the Vistula are high and rising. Great increases in eutrophication and organic pollutants were recorded in the 1960s, and although the rate of increase slowed for several pollutants in the 1980s, nitrogen and phosphorus are still rising at alarming rates (Kajak, 1992). About 70 per cent of the sewage entering the river is untreated and the rest is treated unsatisfactorily. The most important pollutants are nitrogen, phosphorus, phenols, NaCl and SO_4. NaCl comes mostly from coal and sulphur mines, nitrogen and phosphorus from sewage as well as dispersed sources, mostly agricultural. Carter (1993b) also notes that the mercury content of the Vistula below Cracow in 1980 was more than 200 times the permitted norm and that the Vistula is now so poisoned or potentially corrosive that stretches are considered unusable even for factory coolant systems, let alone drinking water.

In places where the economic power and political will are available, however, concerted efforts have been made to improve river water quality with some notable successes. Twenty-one years of records on the Mississippi River, for example, have highlighted the decline in dissolved concentrations of several heavy metals (Table 8.2) and similar success in reducing contamination has been achieved in Europe's Rhine in recent decades. Lelek (1989) studied the health of the Rhine using a three-fold richness classification (abundant, rare, sporadic) for more than 50 of the river's fish species. He showed that 27 species changed to an inferior status of abundance between 1890 and 1950, and 17 between 1951 and 1975, whereas 18 species changed to a superior category of abundance between 1976 and 1985, and just two species diminished during this most recent period.

In terms of organic wastes, the history of the River Thames provides a clear example of the reversibility of many sorry stories of riverine pollution. The quality of the Thames declined in the nineteenth century as London grew and the flushing water closet became widely used, discharging into the river. Five cholera epidemics occurred between 1830 and 1871 and during the

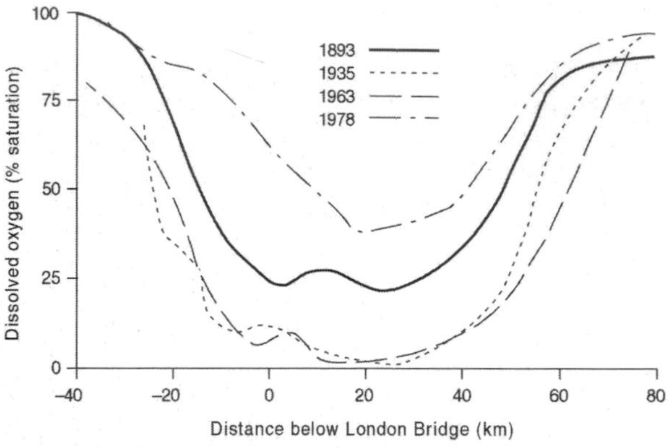

Figure 8.4 Dissolved oxygen levels in the River Thames estuary in four years between 1893 and 1978 (after Wood, 1982).

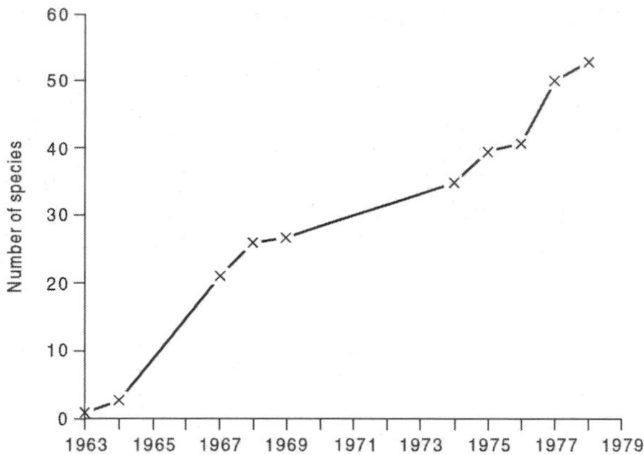

Figure 8.5 Increase of fish diversity in the River Thames estuary as monitored by collections from the West Thurrock power station cooling water intake, 1963–78 (from data in Ellis, 1989).

long, dry summer of 1858, the so-called 'Year of the Great Stink', Parliament had to be abandoned on some days because of the stench. Such a direct impact on the nation's legislators engendered positive action, however, and conditions had improved by the 1890s, when oxygen levels were up to 25 per cent saturation (Fig. 8.4) with the introduction of sewage treatment plants. During the first half of the twentieth century, however, sewage treatment and storage did not keep pace with the growing population, and dissolved oxygen reached zero at 20 km downstream of London Bridge during many summers.

Following tighter controls on effluent and improved treatment facilities introduced in the post-war years, improvement was gradual, with 1976–77 generally taken as the time when water quality reached satisfactory levels. The general improvement in the quality of the Thames' water is indicated by the increased diversity of fish monitored at the water intake of power stations (Fig. 8.5). Much publicity accompanied the landing of the first salmon caught in the Thames since 1833 off the West Thurrock screen in 1974.

River management

River modification and management are rooted in society's view of rivers both as hazards to be ameliorated and resources to be maximised. There are numerous ways in which people have modified rivers: by constructing dams, building levées, widening, deepening and straightening channels (Table 8.3; see also Chapters 9 and 21). The beginnings of direct river modification in the UK, for example, can be traced back to the first century AD with fish ponds and water mills, and changes to facilitate transportation and to effect land drainage (Sheail, 1988). Severe pollution in the mid-nineteenth century led to an organised system of river management. A marked east–west rainfall gradient, combined with an uneven population distribution (a high proportion live in eastern and south-eastern England where the rainfall is low at 600–700 mm/year) means that many rivers in central and southern England, and many urban rivers throughout the UK, are overexploited. Today, most towns and cities rely on interbasin water transfers, and virtually all the major rivers in the UK are regulated directly or indirectly by mainstream impoundments, interbasin transfers, pumped storage reservoirs or groundwater abstractions (Petts, 1988).

Ideally, river management should pursue four objectives (Mellquist, 1992):

1 balancing between users' interests
2 optimisation of resource use

TABLE 8.3 *Selected methods of river channelisation and their US and UK terminologies*

Method	US term	UK equivalent
Increase channel capacity by manipulating width and/or depth	Widening/deepening	Resectioning
Increase velocity of flow by steepening gradient	Straightening	Realigning
Raise channel banks to confine floodwaters	Diking	Embanking
Methods to control bank erosion	Bank stabilisation	Bank protection
Remove obstructions from a watercourse to decrease resistance and thus increase velocity of flow	Clearing and snagging	Pioneer tree clearance; control of aquatic plants; dredging of sediments; urban clearing

Source: after Brookes (1985)

3 inclusion of environmental interests and those of the general public when exploiting water resources
4 cleaning up after past abuses.

In practice, these objectives can be conflicting, and the relative weight given to each by decision-makers is affected by a wide range of influences that include economic, political and environmental considerations. River management is no different from any other natural environmental management issue in that it involves compromises. Some conservationists, for example, argue that river regulation and environmental conservation are intrinsically incompatible since regulation modifies the natural environment in which original communities of organisms became established (e.g. Hellawell, 1988). Indeed, in some cases the ecological requirements of organisms are destroyed or modified beyond the limits of adaptations and the organisms are unable to survive.

Other conservationists, however, adopt a different view of the situation. Moore and Driver (1989) point out that in England and Wales there are more than 500 water supply reservoirs, with a total water surface of 20 000 ha, much of which flooded in the twentieth century. Waterfowl, amphibious and aquatic wildlife have all benefited from this change, one that is especially important after so much wetland has been drained and ploughed for increased agricultural production. The general importance of reservoirs to wildlife conservation is indicated by the designation of 174 reservoirs as Sites of Special Scientific Interest (SSSIs). Modern reservoirs include landscaping and conservation in their planning, construction and operation, while protection from recreational activity, a use of resources that is potentially damaging to wildlife, is an essential part of management at these sites.

The importance of ecological good health in river systems has been increasingly recognised in recent times, both from the purely moral standpoint and from the point of view of human society's use of their resources. There are numerous ways in which damage done in the past can be rectified, many of which are not new in themselves. In 1215, the Magna Carta demanded the removal of numerous weirs along the River Thames so that migratory Atlantic salmon and sea trout could pass upstream to their spawning grounds. There was, however, little response to this edict nor to similar statutes issued in the fifteenth century. The Thames Water Authority has installed fish passes on navigation weirs as part of a salmon rehabilitation programme initiated in 1979 (Mann, 1988). Extensive damage to migratory salmonid fish stocks has also occurred in Finland, caused by the dredging of rivers and brooks both to facilitate boat traffic and, increasingly during the last century, for timber-floating. Timber-floating has now almost completely ceased and Finnish water legislation obliges water authorities to make good any damage caused by dredging. The restoration of rapids and their restocking have been increasingly used to rehabilitate damage caused by dredging (Jutila, 1992).

A critical final point to make with regard to river management, however, is the simple fact that management of the channel alone is usually insufficient. It is important to realise that rivers are an integral part of the landscapes through which they flow. Rivers affect the land in their drainage basins and vice versa. Hence, management of human activities in the entire basin, embodied in 'catchment management plans', is the sensible way forward, as Burt (1993) points out in the UK context.

LAKES

Lacustrine degradation

As with rivers, human use of fresh water from lakes has led to numerous impacts. In some cases, entire lakes have dried up as rivers feeding them have been diverted for other purposes. Owens Lake in California is a case in point. Levels began to drop in the second half of the nineteenth century due to offtakes for irrigated agriculture, but an accelerated decline in water levels occurred with the construction of a 360-km water export system by the Los Angeles Department of Water during the first 20 years of the last century. The lake, which was 7.6 m deep in 1912, had disappeared by 1930. In its place, the dry bed of lacustrine sediments covering 220 km^2 is a source of frequent dust storms, which are hazardous to local highway and aviation traffic, often exceed California State standards for atmospheric particulates and have increased morbidity among people suffering from emphysema, asthma and chronic bronchitis (Reinking *et al.*, 1975). From 1930 to 1969 the Los Angeles system was extended northwards to tap streams flowing into the saline Mono Lake, the level of which had dropped by 14 m between 1941 and 1981, threatening the numerous species endemic to the closed basin lake.

Similar offtakes, in this case for agriculture, have had dramatic consequences for the Aral Sea in Central Asia. Diversion of water from the Amu Darya and Syr Darya rivers to irrigate plantations, predominantly growing cotton, has had severe impacts on the inland sea since 1960. Expansion of the irrigated area in the former Soviet region of Central Asia, from 2.9 million hectares in 1950 to about 7.2 million hectares by the late 1980s, was spurred by Moscow's desire to be self-sufficient in cotton. As a result, inflow to the lake from the two rivers, the source of 90 per cent of its water, was virtually halted in the 1980s. In 1960, the Aral Sea was the fourth largest lake in the world, but since that time its surface area has been halved (Fig. 8.6), it has lost two-thirds of its volume, its water level has dropped by more than 16 m and its salinity has increased to reach that of sea water in the open ocean (Table 8.4).

These dramatic changes have had far-reaching effects, both on-site and off-site. The Aral Sea fishing industry, which landed 40 000 t in the early 1960s, has completely ceased to function as most of its native organisms have died out (Williams and Aladin, 1991). The delta areas of the Amu Darya and Syr Darya rivers have been transformed due to the lack of water, affecting flora, fauna and soils, while the diversion of river water has also resulted in the widespread lowering of groundwater levels (Khakimov, 1989). The receding sea has had local effects on climate, and the exposed sea bed has become a dust bowl from which an estimated 43 million tonnes of saline material is deposited on surrounding areas each year. This dust contaminates agricultural land up to several hundred kilometres from the sea coast, and is suspected to have adverse effects on human

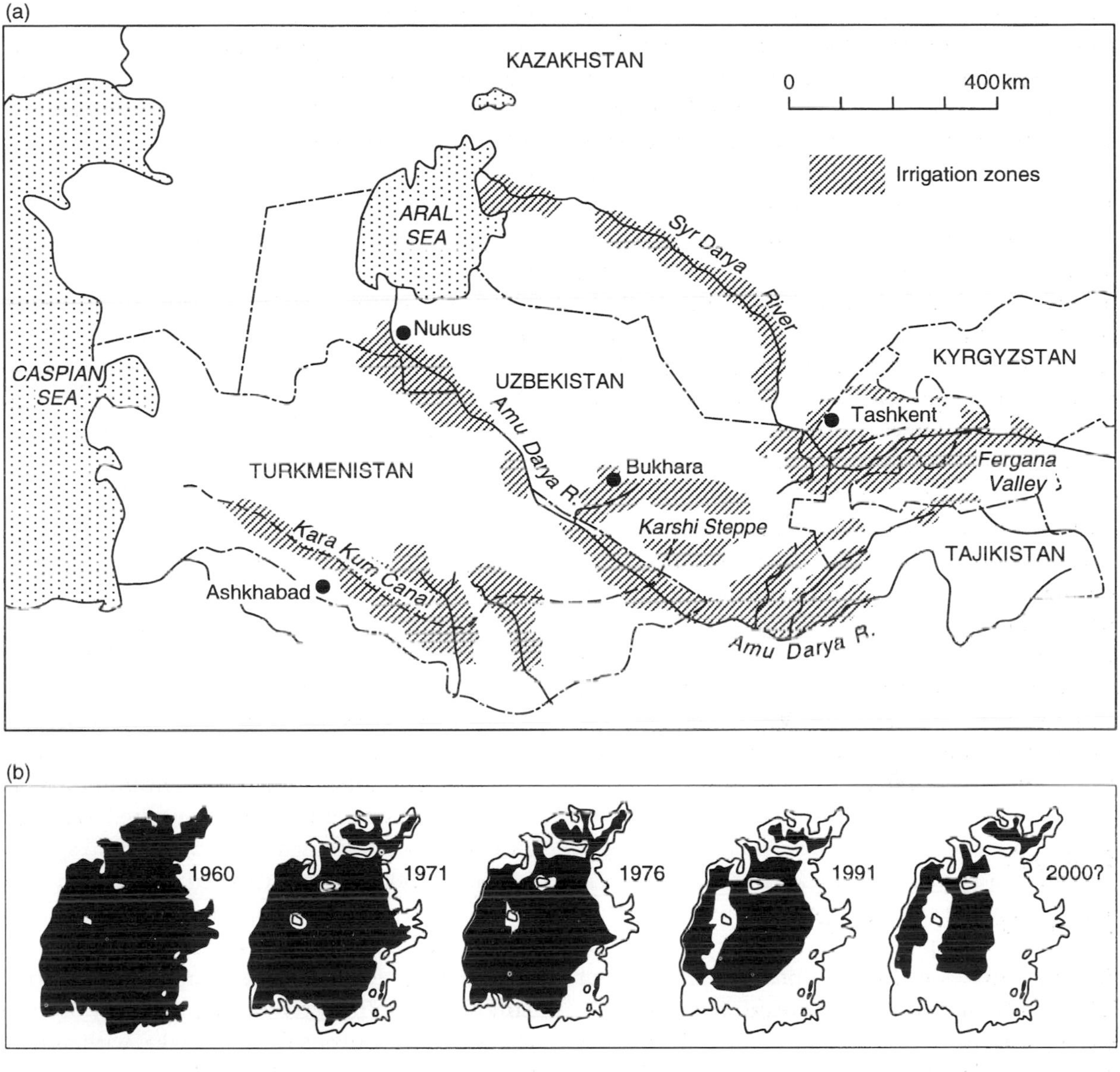

(a)

KAZAKHSTAN
0 400km
Irrigation zones
ARAL SEA
Syr Darya River
Nukus
CASPIAN SEA
UZBEKISTAN
KYRGYZSTAN
Amu Darya R.
Tashkent
TURKMENISTAN
Bukhara
Fergana Valley
Kara Kum Canal
Karshi Steppe
TAJIKISTAN
Ashkhabad
Amu Darya R.

(b)

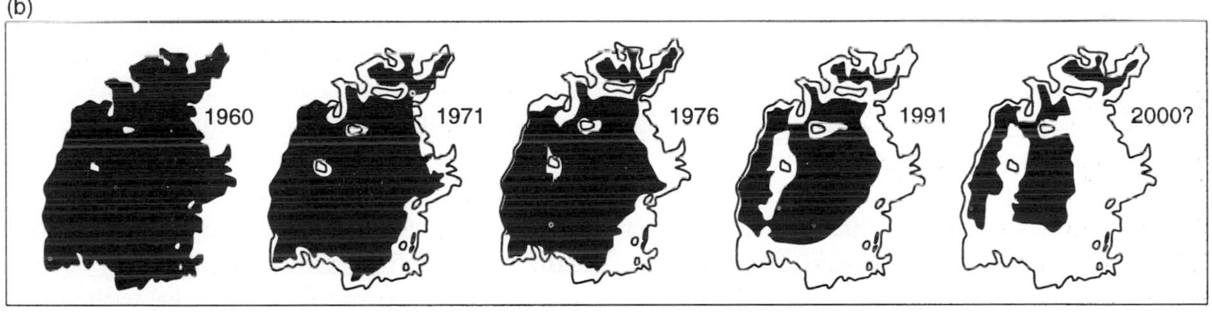

Figure 8.6 (a) Irrigated areas in Central Asia (after Pryde, 1991) and (b) changes in the surface area of the Aral Sea (UNEP, 1992b).

health. The irrigated cropland itself has been subject to problems of salinisation and waterlogging due to poor management, with consequent negative effects on crop yields (Smith, 1992). Drainage water from these schemes is characterised by high salinity and is contaminated by high concentrations of fertiliser and pesticide residues, which have also been linked to poor human health in the region (Glazovsky, 1995).

The ecological demise of the Aral Sea and its drainage basin is one of the late twentieth century's foremost examples of human-induced environmental degradation. But it was not the result of ignorance or lack of forethought, since anecdotal or informal projections of disaster were made in some Soviet quarters well before the situation became serious. Seemingly it is the result of an unspecified cost-benefit analysis in which

TABLE 8.4 *Hydrological parameters of the Aral Sea, 1960–2000*

Year		Average water level (m)	Average surface area (km²)	Average volume (km³)	Average salinity (g/l)
1960		53.4	66 300	1090	10
1970		51.6	60 400	970	11
1980		46.2	52 400	670	18
1990		37.8	34 800	304	33
2000?	Big Aral	32.4	21 000	159	>45
	Little Aral	41.0	3 200	24	30

Source: Glantz *et al.* (1993); Glazovsky (1995)

environmental and health impacts were given little consideration (Glantz *et al.*, 1993). The desire to be self-sufficient in textiles, combined with the Soviet belief that human society was capable of complete domination over nature, overrode any fears of the associated environmental implications. It is an example of the dangers of an extreme technocentric approach to the environment and a complete rejection of the precautionary principle.

The introduction of non-native species to lakes is another way in which, often unwittingly, the human impact has caused serious ecological change (see the case study of the North American Great Lakes on page 271). Deliberately introducing fish species to lakes to provide employment and additional sources of nutrition was a common aspect of economic development projects during the 1950s and 1960s. Predatory species such as largemouth, bass and black crappie, introduced into the oligotrophic Lake Atitlan in Guatemala, succeeded in wiping out many of the smaller fish that had previously been used by indigenous lakeside people. Similar dramatic effects on the pelagic food web resulted from the introduction of the Amazonian peacock bass to Gatun Lake in the Panama Canal. One serious consequence of this predator's elimination of the lake's smaller native fishes was a dramatic increase in malaria-carrying mosquitoes, since the minnows that had previously occupied the shallower lakeside waters had fed off mosquito larvae (Fernando, 1991).

The relative merits, or otherwise, of ecological changes forced by exotic introductions depend to a degree on the stance taken towards environmental issues. In East Africa's Lake Victoria, the predatory Nile perch, introduced in the early 1960s, has consumed to very low numbers (probably in many cases to extinction) most of the lake's endemic openwater Haplochromine cichlid species. Before the introduction of the Nile perch, Lake Victoria contained about 300 cichlid species, but in just 30 years an estimated 200 of these have become extinct (Goldschmidt, 1996). This clear example of a human-induced mass extinction is inexcusable from an ecocentric perspective. However, the Nile perch has significantly boosted the economic output from the lake, with annual fish landings rising from about 40 000 t to 500 000 t over the 20 years to 1992, and the loss of some species may be the price that has to be paid (Viner, 1992). If this more technocentric view is adopted, however, it is sad to note that the Nile perch is already being exploited at its maximum sustainable yield, and with more processing plants planned it looks set to become overexploited, a fate that afflicts virtually all such new commercial fisheries. Lack of finance prevents the countries concerned from supporting adequate fishery regulatory bodies. One of the main lessons to be learned from Lake Victoria's experience, therefore, is that if a radical change is forced upon an ecosystem, the human component of it requires proper management to keep in step with that forced change.

Another range of problems common to many lakes are those stemming from the introduction of plant species, particularly waterweeds. The water hyacinth is a frequent culprit, growing very fast, covering the water surface (so reducing light and oxygen for other plants and for fish), increasing evaporation, blocking waterways and presenting a serious threat to hydroelectrical turbines. It became a major problem in Lake Victoria in the 1990s.

There are numerous examples of lakes being degraded inadvertently, many of which are due to lacustrine sensitivity to pollutants. Some examples of problems caused by pollution, particularly acidification from industrial and mining activity, are covered in Chapters 12 and 19. One of the world's most pervasive water pollution problems, and one that is widespread in all continents, is the eutrophication of standing water bodies (Ryding and Rast, 1989). Eutrophication is a natural phenomenon in lakes, brought on by the gradual accumulation of organic material through geological history. But human activity can accelerate the process, so-called 'cultural eutrophication', by enriching surface waters with nutrients, particularly phosphorus and nitrogen. Such increased nutrient concentrations in lakes have been attributed to wastewater discharge, runoff of fertilisers from agricultural land and changes in land use that increase runoff.

Eutrophication is an important water quality issue that causes a range of practical problems. These include the impairment of the following: drinking water quality, fisheries, and water volume or flow. The main causes of these problems include algal blooms, macrophyte and littoral algal growth, altered thermal conditions, turbidity and low dissolved solids. Health problems range from minor skin irritations to bilharzia, schistosomiasis and diarrhoea.

Monitoring of Lac Léman (Lake Geneva) indicates how the build-up of nutrient pollutants can be rapid (Fig. 8.7). Water quality deteriorated from its relatively clean state during the 1950s as concentrations of phosphorus increased, primarily from point-source discharges, to reach a critical stage in the late 1970s. The situation has improved, however, following the introduction of tertiary wastewater treatment plants and a ban on the use of phosphates in detergents in 1986 (Rapin *et al.*, 1989).

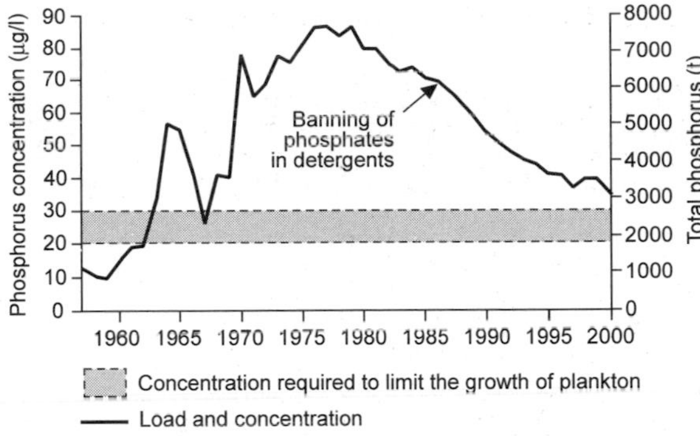

Figure 8.7 Phosphorus concentrations in Lake Léman, Switzerland, 1957–2000 (data from http://www.cipel.org/, accessed March 2002).

Lake management

The restoration and protection of lakes presents all sorts of problems for policy-makers and environmental managers, and the management experience of the North American Great Lakes illustrates the difficulties well. A major restoration effort has been in progress throughout the Great Lakes Basin since the US and Canadian governments committed themselves to the Great Lakes Water Quality Agreement in 1978. The agreement adopted an ecosystem approach, which incorporates the political and economic interests of the 'institutional ecosystem' as well as elements of the natural ecosystem, after two previous management strategies had failed to address fully the dangers faced by the lakes (Hartig and Vallentyne, 1989). Current efforts are focused on reducing pollution entering the lakes, incorporating the 'critical loads' approach adopted in studies of acid rain pollution (see page 202).

Serious pollutants include volatile organic compounds (VOCs), heavy metals, and industrial and agricultural chemicals, which come from both point and non-point sources such as municipal and industrial effluent, rainfall and snowmelt. Special and immediate attention is being given to 43 severely degraded so-called Areas of Concern.

The importance of integrating the ecological needs with those of the lake's users – which are, of course, intimately related – is most acute in lakes that support high population densities. However, the balance of interests also depends upon the nature of the lake and the type of overall regulatory body involved. The example of Lake Baikal in eastern Siberia, the deepest and volumetrically the largest freshwater lake in the world and home to 800 species of plant and 1550 species of animal life, most of which are endemic to its particularly pure waters, illustrates the point well. Theoretically, the state ownership and centralised decision-making of the former USSR was the ideal structure for successfully integrating ecological and user interests on such a unique freshwater body. In practice, however, as numerous examples of environmental problems quoted in this chapter and elsewhere in this book illustrate, the ecological side of the equation was too often given inadequate consideration.

Efforts to protect Lake Baikal against pollution, a struggle that dates from the mid-1960s, was the first major issue in the rise of public environmental awareness in the former USSR (Pryde, 1972). Although there have been small logging and industrial processing operations on the shores of the lake for several decades, concern was raised at the proposal to establish two large wood-processing plants. Great debates occurred over the possible effects of effluents from the plants and the effects of soil erosion resulting from deforestation of the lake shore to supply the plants. Public outcry succeeded in forcing the introduction of new measures to protect the lake's waters, including a ban on logging on surrounding steep slopes, the establishment of several protected areas around the lake, and the eventual conversion of one of the processing plants to a less-damaging manufacturing role. Nevertheless, the success of these measures is still to be evaluated and the real fate of Lake Baikal is still in the balance (Pryde, 1991).

WETLANDS

The term 'wetland' covers a multitude of different landscape types, located in every major climatic zone. They form the overlap between dry terrestrial ecosystems and inundated aquatic ecosystems such as rivers, lakes or seas. The Ramsar Convention (see below) uses a very broad definition, which has been widely accepted:

> *Wetlands are areas of marsh, fen, peatland or water, whether natural or artificial, permanent or temporary, with water that is static or flowing, fresh, brackish or salt, including areas of marine water the depth of which at low tide does not exceed six metres.*

This definition encompasses coastal and shallow marine areas (including coral reefs, which are covered in Chapter 7), as well as river courses and temporary lakes or depressions in semi-arid zones.

It is only in the past 30 years or so that wetlands, which currently cover about 6 per cent of the Earth's land surface, have been considered as anything more than worthless wastelands, only fit for drainage, dredging and infilling. But an increasing knowledge of these ecosystems in academic and environmentalist circles has yielded the realisation that wetlands are an important component of the global biosphere, and that the alarming rate at which they are being destroyed is a cause for concern.

Wetlands perform some key natural functions and provide a wealth of direct and indirect benefits to human societies (Table 8.5). Wetlands form an important link in the hydrological cycle, acting as temporary water stores. This function helps to mitigate river floods downstream, protects coastlines from destructive erosion and recharges aquifers. Chemically, wetlands act like giant water filters, trapping and recycling nutrients and other

TABLE 8.5 *Wetland functions*

Hydrology
Flood control
Groundwater recharge/discharge
Shoreline anchorage and protection
Water quality
Wastewater treatment
Toxic substances
Nutrients
Food-chain support/cycling
Primary production
Decomposition
Nutrient export
Nutrient utilisation
Habitat
Invertebrates
Fisheries
Mammals
Birds
Socio-economic
Consumptive use (e.g. food, fuel)
Non-consumptive use (e.g. aesthetic, recreational, archaeological)

Source: Sather and Smith (1984)

residues. This role is partly a function of the very high biological productivity of wetlands, amongst the highest of any world ecosystem (see Table 1.1). The importance of this function is indicated by the fact that the 6.4 per cent global wetland area is estimated to contribute 24 per cent of global terrestrial primary productivity (Williams, 1990a). As such, wetlands provide habitats for a wide variety of plants and animals. All these functions provide benefits to human societies, from the direct resource potential provided by products such as fisheries and fuelwood, to the ecosystem value in terms of hydrology and productivity, up to a value on the global level in terms of the role of wetlands in atmospheric processes and general life-support systems (Odum, 1979).

Wetland destruction

Estimates of the global area of wetlands drained vary considerably, but a total of 1.6 million km^2 by 1985 is considered by Williams (1990a) to be a reasonable figure. Nearly three-quarters of this total has occurred in the temperate world, and the prime motivation has been to provide more land to grow food. The wetlands of Europe have been subject to human modification for more than 1000 years. In the Netherlands, where particularly active reclamation and drainage periods occurred in the seventeenth, nineteenth and twentieth centuries, more than half of the national land area is now made up of reclaimed wetlands (Fig. 8.8), and Dutch drainage engineers have hired out their skills to neighbouring countries for more than 300 years.

In the USA, where wetlands are very largely concentrated in the east of the country, with major areas around the Great Lakes, on the lower reaches of the Mississippi River, and on the Gulf of Mexico and Atlantic seaboard south of Chesapeake Bay, about half of the pre-settlement area of wetlands had been lost by 1975 (Williams, 1990b). Agriculture has been the major beneficiary from the drainage operations in the USA, while urban and suburban development, dredging and mining account for much of the rest.

Although the rich soils of former wetland areas often provide fertile agricultural lands, conversion to cropland is not always successful. Drainage of the extensive marshlands around the Pripyat River in Belarus during the post-war Soviet period, for example, was largely unsuccessful. The low productivity of reclaimed areas necessitated the application of large amounts of chemical fertilisers, and some former wetland zones also became prone to wildfires (Pryde, 1991).

There have been numerous other reasons for wetland reclamation. On the national scale, in Albania, where about 9 per cent of the country was drained in the period 1946–83, reclamation was as much driven by the fight against malaria as the desire to increase food output, although most

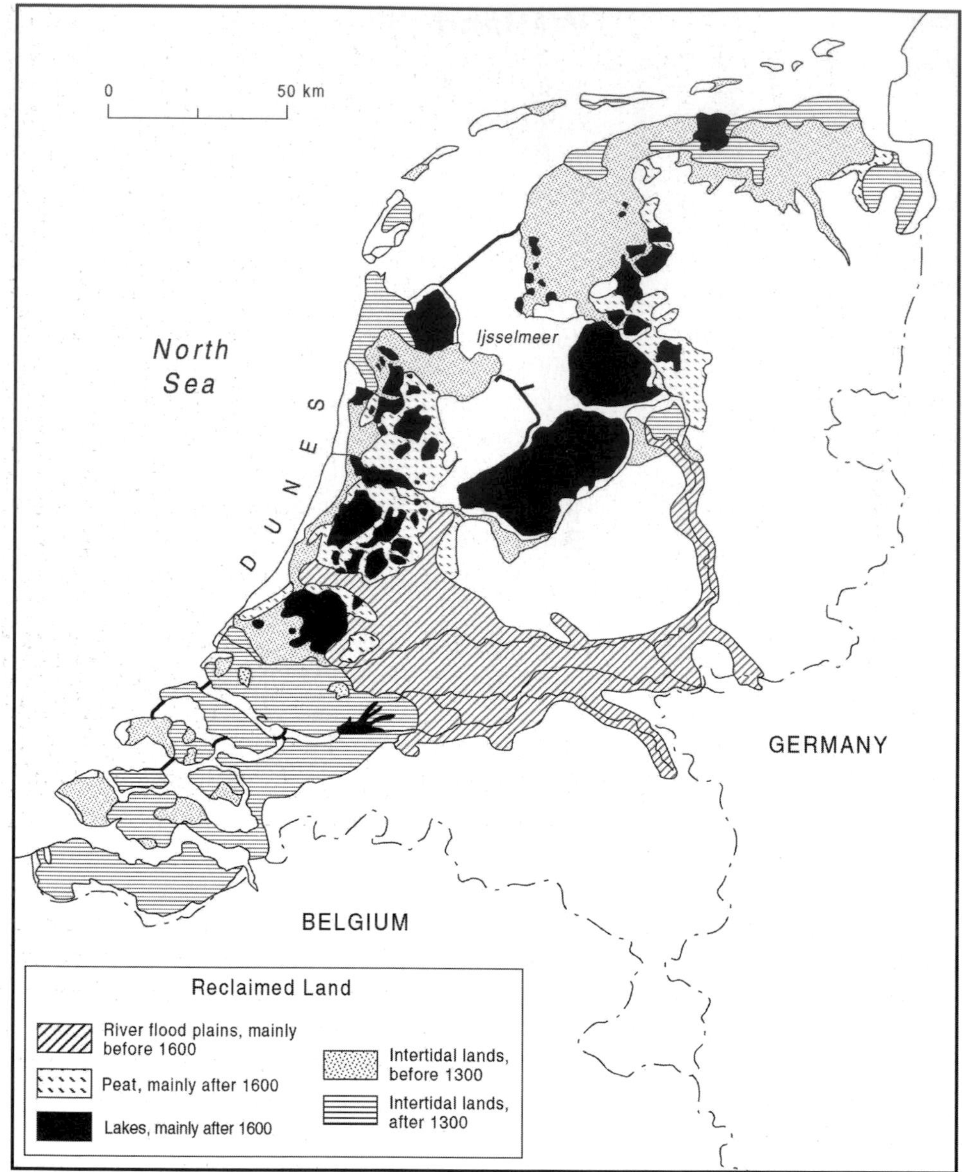

Figure 8.8 Reclamation of wetland types in the Netherlands (after de Jong and Wiggens, 1983).

of the former marshes were converted to irrigated agriculture (Hall, 1993). Urban sprawl has been a major cause in many areas. Mexico City, for example, was surrounded by five shallow lakes when the Spanish first arrived in 1519, but as the city has expanded since then, all but one small wetland area have been desiccated for their water and land. The remaining area at Xochimilco is threatened by the city's declining water table (Fig. 8.9).

In southern Sudan, the Jonglei Canal Project was designed to utilise the large proportion of the White Nile's discharge that currently feeds the Sudd Swamps. The 360-km artificial channel, which aims to bypass and effectively drain the swamps, will provide much-needed additional water to the dryland countries of Sudan and Egypt if it is ever completed – the project has been hindered by civil war in the region. Plans to utilise

Figure 8.9 Mexico City's last remaining area of wetland at Xochimilco, which is in danger of drying up due to the continuing need for water from the world's largest city.

the water resources of wetlands also threaten other sites, such as the inland Okavango delta in semi-arid northern Botswana. The general well-being of the Okavango is also under increasing pressure from cattle encroaching into the lush grazing of the delta, a threat currently held at bay by a 300-km cordon fence; but the lure of resources other than water and land has proved to be a serious cause of wetland degradation in many areas.

The fuel resources harboured in wetlands represent perhaps the most important reason in this respect. In many developing countries fuelwood collection is a prime cause of degradation. It is the major threat to wetlands in Central America, as it is in many coastal mangrove swamps in Africa and Asia (see page 114). In more temperate latitudes, peat is the most important organic fuel of the wetland habitat. Hand-cutting of peat has been practised in rural communities in northern Europe for centuries, and continues today in remote parts of Ireland, Scotland, Scandinavia and Russia, but since the 1950s, peat has been cut from much larger areas with machines. Worldwide production reached 100 million tonnes in the late 1970s (Kivinen and Pakarinen, 1981), most for fuel but with exploitation for horticultural purposes particularly important in some countries. Russia is by far the world's largest producer, with Ireland in second place, deriving 15 per cent of energy consumption from peat burning. In Ireland, 94 per cent of raised bogs and 86 per cent of blanket bogs have been lost, inspiring urgent conservation measures for remaining areas. Raised bogs only occur in eight countries, all in Europe, and while the habitat is close to extinction in the Netherlands and Germany, Eire still has about half of the world's remaining area of oceanic raised bogs worthy of conservation. Their conservation under the EU Habitats

Directive is thus a priority (Irish Peatland Conservation Council, 1998).

The arguments in favour of preserving peatlands take in all the reasons for wetland preservation in general (Table 8.5), and the importance of peatlands for global-scale stability is increasingly being realised. There is a growing body of evidence to suggest that peatlands play a key role in the initiation of ice ages, through feedbacks with geomorphology, surface albedo, atmospheric moisture and the concentration of greenhouse gases (Franzen, 1994). Another possible role of peatlands lies in their important relationship with nearby marine ecosystems. A link has been suggested whereby vital nutrients are exchanged between coastal marine ecosystems and adjacent peatlands. The biological productivity of oceans tends to be limited by a lack of key nutrients, particularly iron. Runoff from peat is rich in iron and probably makes an important contribution to marine productivity in many coastal areas. Productivity in peat ecosystems is often limited by a lack of sulphur, and sulphur from the oceans reaches peat wetlands through the atmosphere (Klinger and Erickson, 1997). One possible feedback resulting from the loss of peatlands is, therefore, decreased marine productivity, and since human populations derive most of their fisheries from coastal waters the consequences for society are clear.

Wetland protection

Realisation of the value of wetlands to human and animal populations has prompted moves to protect this threatened landscape in recent decades, although in many cases conservationists still face a difficult task in protecting wetlands from destructive development projects. In the case of the Göksu Delta on the south coast of Turkey, threatened by a proposed tourist complex, airport and shrimp farm, the Turkish Society for the Protection of Nature enlisted international help in publicising the danger to the delta, and successfully lobbied the Turkish Government, which in 1990 declared the delta a Protected Special Area. The Göksu Delta is a key European site for migrant and wintering birds, holding important populations of no less than 12 globally threatened species, and its beaches are among the main nesting sites for the two Mediterranean species of sea turtle. Following the designation of the delta as a protected area, studies are under way to develop a management plan for the area, but the threats to the Göksu Delta have not all disappeared since its ecology will be altered significantly by the proposed Kayraktepe dam.

A comprehensive national effort towards wetland conservation has been made in the USA, where numerous state and federal laws have a bearing on wetland protection. An attempt is also being made to reverse the loss with the Wetlands Reserve Program, established in 1990, which seeks to restore 405 000 ha of privately owned freshwater wetlands that have been previously drained and converted to cropland.

On the international front, one of the earliest attempts to protect a major world biome was the Ramsar Convention, designed to protect wetlands of international importance. Formally known as the Convention on Wetlands of International Importance especially as Waterfowl Habitats, it was adopted by 18 countries attending its first conference at Ramsar, Iran, in 1971. The convention aims to stem the progressive encroachment on, and loss of, wetlands, and represents one of the first attempts to impose external obligations on the land use decisions of independent states. Each contracting country's main obligation is to designate at least one wetland of international importance for inclusion in a formal list. In early 2003, this Ramsar List had 1263 sites covering 108 million hectares in 136 contracting countries.

The Ramsar Convention has always emphasised that human use on a sustainable basis is entirely compatible with Ramsar listing and wetland conservation in general (Halls, 1997). The Convention emphasises that wetlands should be conserved by ensuring their 'wise use', a term that is synonymous with 'sustainable use' (Table 8.6). Wise use is defined as sustainable utilisation for

TABLE 8.6 The Ramsar Convention's 'wise use' guidelines

Guideline	Action
Adopt national wetland policies	involving a review of existing laws and institutions designed to deal with wetland matters (either as specific wetland policies or as part of national environmental action plans, national biodiversity strategies, or other national strategic planning)
Develop wetland programmes	inventory, monitoring, research, training, education and public awareness
Take action at wetland sites	involving the development of integrated management plans covering every aspect of the wetlands

Source: Davis (1993)

the benefit of humankind in a way compatible with the maintenance of the natural properties of the ecosystem. Sustainable utilisation is understood as people's use of a wetland so that it may yield the greatest continuous benefit to present generations while maintaining its potential to meet the needs and aspirations of future generations. 'Wise use' may also require strict protection. A good example of more than a century of wise use in the Sundarbans mangrove forest of Bangladesh is shown in Table 7.7.

However, many important wetland sites are still not listed and the Convention lacks any legal powers, relying on persuasion and moral pressure to achieve its aims. It is in the developing countries of the world where most concern over the future of wetlands is focused. The rising resource needs of rapidly growing populations, particularly to expand food production, means that many unique wetlands remain under considerable pressure. As with rivers and lakes, sensible wetland management can only be undertaken in these circumstances by incorporating all activities and interest groups present in a particular drainage basin.

The plight of the Azraq Wetland Reserve in Jordan illustrates these points well. This 800 ha mosaic of shallow pools and marshland provides the only permanent water within 12 000 km^2 of desert and began to be degraded shortly after its designation as a Ramsar List site in 1977. The aquifer feeding the two springs that supplied the wetland with water has been heavily overexploited in recent times, exceeding the annual natural recharge rate of about 35 million m^3 (Dottridge and Abu Jaber, 1999) every year since 1983, to supply water to Amman, Zarqa and local irrigation schemes. In 1992, flow from the springs completely ceased and the wetland dried out, allowing slow-burning underground fires to ignite on the peaty soils. Some water has subsequently been restored to Azraq and a management plan has been devised to nurse the area back to health, involving local education and tourist facilities. Plans for a recharge dam have been developed (Al-Kharabsheh *et al.*, 1997) but the acute shortage of water in Jordan remains a threat to the long-term future of the reserve.

FURTHER READING

Adams, W.M. 1993 Indigenous use of wetlands and sustainable development in West Africa. *The Geographical Journal* 159: 209–18. A case study of the inland delta of the River Niger.

Boon, P.J., Davies, B.R. and Petts, G.E. (eds) 2000 *Global perspectives on river conservation: science, policy and practice*. Chichester, Wiley. A collection of 24 papers on conservation.

Glantz, M.H., Rubinstein, A.Z. and Zonn, I. 1993 Tragedy in the Aral Sea basin: looking back to plan ahead. *Global Environmental Change* 3: 174–98. A comprehensive overview of the physical problems and the political history of the Aral Sea disaster.

Gleick, P.H. (ed.) 2000 *The world's water (2000–2001)*. New York, Island Press. A collection of articles on key issues plus tables of data on freshwater resources.

Hall, S.R. and Mills, E.L. 2000 Exotic species in large lakes of the world. *Aquatic Ecosystem Health and Management* 3: 105–35. A review of introductions in 18 lakes on five continents.

Mason, C.F. 1996 *Biology of fresh water pollution*, 3rd edn. London, Longman. A good introduction to the causes and consequences of river and lake pollution, and techniques for pollution abatement.

Smol, J.P. 2002 *Pollution of lakes and rivers: a paleoenvironmental perspective*. London, Arnold. An overview of environmental changes in lakes, rivers and reservoirs with a particular focus on using proxy data in their sediments.

Williams, M. (ed.) 1990 *Wetlands: a threatened landscape*. Institute of British Geographers Special Publication 25. Oxford, Blackwell. A comprehensive appraisal of world wetlands from both physical and human perspectives. The book focuses on the nature of wetlands, the effects of human impacts, and strategies for their management.

Websites

http://www.irn.org/ the International Rivers Network site has information, campaigns and publications (e.g. *World Rivers Review*) on many riverine issues.

http://www.rivernet.org/ the European Rivers Network site has information and news on rivers worldwide, with an emphasis on Europe.

http://www.instreamflow.com/ *Rivers*, an online journal.

http://iksr.firmen-netz.de/icpr/ site for the International Commission for the Protection of the Rhine with information and data on this major European river.

http://www.amrivers.org/ a site covering all aspects of US rivers including annual lists of the most endangered.

http://ilec.or.jp/ site of the International Lake Environment Committee, which promotes sustainable management of the world's lakes and reservoirs; includes news and a large database.

http://ramsar.org/ the Ramsar Bureau's website covers reports, news, documents and archives on all aspects of wetlands.

http://www.wetlands.org/ Wetlands International, a leading conservation group dedicated to wetland conservation, includes the Ramsar database on its site.

POINTS FOR DISCUSSION

- Management of fresh water resources is vital for human society but manipulation of the hydrological cycle for our benefit always causes environmental problems. Discuss with specific examples how these problems can be minimised.
- What are the arguments for maintaining river water quality?
- If the Aral Sea tragedy is the result of a technocentric approach to using the natural environment, is an ecocentric approach the only way of dealing with the problems?
- How and why should wetlands be preserved?

9

BIG DAMS

TOPICS COVERED

The benefits of big dams
Environmental impacts of big dams
Political impacts of big dams

Key words: resettlement, reservoir sedimentation, saline intrusion, decommissioning.

People have constructed dams to manage water resources for at least 5000 years. The first dams were used to control floods and to supply water for irrigation and domestic purposes. Later, the energy of rivers was harnessed behind dams to power primary industries directly, and more recently still, hydroelectricity has been generated using water held in reservoirs, allowing the impoundment and regulation of river flow. The modern era of big dams dates from the 1930s and began in the USA with the construction of the 221-metre Hoover Dam on the Colorado River. But in the past 50 years there has been a marked escalation in the rate and scale of construction of big dams all over the world, made possible by advances in earth-moving and concrete technology. Initially, these dams were for hydroelectricity generation, and subsequently for multiple purposes – primarily power, irrigation, domestic and industrial water supply, and flood control. Some rivers have been intensively manipulated in this way, North America's River Columbia, for example, has, since the mid-nineteenth century, become the site for no fewer than 19 big dams and more than 60 smaller ones, making it the world's largest generator of hydroelectricity (Lee, 1989).

Structures above 15 m in height above their foundations are defined as large by the International Commission on Large Dams, and major if they exceed 150 m. The most active phase of large dam construction was in the period between 1950 and the mid-1980s. About 5000 large dams had been constructed worldwide by 1949, three-quarters of them in industrialised countries. By the end of the twentieth century, there were more than 45 000 large dams in over 140 countries (World Commission on Dams, 2000). The top five dam-building countries (China, USA, India, Spain and Japan) account for nearly 80 per cent of all large dams worldwide; nearly half of the global total are in China. However, the pace at which large dams were built slowed markedly in the last two decades of the twentieth century, especially in North America and Europe where most technically attractive sites are already developed. The average large dam in 2000 was about 35 years old.

In total, the world's major reservoirs were estimated to control about 10 per cent of the total runoff from the land in the 1970s (UNESCO, 1978). That figure had risen to 13.5 per cent by the early 1990s (Postel *et al.*, 1996) and is likely to be more like 15 per cent today. The volume of water trapped worldwide in reservoirs of all sizes (there are roughly 750 000 additional smaller dams globally) is no less than five times the global annual river flow (Davies *et al.*, 2000). Such a large-scale redistribution of water is thought to

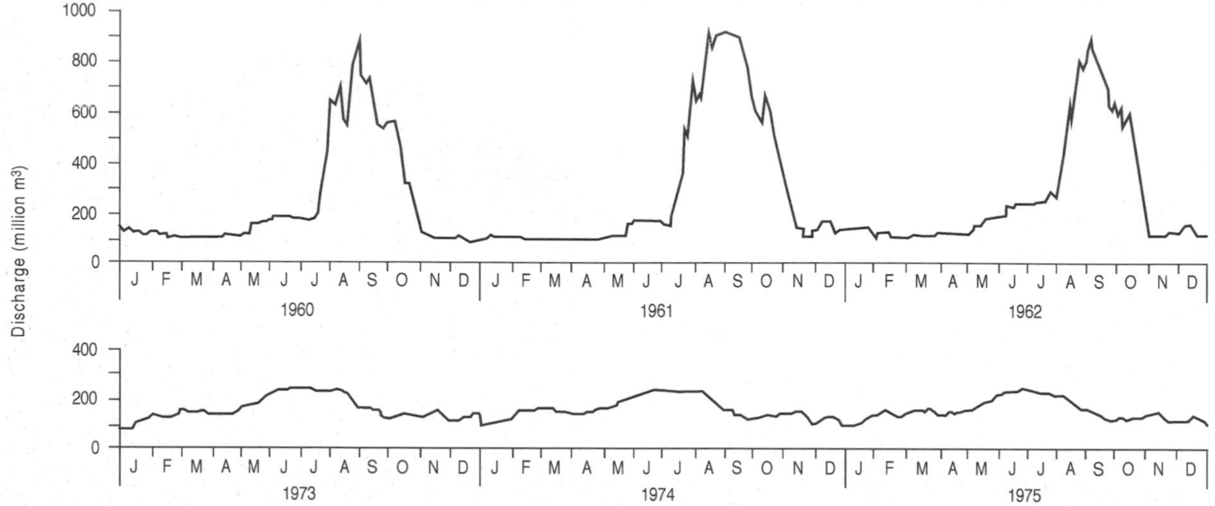

Figure 9.1 Daily discharge regime of the River Nile at Aswan, before and after construction of the High Dam (after Beaumont *et al.*, 1988).

be responsible for a small but measurable change in the orbital characteristics of the Earth (Chao, 1995).

THE BENEFITS OF BIG DAMS

There is no doubt that many big dam schemes have been successful in achieving their primary objectives and in many respects have made substantial contributions to the sustainable use of the world's resources. Half the world's large dams were built exclusively or primarily for irrigation, and about a third of the world's irrigated cropland relies on dams. The World Commission on Dams (2000) identified seven countries, including such populous nations as India and China, as having more than 50 per cent of arable land irrigated by water supplied from dams. Hydropower currently provides about 19 per cent of the world's total electricity supply and is used in over 150 countries. It represents more than 90 per cent of the total national electricity supply in 24 countries.

Egypt's Aswan High Dam, completed in 1970, provides a more detailed illustration of the benefits of big dam construction. Hydroelectricity generated by the dam is a renewable energy source that is cheap to operate after the initially high capital costs of dam construction, and saves on the purchase of fossil fuels from abroad. The Aswan High Dam generates about 20 per cent of Egypt's electricity.

The natural discharge of the Nile is subject to wide seasonal variations, with about 80 per cent of the annual total received during the flood season from August to October, and marked high and low flows depending upon climatic conditions in the main catchment area in the Ethiopian highlands. The dam allows management of the flow of the Nile's discharge, evening out the annual flow below the dam (Fig. 9.1) and protecting against floods and droughts. Management of the Nile's flow has also had benefits for navigation and tourism, resulting from the stability of water levels in the river's course and navigation channels. Irrigation water for cropland is also provided by the dam's reservoir storage, which has allowed 400 000 ha of cropland to convert from seasonal to perennial irrigation and the expansion of agriculture on to 490 000 ha of new land, a particularly important aspect for a largely hyper-arid country with just 3 per cent of its national area suitable for cultivation (Abu-Zeid, 1989).

Large dams are often seen as symbols of

Figure 9.2 The dam on the River Zambezi at Cahora Bassa in northern Mozambique. Just 10 per cent of the dam's electricity-generation capacity would be sufficient for all Mozambique's needs, but the major transmission lines built by the Portuguese serve South Africa, and much of Mozambique is not connected to the dam's supplies.

economic advancement and national prestige for many developing nations (Petts, 1984), but the huge initial capital outlay needed for construction often means that agendas are set to varying extents by foreign interests. A key element in the financing of Ghana's Akosombo Dam, completed in 1965, was the sale of cheap electricity to the Volta Aluminium Company, a consortium with two US-owned companies that produces aluminium from imported alumina, despite the fact that Ghana has considerable reserves of alumina of its own. The construction of the Cahora Bassa Dam in Mozambique (Fig. 9.2) in the 1970s, when the country was still a Portuguese colony, was also largely catering to outside interests. Most of the electricity is sold to South African industry, and part of the original reason for flooding the 250 km-long Lake Cahora Bassa was to establish a physical barrier against Frelimo guerrillas seeking independence. The plan for the dam also envisaged the settlement of up to a million white farmers in the region, who, it was thought, would fight to protect their new lands.

In addition, concerns have often been raised about the main benefits of a new dam being directed towards urban areas. Wali (1988) points out that although the Bayano Hydroelectric Complex in Panama provides 30 per cent of the country's electricity, no less than 83 per cent of national production is consumed in Panama City and Colón, so that the dam is reinforcing the concentration of wealth in the urban areas.

Hence, it is clear that the undoubted benefits of big dams are not always gained solely by the country where the dam is located, and that within the country concerned, the demands of urban populations can outweigh those of rural areas. Many of the drawbacks of such structures, however, are borne by the rural people of the country concerned. Despite the success of many big dams in achieving their main economic aims, their construction and associated reservoirs create significant changes in the pre-existing environment, and many of these changes have proved to be detrimental. It is the negative side of environmental impacts that have pushed the issue of big dams to a prominent position in the eyes of environmentalists and many other interest groups.

ENVIRONMENTAL IMPACTS OF BIG DAMS

The environmental impacts of big dams and their associated reservoirs are numerous, and Goodland (1990) has outlined the main areas that they influence (Table 9.1). The temporal aspect of environmental impacts within a certain area is also important. The river basin itself can be thought of as a system that will respond to a major change, such as the construction of a dam, in many different ways and on a variety of timescales. While the creation of a reservoir represents an immediate environmental change, the permanent inundation of an area not previously covered in water, the resulting changes in other aspects of the river basin, such as floral and faunal communities, and soil erosion, will take a longer time to re-adjust to the new conditions.

TABLE 9.1 *Areas of influence of dam and reservoir projects*

1. The catchments contributing to the reservoir or project area and the area below the dam to the estuary, coastal zone and offshore
2. All ancillary aspects of the project such as power transmission corridors, pipelines, canals, tunnels, relocation and access roads, borrow and disposal areas and construction camps, as well as unplanned developments stimulated by the project (e.g. logging or shifting cultivation along access roads)
3. Off-site areas required for resettlement or compensatory tracts
4. The airshed, such as where air pollution may enter or leave the area of influence
5. Migratory routes of humans, wildlife or fish, particularly where they relate to public health, economics, or environmental conservation

Source: Goodland (1990)

The range of environmental impacts consequent upon dam construction, and their effects on human communities, can be considered under three headings that reflect the broad spatial regions associated with any dam project: the dam and its reservoir; the upstream area; and the downstream area.

The dam and its reservoir

The creation of a reservoir results in the loss of resources in the land area inundated. Flooding behind the Balbina Dam north of Manaus, Brazil has destroyed much of a centre of plant endemism, for example. In some cases the loss of wilderness areas threatened by new dam projects has raised considerable debate, both nationally and internationally. A case in point was the Nam Choan Dam Project on the Kwae Yai River in western Thailand, first proposed in 1982. The suggested reservoir lay largely within the Thung Yai Wildlife Sanctuary, one of the largest remaining relatively undisturbed forest areas in Thailand, containing all six of the nation's endangered mammal species. Debate over the destructive impact of the project resulted in it being shelved indefinitely in 1988 (Dixon *et al.*, 1989).

Some resources, such as trees for timber or fuelwood, can be taken from the reservoir site prior to inundation, although this is not always economically feasible in remote regions. There are dangers inherent in not removing them, however.

Anaerobic decomposition of submerged forests produces hydrogen sulphide, which is toxic to fish and corrodes metal that comes into contact with the water. Corrosion of turbines in Surinam's Brakopondo reservoir has been a serious problem. In a similar vein, decomposition of organic matter by bacteria in the La Grande 2 reservoir in Quebec, Canada, has released large quantities of mercury by methylation. Mercury has bioaccumulated in reservoir fish tissue to levels often exceeding the Canadian standard for edible fish of 0.5 mg/kg (Harper, 1992).

Reservoirs formed by river impoundment typically undergo significant variations in water quality in their first decade or so, before a new ecological balance is reached. Biological production can be high on initial impoundment, due to the release of organically bound elements from flooded vegetation and soils, but declines thereafter. Hence the initial fish yield from Lake Kariba in 1964, its first year of full capacity, was more than 2500 t but by the early 1970s the annual yield had stabilised at around 1000 t (Marshall and Junor, 1981).

Cultural property may also be lost by the creation of a reservoir – 24 archaeological sites dating from AD 70–1000 were inundated by the Tucurúi Dam reservoir in Brazil, for example – although in some cases such property is deemed important enough to be preserved. Lake Nasser submerged some ancient Egyptian monuments, but major ones – including the temples of Abu Simbel, Kalabsha and Philae – were moved to higher ground prior to flooding (Fig. 9.3).

Figure 9.3 The Temple of Philae, which has been moved to higher ground to avoid inundation by Lake Nasser (Photo: Charles Toomer).

Big dams often necessitate resettlement programmes if there are inhabitants of the area to be inundated, and the numbers of people involved can be very large. Global estimates suggest that 40–80 million people have been displaced by reservoirs in the last 50 years (World Commission on Dams, 2000). Some of the biggest projects in this respect have been in China. The Sanmen Gorge Project on the Huang Ho River involved moving 300 000 people, and the Three Gorges Dam on the Yangtze River involves planning for the displacement of up to 1.2 million people from 13 cities, 140 towns and more than 1000 villages by 2009. Some indication of the trade-off between land lost, people displaced and power generated is indicated in Table 9.2 for a selection of big dam projects.

Compensation may be offered to the people who are displaced, but in many remote areas inhabitants do not possess formal ownership documents, slowing or preventing legal compensation. For those who have to resettle, helped by government schemes or not, the move can be a traumatic one. The resettlement of 57 000 members of the Tonga tribe from the area of the Kariba Dam on the Zambezi illustrates some of the adverse effects for the people concerned. Obeng (1978) described the culture shock suffered in moving to very different communities and

TABLE 9.2 *Hydropower generated per hectare inundated, and number of people displaced for selected big dam projects*

Project and country	Approx. rated capacity (MW)	Normal reservoir area (ha)	Kilowatts per hectare	People relocated
Pehuenche (Chile)	500	400	1 250	
Guavio (Colombia)	1 600	1 500	1 067	5 500
Three Gorges (China)*	13 000	110 000	118	1 200 000
Itaipu (Brazil and Paraguay)	12 600	135 000	93	8 000 families
Sayanogorsk (Russia)	6 400	80 000	80	
Churchill Falls (Canada)	5 225	66 500	79	
Tarbela (Pakistan)	1 750	24 300	72	86 000
Grand Coulee (USA)	2 025	32 400	63	
Aswan High Dam (Egypt)	2 100	40 000	53	100 000
Tucurúi (Brazil)	6 480	216 000	30	30 000
Keban (Turkey)	1 360	67 500	20	30 000
Batang Ai (Sarawak, Borneo)	92	8 500	11	3 000
Cahora Bassa (Mozambique)	2 075	266 000	8	25 000
BHA (Panama)	150	35 000	4	4 000
Kariba (Zimbabwe and Zambia)	1 500	510 000	3	50 000
Akosombo (Ghana)	833	848 200	0.9	80 000
Brokopondo (Surinam)	30	150 000	0.2	5 000

*Under construction, 92 000 people relocated by 1998
Sources: Barrow (1981); Dixon *et al.* (1989); Goldsmith and Hildyard (1984); Wali (1988); Goodland (1990); Gleick (1993); McCully (1996)

environments. Drawn-out conflicts over land tenure resulted between the new settlers and previous residents, and since the resettlement area was drier than the Tongan homelands, problems with planting and the timing of harvests were faced. Deprived of fish and riverbank rodents, which traditionally supplemented their cultivated diet, the Tongas faced severe food shortages. When the government sent food aid to relieve the suffering, the food distribution centres became transmission sites for trypanosomiasis.

Development following the construction of big dams can also act as a pull for migrants, bringing associated problems of pressure on local resources. The influx of migrants to the Aswan area led to an increase in population from 280 000 in 1960 to more than 1 million by the late 1980s, mainly due to the increase in job opportunities (Abu-Zeid, 1989).

Over the longer term, other effects of reservoir inundation have become evident. The alteration of the environment can have significant impacts on local health conditions. In some cases these can be beneficial. Onchocerciasis, or river blindness, for example – a disease that is common in Africa – is caused by a small worm transmitted by a species of blackfly. The blackflies breed in fast-running, well-oxygenated waters and dam construction can reduce the number of breeding sites by flooding rapids upstream. This has been the case with Ghana's Akosombo and Nigeria's Kainji dams (Worthington, 1978), although the flies may find alternative breeding sites in new tributary streams.

TABLE 9.3 *Rates of sedimentation in some Chinese reservoirs*

Reservoir	River	Total sediment deposited (million m^3)	Period of record (years)	Storage lost (%)
Sanmenxia	Huang Ho	3391	7.5	35
Qingtongxia	Huang Ho	527	5	84
Yanguoxia	Huang Ho	150	4	68
Liujiaxia	Huang Ho	522	8	11
Danjiangkou	Han	625	15	4
Guanting	Yongdinghe	553	24	24
Hongshan	Laohe	440	15	17
Gangnan	Hutuohe	185	17	12
Xingqiao	Hongliuhe	156	14	71

Source: Biswas (1990)

Malaria, conversely, is likely to increase as a result of water impoundment, since the mosquitoes that transmit the disease breed in standing waters. Local malaria incidence has increased around Tucurúi, Brazil, although management by fluctuating water levels and stranding larvae can help as in the USA's Tennessee Valley Authority water management complex.

Schistosomiasis, also known as bilharzia – a very debilitating though rarely fatal disease, which is widespread throughout the developing world – is transmitted in a different way: by parasitic larvae that infect a certain aquatic snail species as the intermediate host. The incidence of schistosomiasis was considerably increased by the construction of the Akosombo Dam, with infection rates among 5- to 19-year-old children rising from 15 per cent to 90 per cent within four years of its completion (Worthington, 1978). Similar figures have been reported from other large dams, such as Kariba in Zambia (Hira, 1969).

Other biological consequences of large reservoirs include the rapid spread of waterweeds that cause hazards to navigation and a number of secondary impacts, notably water losses through evapotranspiration. Water-fern appeared in Lake Kariba six months after the dam was closed and after two years had covered 10 per cent of the 420 km^2 lake area. More dramatic still was the spread of water hyacinth on Surinam's Brokopondo reservoir, which covered 50 per cent of the lake's surface within two years (UNEP, 1989b). Similar serious difficulties have been encountered at Aswan.

New reservoirs also have effects on geomorphological and, in some cases, tectonic processes. The trapping of sediment is a particularly important aspect of reservoir impoundment. The siltation of reservoirs has a number of knock-on effects downstream of the dam (see below), but it also seriously affects the useful life of the dam itself. Some examples of sedimentation rates in Chinese reservoirs are shown in Table 9.3. An extreme example of rapid sedimentation behind a dam is provided by China's Sanmenxia reservoir. River impoundment began in 1960, but within just 7.5 years of operation the reservoir had lost 35 per cent of its total storage capacity of 9700 million m^3 due to sedimentation (UNEP, 1989b).

The economic impact of reservoir sedimentation has been calculated by Lahlou (1996) in Morocco, where a dam-building programme was

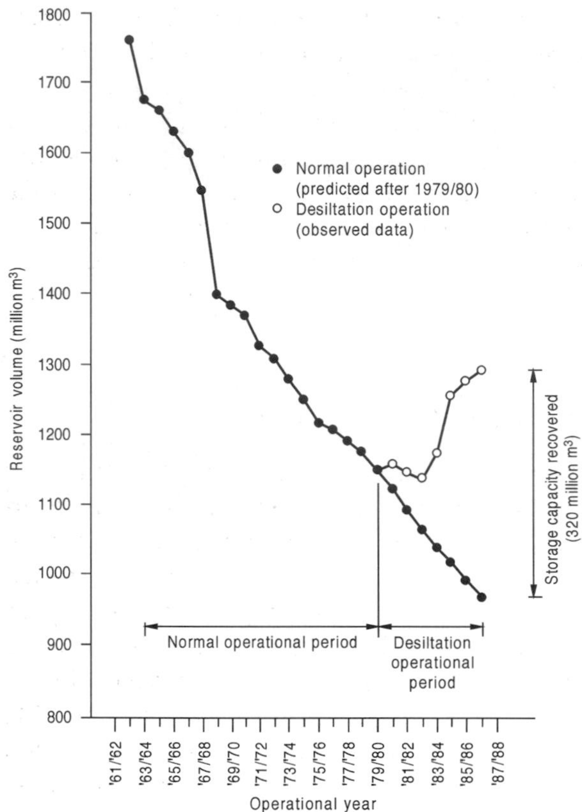

Figure 9.4 Variation of storage capacity in the Sefid-Rud reservoir, Iran during normal and desiltation operation (after Tolouie *et al.*, 1993).

launched in 1975 with the aim of irrigating 1m ha before the year 2000. The country's 1995 total of 84 large dams suffered a combined average annual loss of storage due to reservoir siltation of 0.5 per cent per year. This figure translates to annual losses in the year 2000 of 60m kWh of hydroelectric production, 50m m^3 of potable water supply, 5000 ha of irrigated cropland and 10 000 jobs.

A wide range of techniques is available for reservoir desiltation, the cost of which needs to be budgeted for. Tolouie *et al.* (1993) document the case of the Sefid-Rud reservoir, built to supply 250 000 ha of cropland in north-western Iran, which lost over 30 per cent of its storage capacity in the first 17 years after construction. Desiltation successfully restored about 7 per cent of total capacity in seven years (Fig. 9.4), but the reservoir had to be emptied during the non-irrigation season to enable sediment flushing. Emptying the reservoir released a highly erosive flow downstream of the dam, and hydroelectricity generation was prevented during the operation.

Local heightening of water tables following reservoir impoundment can have deleterious affects on new irrigation schemes through water-logging and salinisation. Waterlogging is an occasional problem around the Kuban reservoir on the River Kuban, near Krasnodar in southern Russia, when the reservoir is filled above its maximum normal level to aid navigation and benefit rice cultivation. The result has been the ruin of over 100 000 ha of crops, and water damage to 130 communities, including 27 000 homes, 150 km of roads and even the Krasnodar airport (Pryde, 1991). Local changes in groundwater conditions have also affected slope stability, causing landslides around some reservoirs. Water displaced by a landslip at the Vaiont Dam in Italy in 1963 overtopped the dam, killing more than 2000 people in the resulting disaster (Kiersch, 1965).

The sheer size of some reservoirs can also create new geomorphological processes. Artificial lakes behind dams on the Volga River are so large that storms can produce ocean-like waves that easily erode the fine wind-blown soils lining the shores. When this process undercuts trees, or creates shoals, navigational hazards result (Pryde, 1991).

The stress changes on crustal rocks induced by huge volumes of water impounded behind major dams have been suspected of inducing earthquakes in some regions. Nurek Dam on the Vakhish River in central Tajikistan is the best-documented example of a large dam, in this case a 315 m-high earth dam, causing seismic activity. Filling of the dam, located in a thrust-faulted setting, began in 1967 and substantial increases in water level were mirrored by significant increases in earthquakes per quarter (three months) during the first eight years of the dam's lifetime (Soboleva and Mamadaliev, 1976; see Fig. 9.5).

The reservoirs behind the Hoover Dam in the

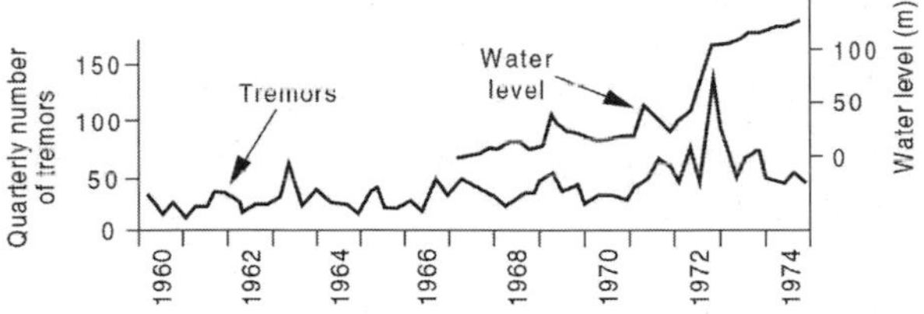

Figure 9.5 Number of earth tremors at the Nurek Dam, Tajikistan, 1960–74 (after Soboleva and Mamadaliev, 1976).

USA and Canada's Manic 3 have also induced local seismic activity, although earthquake incidents suspected to have been caused by other big dams, such as those at the Konya Dam near Bombay in India, Egypt's Aswan High Dam, and at the Kurobe Dam on Honshu Island, Japan, are unlikely to be due to reservoir-induced stresses (Meade, 1991).

The creation of new water bodies with large surface areas is thought by many to affect local climate. Thanh and Tam (1990) suggest that Lake Volta has shifted the peak rainfall season in central Ghana from October to July/August, for example, but few monitoring programmes have proved such effects conclusively. However, changes in the local temperature regime have been observed at the 45 000 ha Rybinsk reservoir north of Moscow in Russia, where the frost-free period has been extended by 5–15 days per year on average in an area of influence that extends for 10 km around the reservoir's shoreline (D'Yakanov and Reteyum, 1965). Evaporation from reservoir surfaces may affect local humidity, and the incidence of fog has been observed to rise in some areas.

Possible effects on global climate have seldom been mentioned in the big dam debate until recently. Indeed, a major argument in favour of big dam construction has been their electricity-generating potential from renewable energy resources without the harmful emissions associated with the burning of fossil fuels. This assumption has been questioned, however, and the impact of hydroelectric dams on the greenhouse effect highlighted by Fearnside (1995) who points out that carbon dioxide and methane are commonly emitted by decaying vegetation in reservoirs. While methane emissions continue slowly after dam closure, carbon dioxide is released in a relatively fast pulse during the reservoir's first decade. Fearnside concludes that carbon dioxide emitted from planned reservoirs in Brazil will greatly exceed emissions avoided by fossil fuel burning for the same amount of electricity generated.

The upstream area

A variety of upstream impacts can be induced or exacerbated by big dam projects. Some of these, in turn, may impact the dam project itself. Notable in this respect is the improved access to previously remote areas. Deforestation in the watershed above the Ambukloo Dam in the Philippines has led to sedimentation of the reservoir, reducing its useful life from 60 years to 32 years (UNEP, 1989b). Conversely, afforestation of catchments above dams has been carried out in many areas specifically to limit sediment accumulation in reservoirs. In the UK, for example, many water authorities have bought land in upper catchments to plant new forests.

An interesting example of some of the upstream changes brought about by reservoir inundation is provided by Wadi Allaqi in Egypt (Dickinson *et al.*, 1994). Prior to the creation of Lake Nasser by the construction of the Aswan High Dam, Wadi Allaqi was a typical hyper-arid wadi: dry for the most part with unpredictable, short-lived storms providing rare flash floods. Water also percolated underground, typically 30 m below the surface, from the Red Sea Hills to the east. Since 1970, however, both surface and subsurface hydrology has been modified fundamentally by Lake Nasser. The lowest 50 km of the wadi are now permanently inundated and another 40 km are periodically submerged, depending upon the height of the lake. The enhanced availability of water along this 40 km stretch has facilitated a significant change in vegetation, effectively transforming a zone with few resources

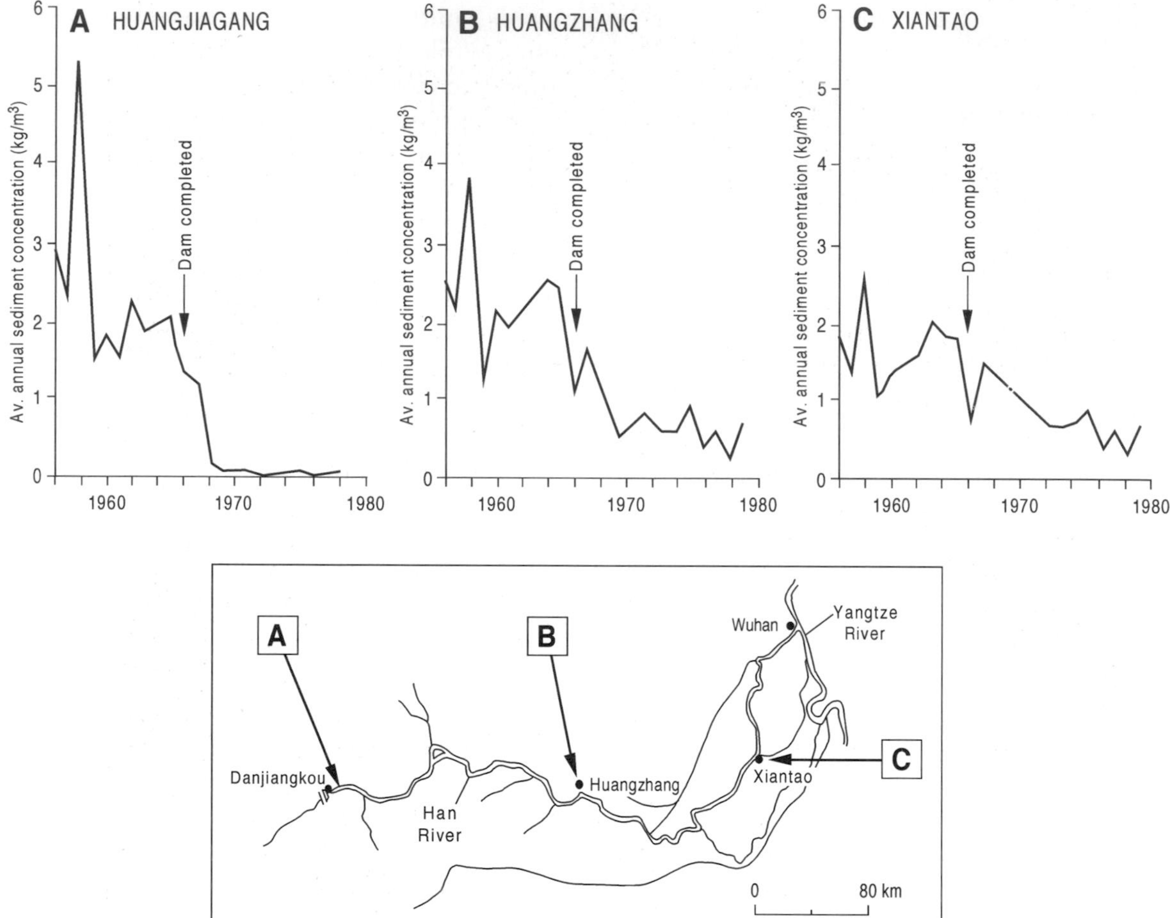

Figure 9.6 Effect of construction of the Danjiangkou Dam on sediment loads of the Han River, China (after Chien, 1985).

into one with significant potential for farming or to provide forage for livestock. Wadi Allaqi has been given conservation status by the Egyptian Government in order to plan carefully for development of its new resources.

The downstream area

Downstream of a reservoir, the hydrological regime of a river is modified. Discharge, velocity, water quality and thermal characteristics are all affected, leading to changes in geomorphology, flora and fauna, both on the river itself and in estuarine and marine environments.

The trapping of sediment behind dams leads to reduced loads in the river downstream. The effects of construction of the Danjiangkou Dam on the sediment load of the Han River, a tributary of the Yangtze, are shown in Fig. 9.6. The resulting flow downstream of the dam is highly erosive, with degradation of the bed and banks observed 480 km below the dam at Xiantao (Chien, 1985). Similar effects on the River Nile have been noted, downstream of the Aswan High Dam, and the lack of silt arriving at the Nile delta has had effects on coastal erosion, salinisation through marine intrusion and a decline in the eastern Mediterranean sardine catch (Abu-Zeid, 1989). In the case of the

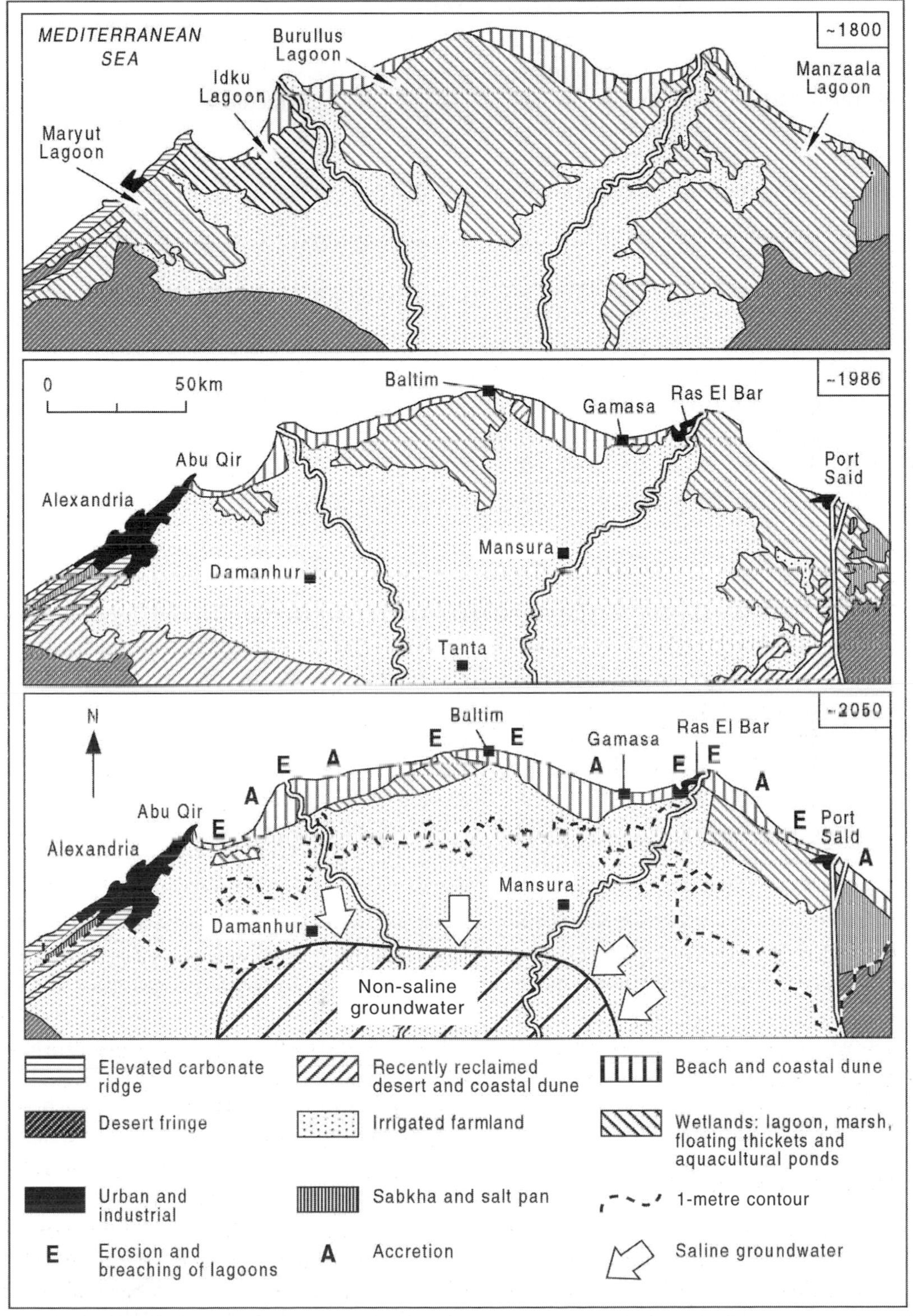

Figure 9.7 Nile delta changes (reprinted with permission from Stanley and Warne, 1993. Copyright, 1993, American Association for the Advancement of Science).

Nile, however, entrapment of sediment behind the High Dam may not be the main cause of coastal changes. Stanley (1996) suggests that deposition of modern and reworked Nile sediment in the extremely dense network of irrigation and drainage channels on the delta plain is more influential in reducing the amounts arriving at the coast. Figure 9.7 illustrates some of the changes that have occurred in the northern Nile delta in the last 200 years, due both to the effects of barrages and dams on the Nile and associated water use, as well as the spread of irrigated farmland and the natural rise in sea level, which can be expected to accelerate with global warming. Expected further changes due to these factors are also shown. To some extent the loss of fisheries off the Nile delta has been offset by a new fishing industry in Lake Nasser, which has provided employment to 7000 fishermen.

The intrusion of sea water into delta areas when river flows have declined due to reservoir impoundment is a common downstream effect. Construction of the Farakka Barrage on the River Ganges in India in 1975 has significantly increased the amount of cropland plagued by salinity problems in Bangladesh as dry-season discharge has declined. Nationally, nearly 350 000 ha of land in Bangladesh was affected by salinity problems before the barrage was built, but by the mid-1990s this total had more than doubled to 890 000 ha (Table 9.4). Downstream changes in salinity due to construction of the Cahora Bassa Dam in Mozambique are also threatening mangrove forests at the mouth of the Zambezi. Mangroves provide the breeding grounds for prawn and shrimp, a major source of foreign currency, but strategic water release from the dam could be used to offset the possible deleterious effects on shrimp and prawn catches (Gammelsrød, 1992). The downstream ecological effects of dam construction are not limited to rivers, estuaries and coastal areas, as the example of the Black Sea illustrates well (see page 98).

Dams can also affect marine and lake fish populations through the barrier they create, which effectively cuts off access to spawning grounds. This effect has been evident on salmon and aloses in the River Garonne and its tributaries in southwestern France since the Middle Ages (Décamps and Fortuné, 1991). In the twentieth century, decline in the landed catches of Caspian Sea sturgeon, the source of caviar, from 40 000 t early in the century to just 11 000 t in the 1970s, is attributable primarily to large hydroelectric dams on the Volga and the consequent loss of spawning grounds. Although catches had largely recovered to pre-dam levels in the early 1980s with the establishment of new sturgeon farms on the Caspian shores (Pryde, 1991), sturgeon stocks have since been seriously reduced again by overfishing, pollution and habitat loss.

TABLE 9.4 *Increase in area affected by salinity in Bangladesh*

	Affected area (thousand ha)
1966–73	344
1979	464
1989–91	578
1994–95	890

Source: Saheed (1995)

Of all the downstream effects of impoundment, the most dramatic have occurred on rivers dammed in several places. The construction of a series of dams in the twentieth century on the Colorado River, today one of the most heavily used waterways in the USA, stemmed the river's heavy sediment load, which had led the Spanish explorer Garces to name it 'Colorado' (Spanish for ruddy). Before 1930, the river carried 125–150m t of suspended sediment each year to the delta of the Gulf of California, but it has delivered neither sediment nor water to the sea since the late 1960s (Schwarz *et al.*, 1990).

A more recent example of large-scale environmental change due to multiple impoundment has occurred in the marshlands of the Tigris–Euphrates delta. Construction of more than 30 large dams on the Tigris and Euphrates in the second half of the twentieth century resulted in a

TABLE 9.5 *Changes in the surface area of Mesopotamian marshlands, 1973–76 to 2000*

Marshland	1973–76 (km²)	2000 (km²)	2000 as % of 73–76
Central	3121	98.0	3.1
AL-Hawizeh	3076	1025.0	33.3
Al-Hammar	2729	173.9	6.4
Total Wetlands	8926	1296.9	14.5

Source: UNEP (2001)

decisive change in the regime and volume of water in the two rivers, considerably reducing water supply and eliminating the floodwaters that fed the wetland ecosystems in the lower basin. The scale of the changes is indicated by the fact that the usable storage capacity of existing dams in the Tigris–Euphrates basin considerably exceeds the annual total discharge of both rivers (UNEP, 2001).

Iraq's political situation in the 1990s limited outsiders' access to the area and hindered monitoring of events in the delta, but analysis of satellite images over the lower Tigris–Euphrates indicates that in less than 30 years at least 7600 km^2 of primary wetlands (excluding the seasonal and temporary flooded areas) disappeared between 1973 and 2000 (Table 9.5). The Central and Al-Hammar marshes in southern Iraq had each been largely desiccated, with more than 90 per cent of their land cover transformed into bare land and salt crusts. While damming of the two rivers in more than 30 places since 1960 has reduced water flow to the entire area, their fate was accelerated by massive drainage works implemented in southern Iraq in the early 1990s following the second Gulf War in 1991. It is interesting to note that these drainage engineering works in Iraq were to a large extent made physically possible by reduced flow thanks to impoundments further upstream.

The Al-Hawizeh marshes, which straddle the Iran/Iraq border, had been reduced to one-third of their previous area by 2000. The remaining Al-Hawizeh marshes remain at high risk of desiccation due to upstream activities, including the Karkheh Dam in Iran, inaugurated in 2001, and its associated water transfer designs to Kuwait, and the planned Ilisu Dam in Turkey.

The lower Mesopotamian marshlands were one of the world's most significant wetlands and a biodiversity centre of global importance, as well as being home to an estimated half-million Marsh Arabs, a distinct indigenous people. About one-fifth of the Marsh Arab population, whose livelihood has been entirely dependent on the wetland ecosystem for about 5000 years, now lives in refugee camps in Iran with the rest internally displaced within Iraq. Several endemic species of marsh mammals, birds and fish have become extinct or are seriously threatened.

Coastal fisheries in the Persian Gulf, dependent on the marshland habitat for spawning migrations and nursery grounds, have also experienced significant reductions. In the Shatt al-Arab estuary, the decrease in freshwater flows has stimulated sea water intrusion and disrupted its complex ecology.

The destruction of the vast Mesopotamian marshlands has been likened to other human-engineered changes such as the desiccation of the Aral Sea and the deforestation of Amazonia as one of the Earth's major and most thoughtless environmental disasters.

POLITICAL IMPACTS OF BIG DAMS

Big dams are significant politically by their very nature as large endeavours in environmental

TABLE 9.6 *History of the Three Gorges Dam Project*

Date	Development
1919	Chinese leader Sun Yat Sen proposes construction of a dam at Three Gorges to improve navigation of the Yangtze River and to make better use of the river's resources
1944	The US Bureau of Reclamation surveys the Yangtze River and drafts a proposal for a dam at Three Gorges
1947	The dam project is suspended by the Chinese Government because of high inflation and an economic crisis
1953	Mao Zedong is presented with a proposal for constructing reservoirs; Mao requests that a dam be built at Three Gorges to control flooding
1954	The Yangtze Valley experiences the second worst floods of the century
1982	Denz Xiaoping pledges to proceed with the Three Gorges Dam project
1984–89	Feasibility studies for the Three Gorges Water Control Project
1992	The People's Congress approves a resolution to proceed with the Three Gorges Dam project
1993	The Three Gorges Project Construction Committee (TGPCC) is formed to represent China's State Council in decision-making
1994	Chinese premier, Li Peng, announces the official launch of construction on Three Gorges Dam
1996	Two major transportation projects, the Xiling Bridge across the Yangtze and an airport in Yichang, are completed. In May, the US Export-Import Bank withdraws support for the Three Gorges Dam project because of failure to follow the bank's environmental guidelines. In December, the Japanese Export-Import Bank announces financial backing for Japanese companies to participate in Three Gorges
1997	The TGPCC issues its first corporate bonds to raise construction funds. By September, the first phase of residents to be relocated from the reservoir region are resettled. In October, construction of transportation infrastructure for the project is completed. In November, the Yangtze River is blocked, signalling completion of the first phase of construction on the Three Gorges Dam project
1998	The Yangtze Valley experiences the worst floods of the century
1998–2006	Phase 2 of the Three Gorges Dam project: construction of the dam, including a flood discharge system and the hydroelectric plant. Between 2003 and 2006, 14 generators are scheduled to go online
1999	4.49 million cubic metres of cement are poured for the foundation of the Yangtze River power plant, a new world record for the most cement poured for the construction of a hydropower project
2004–09	Phase 3 of the Three Gorges Dam project: construction of two five-stage locks and ship lifts. In 2009, another 12 generators will be installed, making the Three Gorges Dam fully operational

Sources: International Energy Outlook (1998) and press reports

management. Summoning the political will to embark on such projects, as well the necessary finance and engineering skills, can take many years. China's Three Gorges Project, one of the most ambitious and controversial big dams in recent times, was first proposed in 1919 but 75 years passed before construction began in 1994 (Table 9.6). It is not due to be fully operational until 2009.

Increasing public awareness of the environmental and social implications of big dams has been a factor in the recent global slow-down in the rate of their construction. From an average of 885 large dams a year in the period from 1950 to the mid-1980s, this rate declined to about 500 a year in the mid-1990s (Postel *et al.*, 1996). Heated debates over the issues have in some cases led to the shelving of construction plans. The Nam Choan Dam Project in Thailand has been mentioned in this respect. Another example is the Tasmanian Franklin River Project, which was stopped on environmental grounds in 1983. India's Narmada Dam has also come under severe criticism over its anticipated impacts.

In such cases, the obvious benefits of dam construction must be carefully weighed against the costs measured in environmental and social terms, and the potential impacts predicted and ameliorated by sensible planning (Goodland, 1986). There is little doubt that many of the adverse impacts of dams can be greatly reduced by good planning and anticipation, and aid agencies that finance such projects now require an Environmental Impact Assessment (EIA) before approval of funding is given. Progress has also been made in the widening scope given to consultation prior to dam construction (Table 9.7). Nevertheless, a sensible operating schedule is also a key factor – many of the problems caused by dams are the result of operators aiming to maximise water use – through releases for hydroelectricity generation and irrigation, for example – to such an extent that other concerns are given too little consideration.

The increasing sensitivity of the big dam issue, and the need to investigate rigorously the benefits and costs to society of dam projects, prompted the establishment of a World Commission on Dams in 1998 by the World Conservation Union (IUCN) and the World Bank. The commission was set up to review the environmental, social and economic impacts of large dams, and to develop new guidelines for the industry. It reported in 2000 on its two main goals:

1. to review the development effectiveness of dams, and assess alternatives for water resources and energy development
2. to develop internationally acceptable standards, guidelines and criteria for decision-making in the planning, design, construction, monitoring, operation and decommissioning of dams.

The Commission's framework for decision-making was based on five core values: equity,

TABLE 9.7 *Broadening of consultation in the design of big dams*

Design team	Approximate era
Engineers	Pre-Second World War
Engineers + economists	Post-Second World War
As above + environmentalists	Late 1980s
As above + affected people	Early 1990s
As above + NGOs	Late 1990s
As above + national consensus	Early 2000s?

Source: after Goodland *et al.* (1993b)

sustainability, efficiency, participatory decision-making and accountability. It proposed:

- a rights-and-risks approach as a practical and principled basis for identifying all legitimate stakeholders in negotiating development choices and agreements
- seven strategic priorities and corresponding policy principles for water and energy resources development – gaining public acceptance, comprehensive options assessment, addressing existing dams, sustaining rivers and livelihoods, recognising entitlements and sharing benefits, ensuring compliance and sharing rivers for peace, development and security
- criteria and guidelines for good practice related to the strategic priorities, ranging from lifecycle and environmental flow assessments to impoverishment risk analysis and integrity pacts.

A late-twentieth century trend relating to large dams noted by the Commission is that of decommissioning dams that no longer serve a useful purpose, are too expensive to maintain safely, or have unacceptable levels of impacts in today's view. The desire to restore rivers to their pre-dam ecological state has accelerated in many countries in recent years, particularly in North America. Indeed, in the USA the decommissioning rate for large dams has overtaken the rate of construction since 1998. Decommissioning dams can enable the restoration of fisheries and riverine ecological processes as experience in North America and Europe has shown, but the removal of dams needs to be preceded by careful planning. Environmental problems are also inherent in the removal of such large-scale structures. These include negative impacts on downstream aquatic life due to a sudden flush of the sediments accumulated in the reservoir.

The building of dams on rivers flowing through more than one country brings international political considerations on to the agenda of big dam issues. Such considerations are particularly pertinent in dryland regions where rivers represent a high percentage of water availability to many countries. The main issues at stake here are those of water availability and quality.

In several international river basins, peaceful cooperation over the use of waters has been achieved through international agreement. One such agreement, between the USA and Mexico over use of the Rio Grande, was signed in 1944 and is operated by the International Boundary and Water Commission. This body ensures equal allocation of the annual average flow between the two countries. Similarly, an international treaty between Egypt and Sudan governs the volume of Nile water allowed to pass through the Aswan High Dam, although none of the other eight Nile Basin countries have agreements over use of the Nile's waters. India and Bangladesh signed a water-sharing accord for the Ganges in 1996. It specifies water allocation in normal and dry periods after 20 years of wrangling over the effects of the Farakka Barrage.

In other international basins, such as the Tigris–Euphrates in the Middle East, the lack of agreement represents significant potential for conflict. While there is currently a water surplus in this region, the scale of planned developments raises some concern (Agnew and Anderson, 1992). Turkey's Southeastern Anatolian Project, a regional development scheme on the headwaters of the two rivers, centres on 22 dams. In early 1990, when filling of the Ataturk Dam reservoir (Fig. 9.8) commenced, stemming the flow of the Euphrates, immediate alarm was expressed by Syria and Iraq, despite the fact that governments in both countries had been alerted and discharge before the cut-off had been enhanced in compensation. Syria and Iraq nearly went to war when Syria was filling its Euphrates Dam. Full development of the Southeastern Anatolian Project, expected in about 2030, could reduce the flow of the Euphrates by as much as 60 per cent, which could severely jeopardise Syrian and Iraqi agriculture downstream. The three Tigris–Euphrates riparians have tried to reach agreements over the water use from these two rivers, and the need for such an agreement is becoming ever more pressing.

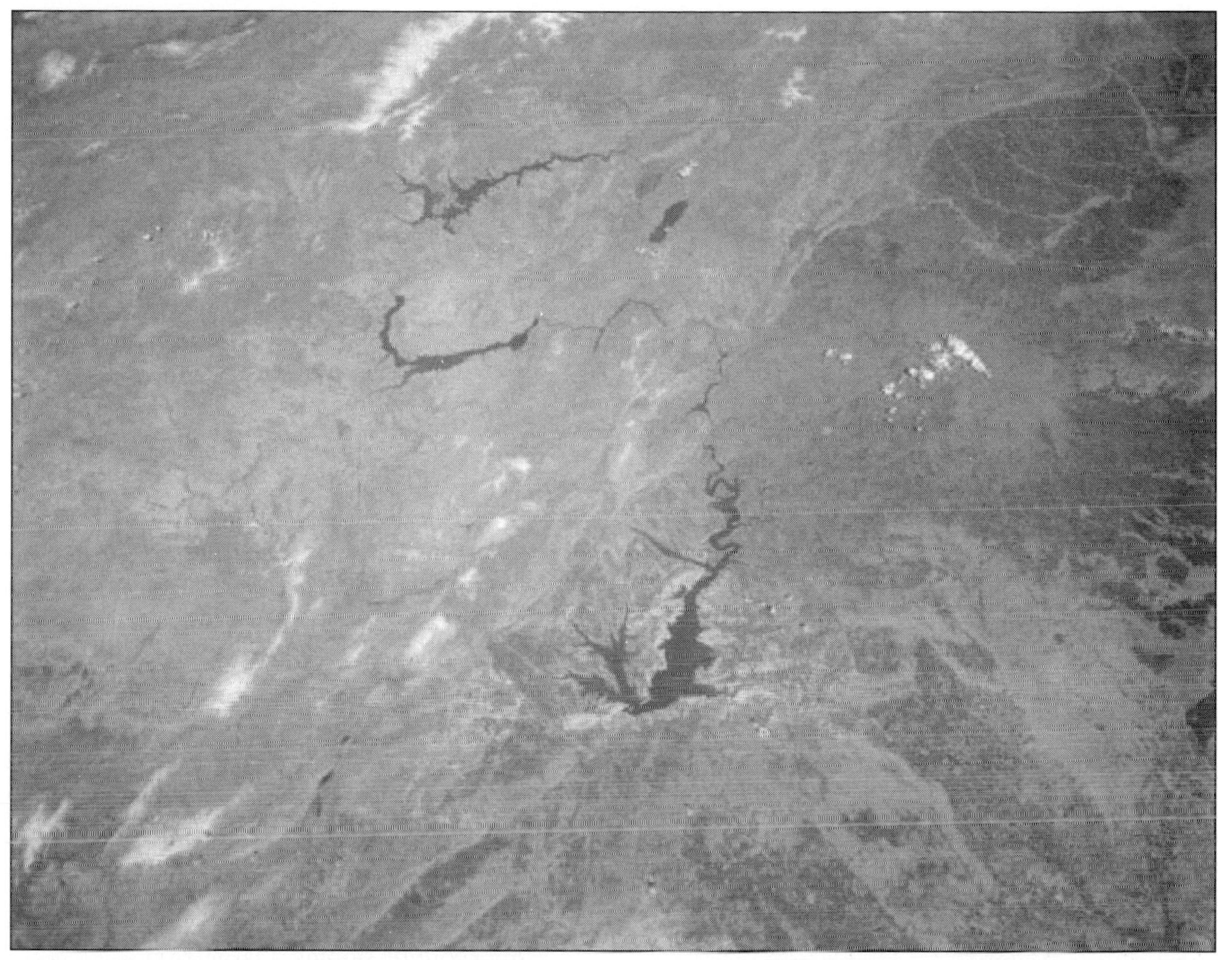

Figure 9.8 Space shuttle photograph of the River Euphrates flowing through south-eastern Turkey. The largest of the three reservoirs in this picture is that behind the Ataturk Dam, completed in 1992. Concern at the scale of water use in Turkey's Southeastern Anatolian Project, which has contributed to the decline of the Tigris–Euphrates marshlands in Iraq, has heightened political tensions in this part of the Middle East (Photo: NASA).

FURTHER READING

Chau, K.C. 1995 The Three Gorges Project of China: resettlement prospects and problems. *Ambio* 24: 98–102. Analysis of the balance between environmental conservation, poverty and economic development in China's largest dam project.

Cummings, B.J. 1990 *Dam the rivers, damn the people*. London, Earthscan. The story of two major hydroelectric developments in the Amazon, and the conflicts between them and local environments and populations.

Goldsmith, E. and Hildyard, N. 1984 (Vol. 1), 1986 (Vol. 2) *The social and environmental effects of large dams*. Wadebridge, Cornwall, Wadebridge Ecological Centre. Volume 1 presents arguments against water development based on large dams, and Volume 2 contains 31 detailed case studies of big dam projects. These books have a very definite bias against the development of big dams (see also Trussell below).

Meade, R.B. 1991 Reservoirs and earthquakes. *Engineering Geology* 30: 245–62. A critical appraisal of the assumption that every earthquake occurring in or near a large reservoir is induced by the reservoir.

Palmieri, A., Shah F. and Dinar A. 2001 Economics of reservoir sedimentation and sustainable management of dams. *Journal of Environmental Management* 61: 149–63. An economic appraisal of managing dams in a sustainable manner.

Petts, G.E. 1984 *Impounded rivers – perspectives for ecological management.* Chichester, Wiley. A comprehensive appraisal of the ecological effects of dams and their reservoirs on world rivers.

Trussell, D. (ed.) 1992 *The social and environmental effects of large dams*, Vol. 3. Wadebridge, Cornwall, Wadebridge Ecological Centre. The third volume in this series reviews the very wide literature on large dams by topic. As with the other volumes, the aim throughout is to present a strong case against big dams.

World Commission on Dams 2000 *Dams and development: a new framework for decision-making.* London, Earthscan. Report of the world body set up to review the environmental, social and economic impacts of large dams.

Websites

http://www.dams.org/ official site of the World Commission on Dams.

http://www.sandelman.ocunix.on.ca/PIRGs/CPHJB/ the Dam and Reservoir Impact and Information Archive.

http://www.irn.org/ the International Rivers Network site has information, campaigns and publications (e.g. *World Rivers Review*) on many riverine issues.

http://www.nwppc.org/ documents, journals and related information on the management of the Columbia River Basin in North America.

http://www.narmada.org/ site that aims to present the perspective of grassroots people's organisations on the construction of large dams on the River Narmada in central India.

http://www.recovery.bcit.ca/ River Recovery – Restoring Rivers Through Dam Decommissioning. This project seeks to restore the ecological health of British Columbia's rivers by identifying dams that require alteration or removal.

POINTS FOR DISCUSSION

- On the whole, do you consider big dams to be good or bad things?
- Are any of the environmental impacts of big dams irreversible?
- Is hydroelectricity a sustainable form of power generation?
- Outline the possible environmental impacts of decommissioning a large dam.

10

URBAN ENVIRONMENTS

TOPICS COVERED

Surface water resources
Groundwater
The urban atmosphere
Garbage
Hazards and catastrophes
Towards a sustainable urban environment

Key words: megacity, ecological footprint, subsidence, post-industrial city

Large numbers of people have lived in close proximity to each other in cities for thousands of years. The first urban cultures began to develop about 5000 years ago in Egypt, Mesopotamia and India, but the size of cities and their geographical distribution expanded dramatically after the Industrial Revolution of the previous millennium. The growth rates of cities in recent decades has been unprecedented. In 1975, five world cities had a population of more than 10 million people; that number of so-called 'megacities' had reached 19 by the year 2000 and was projected to reach 23 by 2015 (UNFPA, 2001). Worldwide, 411 cities had populations of more than 1 million in 2000, compared with 326 in 1990. At the turn of the current century, almost 3 billion people lived in urban areas, nearly half the world population. Fig. 10.1 indicates the growth of 20 megacities. Many cities in the developed world, such as New York and London, grew little in the last 30–50 years of the twentieth century, but cities in the industrialising world show remarkable growth over the same period. The populations of Mexico City, São Paulo, Karachi and Seoul grew by more than 800 per cent in the second half of the twentieth century. The urban area of Mexico City, which is probably the world's largest conurbation, expanded from 27.5 km^2 in 1900 to cover 1250 km^2 in 1990. The phenomenal growth of some cities, and the high concentrations of people they represent (the urban density of Mexico City in 1980 was 14 082 people/km^2), has created some acute environmental problems both outside and within the city limits.

Cities represent a completely artificial environment; they absorb vast quantities of resources from surrounding areas and create high concentrations of wastes to be disposed of. The degree to which cities impinge on their hinterlands, their 'ecological footprint', is indicated by a few examples. About 10 per cent of prime agricultural land has been lost to urbanisation in Egypt. The twentieth-century growth of São Paulo was fuelled by the expansion of coffee plantations in south-east Brazil, which reduced the forest cover of São Paulo State from 81 per cent in 1860 to 6 per cent in the late 1980s (Monteiro, 1989). The demand for water in Tehran spurred the construction of a series of dams and canals in the early decades of the last century, to bring water 50 km from the River Karaj to the west, reducing the water available for rural agriculture. By the 1970s, supplies were again running low, so water was diverted more than 75 km from the River Lar to the north-east (Beaumont *et al.*, 1988). In Rio de Janeiro's Guanabarra Bay, pollution from two oil refineries, two ports, 6000 industries, 12 shipyards, 16 oil terminals, sewage and garbage

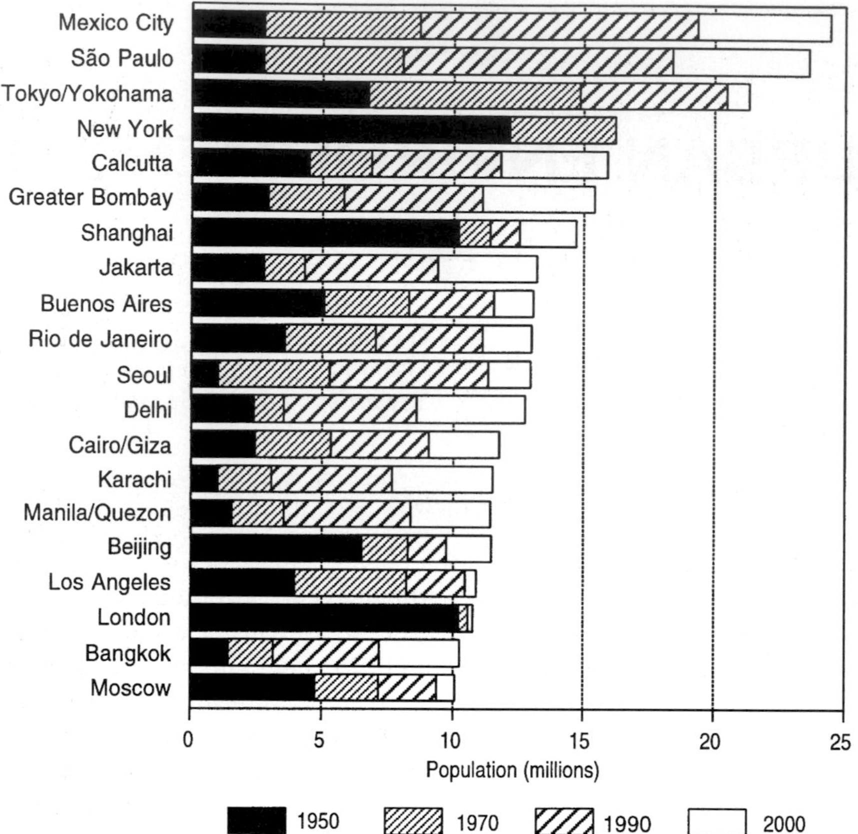

Figure 10.1 Estimated population of 20 megacities in 1950, 1970, 1990 and 2000 (from data in United Nations, 1989a).

dumps has reduced commercial fishing by 90 per cent and mangrove cover by 90 per cent, led to outbreaks of water-borne diseases such as infectious hepatitis and typhoid, and is silting the bay by 81 cm/100 years (Kreimer *et al.*, 1993).

The acute environmental problems that occur within many cities, particularly in the developing world, their underlying reasons, and the scale of the clean-up task faced by urban authorities are well summarised in the case of Manila Metro, capital of the Philippines. A 1990 population of 8 million was projected to rise to 13 million by 2000. All the city's rivers are biologically dead. Each day, 2000 t of solid waste is left uncollected, to be burnt, thrown into waterways or moulder on the ground. Much of the garbage collected is dumped on 'Smokey Mountain', a 23-ha open tip, which represents a severe health hazard to the 20 000 people who reside on its fringes and earn a living by scavenging from the dump (Jimenez and Velasquez, 1989).

About 65 per cent of the country's 1500 recognised industrial enterprises are located in the Manila Metro area, and only one-third to one-half of them are thought to comply with minimal air and water pollution emission standards. One million vehicles, more than half the country's total, operate in the Manila Metro area. Just half of these vehicles are thought to meet even minimal emission standards. The annual cost to the economy due to congestion alone is estimated to be more than US$50 million, which is low by the standards of other Asian capitals, while the economic burden of air pollution may be an order of magnitude higher (Table 10.1).

The basic cause of Manila Metro's severe envi-

TABLE 10.1 *Some estimates of the annual cost of congestion and air pollution in selected Asian cities*

City	Cost (US$ million/year)	
	Congestion	Air pollution
Bangkok Metropolitan Area (1989)	272	380–580
Bangkok Metropolitan Region (1993)	400	1300–3100
Seoul	154	–
Manila	51	–
Jakarta	68	400–800

Source: after Brandon and Ramankutty (1993)

ronmental problems is that 8 million people are using infrastructure much of which dates from the US colonial period, estimated to be adequate for about 2 million people, at most. A large proportion of the solid and liquid wastes are simply inaccessible for collection by virtually any means due to the density of squatter settlements, inappropriate collection systems and the simple lack of services such as septic tank desludging. The problems of physical infrastructure are exacerbated by the government's inability to stop polluters, largely a function of serious understaffing at the metropolitan regulation agency (World Bank, 1989).

SURFACE WATER RESOURCES

Urbanisation can result in numerous impacts on waterways by altering their hydrology, affecting stream morphology, water quality and the availability of aquatic habitats (Table 10.2), effects that are often continuous and synergistic. In hydrological terms, urban areas are typified by 'flashy' discharge regimes caused by the increased area of impervious surfaces such as tarmac and concrete, and networks of storm drains and sewers. The timelag between rainfall and peak discharge of a river is decreased and the peak discharge itself is increased as a result of these impervious surfaces, which has a significant impact on the volume, frequency and timing of floods. Many urban rivers have been subject to structural measures to reduce the flood hazard, such as channel straightening and channelisation, the building of levées, and the stabilisation of banks (see Chapter 21).

One of the most important environmental issues that stems from urban modifications to the hydrological cycle is that of poor water quality. Runoff from developing urban areas is usually choked with sediment during construction phases, when soil surfaces are stripped of vegetation, and a finished urban zone greatly increases runoff. This drastically modified urban drainage network feeds large amounts of urban waste products into rivers and ultimately into oceans.

Many rivers that flow through urban areas are biologically dead, thanks to heavy pollution. Hardoy *et al.* (1992: 73) sum up the state of urban rivers in developing countries as follows: 'Most rivers in Third World cities are literally large open sewers.' They go on to point out that of India's 3119 towns and cities, only 209 have partial sewage treatment facilities and just eight have full facilities. India's Jamuna River, for example, contains 7500 coliform organisms per 100 ml of water on entering New Delhi, a figure that rises to 24 million coliform organisms per 100 ml after flowing through the city. For comparison, the World Health Organization (WHO) guidelines for such microbiological pollution are 10 coliform organisms per 100 ml for drinking water and 1000 per 100 ml for irrigation purposes. Industrial effluents combine with this domestic

TABLE 10.2 *Major impacts of urbanisation on waterways*

Change in stream characteristics	Effects
Hydrology	Increased magnitude and frequency of severe floods
	Increased annual volume of surface runoff
	Increased steam velocities
	Decreased dry weather baseflow
Morphology	Channel widening and downcutting
	Increased erosion of banks
	Stream enclosure or channelisation
Habitat and ecology	Changes in diversity of aquatic insects
	Changes in diversity and abundance of fish
	Destruction of wetlands, riparian buffers and springs
Water quality	Massive sediment pulses
	Increased pollutant washoff
	Nutrient enrichment
	Bacterial contamination
	Increased organic carbon loads
	Higher levels of toxics, trace metals, hydrocarbons
	Increased water temperature
	Trash/debris jams

Source: after Baer and Pringle (2000)

source of riverine pollution to make urban rivers the most polluted freshwater sources on Earth.

All the rivers flowing through Jakarta, Indonesia, are heavily polluted from numerous, mostly untreated, discharge sources: household drains and ditches, overflows and leaks from septic tanks, commercial buildings, and industries. Water-related diseases such as typhoid, diarrhoea and cholera increase in frequency downstream across the metropolitan area (Hardoy *et al.*, 1992). Untreated sewage and discharge from the 20 000 classified water-polluting industries that feed into Bangkok's canal system have created a distinct sag in the dissolved oxygen profile of the Chao Phraya where the canals feed the river (Phantumvanit and Liengcharernsit, 1989). Although the example of the Thames at London (see Fig. 8.4) shows how such near-anaerobic river conditions can be improved, neither the money nor the political will are currently as forthcoming in Thailand.

The local hydrological impact of the Saudi capital, Riyadh, provides a very contrasting example to the depressing catalogue of riverine disaster areas typically associated with large, rapidly growing cities. Discharge of Riyadh's wastewater feeds the Riyadh River, which scarcely existed 20 years ago, but now flows throughout the year down what was the seasonal Wadi Hanifa. The water, which is originally derived from desalinated Gulf sea water, is partially treated before being released to flow down the steep-sided wadi and enter open countryside, eventually disappearing 70 km from Riyadh. The new flow has created an attractive valley lined by tamarisk trees and phragmites, which is becoming an important recreational site for Riyadh's 2.3 million population. Beyond the wadi, significant irrigated agriculture has grown up, drawing on the groundwater around the river. This unique new feature is, however, under some threat from needs to further recycle

Figure 10.2 Buildings that disturb the thermal equilibrium of permafrost can lead to heaving and subsidence, as seen on the left. One solution is to raise buildings and pipelines up above the ground surface on stilts, so ventilating heat-generating structures (right). Both photographs were taken in Yakutsk, eastern Siberia.

the much-needed water resource (Meynell, 1993).

Indeed, the quantity of water available is another critical environmental issue for many cities, as the situation in Tehran mentioned above indicates. Nearly half of the 640 major cities in China face water shortages, with 100 experiencing severe scarcities. The quantity and quality of water resources are related, of course, and severe pollution of many urban stretches of rivers in China reduces the options for their use. Testing of 135 urban river sections in 1994 found 54 to be of such low quality that the water was deemed unsuitable for even industrial or agricultural use (NEPA, 1997).

Different types of environmental problem are encountered in permafrost areas where surface water and soil moisture is frozen for much, and in some places all, of the year. Frozen rivers and lakes mean that many of the uses such water bodies are commonly put to at more equable latitudes, such as sewage and other waste disposal, are not always available. The low temperatures characteristic of such regions also mean that biological degradation of wastes proceeds at much slower rates than those elsewhere. Hence, the impacts of pollution in permafrost areas tend to be more long-lasting than in other environments.

The nature of the permafrost environment also presents numerous environmental challenges to the construction and operation of settlements – challenges that have been encountered in urban developments associated with the exploitation of hydrocarbons and other resources in Alaska, northern Canada and northern Russia.

Disturbance of the permafrost thermal equilibrium during construction can cause the development of thermokarst (irregular, hummocky ground). The heaving and subsidence caused can disrupt building foundations and damage pipelines, roads, railtracks and airstrips. Terrain evaluation prior to development is now an important procedure in the development of these zones, following expensive past mistakes. Four main engineering responses to such problems have been developed: permafrost can be neglected, eliminated or preserved, or structures can be designed to take expected movements into account (Johnston, 1981). Preservation of the thermal equilibrium is achieved in numerous ways, such as by insulating the permafrost with vegetation mats or gravel blankets, and ventilating the underside of structures that generate heat (e.g. buildings and pipelines; see Fig. 10.2).

GROUNDWATER

The water needs of urban population and industry are often supplemented by pumping from groundwater, and pollution of this source is another problem of increasing concern in many large cities. Seepage from the improper use and disposal of heavy metals, synthetic chemicals and other hazardous wastes such as sewage, is a principal origin of groundwater pollution. Some of the major pollutants involved in a selection of cities are shown in Table 10.3. The quantity of such compounds reaching groundwater from waste dumps in Latin America, for example, is thought to be doubling every 15 years (World Bank, 1992a). A serious threat to groundwater quality in Bermuda is posed by those parts of the urban area without sewage systems. Unsewered sanitation provisions, consisting of septic tanks and pit latrines, leak bacteria and nitrogen compounds from excreta. High concentrations of nitrates in groundwater are closely correlated with population density in unsewered areas of Bermuda (Thomson and Foster, 1986). Another regular source of contamination in snow-belt regions of Europe and North America is de-icing agents, usually sodium chloride, applied to roads. Salts washed away from urban highways accumulate in soils as well as groundwater (Howard and Haynes, 1993). Contamination of groundwater beneath any city is a serious long-term issue since aquifers do not have the self-cleansing capacity of rivers and, once polluted, are difficult and costly to clean.

A frequent outcome of overusing groundwater is a lowering of water-table levels and consequent ground subsidence. In Mexico City, use of subterranean aquifers for more than 100 years has caused subsidence of up to 9 m in some central areas (Schteingart, 1989), greatly increasing the flood hazard in the city and threatening the stability of some older buildings, notably the sixteenth-century cathedral.

Marked subsidence episodes in Tokyo have

TABLE 10.3 *Some examples of urban groundwater pollution*

City/region	Country	Major pollutants
Merida	Mexico	Bacteria
Milwaukee	USA	Cl, SO_4, bacteria
Birmingham	UK	Majors, metals, B, P, Si, CN, organics
Narbonne	France	SO_4, NO_3
Cairo	Egypt	NO_3, majors, metals
Bermuda	Bermuda	Micro-organisms, Cl, NO_3

Source: after Lerner and Tellam (1993: 324, Table 1)

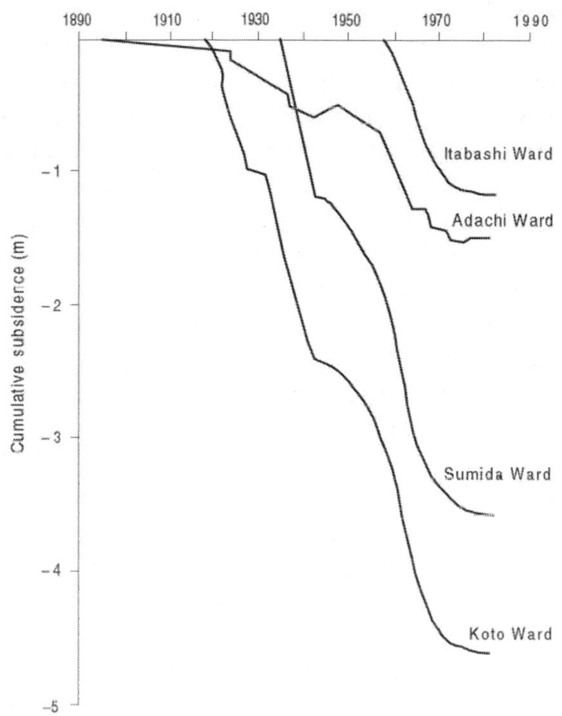

Figure 10.3 Ground subsidence in the Ward Area of Tokyo Metropolis (after TMG, 1985).

mirrored phases of economic and industrial growth. The Tokyo Metropolitan Government suggests that ground subsidence began in the city as economic activity grew after the First World War, and came to a halt for some years in the 1940s following the destruction of industries in the Second World War (TMG, 1985; see Fig. 10.3). Renewed industrial activity during the Korean War accelerated the process once more, but the trend was slowed in the 1960s with the introduction of pumping regulations, and subsidence has been virtually halted since these regulations were strengthened in 1972.

In other coastal cities, depletion of aquifers has created problems of sea-water intrusion. Israel's coastal aquifer, which extends 120 km along the Mediterranean coast, has been heavily exploited for half a century. Overpumping of groundwater in the Tel Aviv urban area depleted groundwater levels to below sea level over an area of 60 km^2 in the 1950s, requiring a programme of freshwater injection along a line of wells parallel to the coast in an attempt to redress the salt water/fresh water balance. Water was brought by the National Water Carrier from the north of the country to supplement natural recharge by precipitation. Despite these efforts, continued pumping for urban and agricultural use has meant a continued decline in the aquifer's water quality: chloride concentrations have increased from 110 mg/litre in 1970 to 190 mg/litre in 1995 and some wells now yield water that neither complies with drinking water standards nor is suitable for unrestricted agricultural irrigation (Gabbay, 1998).

A similar pattern of events occurred in Brooklyn, New York City, although here no attempt was made to prevent sea-water intrusion. By 1947, pumping of the increasingly saline groundwater had ceased and all freshwater supplies were provided by surface sources. Cessation of pumping gradually allowed the water table, which had been reduced to about 11 m below sea level, to rise again. During the half-century of pumping, however, deep basements, building foundations and subways had been sunk, and these were subject to flooding as the groundwater levels rose, necessitating expensive remedial measures.

Rising groundwater levels have become a critical problem for many 'post-industrial' cities as manufacturing industries have given way to service industries, which are much less demanding of water, and legislation has been introduced to control subsidence problems. In London, where loss of water from aged pipes is an additional reason for groundwater levels rising, the period of change from a generally falling to a rising water table occurred in the late 1970s. The potential effects upon the fabric of London's urban environment are now being assessed, with particular interest being shown by the insurance industry. A report issued by the Construction Industry Research and Information Association estimated that the cost of pumping to maintain groundwater levels below the level at which serious damage would occur was up to £30 million (CIRIA, 1989).

TABLE 10.4 *Measures to reduce the impact of rising groundwater beneath urban areas*

Strategy	Examples
Preventative planning	Investigation of development sites to identify hazardous areas Implementation of suitable building regulations on basement depths, construction materials and methods Control water tables by pumping
Water conservation measures	Control of water use and storage Leak detection and maintenance of pipes Wastewater reuse

Source: after George (1992)

The rising groundwater problem has also been reported from many Middle Eastern cities where rainfall is commonly low, potential evaporation high, and natural recharge small and sporadic. Inadvertent artificial recharge from leaking potable supplies, sewerage systems and irrigation schemes has caused widespread and costly damage to structures and services, and represents a significant hazard to public health (George, 1992). A range of measures designed to reduce the impact of rising groundwater prior to and during development is shown in Table 10.4.

THE URBAN ATMOSPHERE

Urban areas have a diverse catalogue of effects on local elements of climate, which are well documented (e.g. Landsberg, 1981), but the most serious environmental issue pertaining to the urban atmosphere is that of quality. The principal sources of air pollution in urban areas are derived from the combustion of fossil fuels for domestic heating, for power generation, in motor vehicles, in industrial processes and in the disposal of solid wastes by incineration. These sources emit a variety of pollutants, the most common of which have long been sulphur dioxide (SO_2), oxides of nitrogen (NO and NO_2, collectively known as NO_x), carbon monoxide (CO), suspended particulate matter (SPM) and lead (Pb). Ozone (O_3), another 'traditional' air pollutant associated with urban areas and the main constituent of photochemical smog, is not emitted directly by combustion, but is formed photochemically in the lower atmosphere from NO_x and volatile organic compounds (VOCs) in the presence of sunlight. Sources of the VOCs include road traffic, the production and use of organic chemicals such as solvents, and the use of oil and natural gas.

These atmospheric pollutants affect human health, directly through inhalation, and indirectly through such exposure routes as drinking water and food contamination. Most traditional air pollutants directly affect respiratory and cardiovascular systems. For example, CO has a high affinity for haemoglobin and is able to displace oxygen in the blood, leading to cardiovascular and neurobehavioural effects. High levels of SO_2 and SPM have been associated with increased mortality, morbidity and impaired pulmonary function, and O_3 is known to affect the respiratory system and irritate the eyes, nose and throat, and to cause headaches. Certain sectors of the population are often at greater risk: the young, the elderly and those weakened by other debilitating ailments, including poor nutrition.

Elements of the natural and built environment can also be adversely affected. Sulphur and nitrogen oxides are principal precursors of acid deposition (see Chapter 12), SO_2, NO_2 and O_3 are phytotoxic – O_3, in particular, has been implicated in damage to crops and forests – and damage

TABLE 10.5 *Cities of the former Soviet Union with the poorest air quality**

City	Main pollutants	Principal sources
ARMENIA		
Yerevan	Benzo(a)pyrene, SPM, NO, NO_2	Chemical industry, power plants, non-ferrous metallurgy, transport
BELARUS		
Mogilev	Benzo(a)pyrene, CS_2, NO_2	Chemical industry, ferrous metallurgy
GEORGIA		
Kutaisy	Benzo(a)pyrene, phenol, SPM	Chemical and petrochemical industries, truck plant, transport
KYRGYZSTAN		
Bishkek	Benzo(a)pyrene, formaldehyde, NO	Power plants, transport
RUSSIA (Asia)		
Angarsk	Benzo(a)pyrene, formaldehyde, SPM	Petrochemical, medical and biological industries, power plants
Bratsk	Benzo(a)pyrene, formaldehyde, CS_2, HF	Non-ferrous metallurgy, pulp and paper mills, power plants
Kemerovo	Benzo(a)pyrene, formaldehyde, CS_2	Fertiliser production, chemical industry, non-ferrous metallurgy
Komsomol'sk-na-Amure	Benzo(a)pyrene, formaldehyde, NH_3, lead	Ferrous metallurgy, power plants, petrochemical industry
Krasnoyarsk	Benzo(a)pyrene, formaldehyde, SPM	Chemical industry, non-ferrous metallurgy, construction materials production, transport
Novokuznetsk	Benzo(a)pyrene, HF, NO_2	Metallurgy, coal mining, power plants
Novosibirsk	Benzo(a)pyrene, formaldehyde, NO_2, NH_3	Transport, power plants, construction materials production
Khabarovsk	Benzo(a)pyrene, formaldehyde, phenol	Power plants, construction materials production, petrochemical industry, road and rail transport
(Europe)		
Berezniky	CS_2, H_2SO_4, NO, NO_2	Chemical industry, fertiliser production
Volzskyi	Formaldehyde, methylmercaptan, NO_2, CS_2	Petrochemical and chemical industries
Gubakha	Sterene, benzo(a)pyrene, formaldehyde	Chemical industry, power plants
Magnitogorsk	Benzo(a)pyrene, sterene, CS_2, NO_2	Ferrous metallurgy
Novocherkassk	Benzo(a)pyrene, formaldehyde, SPM	Metallurgy, petrochemical industry, power plants
UKRAINE		
Kiev	Benzo(a)pyrene, formaldehyde, NH_3	Chemical and petrochemical industries, construction materials production

*Annual mean of three or more pollutants exceeded the 24-hour national limit in 1990
Source: after Shahgedanova and Burt (1994)

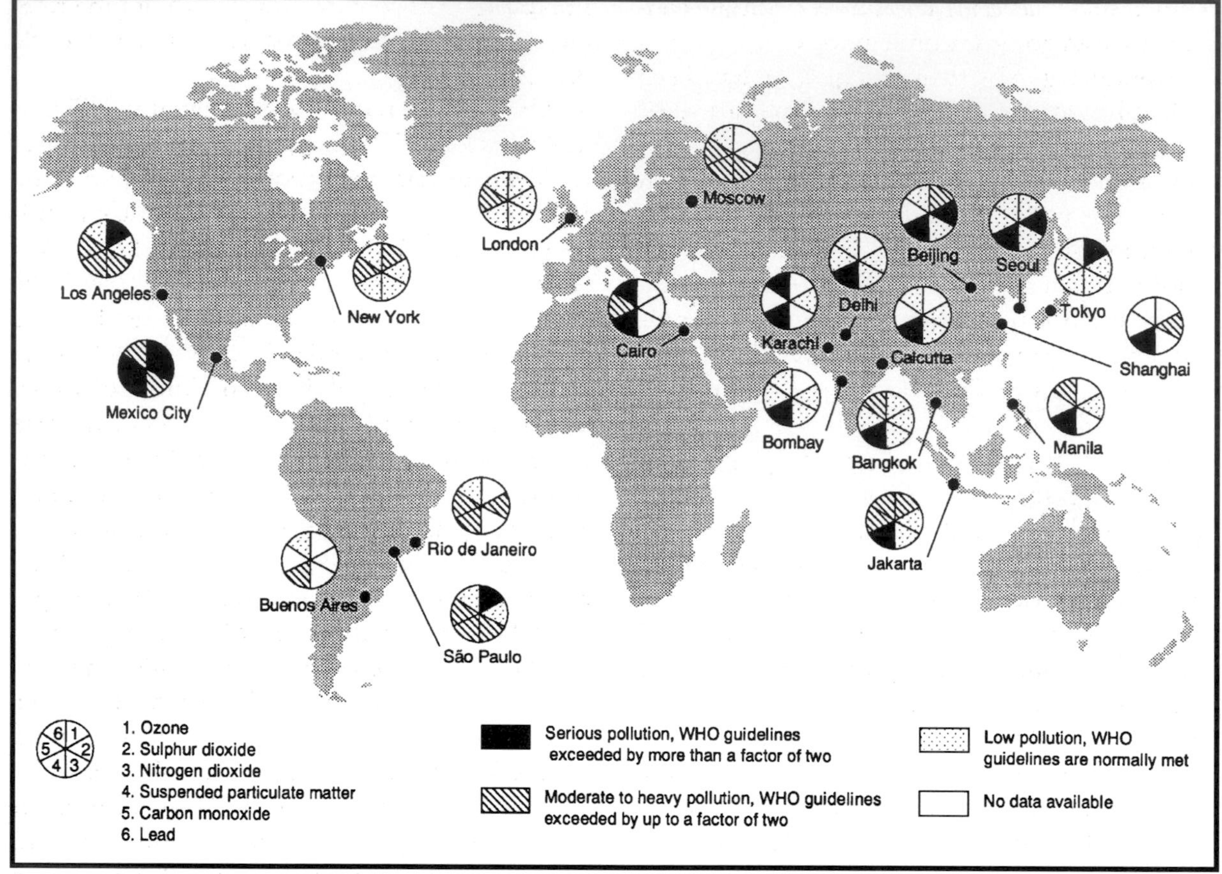

Figure 10.4 Air quality in 20 world megacities (from data in UNEP/WHO, 1992).

to buildings, works of art and materials such as nylon and rubber have been attributed to SO_2 and O_3.

In more recent times, these traditional urban air pollutants have been supplemented by a large number of other toxic and carcinogenic chemicals, which are increasingly being detected in the atmospheres of major cities. They include heavy metals (e.g. beryllium, cadmium and mercury), trace organics (e.g. benzene, formaldehyde and vinylchloride), radionuclides (e.g. radon) and fibres (e.g. asbestos). The sources of these pollutants are diverse, including waste incinerators, sewage treatment plants, manufacturing processes, building materials and motor vehicles. Concentrations of these chemicals are generally low, where they are measured, but this occurs at few sites to date.

Few long-term air quality monitoring programmes have been implemented in cities, and runs of available data are often characterised by changes in the location of sample stations, but data from the former Soviet Union indicate that urban areas with high concentrations of heavy industry using outdated technology have had a 'calamitous' effect on air quality (Shahgedanova and Burt, 1994). Table 10.5 lists cities where the annual levels of three or more pollutants exceeded the 24-hour national limit in 1990. Some of the commonest pollutants present in excessive concentrations are highly toxic: benzo(a)pyrene, a carcinogenic coal tar by-product, phenol and formaldehyde. However, it seems unlikely that the acute air pollution problems of these urban areas will receive much attention as long as financial resources are limited, and more pressing national problems such as housing and food supply continue to head government priorities.

Monitoring of urban air quality has been undertaken at a global network of megacities by UNEP and the WHO since 1974. Available data indicate that while cities in the industrialised countries have made significant reductions in air pollution during the past four decades, rapidly growing urban areas in the industrialising countries pose serious threats to the millions of people who live in them. The clear distinction in urban air quality between rich and poor countries is indicated in Fig. 10.4.

Mexico City emerges as the worst-affected city, with WHO guidelines exceeded by a factor of two or more for levels of SO_2, SPM, CO and O_3. Levels of Pb and NO_2 are almost as bad, exceeding WHO limits by up to two times. The city's poor air quality is exacerbated by its location, in an elevated mountain-rimmed basin where temperature inversions occur, on average, 20 days per month from November to March, which impairs the dispersion of pollutants (Collins and Scott, 1993). Although data are sparse, no particular trends in the six pollutants monitored are discernible in Mexico City, despite the city's rapid growth. This can be attributed to use of cleaner fuels, better emission control, replacement of old industries, and technological improvements (UNEP/WHO, 1992). From 1991, for example, it became obligatory for all new cars to be equipped with catalytic converters.

Not all efforts to control air pollution in the city have been unmitigated successes, however. Concern over rising atmospheric Pb levels, which averaged 8 g/m^3 in 1986 (five times the national standard), resulted in the national oil company reducing the lead content of gasoline sold in the city in September of that year. An unexpected side-effect was a dramatic increase in ozone concentrations, a result of the reaction between atmospheric oxygen and the replacement gasoline additives in ultraviolet sunlight (Ezcurra, 1990). Ozone carried by dominant winds to the Sierra del Ajusco, to the south-west of the Basin of Mexico, is thought to have significantly reduced the chlorophyll content and growth of pine trees on the mountains (Hernández Tejeda and de Bauer, 1986).

Many of the rapidly growing industrial cities in Asia also suffer from poor air quality. Jakarta residents experience particularly high SPM levels and one estimate of the effects on their health suggests that 600 000 asthma attacks, 49 000 emergency room visits and 1400 deaths could be avoided each year if particulate levels were reduced to WHO standards (Ostro, 1994). None the less, the severe pollution conditions observed at several megacities could have been much worse if control measures had not already been introduced. Examples include Beijing, Delhi, Seoul and Shanghai, and the need for such measures is well illustrated at Shanghai where the male lung cancer mortality rate doubled from 21 to 44 per 100 000 men from 1963 to 1985 (Rukang, 1989).

The beneficial effects of tighter legislative controls on air quality are indicated by London's annual mean SO_2 concentrations, which fell by an order of magnitude in four decades: from around 300 g/m^3 in the early 1960s to below 30 g/m^3 in the 1990s (Fig. 10.5). The introduction and enforcement of 'Smoke Control Orders' under

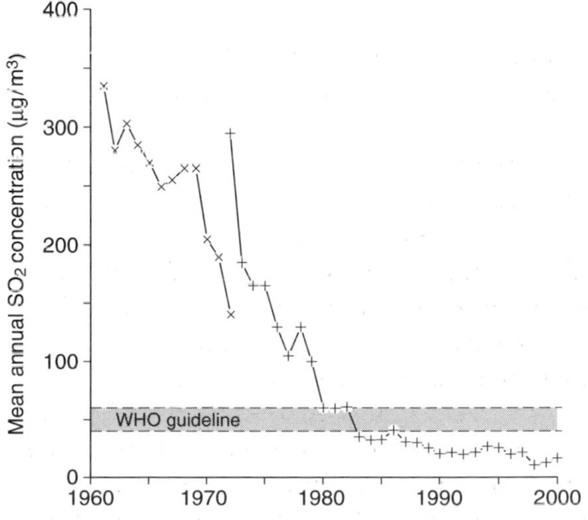

Figure 10.5 Mean annual sulphur dioxide concentrations at two monitoring stations in the City of London, 1961–2000 (data from http://www.aeat.co.uk/netcen/airqual/, accessed April 2002).

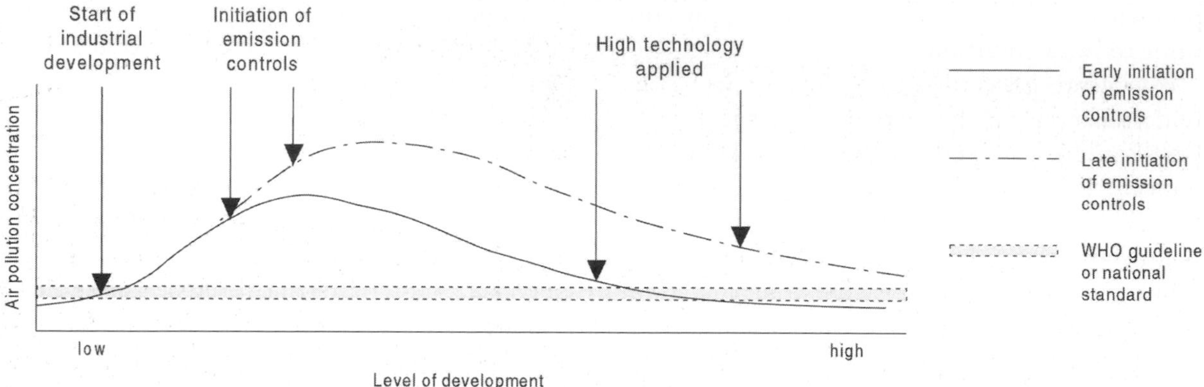

Figure 10.6 Model of air quality evolution with development status (after UNEP/WHO, 1992).

the 1956 Clean Air Act (amended in 1964 and 1968), a response to the infamous London smogs of the 1950s, is the most important factor responsible for this decline in ambient concentrations. Similar successes have been recorded for most of the six pollutants measured in Los Angeles, New York and Tokyo, although Los Angeles still has the most serious O_3 problem in the USA (see page 276). A model for the progression of air pollution problems through time at different levels of development is shown in Fig. 10.6. Pollution rises with initial industrial development, to be brought under control through legislation on emissions. Air quality then stabilises and improves as development proceeds, to be reduced to below acceptable standards by high-technology applications.

GARBAGE

The rapid, and often unauthorised, growth of urban areas has in many cases outpaced the ability of urban authorities to provide adequate facilities, such as the collection of household garbage. Table 10.6 illustrates the scale of the problem in some cities for which reasonable estimates are available. Many other urban areas similarly afflicted are not included due to lack of adequate information. Contrasts are also evident within cities, between more affluent suburbs and poorer areas.

Although the environmental problems associated with garbage do not disappear with its collection (see Chapter 17), uncollected garbage exacerbates many of the environmental hazards covered in this chapter. It can be a serious fire hazard; it attracts pests and disease vectors, creating health hazards; and local disposal by burning or dumping adds to pollution loads and clogs waterways, so increasing the dangers of flooding.

Several animal species, particularly rats, have become adapted to the urban environment by scavenging from urban refuse. Larger species, too,

TABLE 10.6 *Some large cities with poor household garbage collection facilities*

City (Country)	Proportion of garbage not collected (%)
Addis Ababa (Ethiopia)	40
Bogota (Colombia)	50
Dar es Salaam (Tanzania)	65
Guatemala City (Guatemala)	32
Jakarta (Indonesia)	40
Karachi (Pakistan)	65
Kampala (Uganda)	90
Kumasi (Ghana)	70
Lusaka (Zambia)	90
Ouagadougou (Burkina Faso)	70

Sources: after Hardoy *et al.* (1992) and (2001)

Figure 10.7 Marabou storks are a common site in Kampala, Uganda. In the countryside, these scavengers often feed on carrion and were probably first attracted to the cityscape by discarded human corpses during the times of Idi Amin and Milton Obote. They now live on garbage and rodents. Less than 10 per cent of the city's population benefits from a regular collection of household wastes.

have been drawn to garbage bins and dumps, and are also regarded as pests, such as the urban foxes that inhabit many British cities, and polar bears in Churchill, Manitoba, northern Canada. In Uganda's capital city, Kampala, carnivorous Marabou storks strut the streets like normal citizens, living off garbage and doing a useful job in controlling smaller pests (Fig. 10.7).

Some degree of waste recovery occurs in most cities. In many cities of the Third World, large numbers of residents are self-employed in the business of garbage recycling (see the example from Dar es Salaam on page 291). Mexico City and Cairo are just two examples where large squatter communities live and work on official or unofficial rubbish-dump sites. In the case of Cairo, the Zabbalean religious sect has cornered the market in garbage collection, scavenging and recycling, feeding edible portions to their domestic livestock and selling inorganic materials to dealers. Elsewhere, metropolitan authorities run similar programmes. In Beijing, for example, a state-run recycling scheme has been in operation since the 1950s, and in New York City, Local Law 19, brought into force in 1989, requires all residents, institutions and businesses to separate a variety of materials for collection and recycling.

HAZARDS AND CATASTROPHES

The high concentrations of people and physical infrastructure in cities make them distinctive in several ways with regard to hazards. Where money is available, cities are worth protecting because of the large financial and human investment they represent. Adequate provisions of water supply and sanitation are designed to offset the risks of disease; other infrastructure – such as expensive flood protection schemes – protects against geophysical hazards (see Chapter 21).

In developing countries, where escalating urban growth rates and a lack of finance make such provisions inadequate, it is usually the poorest sectors of urban society that are most at risk from environmental hazards. Rapid urban growth and rising land prices have used up the most desirable and safest sites in most cities in the developing world, leaving increasingly hazard-prone land for poorer groups. Such hazards include the pervasive dangers of high pollution levels and the intensive dangers of industrial accidents. The accidental discharge from a pesticide production plant in Bhopal, northern India, in 1984, for example, killed more than 3000 shantytown dwellers. It was primarily caused by inadequate management and lax safety procedures. The general quality of the environment in cities is also a function of infrastructure and services, which tend to be less adequate in poorer areas. A study in Buenos Aires of the distribution of infant mortality, a good indicator of the quality of the environment, showed clear correlations with income levels and basic service provision (Arrossi, 1996). Overall infant mortality rates in poorer suburbs of the city were double those in more prosperous areas, and infant mortality rates by 'avoidable' causes (including tetanus, respiratory infections and perinatal jaundice) were up to three times higher in the poorest zones.

High concentrations of poor housing in Third World cities are built on slopes on hillsides prone to sliding (e.g. Caracas), or in deep ravines (e.g. Guatemala City), on river banks susceptible to flooding (e.g. Delhi), and on low-lying coastlines prone to marine inundation (e.g. Rio de Janeiro). A study of Rio de Janeiro's sister city, Niterói, has shown that its poorest housing has been increasingly vulnerable to landslides in recent years (Smyth and Royle, 2000). The causes of landslides during the rainy season are well known: the undercutting of slopes to build houses and the weight of the houses themselves, along with the deforestation of slopes, which causes local water tables to rise and reduce the strength of the regolith. These factors also threaten more prosperous developments, but the occupants of middle- and high-class housing have greater financial resources, which enable them to adopt better construction techniques (including preventative measures such as slope-retention walls and the installation of storm drains) and thus limit the risk. There is little doubt that the economic poverty of Niterói's shanty dwellers is the key to their increased vulnerability: although often aware of slope stability problems, they can do little to abate them.

Even the destruction caused by citywide hazards, such as earthquakes, can be magnified in unstable sites: in Guatemala, 65 per cent of deaths in the capital caused by the 1976 earthquake occurred in the badly eroded ravines around the city. In other situations, however, the damage and loss of life caused by earthquakes can be greatest in more built-up parts of the urban environment, when buildings themselves become hazardous if they are not constructed to withstand earth tremors. The widespread failure of relatively new constructions in urban areas of Armenia in the 1988 earthquake, in which 25 000 people lost their lives, echoed the experiences of Mexico City in 1985. Seemingly, more sophisticated technology able to withstand tremors had not been incorporated into new buildings for reasons of cost (Krimgold, 1992). Failure of urban infrastructure following earthquakes is one of the commonest causes of damage and loss of life. The most serious earthquake disaster in the USA, in San Francisco in 1906, was largely a function of infrastructural failure. Disruption of gas distribution and service lines caused the outbreak of many fires and interrupted water distribution, so making it difficult to put the fires out.

The most critical environmental problems faced in urban areas of the developing world,

TABLE 10.7 *Countries with particularly poor drinking water and sanitation provision in urban areas, 1990–97*

Country	Access to safe drinking water (% of urban population)	Access to sanitation (% of urban population)
AFRICA		
Benin	46	57
Nigeria	58	50
Sierra Leone	58	17
ASIA		
Afghanistan	39	38
Myanmar	78	56
Viet Nam	47	43
LATIN AMERICA		
Dominican Rep	80	76
Haiti	50	49
Nicaragua	88	34

Source: after WRI (2000: 300–1, Table HD.3)

however, stem from the disease hazards caused by a lack of adequate drinking water and sanitation. Table 10.7 shows some of the countries worst affected. Worldwide, at least 170 million people in urban areas lacked a source of potable water near their homes in 1990, and 375 million did not have adequate sanitation (World Bank, 1992a). The year 1990 marked the end of the WHO's International Drinking Water and Sanitation Decade, which despite nearly US$100 billion in investments, fell far short of its goal of clean water and sanitation for all.

Water-borne diseases (e.g. diarrhoea, dysentery, cholera and guinea worm), water-hygiene diseases (e.g. typhoid and trachoma) and water-habitat diseases (e.g. malaria and schistosomiasis) both kill directly, and debilitate sufferers to the extent that they die from other causes. Again, it is the less well-off sectors of urban society that are most at risk. The effects of improvements to water supply and wastewater disposal on life

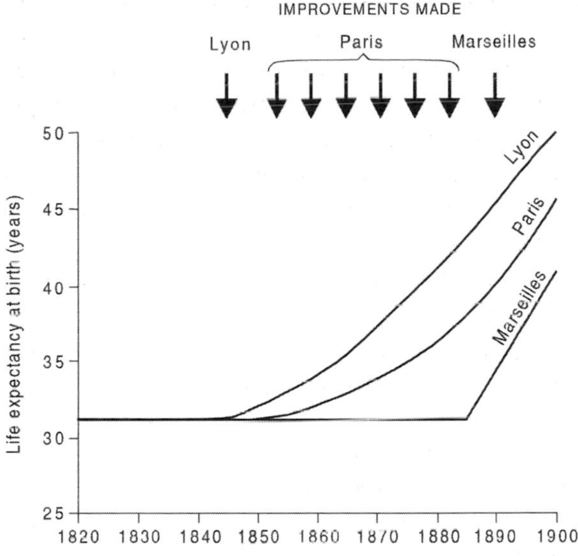

Figure 10.8 Rising life expectancy with improvements in water supply and sanitation in three French cities, 1820–1900 (Briscoe, 1987).

expectancy have been clearly shown in the industrial countries, when services were improved during the nineteenth and twentieth centuries. The trend shown for three major French cities in Fig. 10.8 is typical in this respect, with life expectancy increasing from about 32 years in 1850 to about 45 years in 1900, with the timing of changes corresponding closely to improvements in water and sanitation provision.

However, the implementation of improvements in the needy parts of the world's poorer cities is unlikely to adopt similar approaches to those taken historically in the developed countries. Centralised systems built and maintained by subsidised public agencies will not work in today's developing world simply because the numbers of unserved people are too great and the financial costs too high. Some of the lessons learned during the International Drinking Water and Sanitation Decade give guidance for continuing attempts to provide alternatives (M. Black, 1994). Four key points emerged:

1 systems must respond to local demands, and must be as simple, sturdy and inexpensive as possible

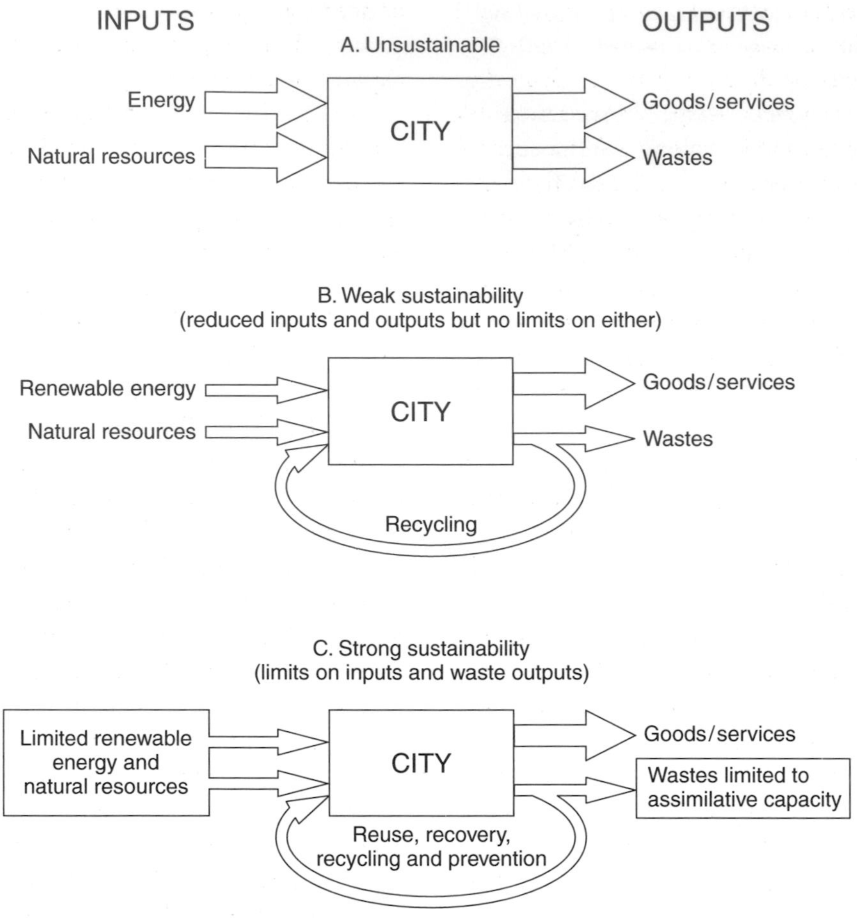

Figure 10.9 Improvements in urban sustainability (after Moffatt, 1999).

2 involving the community and individual households, particularly women, in system design and maintenance is crucial to a project's success

3 governments need to improve the efficiency and sustainability of system operation and maintenance

4 water should be treated as an economic commodity and paid for by users.

TOWARDS A SUSTAINABLE URBAN ENVIRONMENT

It is probably unreasonable to expect that major cities should be supported by the resources produced in their immediate surrounds, and thus urban areas will always leave an ecological footprint on their hinterlands. The idea of increasing the sustainability of urban areas receives a great deal of attention from politicians and urban planners but the fact that cities are not isolated from their surrounds means that improvements to urban sustainability should be undertaken as part of wider programmes of sustainable development everywhere.

A theoretical attempt to trace some of the measures needed to improve urban sustainability is shown in Fig. 10.9. The unsustainable city (A) consumes energy from non-renewable sources and natural resources of all kinds, producing goods and services as well as all sorts of waste products that are released into the environment with varying degrees of management (e.g. motor vehicle emissions, domestic waste buried in landfills). The city depicted in B has improved its sustainability, drawing energy from renewable sources and consuming fewer natural resources as they are used more efficiently. Wastes have also been reduced by introducing recycling measures. Stronger sustainability is shown in C, where limits have been put on the amounts of inputs and outputs, and better waste management has been introduced in the form of reuse, recovery and recycling, as well as improved prevention (see Chapter 17).

While the history of many developed countries' experiences in dealing with urban problems offers numerous indicators as to how the environmental difficulties that plague so many of the developing world's major cities can be solved, the solution to the awesome environmental problems outlined in this chapter appears to depend largely upon available finance and political will. Some of the efforts to combat these problems have been described, but, to conclude, it is useful to examine the experience of Curitiba, a city in south-east Brazil, as a model for other cities where financial resources may be at a premium. Despite a rapidly growing population, rising from 300 000 in 1950 to about 2.3 million people in 1990, Curitiba has greatly improved its urban environment with a series of innovative, relatively low-cost transport, land use and waste disposal measures (Lowe, 1991; Rabinovitch, 1992).

A new public transport system, involving competitively priced bus services using exclusive bus lanes, has been used as a framework for current and future urban development. The system, which carries 1.3 million passengers per day, has reduced congestion on the roads, and lowered air pollution as a typical Curitiba private vehicle now uses 30 per cent less fuel than the average for eight comparable Brazilian cities. The city also boasts one of the lowest motor accident rates per vehicle in the country, and its inhabitants enjoy a very low average transport expenditure. Land use policy has aimed to improve urban conditions, introducing pedestrian precincts and cycleways, concentrating on redeveloping existing sites rather than expanding the urban area, and expanding parks and other green spaces.

Faced with a growing garbage problem, a recycling scheme was introduced in 1989 with education programmes to encourage recycling through municipal collection. Residents were asked to separate garbage before collection, so cutting municipal costs significantly, and more than 70 per cent of households now participate. Effective garbage collection has also meant a significant improvement in the city's squatter settlements (*favelas*). The problem of access to the high-density *favelas* has been tackled by offering bus tickets, food parcels and school notebooks in exchange for household refuse brought to accessible roadsides. The benefits, enjoyed by an estimated 35 000 families, include a marked decrease in litter, rising standards of nutrition and an overall improvement in *favela* quality of life.

Part of the key to success in Curitiba has been the city government's promotion of a strong sense of public participation and its emphasis on low-cost programmes. People are encouraged to build their own houses, with city government loans and assistance from city architects, to offset large municipal outlays on housing projects. Old public buses have been converted into mobile schools that tour low-income neighbourhoods. The creativity displayed in Curitiba shows that many of the environmental problems commonly associated with urban areas can be tackled successfully without large financial resources, given strong leadership and guidance to a population that is willing to participate in solving its own difficulties.

FURTHER READING

Elsom, D. 1996 *Smog alert: managing urban air quality*. London, Earthscan. A comprehensive review of urban air pollution, its effects, particularly those on human health, and methods for its management.

Ezcurra, E. and Mazari-Hiriat, M. 1996 Are mega cities viable? A cautionary tale from Mexico

City. *Environment* 38(1): 6–15, 26–35. A catalogue of Mexico City's urban environmental problems.

Hardoy, J.E., Mitlin, D. and Satterthwaite, D. 2001 *Environmental problems in an urbanizing world*. London, Earthscan. A comprehensive review of environmental issues in rapidly growing cities of the developing world.

Lents, J.M. and Kelly, W.J. 1993 Cleaning the air in Los Angeles. *Scientific American* 269(4): 18–25. Case study of efforts to improve atmospheric pollution in the city with the worst air quality in the USA.

Lowe, M.D. 1991 *Shaping cities: the environmental and human dimensions*. Worldwatch Paper 105. Washington DC, Worldwatch Institute. This pamphlet looks at past problems and future prospects for urban planning from the perspectives of quality of life and environment.

Moffatt, I. 1999 Edinburgh: a sustainable city? *International Journal of Sustainable Development and World Ecology* 6: 135–48. A case study of attempts to make Scotland's capital sustainable.

Warren-Rhodes, K. and Koenig, A. 2001 Escalating trends in the urban metabolism of Hong Kong: 1971–1997. *Ambio* 30: 429–38. An analysis of trends in resource consumption and waste generation.

World Resources Institute 1996 *World resources 1996–97*. New York, Oxford University Press. A special issue on the urban environment.

Websites

http://www.cities21.com/ the International Council for Local Environmental Initiatives site has a wealth of information on programmes for the sustainable development of cities.

http://www.ceroi.net/ Cities Environment Reports on the Internet hosts state-of-the-environment reports from several cities.

http://www.urbanecology.org/ dedicated to sustainability in urban environments.

http://www.gdrc.org/uem/ a virtual library on urban environmental management.

http://www.unchs.org/scp/ the UN's Sustainable Cities Programme.

http://www.unhabitat.org/ the UN's Human Settlements Programme.

http://www.sustainable-cities.org/ the European Sustainable Cities Project.

POINTS FOR DISCUSSION

- Are megacities unsustainable by definition?
- Will air pollution problems in large cities recede as countries develop?
- Most of the environmental problems of urban environments can only be solved with political will and economic resources. Do you agree?
- Why are some areas of cities more hazardous than others?

11

CLIMATIC CHANGE

TOPICS COVERED

Past climatic change
Human impacts on the atmosphere
Greenhouse trace gases
Global warming

Key words: ice age, Intergovernmental Panel on Climate Change (IPCC), Montreal Protocol, Kyoto Protocol, emissions trading, no-regrets initiatives, carbon sequestration

Global climatic change due to increasing atmospheric concentrations of greenhouse gases has dominated the environmental agenda since the mid-1980s and has engendered considerable international political debate. There is no doubt that over the past 100 years or so, human action has significantly increased the atmospheric concentrations of several gases that are closely related to global temperature. It seems likely that these increased concentrations, which are set to continue to build up in the near future, are already affecting global climate, but our poor knowledge and understanding of the workings of the global heat balance make the current and future situation uncertain. Predictions of the climatic nature of the Earth into the next century, and the effects of potential climatic changes on other aspects of the natural and human environment, are thus tentative. However, many of the responses that could ameliorate the impacts of potential climatic change can be argued for on the grounds of their other benefits.

PAST CLIMATIC CHANGE

The Earth's climate has never been static and the human impact on climate has been minor relative to the large-scale perturbations brought about by natural processes. Instrumental or documentary records of climatic parameters are only available for a few thousand years at most, but even on such short timescales some marked variations in climate have been noted (see below). Our knowledge of Earth's climate over longer time periods is derived from 'proxy' indicators. The first evidence of the existence of former 'ice ages' came from glacial landforms, and buried soils and fossils, which indicated that climate in the Pleistocene (1 800 000–10 000 years BP) had alternated between cold glacials and warmer interglacials, with interglacial conditions in most mid- and high-latitude regions being similar to those of today. These lines of evidence have subsequently been supplemented with information gleaned from other sources, most notably deep-sea cores.

Theories developed to explain these changes in the Earth's climate have focused on variations in the amount of solar radiation received by the atmosphere. A composite sequence of these variations due to perturbations in the Earth's orbit around the sun was calculated in the early years of the twentieth century by a Serbian mathematician named Milankovitch. Three types of variation, operating with different periodicities were identified:

1 orbital eccentricity with a periodicity of 100 000 years
2 axial tilt or obliquity with a periodicity of 41 000 years

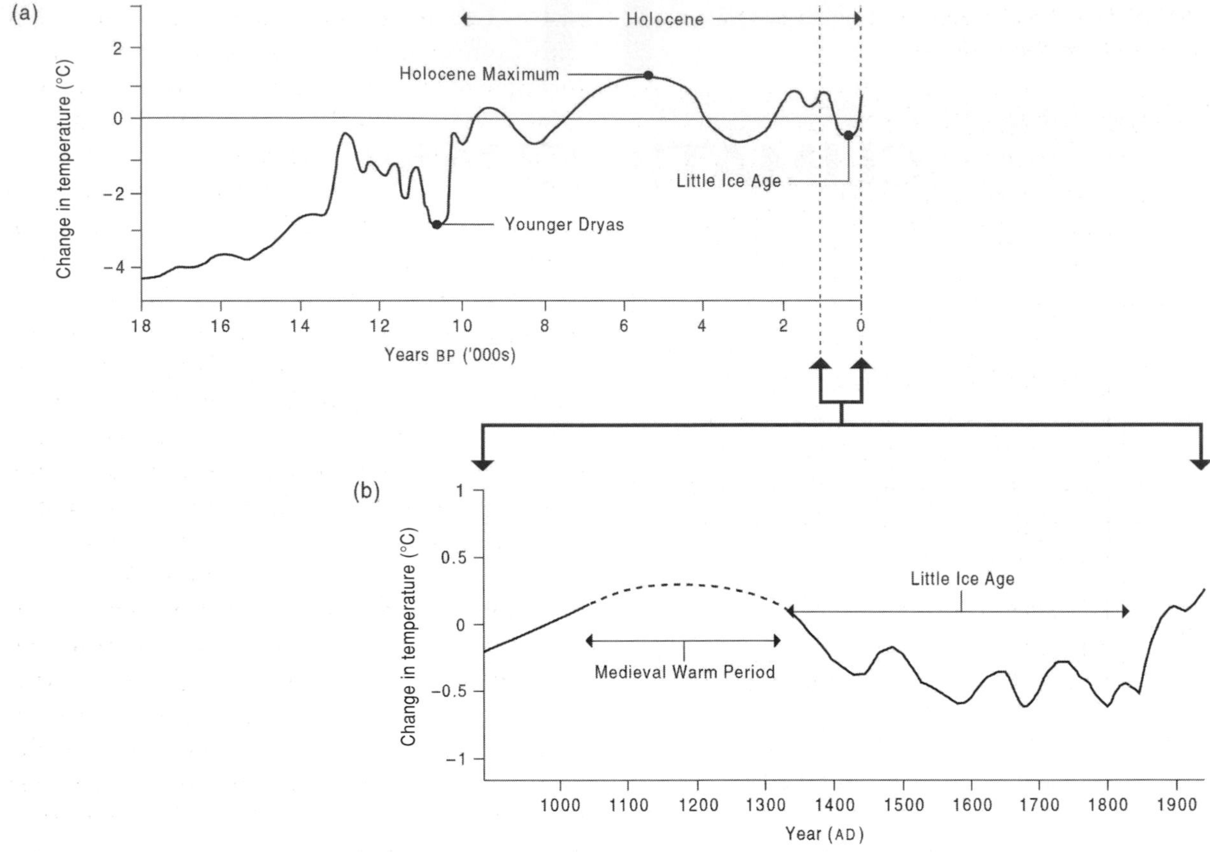

Figure 11.1 Variations in global temperature (after Houghton *et al.*, 1990): (a) over the last 18 000 years; (b) over the last 1000 years.

3 precession of the equinoxes with a periodicity of about 21 000 years.

These periodic variations, which have been confirmed from deep-sea core evidence (Hays *et al.*, 1976), help to explain the occurrence of glacial and interglacial periods on Earth. Over the last 0.8–0.9 million years, for example, eight major glacial–interglacial cycles have occurred, each lasting about 100 000 years, but beginning and ending in quite sudden jumps in temperature. Superimposed on these cyclical fluctuations, there appear to be more frequent and shorter-term climate transitions, most of them on timescales of centuries to decades (Adams *et al.*, 1999).

The last glacial is generally thought to have been, at a maximum, about 18 000 years ago, and since then the Earth's climate has warmed (Fig. 11.1a). The most recent phase of the Quaternary is known as the Holocene, which began 10 000 years ago, and even during this period the Earth's climate has by no means been constant. Indeed, the last 2000 years have seen a relatively warm period during medieval times and a so-called Little Ice Age during the fifteenth to eighteenth centuries (Fig. 11.1b). The end of the Little Ice Age coincided with the era when human activity began to have a significant impact on global temperatures through enhancement of the greenhouse effect (see below).

HUMAN IMPACTS ON THE ATMOSPHERE

Human activities inadvertently affect the workings of the atmosphere in numerous ways, in many cases with possible effects on climatic

TABLE 11.1 *Possible mechanisms for inadvertent human-induced climate change*

DIRECT ATMOSPHERIC INPUTS
- Gas emissions (carbon dioxide, methane, chlorofluorocarbons, nitrous oxide, krypton-85, water vapour, miscellaneous trace gases)
- Aerosol generation
- Thermal pollution

CHANGES TO LAND SURFACES
- Albedo change (deforestation, intensive grazing, dust addition to ice caps)
- Roughness change (deforestation, urbanisation)
- Extension of irrigation

ALTERATIONS TO THE OCEANS
- Current alterations by constricting straits
- Diversion of fresh waters into oceans

Source: after Goudie (1993b)

regimes (Table 11.1). Direct inputs of gases, small particles (aerosols) and heat energy can all affect the operation of climate on various scales. Emissions of aerosols and heat are responsible for local 'heat islands' around urban areas and the creation of photochemical smogs (see Chapter 10), and enhanced inputs of soil aerosols, particularly from agricultural areas in drylands, affect the radiation properties of the atmosphere, possibly resulting in decreased rainfall locally (Bryson and Barreis, 1967). On larger scales, gas emissions are believed to be responsible for an enhanced global greenhouse effect (see below) and depletion of the concentrated layer of ozone present in the stratosphere.

Stratospheric ozone plays a key role in climatic processes through its capacity to absorb incoming solar ultraviolet radiation. This warms the stratosphere and maintains a steep inversion of temperature between about 15 and 50 km above the Earth's surface, affecting convective processes and circulation in the troposphere below. The history of our understanding of and interaction with stratospheric ozone is outlined in Table 11.2. Just how the human-induced depletion of stratospheric ozone, which was first identified as a 'hole' over the Antarctic in 1985 (Farman *et al.*, 1985), may affect global climate is unclear. To date, most concern has focused on the possible ecological effects of increased ultraviolet radiation reaching the Earth's surface. Modifications to the rate of photosynthesis are likely to have significant impacts on many living organisms, reducing productivity in aquatic life such as plankton, and terrestrial plants (some evidence of deleterious impacts on lacustrine flora and fauna is cited on page 212). A direct impact on human health is also likely, through increases in the incidence of skin cancers and cataracts. The issue is no longer confined to the southern hemisphere since stratospheric ozone depletion has now been identified in the northern hemisphere mid latitudes and in the Arctic. Despite prompt international action to reduce chlorofluorocarbons, the compounds mainly responsible for stratospheric ozone depletion, past emissions will continue to cause ozone degradation for decades to come due to the timelag between their production and release into the atmosphere, and their damaging effects. Full recovery is not expected until about 2050 at the earliest.

A number of human modifications to the land surface are suspected of inducing more localised climatic effects by changing the reflectivity or 'albedo' of ground surfaces. Relatively long-lived albedo changes can be brought about by actions that severely modify natural vegetation covers, such as deforestation and heavy grazing, and similar effects have been noted when large amounts of dust particles blown from agricultural soils are deposited on glacier ice. Such albedo changes can also come about through natural processes, however. Drought can alter vegetation cover, and volcanic eruptions can eject huge quantities of dust into the atmosphere. Changes in surface albedo affect the amount of solar energy absorbed by a surface, and hence the amounts of heat energy the surface subsequently releases. This, in turn,

TABLE 11.2 *Human understanding of, and effects on, stratospheric ozone*

Date	Development
1880	Forty years after the discovery of ozone in 1840, its role as a filter to the sun's ultraviolet radiation is recognised, and its concentration in the stratosphere is realised.
Late 1920s	First systematic measurements of the distribution and variability of the ozone layer are made by G.M.B. Dobson, revealing that a maximum occurs during spring at high latitudes.
1950s	Chlorofluorocarbons (CFCs), invented in 1928, come into widespread use particularly for refrigeration and later in air-conditioning, spray cans, foams and in solvents. Hailed as miracle chemicals, CFCs are claimed to be completely harmless to living things.
1970	Concern about the use of supersonic transport planes leads to the discovery of an ozone-destroying catalytic cycle involving nitrogen compounds. Although initial projections of the scale of ozone loss due to these aircraft later prove to be excessive, public awareness regarding the importance and fragility of the ozone layer is raised.
1974	Measurements in the early 1970s reveal a growing abundance of CFCs at ground level, prompting Molina and Rowland (1974) to speculate that CFCs rise to the stratosphere where they break down, producing chlorine, which destroys the ozone layer.
1979	USA, Canada, Norway and Sweden ban CFC use in nearly all spray cans, and the global use of CFCs slows significantly.
Early 1980s	CFC use increases again, due in part to their use as cleaning agents in the rapidly expanding electronics industry.
1985–86	The 'ozone hole' is discovered in 1985 by scientists from the British Antarctic Survey, who report a deepening depletion in the springtime ozone layer above Antarctica of about 40 per cent. Also in 1985, the Vienna Convention for the Protection of the Ozone Layer is signed. The following year, an Antarctic expedition points to chlorine and bromine compounds as the likely cause.
1987	The UN adopts the Montreal Protocol on Substances that Deplete the Ozone Layer, requiring a freeze on annual use of CFCs by 1990 with a 50 per cent reduction by 2000, and a freeze on annual halon production by 1993.
1988	Depletion of the ozone layer in the northern hemisphere mid-latitudes and in the Arctic discovered.
1990	Montreal Protocol amended to accelerate emissions reductions, requiring complete phase-out of CFCs, halons (except for essential uses) and other major ozone-depleting substances by 2010.
2050	Earliest predicted date for full recovery of damage to the ozone layer.

affects such atmospheric processes as convection and rainfall. In simple terms, a greater surface albedo results in a cooler land surface, which, in turn, reduces convective activity and hence rainfall. Such effects may cause a positive feedback, as some researchers have suggested as an explanation for the prolonged drought period in the Sahel since the late 1960s (Charney *et al.*,

1975). In this case, widespread loss of vegetation cover caused by a combination of natural drought and overgrazing could be acting to prolong the drought through the positive feedback induced by albedo change. Warnings that similar feedbacks could operate in areas of widespread tropical deforestation have also been made (Potter *et al.*, 1975), although proving such hypotheses is by no means simple.

Direct human modifications to the hydrological cycle may also have climatic impacts. The rise in the area of irrigated agriculture in many parts of the world (see Chapter 13) and the creation of numerous large water bodies behind dams (see Chapter 9) both modify local albedos, and enhance local evaporation and transpiration rates. The increase in albedo in the area previously occupied by the Aral Sea is thought to have contributed to an enhanced continentality of the climate, with falling relative humidities, rising temperatures, a shift in the seasonality of frosts and a three-fold increase in the number of drought days (Glazovsky, 1995). Impacts on the nature and workings of the oceans might also have local climatic effects. Changes in coastal salinity due to modifications of river regimes, for example, alter the heat capacity of marine waters.

Figure 11.2 Coal-fired power stations like this one at Didcot in southern England are major contributors to the human-induced increase in global atmospheric carbon dioxide concentrations.

GREENHOUSE TRACE GASES

The most worrying type of human-induced climatic change is that brought about through modifications to the natural atmospheric budgets of so-called greenhouse gases. The atmospheric warming caused by greenhouse gases present in the atmosphere in trace amounts is a natural phenomenon caused by the effect of these gases being mainly transparent to incoming short-wave solar radiation but absorbent of re-radiated long-wave radiation from the Earth's surface. The most important greenhouse gases - carbon dioxide, methane, water vapour and nitrous oxide - all occur naturally in the atmosphere, and without their greenhouse properties the Earth's mean temperature would be about 33°C lower than at present (Houghton *et al.*, 1990). However, the atmospheric concentrations of some of these gases have increased dramatically over the past 100 years or so due to modifications of natural cycles by human populations, while new gases with greenhouse properties, notably chlorofluorocarbons, or CFCs, have also been added. Table 11.3 summarises the main atmospheric gases that influence the operation of the greenhouse effect, many of which also effect the destruction of the stratospheric ozone layer.

Most interest has focused on carbon dioxide as the most important greenhouse gas to have been increased by human action, atmospheric concentrations of which have risen by 25 per cent over the last 100 years, with about half of this increase occurring in the past 25 years. The burning of fossil fuels is the most significant source of human additions to atmospheric carbon dioxide (Fig. 11.2), while cement manufacture and land use changes - primarily deforestation - are also important. Global carbon dioxide emissions from fossil fuel combustion and cement manufacture have increased by more than four times since 1950 and these sources now release nearly 7 Gt of carbon a year. As Fig. 11.3 indicates, North America and Europe are by far the largest sources of these industrial emissions, although the trend in these areas has tended to level off or fall in the last decade (precipitously in the centrally planned

TABLE 11.3 *Atmospheric trace gases that are significant to global climatic change*

	Carbon dioxide (CO_2)	Methane (CH_4)	Nitrous oxide (N_2O)	Chlorofluoro-carbons (CFCs)	Tropospheric ozone (O_3)	Water vapour (H_2O)
Greenhouse role	Heating	Heating	Heating	Heating	Heating	Heats in air; cools in clouds
Effect on stratospheric ozone layer	Can increase or decrease	Can increase or decrease	Can increase or decrease	Decrease	None	Decrease
Principal natural sources	Balanced in nature	Wetlands	Soils; tropical forests	None	Hydrocarbons	Evapo-transpiration
Principal anthropogenic sources	Fossil fuels; deforestation	Rice culture; cattle; fossil fuels; biomass burning	Fertiliser; land-use conversion	Refrigerants; aerosols; industrial processes	Hydrocarbons (with NOx); biomass burning	Land conversion; irrigation
Atmospheric lifetime	50–200 years	10 years	150 years	60–100 years	Weeks to months	Days
Pre-industrial concentration (1750–1800) at surface (ppb)	280 000	700	285	0	10	Unknown
Present atmospheric concentration in parts per billion (ppb) by volume at surface	370 000	1770	311	CFC-11: 0.26 CFC-12: 0.53 CFC-113: 0.08	20–40*	3000–6000 in stratosphere
Present annual rate of increase	0.5%	0.9%	0.3%	4%	0.5–2.0%	Unknown
Global warming potential	1	11	270	3400–7100	—	—
Relative contribution to the anthropogenic greenhouse effect	60%	15%	5%	12%	8%	Unknown

*Northern hemisphere

Source: after Earthquest (1991)

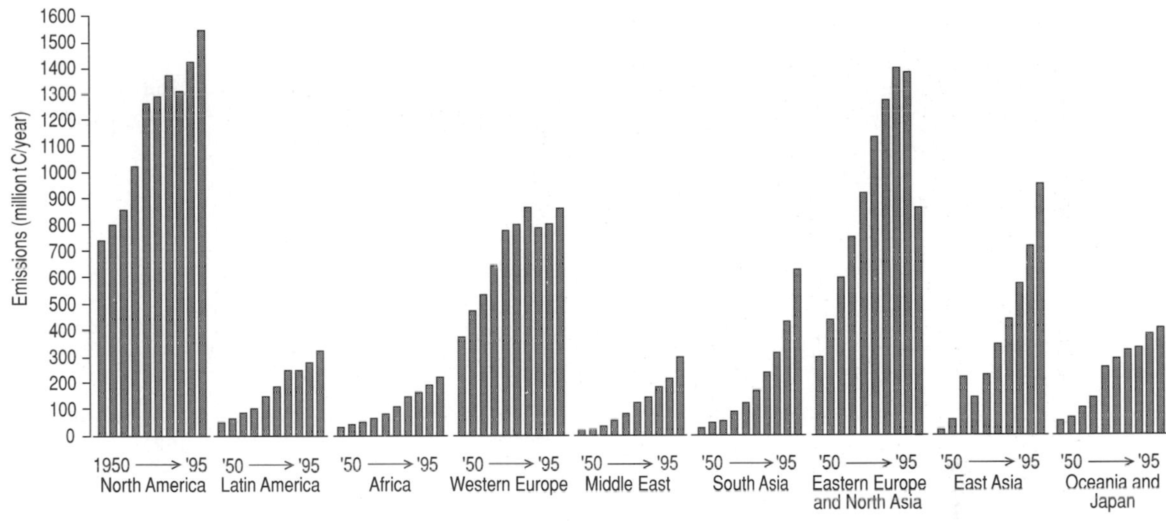

Figure 11.3 Trends in carbon dioxide emissions from industrial sources in major world regions, at five-year intervals, 1950–95 (updated from UNEP, 1993).

economies of eastern Europe and northern Asia due to depressed economies and mild winters), while emissions continue to increase rapidly in other areas, particularly in parts of Asia. On a national per capita basis, emission rates are closely tied to economic prosperity and closely related to energy use. The highest emission rates are from the oil-rich Gulf states (e.g. Kuwait – 29 t of carbon per person per year in 1995), while equivalent figures for the OECD member states average from 10–20 t/year and for many African countries they are 0.5 t/year.

Several estimates of carbon dioxide emissions due to land use changes have been made, ranging from 0.6 to 2.6 Gt/year (Intergovernmental Panel on Climate Change, 1992). The considerable uncertainty reflects the problems of assessing deforestation rates in the tropics (see Chapter 4), which is thought to be the most significant source of increased carbon to the atmosphere from this activity.

The rapid rise in the atmospheric concentration of methane, which is today more than double its pre-industrial concentration, is linked to both industrialisation and increases in world food supply. Estimates of the global methane budget shown in Table 11.4 indicate that methane produced by anaerobic bacteria in the standing waters of paddy fields and the guts of grazing livestock (by so-called 'enteric fermentation'), particularly cattle, is almost equal to the total natural production, an output dominated by rotting vegetation in wetlands. Like carbon dioxide emissions, industrial sources also contain an important fossil fuel element. Overall, anthropogenic sources are currently thought to contribute more than twice as much methane to the atmosphere as natural sources.

CFCs and other halocarbons are compounds that do not occur naturally. Their development, which dates from the 1930s, was for use as aerosol propellants, foam-blowing agents and refrigerants, and their release into the atmosphere has been inadvertent. Although national regulations on the use of CFCs in aerosol sprays during a time of economic recession in the developed world led to a fall in CFC emissions in the late 1970s and early 1980s, emissions climbed again subsequently as economies improved. While aerosol propellants accounted for almost 70 per cent of the market for CFCs in the mid-1970s, by the late 1980s refrigerants and foam-blowing agents accounted for 60 per cent of the market (McFarland, 1989). Concentrations of these

TABLE 11.4 *Estimates of methane source strengths and sinks*

Sources/sinks	Best estimate (million tonnes/ year)	Range (million tonnes/ year)
SOURCES	**515**	
Natural	155	
Wetlands	115	100–200
Termites	20	10–50
Oceans	10	5–20
Fresh water	5	1–25
Methane hydrate	5	0–5
Anthropogenic	360	
Coal mining, natural gas and petrochemical industries	100	70–120
Rice paddies	60	20–150
Enteric fermentation	80	65–100
Animal wastes	25	20–30
Domestic sewage treatment	25	?
Landfills	30	20–70
Biomass burning	40	20–80
SINKS	**500**	
Atmospheric removal	470	420–520
Removal by soils	30	15–45
ATMOSPHERIC INCREASE	**15**	

Source: Intergovernmental Panel on Climate Change (1992)

compounds are far lower than those of other greenhouse gases, but the greenhouse warming properties of CFCs are several thousand times more effective than carbon dioxide. Production of eight halocarbons, including the most abundant CFC-11 and CFC-12, has been severely curtailed by the Montreal Protocol adopted in 1987 (see below), and the stratospheric concentrations of chlorine (the CFC breakdown product that actually destroys ozone) were predicted to peak between 1997 and 1999 (Hofmann, 1996).

GLOBAL WARMING

The theory relating increased atmospheric concentrations of greenhouse gases and global warming is strongly supported by proxy evidence from ice-core data, which show that natural fluctuations in the atmospheric concentrations of greenhouse gases through geological time have oscillated in close harmony with global temperature changes over the past 150 000 years, indicating that the two are almost certainly related (Lorius *et al.*, 1990). Evidence gleaned from a range of other proxy indicators suggests that the twentieth century was the warmest of the last millennium (Jones *et al.*, 1998), and the changes in global mean temperature measured instrumentally since the mid-nineteenth century are shown in Fig. 11.4, which indicates that, overall, the planet has warmed at the surface by about 0.6°C over the past century. Although it is possible that

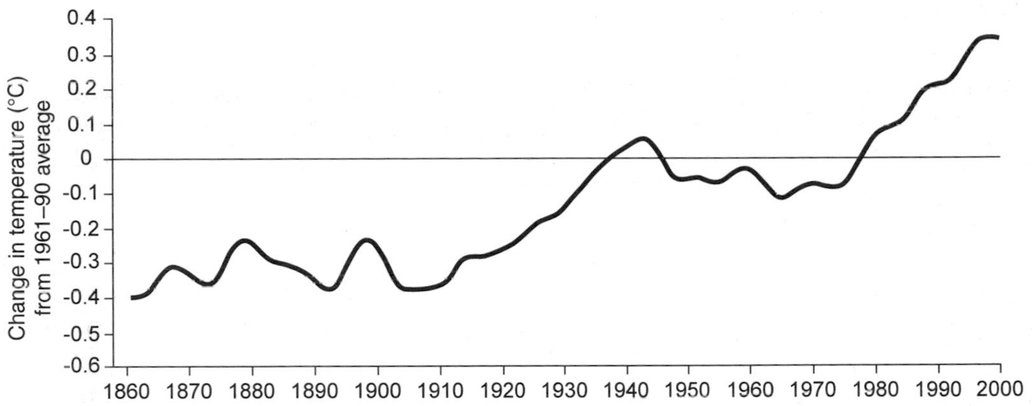

Figure 11.4 Variations in global temperature, 1860–2000 (after Houghton *et al.*, 2001).

this recent warming trend reflects the end of the Little Ice Age (see Fig. 11.1b), most researchers think that the trend shown in Fig. 11.4 is unlikely to be entirely natural in origin. In part, it reflects the operation of an enhanced greenhouse effect due to human pollution of the atmosphere.

Nevertheless, the warming trend over the past 13 or 14 decades has not been continuous through either time or space. Two periods of relatively rapid warming (from 1910 to the 1940s, and again from the mid-1970s to the present) contrast preceding periods, which were respectively characterised by fairly unchanging (1860s to 1900s) and slightly declining (1940s to 1970s) temperature. Spatially, too, global warming has been discontinuous: the two hemispheres have not warmed and cooled in unison and in contrast to most of the planet a few areas of the globe have not warmed in recent decades. These areas include parts of the southern hemisphere oceans and parts of Antarctica (Houghton *et al.*, 2001). Our understanding of the warming effects of enhanced greenhouse gases is further complicated by the observation that highly industrialised areas appear to be warming at a slower rate than less industrial regions. The large emissions of sulphate particles to the atmosphere from industry appear to be retarding warming by reflecting solar radiation back into space (Charlson and Wigley, 1994).

Predicting impacts

The widespread acceptance of the fact that human action has increased the atmospheric concentrations of several greenhouse gases, and the strong probability that these gases are acting to warm the Earth, has engendered a great deal of research into predicting the conditions of a warmer planet. Most of this research uses theoretical numerical models known as general circulation models, or GCMs, run on powerful computers. Higher mean temperatures will, of course, affect many other aspects of the climate, such as winds, evapotranspiration, precipitation and clouds, and by simulating the processes in the atmosphere, the GCMs can be used to predict changes in the distribution of these phenomena through both time and space.

Like all models, GCMs are simplifications of the real world and have numerous deficiencies (Henderson-Sellers, 1994). The models are slow to run, costly to use and their results are only approximate. They can predict changes in climatic conditions on continental scales, but predictions at finer resolutions – for a country the size of the UK, for example – are impossible at present. Part of the problem is the fact that we still do not understand fully all the processes of the climatic system, although we do realise its complexity. Different processes operate on different spatial scales and over varying timescales, and there are numerous feedbacks between different elements of the system, which could enhance or dampen particular environmental responses to change. Present GCMs are poor at incorporating the processes in oceans, which are intimately linked to those of the atmosphere, and can only give a very simple representation of the influences of clouds, for example. Both of these elements may have key feedback effects, which are poorly understood and inadequately modelled at present. Cloud cover, for example, is expected to be increased due to higher levels of atmospheric moisture consequent upon higher surface temperatures. However, whether greater cloud cover will amplify or dampen warming overall is unclear. Clouds reflect incoming solar radiation, so reducing warming on the ground, but they also transfer heat from their upper to lower surfaces, effectively acting like heat pumps (Maddox, 1990).

Other possible feedbacks that are poorly represented in many GCMs are those associated with global vegetation. An increase in CO_2 may make a further contribution to global warming by inducing a physiological response in vegetation. The way in which plant stomata operate is likely to be affected, with the overall result of reducing transpiration. Carbon dioxide also has a fertilising effect on plants, so that a CO_2-enriched atmosphere should result in more vegetation on the ground. This structural change in global vegetation will alter the surface albedo and thus produce another feedback on global climate in some way. One simulation of the overall effect of vegetation changes suggests that structural changes in vegetation will partially offset physiological vegetation–climate feedbacks, but overall climatic feedbacks from global changes in vegetation result in significant regional effects (Betts *et al.*, 1997).

The performance of GCMs is tested against our knowledge of present and past climates. The models are constantly being improved, but there is still a long way to go. Our relatively poor ability to simulate the operation of the climatic system is illustrated by the fact that GCMs predict that global temperatures should have risen by about 1°C since about 1880, yet as Fig. 11.4 shows, they have only risen by half that amount.

Despite the many problems of GCMs, however, most people agree on perhaps the most important aspect of global climatic change from the viewpoint of contemporary human societies: the rate of change will be faster than anything previously experienced. This being the case, the approximate predictions produced by the GCMs are being used to gain some insight into the nature and conditions of the world we will be inhabiting in the next few generations, since pre-industrial CO_2 levels are expected to double some time in the twenty-first century, with consequent significant changes in global climate.

The impacts

Since the atmosphere is intimately linked to the workings of the biosphere, hydrosphere and lithosphere, the projected changes in climate will have significant effects on all aspects of the natural world in which we live. Research into the possible impacts has been one of the central tasks of the Intergovernmental Panel on Climate Change (IPCC), an international body of scientists set up especially to look into the issue of global warming. The potential for disruption to socio-economic systems has been illustrated through history. The collapse of the Subir civilisation in northern Mesopotamia around 2200 BC, for example, has been linked to the stresses imposed by natural climatic change, largely in the form of increased aridity in this case (Weiss *et al.*, 1993). Shifts in the Earth's climatic zones over several hundred kilometres during the next 50–100 years would affect terrestrial ecosystems, with different species able to adapt to or migrate with the changes with varying degrees of success (see, for example, page 366 for the predicted effects on mosquito distribution and malaria incidence). The warming recorded during the twentieth century has already been related to numerous changes among plant and animal communities, some of which are shown in Table 11.5. Shifts in the distribution, or range, of species poleward and towards higher altitudes have been noted, with some species (e.g. butterflies) responding faster to changes in the location of temperature isotherms than others (e.g. alpine plants).

Such climatic shifts also affect patterns of agricultural land use, making the production of certain crops and livestock more, or less, suitable in the changed climate, which would not only affect climatic growing conditions but also the hazards from pests and diseases. In Australia, for example, climate trends have already had a significant effect on the increased average yield of wheat by about 0.5 t/ha since 1952. The most important climatic factor in this case has been the increase in minimum temperatures. Among other things, the observed increase in minimum temperatures has meant fewer frosts, causing less damage to wheat harvests (Nicholls, 1997).

Similar good news for farmers has been detected in Europe where the growing season has increased by 11 days on average since the early 1960s because of the warming climate. Researchers analysed data from 1959 to 1993 from more than 40 gardens containing genetically identical trees and shrubs in a network that extends roughly between Scandinavia, Macedonia and Ireland. From records for events such as the appearance of leaves or needles, blossoming, and leaves turning colour and falling, they found trends suggesting that over 30 years, on average, the season has begun about six days earlier and ended about five days later (Menzel and Fabian, 1999). This finding is in line with a study of satellite data showing that in the last two decades of the twentieth century there was a gradual greening of the northern latitudes as plant life grew more vigorously above 40°N, representing a line stretching from New York to Madrid to Beijing (Zhou *et al.*, 2001).

Globally, however, one study (Fischer *et al.*, 1994) suggests that a doubling of pre-industrial CO_2 levels would decrease world cereal production by up to 5 per cent, depending upon the climatic scenario, even assuming some adaptation to the changing conditions by farmers and allowing for the fact that greater atmospheric carbon dioxide will increase global plant productivity. The largest falls in output would occur in developing countries and this loss of production,

TABLE 11.5 *Evidence of observed ecological changes linked to the recent warming of climate*

Ecology	Location	Observed changes	Climate link
Treeline	Europe, New Zealand	Advancement towards higher altitudes	General warming
Lowland birds	Costa Rica	Extended distribution from lower mountain slopes to higher areas	Dry season mist frequency
Butterflies	North America, Europe	Northward range shifts	Increased temperatures
Wheat	Australia	Increased average yields	More frost-free days
Numerous plant species	Europe	Longer growing season	General warming
Numerous bird species	North America, Europe	Earlier spring migration and breeding	General warming

Source: after Walther *et al.* (2002)

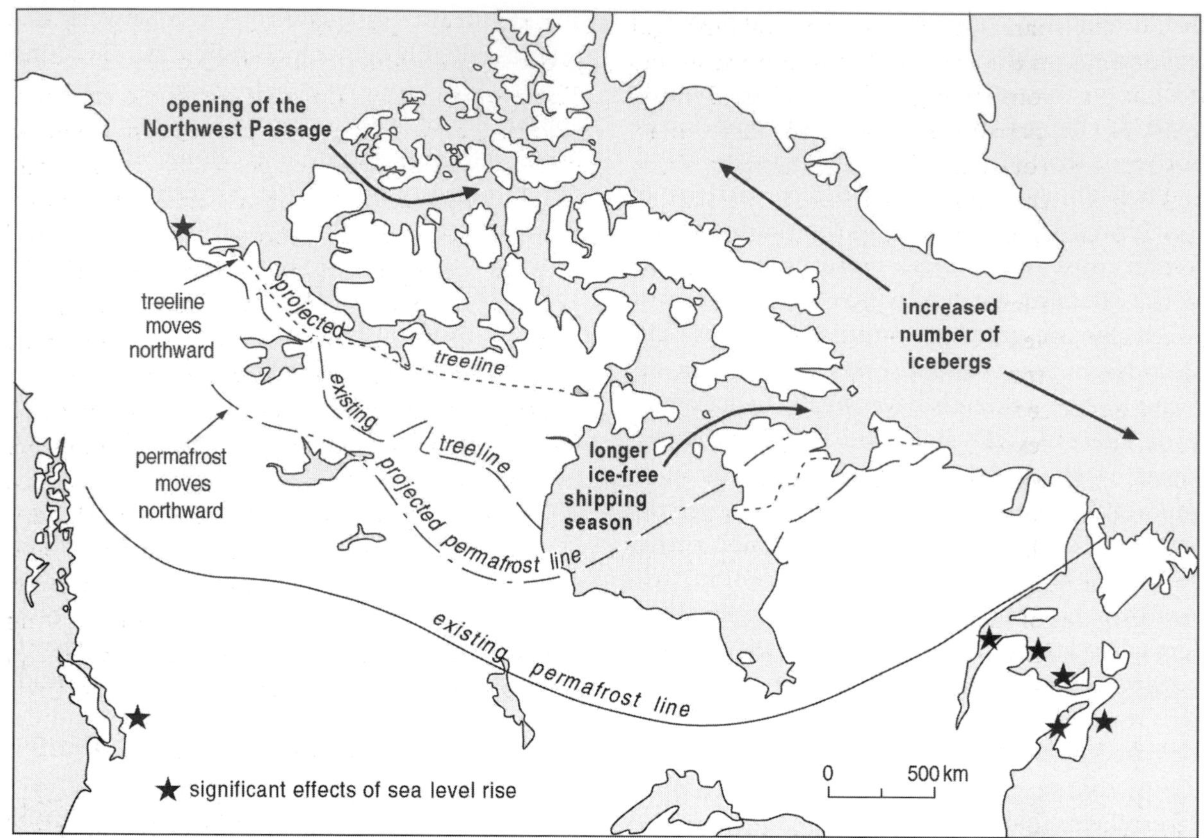

Figure 11.5 Projected changes in northern Canada following climate warming (Slaymaker and French, 1993).

combined with rising agricultural prices, is predicted to increase the number of people at risk from hunger in the order of 5–50 per cent depending on the climatic scenario.

Relatively small changes in climate can also influence the availability of water. Consequent problems would be particularly acute in semi-arid regions and more humid areas where demand or pollution has already created water scarcity. The Mediterranean Basin is one example here (Smith, 1997). Decreasing trends in precipitation totals have been identified in western-central parts of the basin in recent decades, as well as marked changes in seasonality. A clear tendency for rainfall to be concentrated into a shorter period of the year has been noted in the Alentejo region of southern Portugal, with the proportion of annual precipitation falling in autumn and winter increasing at the expense of spring totals. Springtime rainfall has also been decreasing in southern Spain (Corte-Real *et al.*, 1998). Such changes in the seasonality and intensity of rainfall may also be expected to have an impact on the flood regimes of rivers. Work on a number of the world's large drainage basins has already identified a significant rising trend in the risk of great floods (those with a return period of 100 years) in the twentieth century (Milly *et al.*, 2002).

In high latitudes, where global greenhouse warming is expected to be greatest, significant changes are predicted in glacial and periglacial processes, affecting glacier ice, ground ice and sea-ice, which, in turn, would affect vegetation, wildlife habitats, and human structures and facilities. A prediction of the pattern of changes around the Arctic Circle, in northern Canada, is shown in Fig. 11.5. There is a strong possibility that the Arctic Ocean's ice cover will disappear, facilitating marine transport and oil and gas exploration on the one hand, but also increasing the dangers from icebergs. The northward movement of the permafrost line would have many implications for roads, buildings and pipelines now constructed on permafrost, necessitating reinforcement, and new engineering design and construction techniques. A serious feedback aspect of high-latitude permafrost melting is the consequent release of methane, a greenhouse gas. The warming trend documented for the Antarctic Peninsula since the 1940s has already affected the frequency of extensive sea-ice in the Southern Ocean. Cold winters with extensive sea-ice cover occurred on average in four out of five years in the middle of the last century, but have decreased to just one or two years in five since the mid-1970s. One function of this trend appears to be a decline in the abundance of krill in the Southern Ocean (Loeb *et al.*, 1997; see also Chapter 6).

Numerous effects of global climatic change on geomorphological processes can be expected. These would occur through the direct effects of warming, and as a consequence of related changes in precipitation and temperature regimes, and their impacts on such geomorphologically significant variables as vegetation. In many semi-arid areas, for example, soil moisture is predicted to be reduced by larger losses to evapotranspiration and decreased summer runoff, and increased rates of soil erosion by wind can be expected as a consequence. In many parts of the world, dry periods are normally associated with high frequencies of dust storms and this natural effect has been exacerbated in some cases by human disturbances to vegetation and soil surfaces (Goudie and Middleton, 1992). Many cases of such synergy between climatic change effects and pre-existing human impacts can be predicted and Table 11.6 gives a number of examples. Such synergistic

TABLE 11.6 *Examples of synergy between climate change and more direct human impacts on geomorphology*

Phenomenon	Current human abuse	Potential global warming impact
Groundwater reduction in High Plains, USA	Overpumping by centre pivot irrigation	Increased moisture deficit
Desiccation of Aral Sea and associated duststorms	Excessive irrigation offtake and interbasin water transfers	Increased moisture deficit
Permafrost subsidence	Vegetation and soil removal, urban heating, etc.	Warming
Quaternary dune reactivation	Intensive grazing and agricultural activities	Further vegetation depletion
Coastal retreat	Sediment starvation by dam construction and coastal engineering structures	Sea level rise
Coral reef stress	Pollution, siltation, mining, overexploitation	Overheating, more hurricanes, rapid sea level rise
Coastal flooding	Groundwater and hydrocarbon mining	Sea level rise and more frequent storms

Source: after Goudie (1993c)

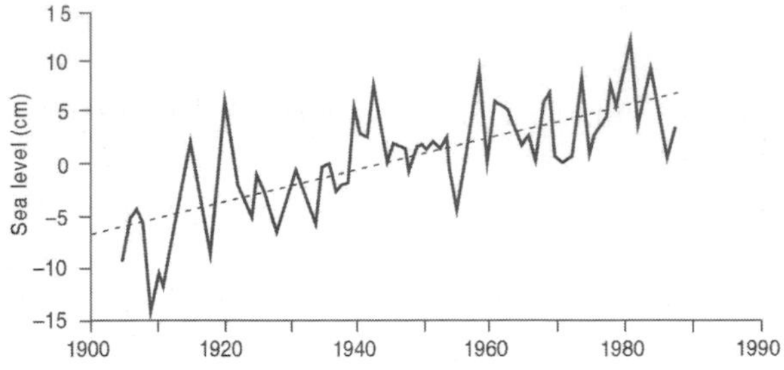

Figure 11.6 Variations in sea level at Oahu, Hawaii, USA, 1905–90 (Nunn, 1994).

effects may also work to ameliorate environmental problems, however. An assessment of the success of soil conservation measures on the highly erodible loess farmlands of the middle reaches of the Huang Ho River found that sediment yields in one of the Huang Ho's tributaries had declined by 74 per cent over the period 1957–89 (Zhao *et al.*, 1992). Just half of this reduction was attributed to improvements in soil and water conservation measures, while the other 50 per cent was a reflection of the shift to a drier climate.

Synergies between climate change and other human effects are also likely to occur in many other areas of the physical environment. Higher temperatures mean that ground-level ozone forms faster and high atmospheric concentrations of ozone persist for longer. Global warming is thus likely to accelerate the photochemical reaction rates among chemical pollutants in the atmosphere, increasing pollution in urban areas.

One aspect of a warmer world that has received considerable attention is that of higher sea levels caused both by thermal expansion of the oceans and the added input from melting ice. As Schneider (1989a) suggests, sea level rise will undoubtedly be the most dramatic and visible effect of global warming into the greenhouse century. Given the very large concentration of human population in the coastal zone (see Chapter 7), the consequences for low-lying and subsiding

Figure 11.7 The Piazza San Marco in Venice is barely above mean sea level and is frequently flooded. Any rise in sea level due to global warming would increase the city's flood frequency.

TABLE 11.7 *Threats posed to the Maldives by projected global climatic change*

1. Increased rates of coastal erosion and alteration of beaches with increased impacts from high waves
2. Changes in aquifer volume associated with increased saline intrusion
3. Increased energy consumption (e.g. air-conditioning)
4. Coral deaths as a result of increased sea water temperature
5. Accelerated interisland migration due to declining stability and habitability of islands
6. Loss of capital infrastructure on smaller tourist resort islands
7. Changes in reef growth and current patterns
8. Increased vulnerability of human settlements due to aggregation and increasing size

Source: Pernetta and Sestini (1989)

regions are potentially very severe. Although sea levels change under the influence of natural processes, the recent rise, which averages about 1.5 mm/year over the past 80 years in the Pacific (Fig. 11.6), may herald the early stages of a greenhouse rise, which is predicted to be up to 88 cm between 1990 and 2100 (Houghton *et al.*, 2001).

Increased flooding and inundation are the most obvious threats and many low-lying settlements on continental coasts, such as Bangkok, Lagos, London and Tokyo to name but a few, would be affected. The Mediterranean city of Venice is particularly at risk because most of the urban area is only 80 cm above mean sea level (Fig. 11.7). The city has also experienced human-induced subsidence in recent decades. While prior to the 1950s the natural subsidence of deltaic coastal plains was compensated for by active sedimentation, the period since the 1950s has been characterised by accelerated subsidence due to the draining of marshes, reduced sedimentation and the overabstraction of water from aquifers (Sestini, 1992).

A large number of the islands that existed in the Pacific 18 000 years ago have since been drowned (Gibbons and Clunie, 1986) and a continued rise of just a few metres would greatly alter the present map of the central Pacific and Indian Oceans. Several small island countries in the Pacific – such as Tokelau, the Marshall Islands, Tuvalu, the Line Islands and Kiribati – could cease to exist if worst-case scenarios for sea level rise are realised (Pernetta, 1989). A number of other physical impacts can also be expected for other oceanic islands. As sea levels rise, coastal erosion rates are likely to be affected although the exact nature of these changes will not be straightforward. They will depend on many other factors, including the tectonic history of the island, the rate of sediment supply relative to submergence, the width and growth rate of existing coral reefs and their health, whether islands are anchored to emergent rock platforms, and the presence or absence of natural shore protection such as beachrock or conglomerate outcrops, or vegetation such as mangroves (Nurse *et al.*, 1998). The wide range of physical and consequential socio-economic impacts for small islands is illustrated in Table 11.7 for the Maldives.

Sea level rise poses a particular problem to much larger, more populous low-lying countries such as Bangladesh. Inundation and salinity intrusion were identified as two of the most serious threats to the national economy (the other was drought) in a recent climatic change projection for the country that analysed the impacts under two scenarios:

1. a lower limit of 2 °C change associated with 30 cm sea level rise
2. an upper limit of 4 °C change associated with 100 cm sea level rise.

Other significant primary physical effects were found to be increased incidence of cyclonic storm surges, flashfloods and low water flow. The impact on agricultural production, for example, was predicted to be a fall in foodgrain production due to a combination of drought-induced stress on crops, caused by increased evapotranspiration, and lower water availability. Most development efforts planned for the coming three decades would be negated by sea level rise and further threatened by the lack of a water-sharing agreement with India. The study also highlighted the danger to the Sundarbans mangrove forest: the ecosystem's ability to adapt to the pace of projected sea level rise is limited, not least by a lack of space inland (BCAS, 1993).

The threat from increased frequencies of tropical storms highlighted in the Bangladesh study is another serious facet of recent warming trends in tropical latitudes where tropical cyclones develop only in areas with sea-surface temperatures exceeding 26.5 °C. Tropical cyclones can cause severe disturbances to coastal and island ecosystems, quite apart from the damage to social structures, but although a warmer planet would seem logically to result in more frequent tropical cyclones, no significant trends in frequency have been identified over the twentieth century (Houghton *et al.*, 2001). Other changes in tropical cyclones are thought to be likely, however: increases in the intensities of their peak winds and their precipitation. One computer simulation of the effects of a sea-surface temperature warming of 2.2 °C in the north-west Pacific yielded hurricanes that were up to 10 per cent more intense in terms of wind speed (Knutson *et al.*, 1998). The extreme intensities of super-hurricanes Gilbert (1988), Hugo (1989) and Andrew (1992), as well as the series of very severe gales in Europe in early 1990, are regarded by some as indications of global warming beginning to have a noticeable effect on such extreme climatic events. The implications for society and the insurance industry are clear (Berz, 1991).

Responses to global climatic change

The formidable economic, social and political challenges posed to the world's governments and other policy-makers by impending global climatic change are unprecedented. Policy responses can be categorised broadly into those that aim to prevent change and those that accept the changes and focus on adapting to them. While the issue is a truly global one, since all greenhouse gas emissions affect climate regardless of their origin, the costs and benefits of measures to mitigate the effects of global climatic change are likely to be spread unevenly across countries. The issue raises important questions of international equity since, at present, the major proportion of greenhouse gas emissions comes from the industrialised countries, which contain only about 25 per cent of the world's population. Developing-world leaders have called for reductions in emissions from the industrialised countries to make more of the planet's capacity for assimilation of greenhouse gases available to those countries that are industrialising now, a plan that should be facilitated by transfers of finance and technology from the North to the South (Tolba and El-Kholy, 1992). It has also been pointed out that some of the countries most at risk, the small island states, are effectively subsidising the economies of industrial countries that are net producers or exporters of carbon dioxide, since more natural carbon dioxide is fixed by the tropical rain forests, oceans, coral reefs and mangroves of small islands than is emitted locally from these islands (Pernetta, 1992).

Most countries have accepted the need to make some effort to prevent change, or at least slow its pace, by reducing greenhouse gas emissions. A contribution has been made in this respect by the Montreal Protocol, which was signed in 1987 and amended in 1990. Governments have committed themselves to reduce consumption and production of substances that deplete the stratospheric ozone layer, many of which also contribute to global warming. (CFCs and other halocarbons are due to be

TABLE 11.8 *Examples of 'no regrets' initiatives to combat global warming by mitigating the causes*

Initiative	Effect on greenhouse gases	Other benefits
Energy conservation and energy efficiency	Reduced CO_2, methane and N_2O emissions	Conservation of non-renewable resources for current and future generations, reduction of other problems such as acid rain
CFC emission control	Reduced CFC emissions	Reduced stratospheric ozone layer depletion, reduced surface skin cancer and blindness
Tree planting	Increased biosphere carbon sink capacity	Improved microclimate, reduced soil erosion, reduced seasonal peak river flows
Prevent deforestation	Maintain biosphere carbon sink capacity	Biodiversity conservation and maintenance of environmental services
Reduce motor vehicle use	Reduced CO_2 emissions	Improved air quality and congestion
Improve efficiency of N fertiliser use	Reduced N_2O emissions	Improved water quality

phased out by 2010.) Most attention has been focused on carbon dioxide, however, an initiative that has a global forum in the UN Framework Convention on Climate Change (UNFCCC). The Convention was signed by more than 150 governments at the Earth Summit in 1992 and came into force in March 1994 following the fiftieth country ratification. In effect, signatory states have agreed to reduce emissions to earlier levels, in many cases the voluntary goal being a reduction of carbon dioxide emissions to 1990 levels. An attempt to make agreed reductions legally binding was made in 1997 at the Kyoto Protocol, a follow-on to the original climate treaty. Kyoto also focused on a wider range of greenhouse gas emissions, to include methane, nitrous oxide, hydrofluorocarbons, perfluorocarbons and sulphur hexafluoride, in addition to carbon dioxide. The declared aim of the Protocol is to cut the combined emissions of greenhouse gases by about 5 per cent from their 1990 levels by 2008–12, specifying the amount each industrialised nation must contribute towards this overall aim. Those countries with the highest carbon dioxide emissions, including the USA, Japan, the EU and most other European states, are expected to reduce their emissions by 6 to 8 per cent. In practice, individual country reductions can be greater or less than those agreed since the Kyoto Protocol also officially sanctioned the idea of 'emissions trading' between industrialised nations. Hence, if a state's emissions fall below its treaty limit it can sell credit for its remaining allotment to another country to help the buyer meet its treaty obligation.

The practical implementation of this pledge, however, will entail large financial investments to insure against future events that are far from certain. This being the case, the additional beneficial aspects of these and other mitigating policies are being widely promoted: these 'no regrets' initiatives are sensible anyway (Table 11.8). For example, planting trees to lock more carbon in the biosphere, so-called 'carbon sequestration', and reducing deforestation also have the additional benefits of conserving biodiversity and maintaining environmental services such reducing soil erosion. Such initiatives are also, from the global warming perspective, examples of the precautionary principle in operation.

Realistically, however, no government is likely to sacrifice significant economic growth for reductions in carbon dioxide emissions, so that long-term strategies to reduce emissions must uncouple economic growth from growing fossil fuel consumption. Reducing the amount of energy used per unit of GDP will be one element in such a strategy, but there is also a need for a significant shift away from fossil fuels to using more renewable energy sources (see Chapter 18). Many industrialised countries do have experience of economic growth with declines in energy consumption. It occurred during the late 1970s and early 1980s, sparked by the 1970s oil crisis (Goldemberg *et al.*, 1988). Unfortunately, virtually all governments have reinstated their faith in the belief that economic growth must be based on an increase in energy consumption. But the lessons of global warming make it clear that this kind of industrialisation, based on an inefficient use of fossil fuel resources, is not a sustainable form of development.

Two scenarios based on projections for world energy demand are shown in Fig. 11.8. If fossil fuels continue to be used for more than 80 per cent of world energy, as they were in 1990 (the 'business as usual' scenario), carbon dioxide emissions are projected to more than triple by the year 2050, but the phasing in of more renewable energy sources is projected to bring carbon dioxide emissions back to 1990 levels soon after 2050. Even this latter scenario, which envisages the contribution of renewable sources to world energy demand to rise from 10 per cent to 60 per cent in 60 years, would represent an unprecedentedly rapid shift (World Bank, 1992a). It would also still entail a significant increase in atmospheric carbon dioxide concentrations. There is no doubt, therefore, that CO_2 emissions need to be reduced to levels far below the modest proposals agreed at Kyoto (Wigley *et al.*, 1996).

The projections in Fig. 11.8 suggest, therefore, that warming is still likely to take place in the next few decades, and thus a need to prepare for adaptive responses remains. An integral part of this strategy is to continue the funding of scientific research and data collection, which can help to reduce the uncertainty surrounding possible climate change impacts. Nevertheless, the many question marks that remain over the nature, magnitude, patterns and timing of change make it

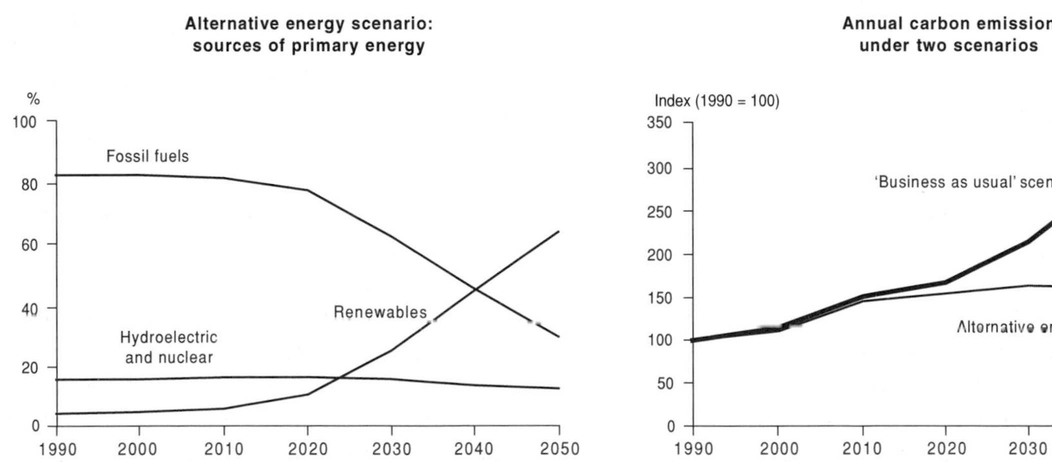

Figure 11.8 Scenarios for projected global carbon emissions, 1990–2050 (after World Bank, 1992a).

TABLE 11.9 *Canada Climate Change Development Fund's four programme areas for aid projects in developing countries*

Programme area	Expected results	Examples of suggested indicators
Emissions reduction	Reduction in growth rate of greenhouse gas emissions	• Degree of efficiency of fossil fuel-based power sources • Degree of transmission and distribution efficiencies in power systems • Level of emissions avoided • Rate of introduction of renewable energy sources • Level of knowledge and capacity to undertake potential emission reduction initiatives (feasibility study results)
Carbon sequestration	Increase in sequestration of carbon in sinks	• Area of land reforested, afforested or preserved from deforestation • Level of soil organic matter and decreased erosion • Level of knowledge and capacity to undertake potential carbon sequestration initiatives (feasibility study results)
Adaptation	Reduction in vulnerability to the adverse effects of climate change	• Area of land protected from soil erosion • Degree of reliance on crop monocultures that are not drought-resistant • Degree of irrigation efficiency • Level of water conservation (for industry, agriculture, domestic use) • Quality of coastal zone protection and management • Quality of early warning systems • Level of capacity to monitor and project vulnerability
Core capacity building	Increase in capacity to participate in global efforts to combat climate change	• Creation of national emission inventories • Level of market barriers to more sustainable energy use • Quality of regulatory frameworks and policies related to climate change • Level of capacity within civil society to influence decision-making on climate change

Source: after CIDA (2000)

necessary to emphasise again the 'no regrets' aspects of preparing for adaptive responses:

> *Developing alternative energy sources, revising water laws, searching for drought-resistant crop strains, negotiating international agreements on trade in food and other climate-sensitive goods – all these steps could also offer widespread benefits even in the absence of any climatic change.*
>
> *(Schneider, 1989b: 46)*

These adaptive responses are being undertaken all over the world as individual countries develop national climate change strategies. The UNFCCC also committed industrialised countries to providing technological and financial resources to help developing countries address climate change, and hence many aid programmes have now incorporated climate change objectives into their projects. The Canadian International Development Agency (CIDA) has established the

Canada Climate Change Development Fund (CCCDF) with the specific aim of promoting activities to combat the causes and effects of climate change in developing countries while helping to reduce poverty and promote sustainable development. The CCCDF finances projects under four programmes, adopting an approach that combines technology transfer and capacity building. The four programmes focus on emissions reduction, carbon sequestration, adapting to the effects of climate change, and enhancing basic capacities for dealing with climate change. Further details of these four programmes are shown in Table 11.9.

FURTHER READING

Adams J., Maslin M. and Thomas E. 1999 Sudden climate transitions during the Quaternary. *Progress in Physical Geography* 23: 1–36. A review of climate changes in the past few million years, which are typically quite rapid.

Barnett, J. 2001 Adapting to climate change in Pacific island countries: the problem of uncertainty. *World Development* 29: 977–93. This paper outlines possible climate change problems faced in Pacific islands and assesses ways of improving resilience.

Houghton, J.T. 1997 *Global warming: the complete briefing*, 2nd edn. Cambridge, Cambridge University Press. A good introduction to global warming by one of the issue's leading researchers.

Houghton, J.T., Ding, Y., Griggs, D.J., Noguer, M., van der Linden, P.J. and Xiaosu, D. (eds) 2001 *Climate change 2001: the scientific basis*. Cambridge, Cambridge University Press. A key IPCC report.

Kemp, D.D. 1994 *Global environmental issues: a climatological approach*, 2nd edn. London, Routledge. An accessible textbook covering the effects of climate and climatic change on many environmental issues.

Mirza, M.M.Q. 2002 Global warming and changes in the probability of occurrence of floods in Bangladesh and implications. *Global Environmental Change* 12: 127–38. A case study of the possible flood implications of global warming on the Ganges, Brahmaputra and Meghna rivers.

Walther, G.-R., Post, E., Convey, P., Menzel, A., Parmesan, C., Beebee, T.J.C., Fromentin, J.-M., Hoegh-Guldberg, O. and Bairlein, F. 2002 Ecological responses to recent climate change. *Nature* 416: 389–95. A review of studies showing evidence of the ecological impacts of recent climate change.

Wigley, T., Richels, R. and Edmonds, J. 1996 Economic and environmental choices in the stabilization of atmospheric CO_2 concentrations. *Nature* 379: 240–3. An appraisal of the pros and cons facing governments if reductions in CO_2 emissions are to stand a realistic chance of stabilising atmospheric concentrations.

Wyman, R.L. (ed.) 1991 *Global climate change and life on Earth*. London, Chapman & Hall. A collection of papers that focuses on the greenhouse effect and its relationship to other environmental issues such as deforestation, overpopulation and hunger, pollution, sea level change and biodiversity loss.

Websites

http://www.ipcc.ch/ information on the Intergovernmental Panel on Climate Change's work assessing the scientific, technical and socioeconomic aspects of human-induced climatic change.

http://cdiac.esd.ornl.gov/ run by the Carbon Dioxide Information and Analysis Center, features regularly updated information on CO_2 papers and databases.

http://www.cru.uea.ac.uk/tiempo/ the Tiempo Climate Cyberlibrary is an information service with publications, data and educational resources on global warming and climate change, with a focus on the developing world.

http://www.globalchange.org/ an interdisciplinary guide to a vast range of information on human-induced climate change, particularly global warming and stratospheric ozone depletion.

http://unfccc.int/ site of the UN Framework

Convention on Climate Change including reports, data and information on progress towards convention targets.

http://www.climatechangesolutions.com/ a Canadian site with numerous ideas on combating climatic change.

http://www.wmo.ch/ the World Meteorological Organization site has a wide range of information on issues in meteorology, weather and climate.

http://www.ukcip.org.uk/ the UK Climate Impacts Programme helps organisations assess how they might be affected by climate change; with research and impact assessments.

POINTS FOR DISCUSSION

- Have we done enough to solve the problems associated with the hole in the ozone layer?
- Since global warming is predicted to bring benefits as well as problems, should we try to prevent it at all?
- How might actions taken to curb global warming affect other environmental issues?
- Outline the importance of tackling global warming at global, national and local scales.

12

ACID RAIN

TOPICS COVERED
Nature of acid rain
Geography of acid deposition
Effects of acid rain
Combating the effects of acid rain

Key words: buffering capacity, critical load, 30% Club, 'waldsterben', memory effect

The effects of atmospheric acidity on biology and the urban fabric were first noted in England about 150 years ago when a Scottish chemist found that local rainfall in Manchester was unusually acidic. Smith (1852) suspected a connection with sulphur dioxide from local coal-burning factories, and commented on the effects the deposits had on buildings and vegetation. It is Smith who is credited with the first use of the term 'acid rain'.

Interest in the phenomenon was awakened a century later when Gorham (1958) observed that air masses passing over industrial regions influenced the acidity of rain downwind over the English Lake District, and that the consequent acid deposition could affect the ecology of upland lakes. The effects of acidification on lake and river ecology in Sweden pushed the issue on to the international agenda in the 1960s when Swedish researchers suggested that much of the enhanced acidity of precipitation in Sweden was due to the long-range transportation of pollutants from other countries, including Britain (Odén, 1968). Although these claims were rejected by the British Government for many years, the concept of trans-boundary transportation of pollution is now widely accepted, and the widespread and extensive acidification that has occurred in areas such as southern Scandinavia, northern Britain and parts of north-eastern North America is generally agreed to have chiefly been caused by atmospheric acidic deposition. Nevertheless, the effects of this deposition on various aspects of aquatic, terrestrial and built environments are still the subject of some controversy.

THE NATURE OF ACID RAIN

Acid rain is a misleading term. In practice, it refers to acidic deposits from the atmosphere both by wet deposition (rain, snow, sleet, hail, mist and dew) and dry deposition (by gravitational settling and through contact with surfaces). Hence, some researchers prefer the term 'acid deposition'.

Acidity is commonly expressed using the pH scale, which reflects the concentration of hydrogen ions (H^+) in a solution. The scale ranges from 0 to 14, with a value of 7 indicating a neutral solution. Values less than 7 indicate acid solutions (e.g. lemon juice is pH 2.2), while values greater than 7 indicate basic solutions (e.g. baking soda is pH 8.2). The pH scale is logarithmic, so that a solution of pH 4 is ten times more acidic than one with a pH of 5.

The chemistry of the atmosphere is affected by numerous sources of gases and particulate matter that can alter pH. Volcanic emissions, fires, deflation of soil particles and various biological sources all affect atmospheric chemistry as part of the natural cycles of the Earth's matter, but

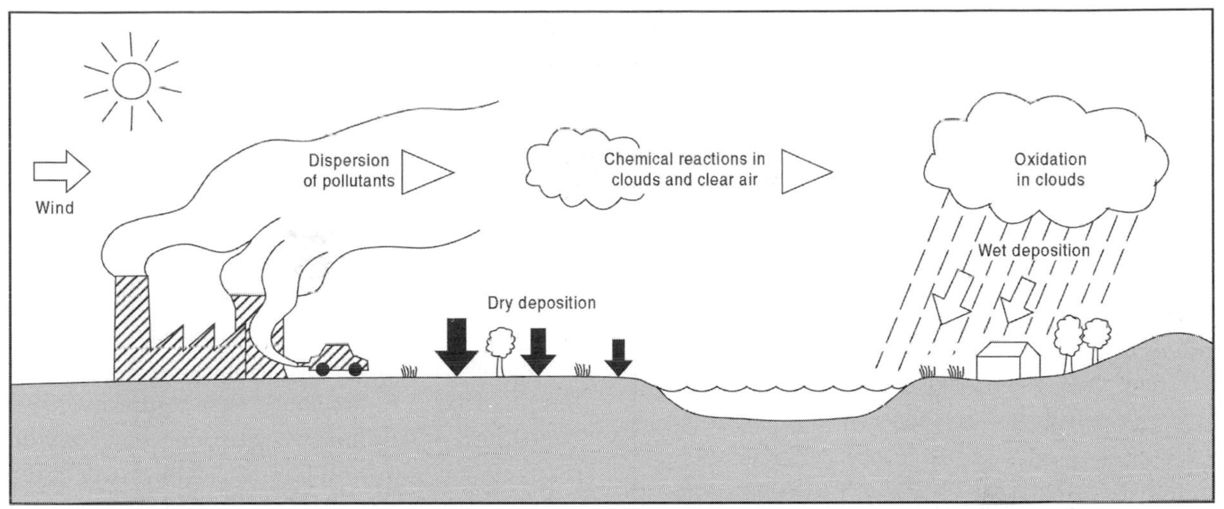

Figure 12.1 Emission, transport, transformation and deposition of pollutants known as acid rain.

precipitation in an 'unpolluted' atmosphere commonly has a pH of 5.6. This slight acidity is due to the ubiquitous presence of carbon dioxide in the atmosphere, which forms a carbonic acid. As with numerous forms of human-induced pollution, the acid rain debate centres around the role of human action in accelerating some of the pathways in the cycles of certain elements: primarily sulphur and nitrogen. The most important emissions are sulphur dioxide (SO_2) and nitrogen oxides (NO and NO_2, which are collectively referred to as NO_x). Large quantities of these oxides of sulphur and nitrogen are emitted into the atmosphere by the combustion of fossil fuels (coal, oil and natural gas) and industrial processes (principally primary metal production), and it is these gases that are either deposited close to the source, as dry deposition, or are converted into sulphuric and nitric acids, and deposited as wet deposition at distances up to thousands of kilometres from the source (Fig. 12.1). While these industrial sources are the largest emitters of acid rain pollutants, a range of other sources also contributes. One of the most widespread is agriculture, which makes important contributions in some areas when ammonia from nitrogenous fertilisers and manure escapes to the atmosphere by volatilisation.

GEOGRAPHY OF ACID DEPOSITION

The absolute quantities of atmospheric sulphur and nitrogen produced by human action have risen dramatically over the past 100 years or so, to the point at which human-induced emissions are now of a similar order of magnitude to natural sources. However, the uneven global distribution of industrialisation means that the concentration of human-induced acid deposition is highly uneven. Data for the mid-1980s, for example, indicated human-induced sulphur emissions to the atmosphere to be about 60 per cent of the total, but over the industrialised regions of Europe and North America the human-induced flow was thought to be 12 times the natural flow (Bates *et al.*, 1992). Globally, human emissions of sulphur have been roughly constant since 1980, but their geographical pattern has changed significantly. In 1980, 60 per cent of global emissions were from the North Atlantic basin (USA, Canada and Europe) but by 2000 this region was thought to make up less than 35 per cent of sulphur emissions worldwide thanks to concentrated efforts to reduce this form of pollution (see below). The area of maximum emissions has shifted towards Asia over this period. Asian sulphur

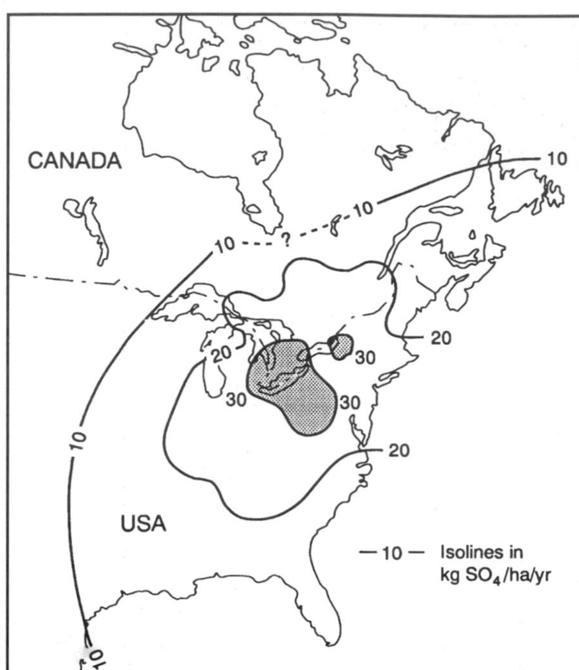

Figure 12.2 Mean annual deposition of SO_4 over eastern North America for the period 1982–87 (after Tolba and El-Kholy, 1992).

emissions, dominated by China, were 26 per cent of the global total in 1980 but had risen to 46 per cent in 2000 (Smith *et al.*, 2001). This shift in the sulphur emissions pattern, from a centre around the North Atlantic to one dominated by Asia, is expected to continue over the coming decades.

Maps such as the one of North America shown in Fig. 12.2 are frequently cited to highlight the seriousness of the acid rain problem, and although such maps have been criticised for the way in which they generalise patterns and give no indication of changes through time (e.g. Kallend *et al.*, 1983) they none the less give an impression of the scale of the problem. Changes in the patterns of emissions and deposition of sulphur in Europe have been described by Mylona (1993), who notes that areas of maximum emissions have shifted over the last century away from coalfields to oil-refining areas and regions with high densities of motor vehicles, and the traditional core areas of western and central Europe have been augmented by new sources in the north, east and south. In 1880, the maximum concentrations of SO_2 (4–7 g S/m^3) were confined to the English Midlands, moving over to the western and central parts of the continent in the early twentieth century. Since the 1950s, maximum emissions are found in a core region that comprises South Saxony and North Bohemia, where emission rates range from 25 to 40 g S/m^3. Areas of maximum deposition have also shifted, from north-western Europe to the German/Czech/Slovak/Polish border region where deposition of about 2 g S/m^2 in 1880 had risen to six times that figure by 1991. Deposition of sulphur over Europe for the period 1880–1991 is shown in Fig. 12.3, indicating that a broad swath from Wales to Poland has received more than 200 g S/m^2 over that period (Mylona, 1993).

While Europe and North America remain the world's most severely affected large-scale regions because of their long histories of emissions, an increasing body of evidence indicates the presence of smaller acid rain 'hotspots' in many other areas with concentrations of particularly polluting industry. Pape (1993), for example, cites several huge metal-smelting centres in Siberia responsible for large emissions of sulphur dioxide, and emissions of both SO_2 and NO_x are high in many of the world's largest urban areas (see Chapter 10). Indeed, while Western European anthropogenic emissions of sulphur decreased by more than 60 per cent between 1980 and 2000, and those from Canada and the USA fell by about 30 per cent, emissions elsewhere have continued to rise, particularly in Asia where those from China and South-East Asia more than doubled over the same period (Smith *et al.*, 2001). Parts of Japan, North and South Korea, southern China and the mountainous portions of South-East Asia and south-western India have all been identified as currently vulnerable to acid rain (Bhatti *et al.*, 1992). Recent steady rises in NO_x emissions have also been recorded in all continents except North America and Europe where they have remained relatively constant.

Another important aspect of the acid rain issue is its movement across national political

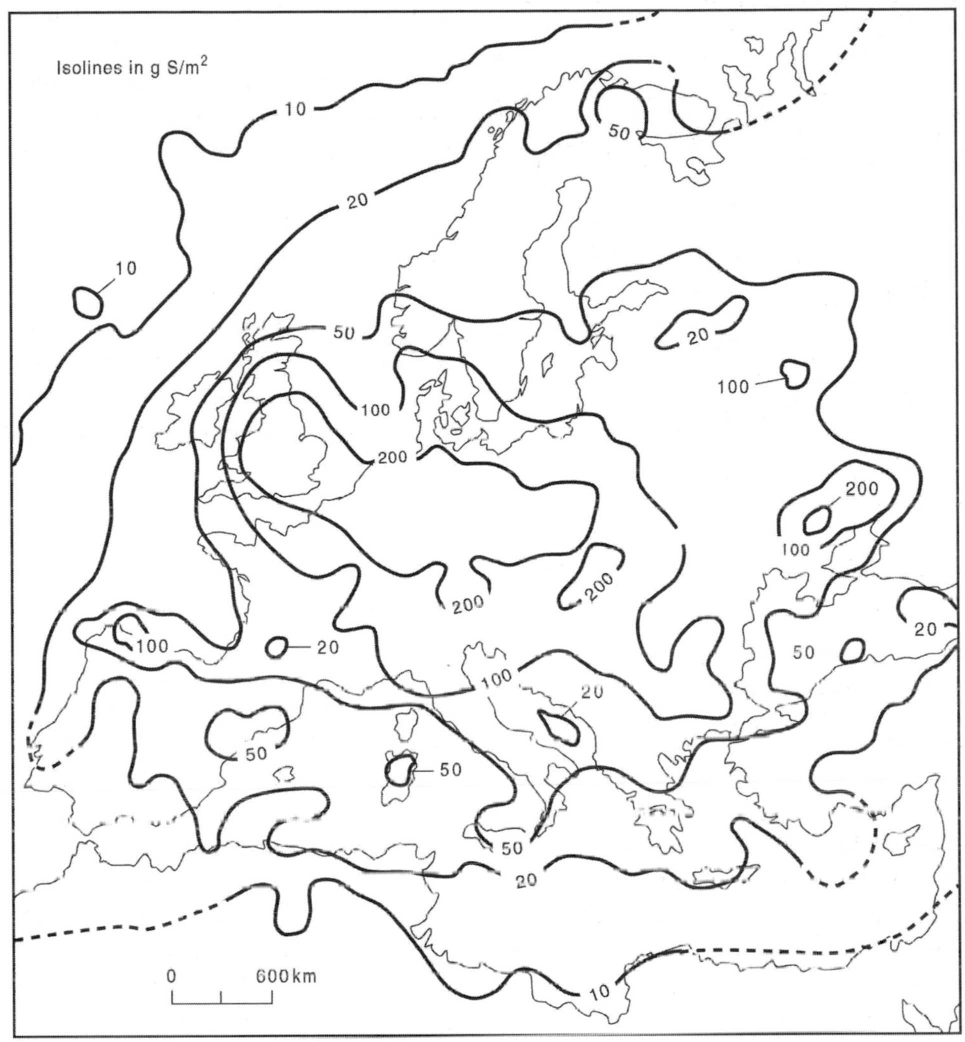

Figure 12.3 Total deposition of oxidised sulphur in Europe, 1880–1991 (after Ågren, 1993).

boundaries; indeed it has been noted previously that the problem first became an international issue when links between acid rain in Sweden were made with polluting industries in other parts of Europe. The long-range transportation of acid rain pollutants is a relatively recent phenomenon. In Britain, for example, reaction to increasing urban air pollution, including the infamous London 'pea souper' smogs of the early 1950s, resulted in the building of higher industrial and power-generation smoke stacks in an attempt to combat local pollution problems. Higher chimneys were indeed successful in reducing pollutants locally, but in doing so they transformed a local pollution problem into a regional and international one.

The acid rain pollutant budget of Sweden is shown in Table 12.1, which has been compiled from data recorded as part of the European Monitoring and Evaluation Programme (EMEP). As the table indicates, only about 7 per cent of the sulphur deposited in Sweden in the late 1990s came from domestic sources, while significantly larger contributions were transported from Poland and Germany to the south, shipping in the Baltic Sea to the east, and from the United

TABLE 12.1 *Sources of sulphur deposited (wet and dry) in Sweden in 1998*

Source	Annual deposition (hundred tonnes)
Poland	196
Germany	150
Baltic Sea	147
UK	143
Exterior sources*	119
Sweden	101
Russia	97
Denmark	48
North Sea	48
All other sources	386
Total	1435

*Sources outside the EMEP area
Source: data from EMEP (2000)

Kingdom to the south-west. The great distances travelled by secondary pollutants, which fall in acid precipitation, is indicated by the fact that nearly 12 000 t of the sulphur deposited in Sweden originates outside the EMEP area, some from as far away as North America and Asia.

EFFECTS OF ACID RAIN

The actual effects of acid rain on the Earth's surface have been, and continue to be, a subject for some debate. It is important to note, for example, that some environments have naturally acidic pH levels due to podsolic soils and peats, and that soil acidification itself is a naturally occurring process. Several studies have illustrated the gradual decrease of soil pH with primary succession, such as that conducted by Crocker and Major (1955) in Glacier Bay, Alaska. Fine morainic till had a pH of 8.0–8.4 when first exposed by glacier meltback, but had fallen to about pH 5.0 after 70 years due to leaching by rainwater and modification by developing vegetation.

The ecological effects of acid rain depend very largely upon the ability of an ecosystem to neutralise incoming acids. This ability – the so-called 'buffering capacity' – is largely determined by the nature of the bedrock: environments on hard, impervious igneous or metamorphic rock, where calcium and magnesium content is low, are most at risk of acidification due to inputs of acid rain. Hence, attention has necessarily focused on areas with low buffering capacity where recent changes in acidification have occurred. Even in these areas, however, different degrees of acidification will have different effects and so the concept of 'critical loads' is now commonly used to determine the susceptibility of a particular environment. A critical load is a measure of the amount of acid deposition that an ecosystem can take without suffering harmful effects – in other words the threshold beyond which damage occurs. To date, critical loads have been defined for soil and fresh water.

The fact that numerous ecosystems have experienced large decreases in their pH is now well documented. Some of the most convincing long-term evidence comes from work on lake sediments, which analyses the species composition of fossil diatoms (single-cell algae that are sensitive to pH) found in sediment cores. Dramatic increases in acidity over the past 100 years or so have been documented from numerous lakes in North America, northern Britain and Scandinavia.

Linking cause and effect is still by no means straightforward, however, since other human impacts may also affect the pH of soils and water, for example. Changes in land use are particularly important here. Human-induced acidification of soils occurs through agricultural practices such as the liberal use of ammonia fertilisers and the production of forage legumes, and through afforestation practices, particularly where conifers are concerned. In South Africa, for example, soil acidification caused by afforestation may proceed faster than that due to the worst-recorded acid rain on the eastern Transvaal highveld (Fey *et al.*, 1990). Long-term diatom data for Swedish lakes indicate the occurrence of a pH increase from about 5.5 to 6.5 dating from the Iron Age (c. 2500 BP). This alkalinisation was due to extensive burning, agricultural forestry, grazing and other agricultural practices, which increased the base saturation and pH of soils, leading to increased transportation of base cations and nutrients from soils to surface waters. Following this period of alkalinisation, the pH of many lakes has fallen to about 4.5 this century and the cessation of many of these practices could account for some of this change (Renberg *et al.*, 1993). Nevertheless, these authors consider acid rain to be more important in producing the unprecedentedly low pHs this century, and in a similar vein, Patrick *et al.* (1990) have looked at the land use and management history around six Scottish lakes over the past 200 years and concluded that acidification can only be explained by acid rain.

Another facet of the acid rain debate stems from the fact that both sulphur and nitrogen are essential minerals for life, so that additional inputs can, in certain circumstances, be beneficial for organic growth. Moderate inputs of acid rain can stimulate forest growth, for example, in soils where cation nutrients are abundant and sulphur or nitrogen is deficient. Similarly, experiments on peanuts indicate that acid rain can increase the dry matter weight per plant and enhance economic yields (Chen *et al.*, 1990). A further, arguably positive, aspect of the large quantities of sulphate particles emitted to the atmosphere by human populations is the suggestion that they may act to offset the atmospheric warming associated with the greenhouse effect over some parts of the Earth (Charlson and Wigley, 1994). This occurs because sulphate particles reflect solar radiation back into space and also act as condensation nuclei, helping clouds to form, making them more reflective and changing their lifetimes.

Nevertheless, despite the difficulties and provisos outlined above, there is a large body of evidence that implicates acid rain in numerous deleterious effects upon ecosystems, human health and the built environment.

Aquatic ecosystems

Much of the work on acid rain's impact on aquatic ecosystems has focused on fish, and numerous studies have documented the loss of fish from acidified waters. One survey of 1679 lakes in southern Norway, for example, showed that brown trout were absent or had only sparse populations in more than half the lakes, and the proportion of lakes with no fish increased with declining pH (Sevaldrud *et al.*, 1980). Thresholds of tolerance to reduced pH can be identified for particular species of fish. A survey of sport-fish populations in Canadian rivers found that Atlantic salmon had been extirpated from seven acidic (pH 4.7) rivers in Nova Scotia that had previously supported the species (Watt *et al.*, 1983). Atlantic salmon were on the decline in other rivers with pH 4.7–5.0, but were stable in rivers with pH 5.0. Different species of fish are affected to varying degrees by reduced pH levels: many cold-water salmonids are generally more sensitive than other species, and among the salmonids, the rainbow trout is the most sensitive.

Fish have been affected both by gradual long-term changes in pH and the sudden 'acid shock' that follows the first heavy autumnal rains or the early spring snowmelt release of acids accumulated over winter. Another factor affecting fish mortality is the increased mobilisation of toxic metals by acid deposition. Hydrogen cations introduced into soils by acid rain may exchange with heavy metal cations formerly bound to colloidal particles, thereby releasing the metals into the soil and watercourses. Aluminium is a particular problem and affects fish by obstructing their gills. A gradual loss of fish stocks in many freshwater systems is often more a result of recruitment failure as opposed to outright mortality, as evidenced by higher proportions through time of more mature individuals. The loss of fish from the top of the food chain consequently has knock-on effects at lower trophic levels.

Other aquatic organisms are also affected by acidification. Battarbee (1994) notes that populations of algae, zooplankton and benthic invertebrates are all affected, and that at higher trophic levels, studies have documented the negative effects on amphibians (e.g. the UK's rare natterjack toad) and birds (e.g. the dipper). Species

TABLE 12.2 *Sequence of biological changes during experimental acidification of the Ontario Lake 223*

Year	pH	Changes observed
1976	6.8	Normal communities for an oligotrophic lake
1977	6.13	Increased chironomid emergence
1978	5.93	Fathead minnow reduced abundance; *Mysis relicta* near to extinction; further increased chironomid emergence
1979	5.64	*Orconectes virilis* recruitment failure; further increased chironomid emergence
1980	5.59	Lake trout recruitment failed; *Orconectes* population decline, parasitism increase, egg attachment failure
1982	5.09	White sucker recruitment failed; pearl dace declined; *Orconectes* near to extinction; chironomid emergence falls back to 1976 levels
1983	5.13	*Orconectes* extinct; mayfly (*Hexagenia* sp.) appears to be extinct

Source: modified after Howells (1990: 163, Table 10.1)

higher up the food chain are more likely to be affected indirectly, through changes to their habitats and food sources. The sequence of biological changes caused by acidification monitored in a Canadian lake deliberately dosed with sulphuric acid over an eight-year period is shown in Table 12.2. The crustacean *Mysis relicta* was an early casualty, while several species of fish were seen to decline and eventually disappear, and an increasing growth of filamentous algae was also noted at about pH 5.6

Although there are few studies, to date, on acid rain effects on marine ecology, there is some evidence to suggest that increased leaching of trace metals from acidified areas can influence phytoplankton growth in coastal waters (Granéli and Haraldson, 1993).

Terrestrial ecosystems

Acid rain may damage vegetation communities in a number of ways:

- increasing soil acidity
- decreasing nutrient availability
- mobilising toxic metals
- leaching important soil chemicals
- changing species composition and decomposer micro-organisms in soils.

The widespread loss of lichens across many of Europe's industrial areas has been attributed to atmospheric pollution, chiefly by sulphur dioxide, to which nearly all lichens are very sensitive. Particularly severe damage to lichens and mosses has been reported from the most polluted sites, such as in Russia's Kola peninsula, site of two of the world's largest SO_2 and heavy metal emission sources. The moss–lichen layer performs many ecological functions, including biomass production, the regulation of soil temperature and water-holding capacity, and protection of soil from erosive processes. One knock-on effect of the widespread loss of this important layer in Kola is extensive damage to trees and dwarf shrubs caused by frost, as well as extensive soil erosion (Kashulina *et al.*, 1997).

The effects of acid rain on trees and forests has been one aspect of terrestrial ecosystems that has attracted particular interest, and also controversy, since trees are subject to many natural stresses, including diseases, pests and climatic variations

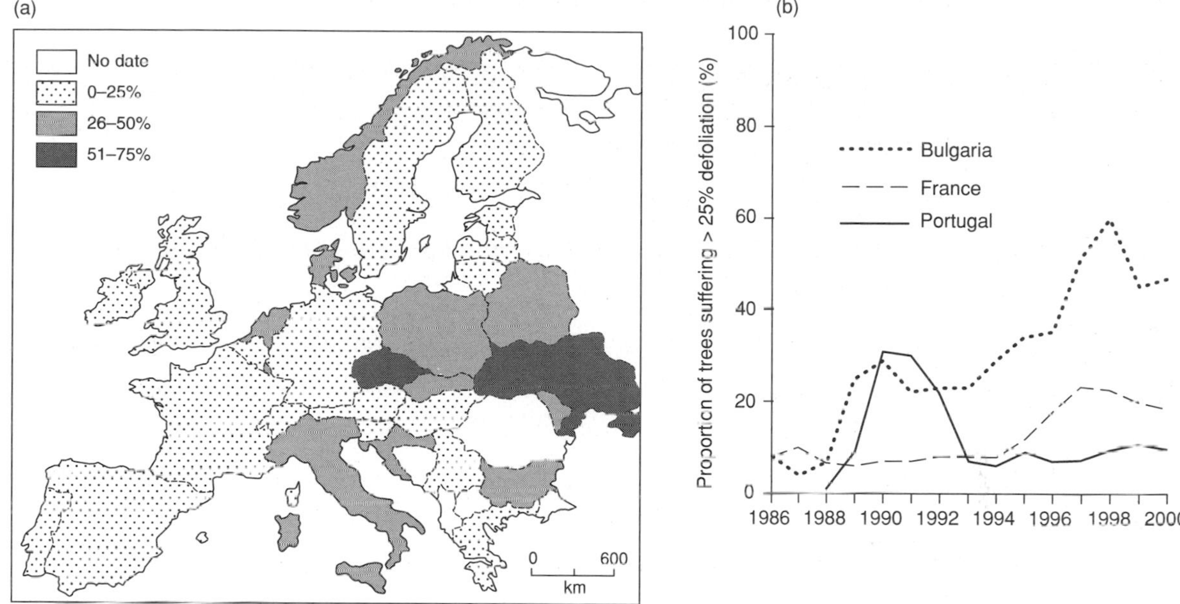

Figure 12.4 Forest damage in Europe, 2000 (from data in Elvingson, 2001): (a) proportion of trees suffering >25 per cent defoliation by country (the data refer to all species except in Eire where they refer to conifers only); (b) time series of defoliated trees in selected countries.

such as hard winters and high winds, as well as other forms of pollution (Innes, 1992). Hence, distinguishing the effects of acid rain from other potential causes of deteriorating tree health is by no means easy, and acid rain pollutants may be just the first link in a chain of degradation as the case of the Kola peninsula described above shows. Indeed, to complicate the issue further, very similar damage to that described for the Kola peninsula is more widespread in neighbouring regions of Finland and Norway, where the cause of moss–lichen layer loss is quite different. Acid rain damage appears to be very limited in Norwegian and Finnish parts of the Barents region, but more widespread loss of the moss–lichen layer is attributed to heavy grazing by reindeer herds.

The symptoms commonly associated with poor tree health – which may be due, at least in part, to acid rain – include thinning of crowns, shedding of leaves and needles, decreased resistance to drought, disease and frost, and direct damage to needles, leaves and bark. A systematic programme monitoring damage to European forests has been carried out since 1986. The programme measures defoliation according to five classes on more than 6000 permanent forest sites across Europe. In the survey for 2000, about one in every four trees was found to be suffering from abnormal thinning of the crown, having lost more than 25 per cent of their leaves or needles. The intensity of defoliation for all species by country is shown in Fig. 12.4a. The percentage varies from one species to another, however, as well as between regions. Damage is lower than the average, for instance, in Scandinavia and on the eastern side of the Baltic Sea while some of the highest percentages of damaged trees are found in central and eastern Europe. Trends through the period 1986–2000 are by no means obvious or constant for all countries (Fig. 12.4b) and country-scale data obscure regional variations; overall, however, the annual surveys show that there has been a steady increase in total damage up to 1995, after which the situation stabilised at a relatively high level (Elvingson, 2001).

The causes of damage include drought, wind,

frost and disease, but air pollution is widely recognised as a contributing factor in those countries where the critical loads for many pollutants are being markedly exceeded (Elvingson, 1993; 1997; 2001). The monitoring surveys indicate that in most places the deposition of nitrogen is now greater than that of sulphur, but sulphur still accounts for most of the acidification because nitrogen is mostly taken up by plants and/or fixed in the soil. The complexities of recognising causes and effects are well illustrated for a number of European forest regions in Table 12.3.

Such impacts can have knock-on effects for faunal species that depend upon forests for food and breeding. In the case of birds, Goriup (1989) suggests that while some species may face population declines and/or range contractions, others may benefit from the superabundance of dead and decaying standing timber. Some forest-dwelling birds may be affected more directly by the acidification of soils. The reproductive capacity of great tits and other species has been reduced in European forests as snail populations have declined on soils with falling pH, leading to a calcium deficiency in birds, which, in turn, is manifested in a number of egg-shell defects (Graveland *et al.*, 1994).

Human health

The affects of acid rain pollutants on human health occur directly through inhalation, and indirectly through such exposure routes as drinking water and food contamination. Most air pollutants directly affect respiratory and cardiovascular systems, and high levels of SO_2 have been associated with increased mortality, morbidity and impaired pulmonary function, although distinguishing between the effects of SO_2 and other pollutants, such as suspended particulate matter, is not always easy. Atmospheric concentrations of SO_2 have been shown to have a positive association with daily mortality in Athens, for example, but carbon monoxide and smoke levels are similarly related (Touloumi *et al.*, 1994). The best-documented effects relate to strong acids in aerosol form (i.e. sulphuric acid,

TABLE 12.3 *Some examples of recent forest decline in Europe and the possible causal sequences*

Region	Causes
Central Europe	Dieback of silver fir, a well-documented historical phenomenon that peaked most recently in the 1970s in association with a series of dry summers
Fichtelgebirge (German–Czech border)	Wet deposition of sulphur and nitrogen, SO_2, Ca/Al and Mg/Al ratios in soil affecting tree nutrition via mycorrhizae
Erzgebirge-Silesia (confluence of German–Czech–Polish borders)	SO_2, heavy metals, fluorine, other gaseous pollutants, frost
Inner Bavarian Forest (Germany)	Acid mist, frost shocks, some soil nutrient problems
Athens Basin	O_3 and other oxidant damage to pines
Netherlands	Nitrogen inputs (NH_3 and NO_3 – enhancing sensitivity to other stresses), drought, perhaps O_3

Source: after Freer-Smith (1998)

H_2SO_4 and ammonium bisulphate, NH_4HSO_4), and the areas of major concern with respect to inhalation of these aerosols are exacerbation of the effects of asthma and the risk of bronchitis in all exposed persons. Inhalation of highly water-soluble acidic vapours such as SO_2 can also cause breathing difficulties, particularly in asthmatics.

The release of heavy-metal cations, such as cadmium, nickel, lead, manganese and mercury from soils into watercourses can eventually reach human populations through contaminated drinking water or fish. Such toxic heavy metals have been associated with a number of health problems including brain damage and bone disorders. Acidified water can also release copper and lead from water-supply systems with similar adverse effects on human health. High levels of lead in drinking water, for example, have been linked to Alzheimer's disease, a common cause of senile dementia.

Materials

Acid deposition has been associated with damage to building materials, works of art and a range of other materials. In all cases, the presence of water is very important, either as a medium for transport and/or in its role in accelerating a vast array of chemical and electrochemical reactions and in creating mechanical stresses through cycles of solution and crystallisation.

Acid rain can accelerate the corrosion of metals and the erosion of stone. Sulphur compounds are the most deleterious in this respect, although rain made acidic to below a pH of 5.6 by CO_2, SO_2 or NO_x can dissolve limestone. Iron and its alloys, which are widely used in construction, are among the most vulnerable construction materials. The metals are attacked by corrosion, particularly by SO_2. Atmospheric corrosion of iron and steel reached rates of up to 170 mm/year in the most heavily polluted areas during the 1930s, while rates in present-day Britain for plain carbon steels range from 20–100 mm/year (Lloyd and Butlin, 1992). Sulphur corrosion was implicated in the collapse of a steel highway bridge between

Figure 12.5 Acid rain damage to a statue in one of the courtyards of the Louvre in central Paris, France.

West Virginia and Ohio in the USA in 1967, a disaster that killed 46 people (Gerhard and Haynie, 1974).

Limestone, marble, and dolomitic and calcareous sandstones are vulnerable to SO_2 pollution since in the presence of moisture, dry deposition of SO_2 reacts with calcium carbonate to form gypsum, which is soluble and easily washed off the stone surface. Alternatively, layers of gypsum can blister and flake off when subjected to temperature variations. Soluble sulphates of chloride and other salts can also crystallise and expand inside stonework, causing cracking and crumbling of the stone surface. Such processes have caused particularly noticeable damage to statues in many urban and industrial areas (Fig. 12.5) and some of the world's most important architectural monuments, including the Parthenon in Athens and the Taj Mahal in Agra, have been adversely affected. The results of one study, in which weathering rates were measured on more than 8000 century-old marble tombstones across North America, suggest that SO_2 pollution was probably responsible for more deterioration than other weathering processes (Meierding, 1993). Upper stone faces in heavily polluted localities with mean SO_2 concentrations of 350 g/m^3 receded at a mean rate >3 mm/100 years due to granular disintegration induced by the growth of gypsum crystals between calcite grains.

Although modern glass is little affected by SO_2, medieval glass, which contains less silica and higher levels of potassium and calcium, can be weakened by hygroscopic sulphates, which remove ions of potassium, calcium and sodium. Paper is affected by pollutants of SO_2 and NO_2 because these are absorbed, making the paper brittle, and fabrics such as cotton and linen respond to SO_2 in a similar way. Damage to materials such as nylon and rubber have also been attributed to SO_2 as well as low-level ozone.

COMBATING THE EFFECTS OF ACID RAIN

Working on the premise that prevention is better than cure, the most obvious method for combating the effects of acid rain is to reduce the emissions of sulphur and nitrogen oxides from polluting sources. There are numerous pathways that can be taken towards achieving this aim. Reducing emissions from power stations can be approached in a number of ways. Energy conservation is the simplest and arguably the most sensible method of reducing emissions from the burning of fossil fuel, while increasing the use of renewable energy sources also helps (Fig. 12.6), although such alternatives also come with an environmental price (see Chapter 18). Alternatively, there are many existing technologies available that can reduce acid rain pollutant emissions from power stations and other industrial sources. These include:

- fuel desulphurisation, which removes sulphur from coal before burning
- fluidised bed technology, which reduces the SO_2 emissions during combustion
- flue gas desulphurisation, which involves removing sulphur gases before they are released into the air.

Similar approaches can also be taken to reduce emissions of NO_x from power stations, while catalytic converters and lean-burn engines have been applied to NO_x emissions from motor vehicle engines (see Chapter 16). Some of these technocentric approaches can simply transform one type of environmental problem into another, however. Flue gas desulphurisation involves using a scrub-

Figure 12.6 Solar-powered water heaters on rooftops in Athens help to reduce acid rain precursors from the burning of fossil fuels in a city where acid rain threatens human health and numerous ancient monuments.

ber to remove SO_2, but the resulting by-products are often disposed of in landfill sites. More sustainable uses of the by-products are being developed, such as that described on page 328 for rehabilitating areas covered by minespoil.

Political aspects of emissions reduction

Many of these actions are costly in economic terms, however, and the uncertainties that have surrounded some of the suggested effects of acid rain have been the reason for some governments' unwillingness to initiate expensive acid rain reduction programmes.

Britain has been a case in point here. The Scandinavian countries were successful in persuading the UN Economic Commission for Europe to set up a Convention on Long-Range Transboundary Air Pollution (CLRTAP), which was formally signed by 35 states in 1979. Its members include all European countries, the USA and Canada. However, when the Convention's protocol on sulphur emissions reduction was adopted in 1985 – the '30% Club', so-called because the members agreed that their sulphur emissions in 1993 would be at least 30 per cent less than their 1980 levels – Britain, amongst others, refused to join. The case is an interesting one for the light that it shines upon the interactions between politics and science.

In the early 1980s, Britain became increasingly isolated politically in its reticence to take the issue of transboundary transportation of acid rain pollutants seriously enough to take preventative action. Initially, an influential ally was found in the former West Germany, but domestic pressure turned the Germans in favour of sulphur reductions following widespread fears over the deteriorating health of German forests: so-called 'waldsterben'. While other European countries, particularly those convinced that their ecology was being adversely affected by pollution from their neighbours, were clamouring for concerted action, Britain's response was to call for more research (Dudley, 1986). The British Government's logic was contrary to the precautionary principle but was based on the fact that action to reduce emissions would cost a great deal of money, and therefore they had to be convinced that the money was being spent correctly. Evidence that was convincing enough for some countries was not convincing enough for others. Other countries that refused to join the 30% Club, such as Spain and Poland, also did so primarily for economic reasons. In 1986, however, Britain did concede the need to reduce SO_2 emissions and began to introduce flue gas desulphurisation equipment to a number of large British power stations.

Irrespective of the political tensions engendered by the 30% Club, its approach had inadequacies, both absolutely and relatively. In absolute terms, many believed that reductions needed to be much greater to make any significant improvements, while it was also recognised that reductions needed to be more specifically targeted at the large emitters and at the worst-affected environments. Hence, the critical load idea was adopted for the new sulphur emission protocol, signed in 1994, to replace the 30% Club. The new protocol marks a significant milestone in international pollution control since it is the first time that different targets have been set for each country. The targets take the form of maximum permissible emissions per target year, which are based upon the ability of the environment to withstand pollution. Great Britain has committed to an 80 per cent reduction in emissions by 2010 from the baseline year of 1980. Germany's

TABLE 12.4 *Excess deposits of sulphur in Norway*

	Area where critical load exceeded (% national land area)
1985	30
1990	25
2010	16

Source: Norwegian Institute for Water Research

TABLE 12.5 *Changes in national emissions of sulphur dioxide and nitrogen oxides in selected countries, 1980–2000*

Country	Annual emissions (thousand tonnes) 1980	1990	2000	Change 1980–2000 (%)
SULPHUR DIOXIDE				
China and centrally planned Asia	7 800	13 000	18 000	131
South and South-East Asia	4 000	6 400	9 400	135
USA	23 500	21 481	16 483	-30
Canada	4 643	3 236	2 534	-45
Poland	4 100	3 210	1 511	-63
UK	4 880	3 754	1 165	-76
NITROGEN OXIDES*				
USA	22 121	21 927	21 713	-2
Canada	1 959	2 104	2 058	5
UK	2 580	2 756	1 512	-41
Poland	1 229	1 280	838	-32

*Note: Emissions for nitrogen oxides are given as nitrogen dioxide equivalents
Source: UNECE/EMEP emission database at http://webdab.emep.int/accessed August 2002 for all except Asian data from Smith *et al.* (2001)

commitment is 87 per cent, that of Poland 66 per cent and Russia 40 per cent. The benefits of these reductions are already being felt. While 30 per cent of Norwegian territory received amounts of sulphur that exceeded the critical load in 1985, calculations based on the sulphur protocol indicate that the proportion of the national territory where critical loads will be exceeded will be reduced to about 16 per cent by 2010 (Table 12.4).

Some of the largest emitters of sulphur and nitrogen are shown in Table 12.5. This table illustrates the success of several industrialised countries in reducing sulphur dioxide emissions over the period 1980–2000. Decreases have been achieved in many cases through pollution control strategies, energy conservation and fuel switching. Even greater reductions have been achieved in other countries, such as Sweden, where sulphur dioxide emissions have been cut from 491 000 t in 1980 to 58 000 t in 2000, an 88 per cent reduction, by a combination of regulatory measures and the introduction in 1988 of a 'sulphur tax' levied on combustion plants. The effect on atmospheric concentrations in many cities in the more developed countries is illustrated by the trend shown for Tokyo in Fig. 12.7. Worrying

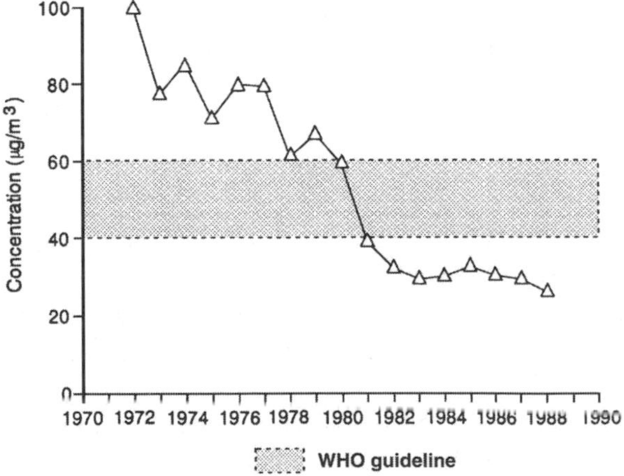

Figure 12.7 Decline of mean annual sulphur dioxide concentration at Omori-Minami, an industrial city centre site in Tokyo, 1972–88 (after UNEP/WHO, 1992).

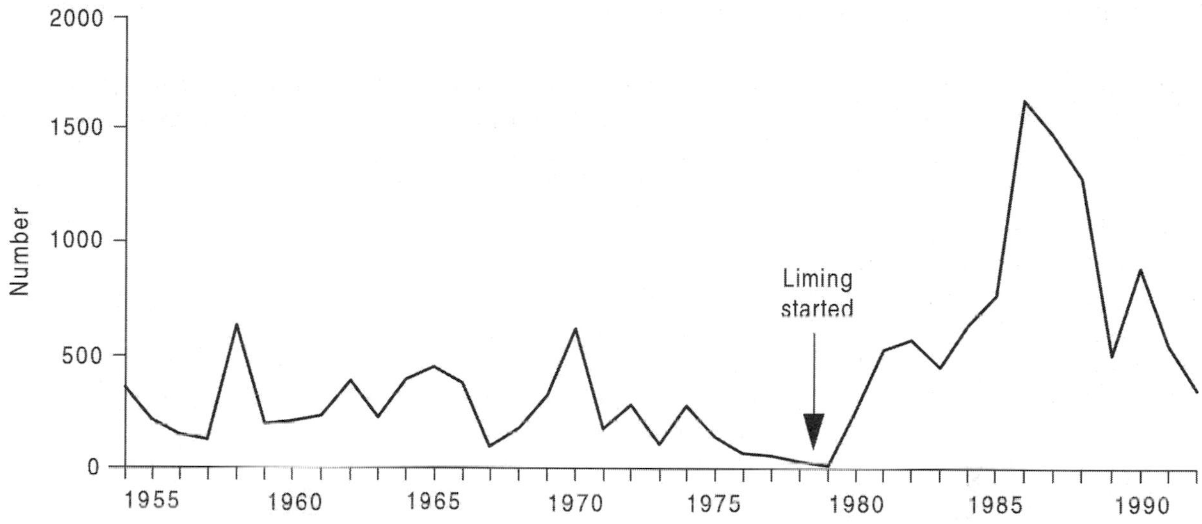

Figure 12.8 Counts of upstream migrating salmon in the River Högvadsån, south-western Sweden, 1954–92 (after Fleischer *et al.*, 1993).

trends are illustrated for most of Asia, however, where emissions have more than doubled over the 20-year period.

Recent trends for nitrogen oxides in North America are less encouraging, with little change over the period, but significant decreases have been achieved in much of Europe. China and India increased emissions by half over the decade 1980–90, and although similar data for 1990–2000 are not available, this trend is likely to have continued. Motor vehicles are a major source of nitrogen oxides and larger emissions probably reflect increases in national fleets.

Environmental recovery

Even given the actual and planned reduction in acid rain emissions, the response of affected environments is not necessarily known or guaranteed. There is some evidence for fairly rapid recovery following emissions control, such as that reported for lake fauna downwind of the Sudbury metal smelters in Canada after a 50 per cent reduction in sulphur emissions (Gunn and Keller, 1990), but not all organisms may respond as immediately. In many instances there is likely to be a timelag between emissions reduction and a detectable reduction in environmental effects, the so-called 'memory effect' as far as building stone weathering is concerned (Cooke, 1989). None the less, 20 years of micro-erosion meter measurements on the balustrade of St Paul's Cathedral in London have shown a reduction in the mean erosion rate on horizontal surfaces from 0.045 mm/yr in the period 1980–90 to 0.025 mm/yr in 1990–2000, while a decline in atmospheric SO_2 was recorded from 20–25 ppb in 1980–82 to around 10 ppb in 1990–2000 (Trudgill *et al.*, 2001; see Fig. 10.5).

In ecosystems, it is thought that recovery from acidic deposition is probably a complex, two-phase process in which chemical recovery precedes biological recovery. Work in the English Lake District has shown improvements in freshwater ecology following two decades of declining acid emissions. Tipping *et al.* (2002) found acid-sensitive stoneflies that had been absent in surveys conducted during the 1960s and 1970s had returned by 1999. Diatoms found in lake sediments had also responded to the increasing pH in some of the lakes surveyed, though not in all.

Generally the time for biological recovery is better defined for aquatic than terrestrial ecosystems. Evidence from the Hubbard Brook Experimental Forest, a long-term research site in the White Mountains of New Hampshire in north-eastern USA, suggests that for acid-impacted aquatic ecosystems, stream macroinvertebrate and lake zooplankton populations would recover in three to ten years after favourable chemical conditions are re-established. Fish populations would follow within a decade, following the return of the zooplankton on which they depend. For terrestrial ecosystems, trees would probably respond positively to favourable atmospheric and soil conditions over a period of decades (Driscoll *et al.*, 2001). Soils heavily leached by acid rain, in some cases losing as much as half their calcium and magnesium contents, may take many decades to recover (Kaiser, 1996).

However, in the case of acidified lakes, some researchers believe that it might take more than 100 years before fresh waters regain something close to their original species composition with maintained functions. This view emphasises the continued need for complementary local measures such as appropriately managed land use and the continued implementation of liming programmes designed to increase water pH. Liming is carried out in several European countries and in North America. The most comprehensive operation is that in Sweden, where a large-scale freshwater liming programme initiated in 1977 had treated about 6000 lakes up to 1992 (Fleischer *et al.*, 1993). The programme has been chemically successful and in many cases considerable biological improvements have resulted, as the Atlantic salmon counts for a river in south-west Sweden, shown in Fig. 12.8, indicate. In this river, salmon counts rose as upstream migration increased after liming in the late 1970s; in other rivers and lakes, fish populations may require help by stocking.

Recovery of lacustrine ecosystems is also likely to be slowed further by the combined effects of acid rain and other major environmental issues. A study of North American lakes related increased rates of death and disease among fish and aquatic plants to both global warming and the depletion of the ozone layer (Schindler *et al.*, 1996). The lakes' levels of dissolved organic carbon were found to have fallen due to both climate warming and acidification. Since carbon absorbs ultraviolet radiation, which has increased due to stratospheric ozone depletion, ultraviolet radiation has penetrated lake waters to much greater depths, adversely affecting the health of aquatic organisms. No less than 140 000 lakes in North America are estimated to have carbon levels low enough to be at risk from such deep ultraviolet penetration. The deleterious effects can also be compounded by another aspect of climate, when sulphur compounds stored in lake sediments are oxidised in response to receding lake levels during drought periods (Yan *et al.*, 1996).

FURTHER READING

Hodgson, D.A., Vyerman, W., Chepstow-Lusty, A. and Tyler, P.A. 2000 From rainforest to wasteland in 100 years: the limnological legacy of Queenston mines, Western Tasmania. *Archiv für Hydrobiologie* 149: 153–76. A southern hemisphere example of historical acidification around a copper smelter.

Howells, G. 1990 *Acid rain and acid waters.* London, Ellis Horwood. A concise and comprehensive critical review of the acid rain issue, particularly the effects on water, soils and vegetation.

Innes, J.L. 1992 Forest decline. *Progress in Physical Geography* 16: 1–64. A comprehensive review of the many factors that can effect forest decline.

McCormick, J. 1997 *Acid earth: the politics of acid pollution*, 3rd edn. London, Earthscan. A global overview of the science, politics and economics of acidification.

Mason, B.J. 1992 *Acid rain: its causes and effects on inland waters.* Oxford, Clarendon Press. This book focuses on acidification of lakes and streams, and the effects on aquatic animals. It draws largely on experimental and modelling work conducted in the UK and Scandinavia.

Smith, S.J., Pitcher, H. and Wigley, T.M.L. 2001 Global and regional anthropogenic sulfur dioxide emissions. *Global and Planetary Change* 29: 99–119. An examination of the geographical shift in human emissions of sulphur since 1980.

Steinberg, C.E.W. and Wright, R.F. (eds) 1994

Acidification of freshwater ecosystems: implications for the future. Chichester, Wiley. A collection of 24 papers on all aspects of acidification, including the recovery of ecosystems.

Websites

http://www.acidrain.org/ the Swedish NGO Secretariat on Acid Rain site includes the publication *Acid News*.

http://www.epa.gov/airmarkets/ the US Environmental Protection Agency's site contains reports, data and maps of deposition in the USA.

http://www.ngu.no/Kola/ the Kola Ecogeochemistry Project site has details on this environmental survey in the Barents region, which includes two of the world's largest SO_2 emission sources.

http://www.hbrook.sr.unh.edu/hbfound/hbfound.htm the Hubbard Brook Research Foundation carries out important research into acid rain in the USA.

http://www.emep.int/ the site of the programme for monitoring and evaluation of the long-range transmission of air pollutants in Europe gives access to numerous publications and data.

POINTS FOR DISCUSSION

- Do you agree that acid rain is as much a political issue as an environmental one?
- Why is it so difficult to attribute forest decline directly to any one cause?
- How has the concept of thresholds in the physical environment been used to help solve the acid rain problem?
- Examine the reasons for the timelag between reductions in acid rain emissions and biological recovery.

13

FOOD PRODUCTION

TOPICS COVERED

Agricultural change
Fertiliser use
Irrigation
Agricultural pests
Biotechnology
Aquaculture

Key words: domestication, Green Revolution, integrated pest management, GM food, terminator technology, Blue Revolution

The origins of agriculture, which includes both crop and livestock production, are traced back 10 000 years to the start of the Holocene. Before that, the genus *Homo* had lived for more than 2 million years by gathering plants and hunting wild animals for food, and *Homo sapiens* continued this practice for 100 000 years. For the last 5000 years, virtually the entire world's population has been reliant on farmers and herders who consciously manipulate natural plants and animals to provide food. As global population has grown, the area used for food production has expanded and production techniques have become increasingly sophisticated. Both the extensification and the intensification of food production have thrown up numerous environmental issues. Many of these are covered in this chapter, but other environmental impacts of agriculture are dealt with elsewhere. These include the problems associated with the clearance of forests for farming (Chapter 4), the environmental impacts of heavy grazing in drylands (Chapter 5), soil erosion on agricultural land (Chapter 14) and some of the impacts associated with irrigation (Chapters 5, 8 and 9). Most of the world's food supply is produced on land, but fish farming has grown in importance in recent years and is detailed in this chapter. Nevertheless, the majority of fish are still hunted, and environmental issues surrounding this form of food production are covered in Chapter 6.

AGRICULTURAL CHANGE

Agricultural management enables more calories of food to be obtained from a given area of land over a more predictable period of time than the simple collecting of wild foods. Some people (e.g. Cohen, 1977) believe that agriculture developed in response to rising human populations, while others have argued that new methods of food production were a cause rather than a consequence of population growth (e.g. Hassan, 1980). It is likely that social, economic, technological and environmental factors all contributed in varying combinations to the early transition from hunters and gatherers to herders and farmers, and this 'Neolithic Revolution' was associated with the emergence of urban civilisations on some of the world's great rivers (the Nile, Tigris–Euphrates, Indus and Huang Ho) between 3500 and 5500 years ago, and in MesoAmerica and the Central Andes around 3000 years ago.

Part and parcel of the start of agriculture was the process of domestication, the deliberate selection and breeding of wild plants and animals that

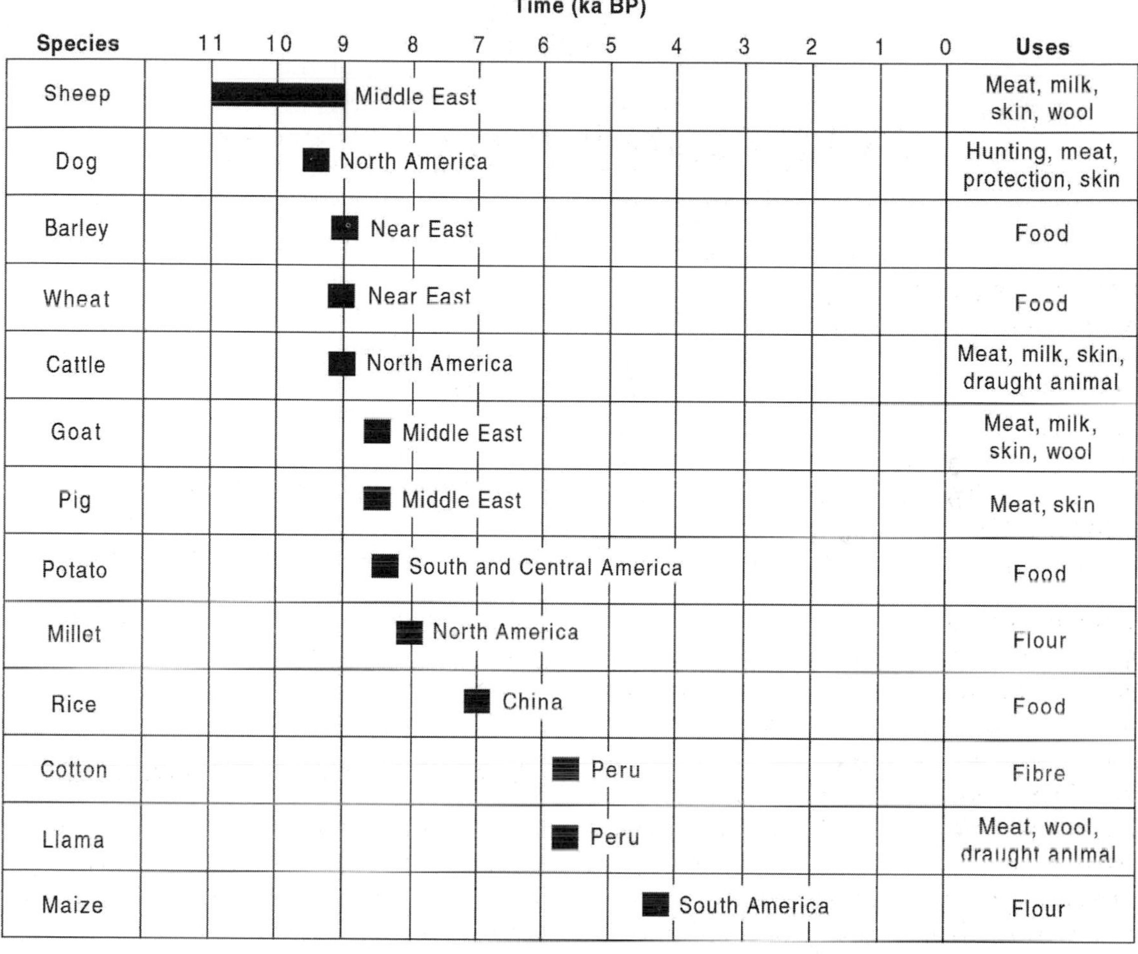

Figure 13.1 Origins of domesticated plants and animals (after Williams *et al.*, 1993).

leads to genetic change. By cultivating plants, for example, farmers chose particular seeds that offered the best results and by planting more of them increased their chances of survival. The dominance of particular characteristics was thus deliberately increased. In crops valued for their edible seed, such as wheat, a shift towards non-shattering occurred quickly, and hence the natural mechanism for seed dispersal was lost and the plant became dependent on humans for survival. The origins of domestication, which can be seen as one of the most important human impacts on the natural environment, have been traced to diverse parts of the world using archaeological evidence (Fig. 13.1). The process of selection and breeding continues today in a more sophisticated manner using such techniques as tissue culture and genetic engineering (see below).

Irrespective of the original causes of the emergence of agriculture, there is no doubt that the ability to provide more food from a unit area of land per unit time has been associated with an exponential growth of the world's human population. A global population thought to have been around 5 million 10 000 years ago increased by a factor of 20 to 100 million in 5000 years (May, 1978). The 500 million mark was passed about 300 years ago and by 1850, 1000 million (1 billion) people inhabited the planet. By this time, the exponential rate of population growth was

Figure 13.2 A landscape transformed by agriculture on the Portuguese North Atlantic island of Faial in the Azores.

clear and world population had surpassed 4 billion by 1978, 5 billion by 1990 and 6 billion by 2000.

Until the early 1950s, world agricultural production had kept pace with population growth chiefly through expansion of the cultivated area. Expansion subsequently slowed and had all but stopped by the end of the 1980s. Today, after 10 000 years of agriculture, cropland occupies about one-sixth of the world's biologically productive land area (Gilland, 1993), while the area used as pasture for livestock is much greater. By clearing natural vegetation and replacing it usually with fewer domesticated species, or by grazing livestock, agriculturalists have been responsible for one of the most widespread changes wrought by humans on the natural environment. Forests and grasslands of the temperate latitudes have been the most severely altered, but all the world's major biomes, with the exception of the polar ice caps, have been affected (Fig. 13.2).

Continued increases in food production since the 1950s – to keep up with population growth, which has continued at unprecedented rates – has been achieved chiefly by increasing crop yields. The average cereal yield in Britain, for example, has risen steadily from about 2 tonnes per hectare in the mid-1940s to more than 7 tonnes per hectare in 2000 (Robinson and Sutherland, 2002). In fact, the rising trend in yields has a long history in many parts of the world and has been the result of numerous improvements in agricultural efficiency. The tonnage of wheat harvested from an average hectare of English cropland in the 1980s was three times that in the 1930s, five times that in 1800 and ten times that of 1300. The increases were achieved first by the replacement of open-field farming by mixed farming, and then by the replacement of mixed farming by increasingly mechanised chemical farming (Grigg, 1992). One of the outcomes of using more machinery on arable land has been the removal of hedgerows to enlarge field sizes, which in England is akin to a reversion to the large open fields of the Medieval period as illustrated in Table 13.1. The average field size in arable Cambridgeshire has more than

TABLE 13.1 *Changes in the length of hedgerows in three parishes in Huntingdonshire, England, since the fourteenth century*

Date	Length of hedges (km)	Date	Length of hedges (km)
Pre-1364	32	1780	93
1364	40	1850	122
1500	45	1946	114
1550	52	1963	74
1680	74	1965	32

Source: Moore *et al.* (1967)

doubled since 1945 (Robinson and Sutherland, 2002), but modern cultural attachment to hedgerows with their flora and fauna, which only became established in the eighteenth century, has caused strong reactions.

Worldwide, it is the adoption of modern chemical farming techniques, both in the developed countries and in parts of the developing world during the so-called Green Revolution, which has enabled the continued increase in yields. Large fields and mechanisation have gone hand in hand with several other key components of these techniques:

- application of increasing amounts of synthetic fertilisers to supply plant nutrients (nitrogen, phosphorus and potassium)
- expansion and improvement of irrigation systems
- protection of crops from diseases, pests and weeds with synthetic chemicals
- development of new varieties of crop plants that are more responsive to fertilisers, more resistant to pests and with a higher proportion of edible material.

The environmental impacts that have come with these developments will form the focus of this chapter, but parallel intensification has occurred over the past 50 years in the livestock industry, particularly in the industrialised world where livestock products provide 30 per cent of the calorie intake, in contrast to the less-developed countries where they provide less than 10 per cent (Grigg, 1993). Factory-style farms are commonplace as pigs and poultry have been moved from the farmyard to indoor feeding facilities and cows are fattened with special feeds. Abundant feed grain, drugs to prevent disease, and improved breeds have enabled animals to be raised faster and leaner as meat production for consumers rich enough to eat it regularly has become the focus of livestock production, taking over from the multiple benefits of manure, draught power, milk and eggs.

The intensification of modern livestock farming has raised a number of environmental issues, not least of which are the concerns over animal welfare, which has engendered a minor shift back towards 'free range' animal products in some western countries (Fig. 13.3). An increasingly serious issue is that of pollution from farm wastes. While the traditional mixed farm operated as a closed system, with manure being spread on cropland to provide valuable nutrients, specialisation in the livestock industry has meant a trend towards concentration in areas distant from centres of arable production. Leakage of manure, with a high biochemical oxygen demand, from storage tanks has become an increasingly serious point-source pollutant, depleting the dissolved oxygen in groundwater, wells and rivers (see Chapter 8). Nitrogen in manure is also lost to the atmosphere as ammonia, contributing to acid deposition: the livestock industry is the single largest source of acid deposited on soils in the Netherlands, for example (Durning and Brough, 1991). Nitrogen from the intensive livestock industry, cropland and industry gives the Netherlands the dubious distinction of the world's highest deposition rates (4–9 $g/m^2/yr$). Excess nitrogen can also affect ecosystem composition and biodiversity, and these consequences in the Netherlands are well documented: the conversion of heathland to species-poor grasslands.

Figure 13.3 Pigs on a free-range farm in the English Midlands.

FERTILISER USE

Although manure and other materials have long been used as natural fertilisers, and are still applied to cropland in many parts of the world, the rise of modern farming techniques has been underpinned by a rapid increase in the use of manufactured chemical, or synthetic, fertilisers to supply plant nutrients. Specially manufactured chemical fertilisers date back to the mid-nineteenth century when a factory producing superphosphates by dissolving bones in sulphuric acid was opened in the English county of Kent, although their use has become globalised only in the last 50 years. World production of nitrogenous fertilisers in 1913 was 0.77 million tonnes nitrogen content and had risen to 3.6 million tonnes in 1950. By 1975, production had increased more than ten times to 42 million tonnes, and continued to rise to 85 million tonnes in 1990 when its manufacture absorbed about 1.3 per cent of world energy consumption. Similarly rapid increases have occurred in the quantities manufactured of the other two main fertilisers: phosphates and potash. This rise in supply has been a key component of the increase in average world cereal grain yields from 1.13 t/ha in 1948–52 to 2.82 t/ha in 1994, and the maintenance of cereal production per head at the present level to 2030 will require a further doubling of chemical nitrogen consumption (Gilland, 1993).

The change in use of the three main types of fertiliser over a 30-year period for a selection of countries is shown in Table 13.2. The global pattern is reflected in most cases, with two notable exceptions. In the Netherlands, the intensity of use rose to a peak in the early 1980s and then declined to a level similar to that in the mid-1960s in response to environmental concerns and the introduction of the National Environmental Policy Plan (see page 379). In Bulgaria, fertiliser use fell sharply in the 1990s due to economic factors related to the change from a centrally planned to a market-based economy.

TABLE 13.2 *Change in use of chemical fertilisers for selected countries*

Country	Fertiliser used on cropland (kg/ha/year)	
	1964–66	1994
Chile	26	95
China	24	309
Bulgaria	82	52
Egypt	117	264
India	5	80
Netherlands	582	592
Nigeria	0	9
Mexico	15	62
Philippines	11	64
Syria	3	59
UK	288	381
USA	63	103
Venezuela	10	68

Source: from data in WRI (1986, 1998)

The recent increase in fertiliser use in the large majority of countries, in combination with several other human activities such as fossil fuel burning and the modification of natural ecosystems through land clearance and wetland drainage, has led to unprecedented effects on the nitrogen cycle. Human activities now contribute about 210 million tonnes to the global supply of biologically available or 'fixed' nitrogen, which is considerably more than the 140 m tonnes made available through natural processes (Table 13.3). The rising trend in the use of nitrogenous and other fertilisers has fuelled concerns over the environmental effects of nutrients that are not taken up by crops. In practice, fertilisers are lost from a field because of overexcessive or poorly timed applications, and they can become pollutants in three ways:

1. excess nitrate and phosphate is lost by runoff and leaching, and enters rivers, lakes, groundwater and coastal waters where it can cause eutrophication and, in the case of nitrate, may cause 'blue baby syndrome' and stomach cancer

2 excess nitrate can be converted by denitrification into nitrous oxide, which is released into the atmosphere (nitrous oxide is a greenhouse gas and also contributes to stratospheric ozone depletion)

3 excess nitrogen can be converted by volatilisation into ammonia, which is released into the atmosphere where it can contribute to the effects of acid rain (see Chapter 12).

Rising levels of nitrates in surface waters have been reported from many areas where modern intensive farming is practised (see Chapters 6 and 8), and the excess of nitrogen is particularly severe in areas where intensive agriculture and fossil fuel use coincide. Some coastal rivers in the north-east USA and northern Europe are thought to be receiving as much as 20 times the natural amount from agricultural and airborne sources, and nitrate levels in many Norwegian lakes have doubled in less than a decade (Vitousek *et al.*, 1997a).

Groundwater levels are also rising. Sampling in Denmark indicated that mean groundwater concentrations of around 1 mg/l NO_3-N in the 1940s and 1950s had risen to 3 mg/l in 1980, with just under 10 per cent of waterworks tested exceeding the maximum limit for drinking water of 11 mg/l (Forslund, 1986). In the USA, the US Geological Service found that in a third of all counties surveyed, 25 per cent of wells had nitrate concentrations that exceeded background levels and in 5 per cent of counties 25 per cent of wells exceeded the Federal drinking water standard of 10 mg/l (Goodrich *et al.*, 1991). Even if field losses are reduced by changes in land use and fertiliser application, groundwater concentrations are likely to continue to rise in the short to medium term in response to past losses that reach the water table with a timelag dependent upon the depth of the unsaturated zone. Analysis of pore water in the unsaturated zone of chalk in southern England has revealed a front of nitrate-rich water moving towards the water table at 1–2 m per year (Burt *et al.*, 1993).

TABLE 13.3 *Estimates of global sources of fixed nitrogen*

Source	Release of fixed nitrogen (million tonnes/year)
ANTHROPOGENIC	
Fertiliser	80
Legumes and other plants	40
Fossil fuels	20
Biomass burning	40
Wetland draining	10
Land clearing	20
Total: human sources	210
NATURAL	
Soil bacteria, algae, lightning, etc.	140

Source: after Vitousek *et al.* (1997a)

IRRIGATION

Irrigation probably first began in the sixth millennium BC when two river-basin civilisations, on the Tigris–Euphrates of Lower Mesopotamia and on the Nile in Egypt, first began to sustain settled agriculture. In modern times, irrigation is often one of the key tools used to intensify agriculture and the global area of irrigated cropland more than doubled in the second half of the last century (Fig. 13.4); it stood at 272m ha in 2000. In many dryland countries, the national area under cultivation would be considerably smaller without irrigation, and in the extreme case of Egypt it would be almost zero (Fig. 13.5).

Unfortunately, however, poorly managed irrigation systems can lead to a wide range of environmental problems. Excessive applications of surface water can result in rising water tables, which cause salinisation and waterlogging. These processes reduce crop yields on irrigation schemes all over the world (see Table 5.5) and can accelerate the weathering of buildings as Akiner *et al.* (1992) have documented for numerous examples of ancient Islamic architecture in Uzbekistan. Conversely,

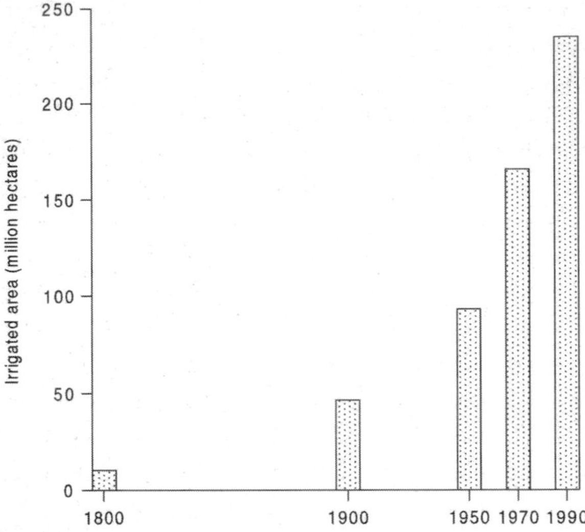

Figure 13.4 Rise in the global area of irrigated cropland, 1800–1990 (from data in FAO, 1993; Gleick, 1993).

overpumping of groundwater resources can result in rapid depletion of supplies, which can threaten the long-term viability of the irrigation schemes themselves. Many examples involving centre-pivot irrigation technology can be quoted, such as the depletion of the Ogallala aquifer beneath the High Plains in south-west USA (Riebsame, 1990) and the rapid drop in groundwater level of the Minjur aquifer, in central Saudi Arabia, where rates of 1.8 m per year were recorded between 1984 and 1990 as irrigated wheat cultivation expanded in the Tebrak area (Al-Saleh, 1992; Fig. 13.6). The problem of receding water tables is particularly prevalent in Iran, where the widespread use of tubewell irrigation has overextracted groundwater from beneath nearly 20m ha, or one-third of all agricultural land (FAO, 1994). Overpumping of groundwater in coastal locations may result in the contamination of freshwater aquifers by sea-water

Figure 13.5 Cropland irrigated with water from the River Nile in Egypt. Irrigation first began on the Nile 6000 years ago, and agriculture in modern Egypt is no less dependent upon the river (Photo: Stephen Stokes).

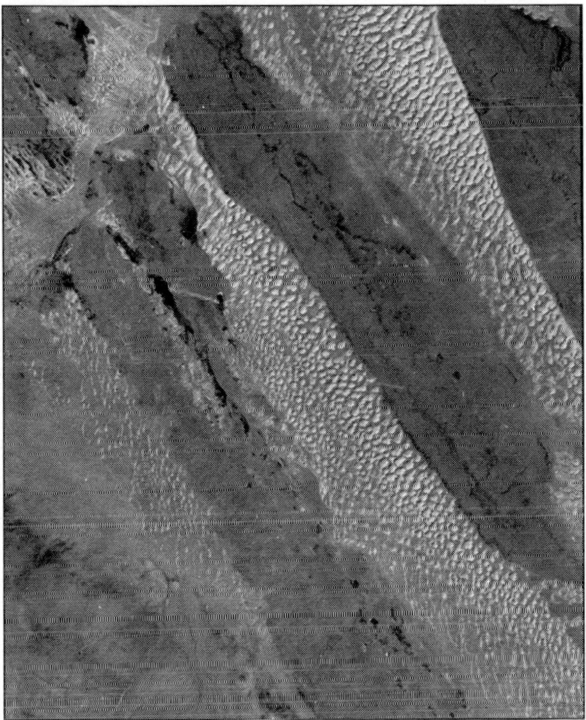
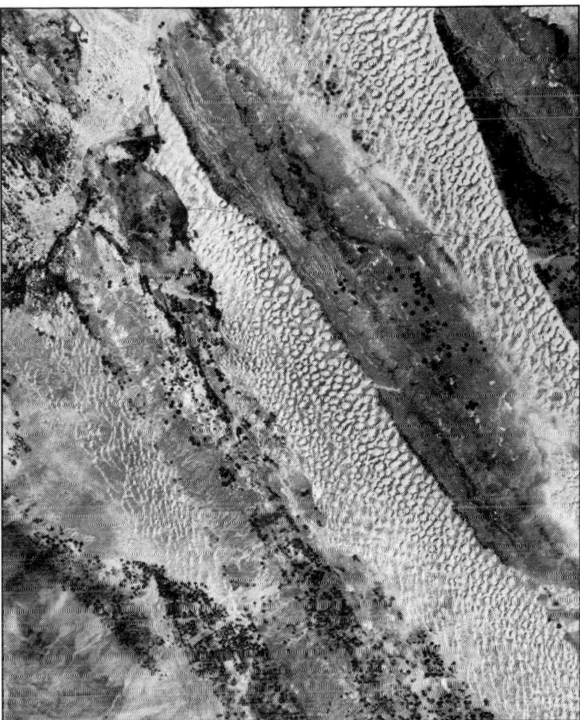

Figure 13.6 The dramatic increase in use of centre-pivot systems to irrigate wheat with water from the Saudi Arabian Minjur aquifer can be seen by comparing these two satellite images of the Tebrak area taken in 1972 and 1986. Bare desert areas in 1972 either side of the sand dune belt running from top left to bottom right are dotted with about 40 distinctive round irrigated areas to the left of the dunes in 1986 and many more to the right. The 1986 image also shows a new highway crossing the dunes just below centre (Images: UNEP GRID).

intrusion, as Mistry (1989) describes in the Saurashtra Peninsula in Gujarat, India.

An increased incidence of water-associated diseases is another problem often associated with the introduction or extension of irrigation schemes. Amin (1977) reports that the prevalence of schistosomiasis in Sudan's Gezira Project rose from 5 per cent in 1945 to 60 per cent in 1973 (see also page 147). Other detrimental health effects can stem from the overexcessive use of pesticides on irrigation schemes, as exemplified by Central Asian cotton plantations (see below), an example that also illustrates the scale of potential off-site environmental effects on the Aral Sea (see page 130).

AGRICULTURAL PESTS

Losing a portion of a field's crop to pests has been a problem for farmers since people first started cultivating soils. Worldwide, an estimated 67 000 different pest species attack agricultural crops and about 35 per cent of crop production is lost to them each year. Insects cause an estimated 13 per cent of the loss, plant pathogens 12 per cent, and weeds 10 per cent (Pimental, 1991). Hence, pests represent a serious problem for farmers and for human society in general.

Pesticides

The use of synthetic chemical sprays to control agricultural pests and diseases dates back to the 1860s when they were used to control Colorado beetle, which damaged potato crops in the USA, and vine mildew in France. Pesticides (which include insecticides, herbicides and fungicides) entered general use in many of the world's agricultural areas along with synthetic fertilisers in the 1950s, following the discovery of DDT in Switzerland in 1939 and 2,4-D in the USA during the Second World War. This development has been a response to an age-old problem but also an attempt to limit the escalation of the pest problems that came about as a result of changing cultivation practices, such as shortened fallow periods, narrow rotation and the replacement of mixed cropping by large-scale monocultures of genetically uniform varieties.

Worldwide, annual sales of pesticides grew from around US$7 billion in the early 1970s to US$25 billion in 1990 (Tolba and El-Kholy, 1992). Most are used in the industrialised countries, but while consumption in these countries has levelled off in recent years, sales to developing countries continue to grow. Most pesticides are used in agriculture, with about 10 per cent used in public health campaigns such as the fight against mosquitoes, which transmit malaria.

One unfortunate aspect of the increasing use of pesticides is the development of resistance to the chemicals. While most individuals of a pest population die on exposure to a pesticide, a few individuals survive by virtue of their genetic make-up and pass on this resistance to future generations. Increasing resistance is often countered by stronger doses of chemicals or more frequent applications, at increasing cost to the farmer, which also speeds up the trend towards resistance, so encouraging the development of new chemicals. Elimination of one pest can also result in the rapid growth of secondary pest populations, however. Widespread applications of DDT in Egypt, for example, to control the American bollworm in cotton plantations, caused a species of whitefly, which had previously been a secondary pest, to explode in numbers in the late 1970s. DDT treatment actually stimulated whiteflies to produce a higher number of eggs than unsprayed flies and the whitefly supplanted bollworm as the country's worst cotton pest (Dittrich *et al.*, 1985).

Pesticides are designed to kill living things, whether they be insects, weeds or plant diseases, but any amount of chemical that misses its intended target can affect numerous other aspects of the environment. In practice, the proportion may be large: some estimates suggest that for many insecticides as little as 0.1 per cent of the amount applied to a particular field may reach the target organism (Pimental and Levitan, 1988); the remainder becomes a contaminant. The resultant effects on non-target plants, animals, soil and water organisms, and the pollution of land, air and water, which can also affect human populations, has caused considerable concern.

TABLE 13.4 *Bioaccumulation of DDT residues up a salt marsh food web in Long Island, USA*

Organism	DDT residues (parts per million)
Water	0.000 05
Plankton	0.04
Silverside minnow	0.23
Sheephead minnow	0.94
Pickerel (predatory fish)	1.33
Needlefish (predatory fish)	2.07
Heron (feeds on small animals)	3.57
Herring gull (scavenger)	6.00
Fish hawk (osprey) egg	13.8
Merganser (fish-eating duck)	22.8
Cormorant (feeds on larger fish)	26.4

Source: after Woodwell *et al.* (1967)

Natural predators of the problem pest may be destroyed, for example, thereby increasing the need for more pesticide applications. Other creatures useful to the farmer, such as bees and other pollinators may be affected. The concentration of some compounds has been seen to accumulate up the food chain, reaching damaging levels in higher species. DDT, a particularly persistent compound, is the classic example (Table 13.4), which at high concentrations has caused considerable increases in mortality among seabirds and birds of prey (see page 254), and has also been reported at dangerous concentrations in human mother's milk. Similar bioaccumulation has been documented for methyl mercury.

On the national scale, the excessive use of pesticides has left large agricultural areas poisoned. In Romania, more than 900 000 ha are chemically polluted, of which 200 000 ha have been rendered totally unproductive through the excessive use of pesticides, consumption of which rose from 5900 t in 1950 to 1.2 million tonnes in 1985 (Turnock, 1993). Contamination of groundwater was first recognised in the USA in 1979 when pesticide residues were found in 96 wells on Long Island and in more than 2000 wells in California (Goodrich *et al.*, 1991). The dangers posed to human health by high concentrations of residues are illustrated in Central Asia, where Glantz *et al.* (1993) suggest that pesticide applications have been on average up to ten times the levels used elsewhere in the former USSR or USA. These high levels have been related to deteriorating human health via contaminated surface water and groundwater. A sharp increase in oesophageal cancers, high rates of congenital deformities, outbreaks of viral hepatitis and a life expectancy in some areas of about 20 years fewer than that for the former USSR in general, have been cited for Central Asian regions of cotton monoculture (Feshbach and Friendy, 1992); these impacts upon human health have been exacerbated by the dearth of medical and health facilities in the Aral Basin.

More direct exposure to dangerous levels of pesticides occurs in numerous ways. Agricultural workers are particularly at risk if not adequately protected during pesticide applications, as is often the case on plantations in the developing world. In one tragic incident in the 1970s, 1500 male banana-plantation workers in Costa Rica became sterile after repeated contact with dibromochloropropane (Thrupp, 1991). A number of cases have also been documented in which seeds treated with pesticides have been eaten rather than planted. Wheat seeds treated with mercury-based fungicides and distributed as food in Iraq caused the hospitalisation of 370 people in 1960 and 6530 in 1972, with 459 deaths in the latter case (Bakir *et al.*, 1973). These authors note similar cases in Guatemala in the early 1960s and Pakistan in 1969. One estimate suggests that as many as 25 million people in developing countries may be affected by pesticide poisonings each year, although this number also includes suicide attempts (Jeyaratnam, 1990).

Realisation of the dangerous effects of many pesticides has encouraged some governments to introduce programmes to reduce their use in agriculture. Some of the countries most heavily dependent upon synthetic chemicals, such as Denmark, the Netherlands and Sweden, have been the most active in this regard (Hurst, 1992). These moves follow the introduction of complete bans on the use of some of the more toxic and persistent pesticides in many developed countries. Long-lasting organochlorine insecticides, such as DDT, aldrin and dieldrin, and herbicides such as 2,4-D and 2,4,5-T, are among the pesticides now rarely used in the more developed countries, although these compounds are still common inputs to agriculture in Latin America (UNECLA, 1990) and many other developing countries.

In some cases, however, such bans have not been without their critics. The use of dieldrin in locust-control programmes is a good example. It was successfully used in Africa throughout the 1960s and 1970s by spraying strips of land near locust hatching areas. Although dieldrin breaks down quickly in tropical sunlight, the strips remained lethal to hoppers for six weeks. Concern over dieldrin's lethal effects on birds, its persistence in the food chain, and its toxicity to people, led to bans in developed countries, which also refused to donate the pesticide to African countries. A much shorter-lived replacement, fenitrothion, was advocated, but fenitrothion's three-day active period means that spraying has to be delayed until mature locusts are swarming, and then they must be sprayed directly. This approach requires precise monitoring and timing. Swarms also need to be sprayed much more frequently, making the environmental advantages over dieldrin marginal. Using fenitrothion also takes more work, more spray and more money to stop a locust swarm (Skaf, 1988).

Alternatives to pesticides

Despite the widespread use of pesticides, about one-third of total crop production is still lost to insects, weeds and plant diseases, a similar proportion to that lost in pre-synthetic chemical times (Pimental, 1991). Although this percentage has been maintained during a period of rapidly increasing crop production, the success of chemicals in controlling pests has not been as great as was once hoped. Alternatives to synthetic chemicals have long been used and, given the problems of resistance and contamination associated with chemical use, these alternatives are being given more attention in both developing and developed countries.

There are five main alternative approaches to chemical pest control (El-Hinnawi and Hashmi, 1987):

1. environmental control
2. genetic and sterile male techniques
3. biological control
4. behavioural control
5. resistance breeding.

A combination of these approaches is often employed as a form of 'integrated pest management', which acknowledges that both crops and pests are part of a dynamic agricultural ecosystem, and aims to limit pest outbreaks by enhancing the effects of natural biological checks on pest populations and only resorting to chemicals in extreme cases.

Environmental control measures incorporate all human alterations to the environment, which range from simple measures, such as digging up the egg pods of pests and the planting of trap crops to lure pests away from the main crop, to larger-scale efforts such as the draining of wetlands that harbour pests. The special breeding for release of large numbers of genetically altered, often sterile, individuals has been used in a number of cases, the aim being to decimate pest populations by preventing reproduction. The technique was pioneered in the control of the screw-worm fly, a pest affecting cattle. Behavioural control techniques that aim to control pests by use of sex pheromones and by attractant or repellent chemicals are in the early stages of development, while the selected breeding of organisms resistant to pests has long been practised and is now a more precise science thanks to modern biotechnology (see below).

The most widely practised and successful method of biological control, which uses living organisms as pest control agents, has been the introduction and establishment of appropriate species to provide permanent control of a pest. A well-known historical example is the introduction of a South American moth to Australia in the 1920s to combat the explosive spread of prickly pear across the east of the country. The prickly pear had itself been introduced from South America in the late eighteenth century by the first Governor of Australia in the mistaken belief that it might form the basis for a dye industry. Having become established on a new continent, the prickly pear spread very rapidly through eastern Australia in the early decades of the twentieth century, and at the peak of infestation occupied an area larger than Great Britain, about 24 million hectares (Fig. 13.7), ruining many farmers and ranchers, and threatening the rural economy of the entire continent.

Having identified the Cactoblastis moth larvae as a natural predator of the prickly pear, eggs were first placed on plants in September 1927 with dramatic effect. After a temporary decline in moth numbers in 1931–32, when the prickly pear regained some of its former territory, the Cactoblastis population resurged to contain the pest. Since 1940, a balance between pear biomass and insect numbers has been established, at a level slightly less than 1 per cent of the peak prickly pear biomass (Freeman, 1992).

Introductions for biological control can have unwanted effects, however. A particularly damaging example occurred on the French Polynesian

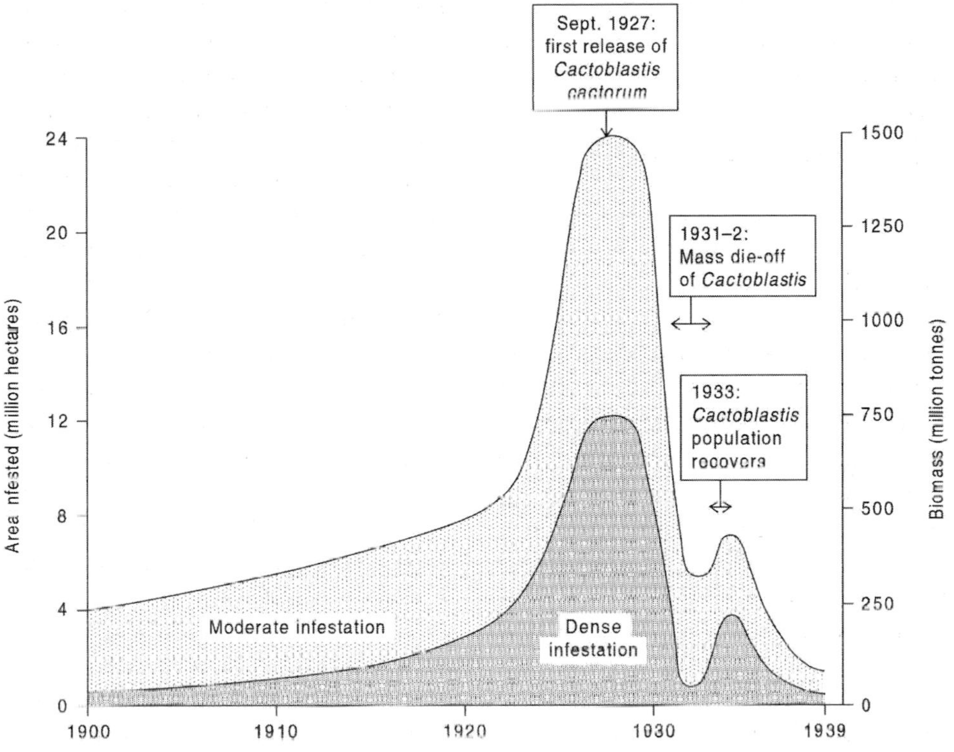

Figure 13.7 Growth and collapse of prickly pear infestations in eastern Australia, 1900–39 (after Freeman, 1992. Reproduced with modifications by permission of the American Geographical Society).

island of Moorea, where the giant African snail was introduced in the nineteenth century by colonial governors with a yearning for snail soup. However, the snail spread across the island and caused widespread destruction to crops, so a predaceous snail, the Ferussac, was brought in to control the giant African snail in 1977. Unfortunately, the Ferussac found native snail species to be more palatable targets and, by 1987, seven species endemic to Moorea had become extinct in the wild, although six survive in captivity (Murray *et al.*, 1988).

BIOTECHNOLOGY

Although the term biotechnology is a recent one, people have manipulated the genetic make-up of crops and livestock for thousands of years by selecting and breeding according to their needs. Similarly, the production of bread, cheese, alcohol and yoghurt are age-old examples of the deliberate manipulation of organisms or their components to the benefit of society and are, therefore, examples of biotechnology. Some of the recent steps in the cross-breeding of wheat strains are shown in Fig. 13.8. Higher yields have been attained by cross-breeding with varieties that had more ears per stem, but the extra weight often caused the stem to break, making it impossible to harvest. Cross-breeding with short-stemmed varieties solved this problem. Diseases and pests are a constant problem which can be combated with pesticides and further cross-breeding.

For the last five millennia, virtually the entire world's population has been reliant on farmers and herders who consciously manipulate natural plants and animals to provide food. This means that the crops we grow for food and the animals we raise to eat are radically different from those that existed in the 'natural' state. The same even

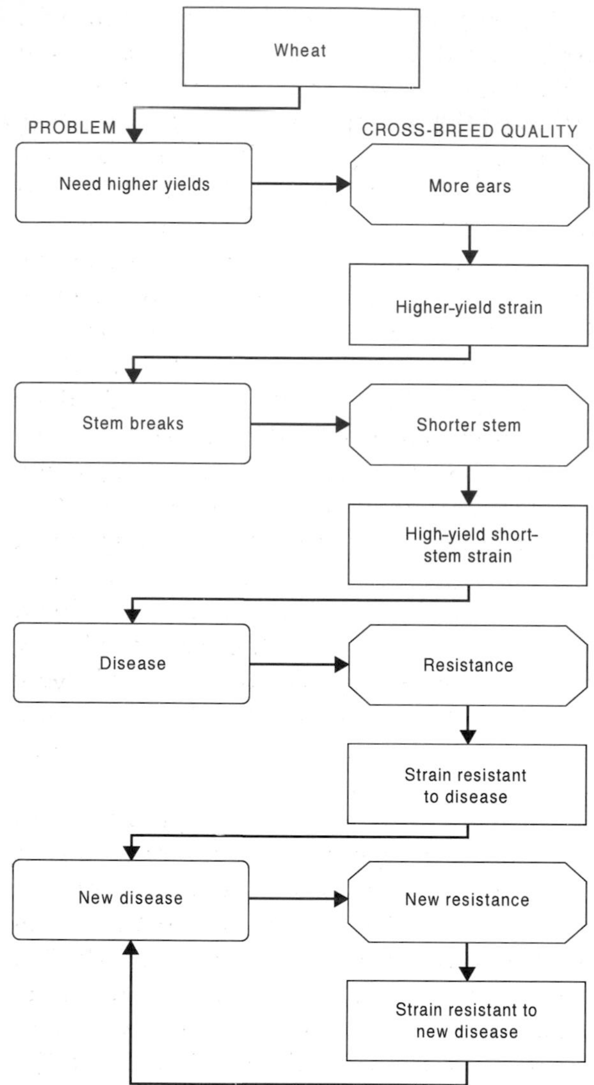

Figure 13.8 Steps in the breeding of new wheat strains.

applies to our pets and the plants in our gardens. Thousands of years of selective breeding have boosted the yields of crops, the milk production of cows, the amount of meat on cattle, and the colours and sizes of our flowers and dogs. Virtually every living thing that humankind exploits has been genetically modified in a major way.

Such procedures can now be conducted much more precisely using genetic engineering, which transfers useful DNA material from one organism into the genetic make-up of another organism. Today we use the term 'genetically modified', or GM, to refer to food produced from plants or animals that have had their genes changed by scientists in the laboratory rather than farmers in the field. In many respects, genetic engineering is no different in principle from conventional breeding, except that it is faster. However, genetic engineering can go further than conventional breeding. Scientists can now select a single gene for a single characteristic and transfer that stretch of DNA from one organism to another. The implanted genes may come from other species and can cross between the animal and plant kingdoms. For example, inserting genes for scorpion toxin or spider venom into maize and other food crops can act as a 'natural pesticide' to deter insects and birds from feeding on them.

The scope for modifying the properties of commercial food sources is very wide. Among the many applications, DNA can be transferred to increase resistance to pests and diseases, to modify the timing of fruit ripening, to enhance nutritional value, and to increase tolerance of altitude or temperature regimes. These techniques are now beginning to yield new commercial food products. The first to be marketed, in 1994, was a tomato developed by a US biotechnology company. The tomato has been genetically manipulated to ripen more slowly than natural tomatoes, so enhancing its flavour.

Once a new variety of plant has been created using recombinant DNA techniques, it can be mass-produced using cell or tissue culture, another important biotechnology technique. The approach enables a complete plant to be regenerated in the laboratory from a single cell exposed to nutrients and hormones, and is commonly used for clonal propagation, the mass production of clones or genetic duplicates that can then be planted like conventional seedlings.

Although biotechnology offers great potential for world food production, there are fears that many of the more expensive techniques will continue to be used predominantly on high-value crops grown in industrialised countries, with limited application in more needy areas of the developing world. Developing countries may be disadvantaged in other ways: some see the virtual monopoly over biotechnology held by transnational companies as inherently unfair to poorer countries, and those whose economies are heavily reliant on such crops as sugar and cocoa may suffer as substitutes are found (Hobbelink, 1991). Even in developed countries, there may be resistance from consumers against GM foods in the same vein as the backlash against the inhumane treatment of animals in intensive factory farms. Already the use of some growth hormones, such as BST in cattle, has been banned in the USA, for example, due to safety fears from consumers but also due to milk producers' worries that increased production might depress prices.

Developments in biotechnology have also raised environmental concerns. The notion that introducing a single gene (that releases a toxin, for example, to endow resistance to a pest) to a plant will produce just a single result (resistance to the pest) is naive since any human-induced change is bound to have other knock-on effects. Another difficulty of permanently inducing resistance to a pest by genetic modification is that no control over the expression of the new gene is yet possible. Hence, the pest population will be exposed to the toxin continually and eventually the pest will develop a resistance to the toxin. Further, genetically altered plants or microorganisms might have a competitive advantage that could disturb natural ecology if they were released into the wild. The potential effects could echo those caused by unintended releases of exotic insects and other organisms into environments where they have proliferated to the detriment of many other species, as the above examples of the prickly pear in Australia and the Ferussac snail in Moorea illustrate (see also page 270). Many feel that the current pace at which GM foods are being introduced is too fast. Adoption of the precautionary principle is urged, running indoor laboratory trials first, only gradually moving toward unrestricted outdoor release once the other trials have proved satisfactory.

The risk of GM genes escaping into the wild could be reduced by incorporating 'terminator technology'. These are genes that are expressed in embryos at a very late stage of development allowing crops to develop normally but killing them when they are mature, so effectively making the harvested seed sterile. This modification helps to ensure that there should be no offspring, although it does not eliminate the risk completely. Terminator technology, however, has raised a storm of protest over its ethical implications. This type of genetic manipulation would effectively place the company owning the GM crop in a very powerful position: the farmer has to buy a new batch of seeds from the company each year. Many people fear that this would stifle competition, and operate against the common good interests of crop producers and consumers. The feared impact would be particularly great in developing countries. Here the age-old practice of seed saving from one year to the next, which enables hundreds of millions of poor farmers to subsist, would effectively cease.

The new techniques also highlight the need to maintain the reservoir of genetic diversity in the wild, to provide new genetic material that could be of use to humankind for food production and in other applications such as pharmaceuticals. The dangers of our increasing reliance on a small number of food crops – 75 per cent of food comes from just 12 crops (Groombridge and Jenkins, 2000) – are enhanced by their limited genetic diversity, a situation exacerbated by crop cloning, as examples of major crop failures through history illustrate (Table 13.5). The constant need for new genetic material, much of it found in wild relatives of domesticated crops, can be shown by just one example: Plucknett and Smith (1986) demonstrated that a new variety of sugar-cane is required in Hawaii about every ten years to adapt to pests and to maintain yields. While realising the need to protect and maintain genetic diversity in the wild, regulating the use of such material, thereby enabling compensation to be paid to the countries of origin, is another pertinent issue, particularly for developing countries. It is an aspect of the

TABLE 13.5 *Crop failures attributed to genetic uniformity*

Date	Location	Crop	Cause and result
900	Central America	Maize	Anthropologists speculate that collapse of Mayan civilisation may have been due to a maize virus
1846	Ireland	Potato	Potato blight led to famine in which 1 million people died and 1.5 million emigrated
Late 1800s	Sri Lanka	Coffee	Fungus wiped out homogeneous coffee plantations on the island
1943	India	Rice	Brown spot disease, aggravated by typhoon, destroyed crop, starting 'Great Bengal Famine'
1960s	USA	Wheat	Stripe rust reached epidemic proportions in Pacific north-west
1974–77	Indonesia	Rice	Grassy stunt virus destroyed over 3 million tonnes of rice (the virus plagued South and South-East Asian rice production from 1960s to late 1970s)
1984	USA	Citrus	Bacterial disease caused 135 nurseries in Florida to destroy 18 million trees

Source: modified after WCMC (1992: 428, Table 27.22)

biodiversity debate that the Convention on Biological Diversity signed at the Earth Summit in 1992 specifically recognises and aims to tackle.

Sustainable agriculture

Having looked at some of the key recent technological advances that have allowed world agriculture to increase output, it is appropriate to outline some of the basic tenets of a more sustainable approach to farming. Many of the environmental issues spawned by these agricultural advances have already been outlined, but Table 13.6 provides further evidence of the need for a move towards more sustainable options. Modern intensive farming requires high inputs of energy in the form of fertilisers, pesticides, and so on, while much more efficient production is achieved with less intensive methods. As Table 13.6 shows, a highly intensive spinach farm in the USA not only needs nearly 19 times more energy per hectare than a Mexican peasant to produce its crop, but also does so 50 times less efficiently. According to this criterion, the Mexican peasant's approach is much more sustainable than that of the US spinach farmer.

Sustainable agriculture aims to produce crops using methods that are more closely based on the functioning of natural systems and the use of ecological processes, as opposed to a large-scale manipulation of nature and high inputs of energy. A checklist of criteria for more sustainable agriculture, based on alternative systems devised in the USA to avoid the forms of environmental degradation described in this chapter and in Chapter 14, is provided by Lockeretz (1988):

- diversity of crop species
- selection of crops and livestock that are well adapted to the environment in question

TABLE 13.6 *Energy efficiency of some agricultural production systems*

Agricultural system	Energy input (kcal/ha/year)	Energy production (kcal/ha/year)	Energy efficiency (production/input)
Hunter-gatherer	2 685	10 500	3.9
Pastoralism (Africa)	5 150	49 500	9.6
Peasant farming (Mexico)	675 700	6 843 000	10.1
Estate crop production (Mexico)	979 400	3 331 230	3.4
Estate crop production (India)	2 837 760	2 709 300	0.9
Maize (USA)	1 173 204	3 306 744	2.8
Wheat (USA)	4 796 481	8 428 200	1.7
Apples (USA)	18 000 000	9 600 000	0.5
Spinach (USA)	12 800 000	2 900 000	0.2

Source: after Pimental (1984)

- preference for farm-generated resources rather than purchased materials
- use of nutrient cycles to minimise nutrient losses
- livestock housed and grazed at low stocking densities
- enhancement of nutrient storage in the soil
- maintenance of protective cover on soil
- rotations that include deep-rooted crops and that help to control weeds
- use of soluble inorganic fertilisers
- use of pesticides only as a last resort.

Most proponents of alternative, low input agriculture would agree with these recommendations, although a deep ecologist would no doubt omit the final two practices on this list.

AQUACULTURE

A notable recent development in global food production is the rise in importance of aquaculture, which is dominated by fish farming but also includes the production of other aquatic organisms such as seaweeds, frogs and turtles. Between 1987 and 1997, the global production of farmed fish and shellfish more than doubled in weight and value, and farmed fish currently accounts for more than a quarter of all fish directly consumed by people (Naylor *et al.*, 2000). Asia accounts for about 90 per cent of world aquacultural output, and China is by far the world's largest producer. Most of the Chinese production comprises freshwater species such as carp and tilapia, and globally aquacultural output is split about 60/40 between inland and marine production.

Fish farming has a long history in parts of East and South Asia, dating back to 3000 BC in China, and has been practised in Europe since before Medieval times. At its simplest level, fish are raised in ponds, where they find their own food, from eggs collected in the wild, and are harvested when grown to a certain size. More sophisticated approaches involve a series of ponds or cages for different levels of development, including special nursery ponds and/or hatcheries. Provision of feed and control of natural predators, competitors and diseases is also common practice. Particular advances in fish-farming techniques have occurred in the last 50 years: selective breeding of stock is now widely practised in more advanced farms, reproduction is controlled by hormonal injection and disease by antibiotics. Such intensive management has been used mainly in more developed countries for species that command a high market price. The spectacular

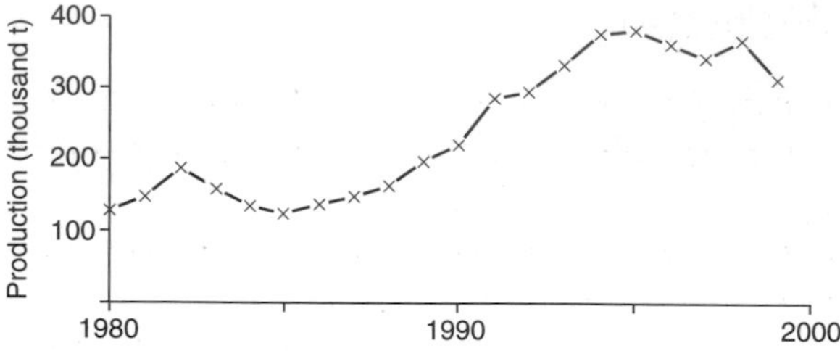

Figure 13.9 Production of shrimps and prawns in Thailand, 1980–99 (FAO data).

development of the Atlantic salmon farming industry in Norway, initiated in the early 1970s and reaching a 1990 production of 140 000 t, is an important example (Coull, 1993). Such intensive fish farming is, however, associated with a number of adverse impacts on the environment. These include the displacement of native fish species, control of wildlife that might prey on the farmed species, and local contamination of waters. In a review of the issues, Iwama (1991) highlights the output of organic particulate wastes from feed and faeces, which can cause eutrophication and can smother sea or lake bed organisms, as the most pressing area of concern.

On a wider scale, the conversion of natural habitats for fish and prawn farming has been a primary cause of mangrove destruction in many coastal areas of the tropics. The rapid rise in the production of shrimps and prawns in Thailand in the early 1990s (Fig. 13.9) has largely been at the expense of mangroves (see page 118) and this loss of habitat has a feedback effect on marine production. It has been estimated that for every kilogram of shrimp farmed in Thai shrimp ponds developed in mangroves, 400 g of fish and shrimp are lost from capture fisheries (Naylor *et al.*, 2000).

Some predictions suggest that global production from aquaculture could reach 50 million tonnes in the next 25 years, which has engendered talk of a 'Blue Revolution' to rival the agricultural Green Revolution (Doumenge, 1986). In essence, this would amount to a major shift from capture to farming or husbandry in water, which would parallel that which occurred when hunting and gathering gave way to agriculture on land. Some observers believe the recent rise in aquacultural output is relieving pressure on ocean fisheries, but the farming of carnivorous fish species actually has the opposite effect because it is based on large inputs of wild fish for feed. Eel, salmon, trout and tilapia produced in intensive aquacultural systems are fed two to five times more fish protein, in the form of fish meal, than their actual weight (Tacon, 1996).

Cheaper and more ecologically sound management practices are long established in China, however, where carp are reared alongside pigs with reciprocal recycling of nutrients between land and water. Other species, such as tilapia, a prolific breeder that can also be fed mainly on wastes, are widely farmed in South Asia due in part to the traditional dietary importance of fish. Here, and in many other parts of the world, aquaculture has the additional benefit of helping to deal with problems of waste disposal. The largest of many such systems that are based on sewage is in Calcutta, where 7000 t of fish are produced annually for the local market. Water and sewage are channelled into two lakes covering about 3000 ha, and carp and tilapia are introduced after an initial bloom of algae. Thereafter, the lakes are fed with additional sewage on a monthly basis. Health concerns over fish raised in this manner can be eliminated by retaining sewage in settling ponds for 20 days before introducing it into fish ponds, or by moving fish to clean-water ponds before harvesting. Despite the importance of this fishery to sewage treatment and pollution control, income generation and employment for the poor, the Calcutta lakes are gradually being depleted in the face of increasing demands for housing land (Mukherjee, 1996).

FURTHER READING

Conway, G.R. and Pretty, J.N. 1991 *Unwelcome harvest: agriculture and pollution*. London, Earthscan. A comprehensive look at the main forms of agricultural pollution (pesticides, fertilisers and farm wastes) and the impacts of all types of pollution on agriculture.

Dierber, F.E. and Kiattisimkul, W. 1996 Issues, impacts, and implications of shrimp aquaculture in Thailand. *Environmental Management* 20: 649–66. National case study illustrating the problems of a rapid rise in aquacultural production and suggestions for management solutions.

Gibbon, D., Lake, A. and Stocking, M. 1995 Sustainable development: a challenge for agriculture, in Morse, S. and Stocking, M. (eds) *People and environment*. London, UCL Press: 31–68. A thorough review of how sustainable development can be applied to agriculture.

Goodman, D. and Redclift, M. 1991 *Refashioning nature: food, ecology and culture*. London, Routledge. This book takes an interesting look at food as one of the chief links between humankind and nature, and how modern farming has tended to further widen the gap between the two.

Lappé, M. and Bailey, B. 1999 *Against the grain: the genetic transformation of global agriculture*. London, Earthscan. A good assessment of the environmental impacts of agricultural biotechnology.

National Research Council 1993 *Sustainable agriculture and the environment in the humid tropics*. Washington, DC, National Academy Press. A practical discussion of 12 major land use options for boosting food production while protecting the natural resource base. The book also covers technology and policy, and contains seven country profiles.

Robinson, R.A. and Sutherland, W.J. 2002 Post-war changes in arable farming and biodiversity in Great Britain. *Journal of Applied Ecology* 39: 157–76. A comprehensive review.

Thrupp, L.A. 2002 *Fruits of progress: growing sustainable farming and food systems*. Washington, World Resources Institute. This book documents the 'green' transformation currently sprouting in the food and agriculture industry in many parts of the world.

Websites

http://www.fao.org/ the FAO site has a very wide range of information, data and reports on all aspects of food production.

http://cipm.ncsu.edu/agVL/ the World Wide Web virtual library for agriculture.

http://www.pan-international.org/ the Pesticide Action Network is a network of non-governmental organisations (NGOs), institutions and individuals working to replace the use of hazardous pesticides with ecologically sound alternatives.

http://www.monsanto.com/ Monsanto is a major agrochemical company.

http://www.icgeb.trieste.it/ the site of the International Centre for Genetic Engineering and Biotechnology, which has a focus on the developing world, has a wide range of information.

http://ipmwww.ncsu.edu/biocontrol/biocontrol.html Biological Control Virtual Information Center.

http://cipm.ncsu.edu/ Center for Integrated Pest Management.

http://biosafety.ihe.be/ this Belgian site is devoted to biotechnology issues in Europe and elsewhere.

http://www.was.org/ the World Aquaculture Society site.

POINTS FOR DISCUSSION

- Why should we be concerned about the human impact on the global nitrogen cycle?
- Is biotechnology the answer to problems of world food supply?
- Is it reasonable to expect all farmers to practise sustainable agriculture in a world of more than 6 billion people?
- Outline the probable benefits and possible problems that might arise from a 'Blue Revolution'.

14

SOIL EROSION

TOPICS COVERED

Factors affecting soil erosion
Measuring soil erosion
Effects of erosion
Accelerated erosion
Soil conservation

Key words: erosivity, erodibility, sediment delivery ratio, Universal Soil Loss Equation (USLE), Global Assessment of Human-Induced Soil Degradation (GLASOD), mulching

Soil erosion is a natural geomorphological process that occurs on most of the world's land surface, with the principal exception of those areas on which soil eroded from elsewhere is deposited. Since soil erosion rarely occurs in dramatic events, perception of its seriousness, measurement of its progress and persuasion of the need to take ameliorative action has been a long process. A fundamental distinction is made between natural or geological soil erosion and rates that are accelerated by human activity, and it is accelerated erosion that has received most scientific attention. Such attention is not new, however. Ancient Greek and Roman observers of the natural world commented on the effects of such human activities as agriculture and deforestation on soil loss.

Soils are an integral part of the support system for ecosystems and human communities, and are important to society in several ways:

- as a medium in which crops, forests and other plants grow
- for their filtering, buffering and transformation activity, between the atmosphere, groundwater and plant cover, servicing the environment and people by protecting food chains and drinking-water reserves
- as biological habitats and gene reserves
- by serving as a spatial base for society's structures and their development (e.g. the construction of buildings and dumping of refuse)
- as a source of raw materials (e.g. clay, sand and gravel for construction), and also as a reserve of water and energy.

Hence the erosion of soil, the most widespread form of soil degradation, is an important environmental issue. This importance is all the more pressing when we consider the fact that the world's human population is increasing, along with its use of resources, and we know that natural rates of soil formation are slow, essentially making it a finite resource.

FACTORS AFFECTING SOIL EROSION

Soil is principally eroded by the forces of water or wind acting on the soil surface, and on steep slopes by mass movement, although other agents, such as animals and humans, can also contribute (e.g. footpath erosion). Water erosion occurs both by the action of raindrops, which detach soil particles on impact and move them by splashing, and

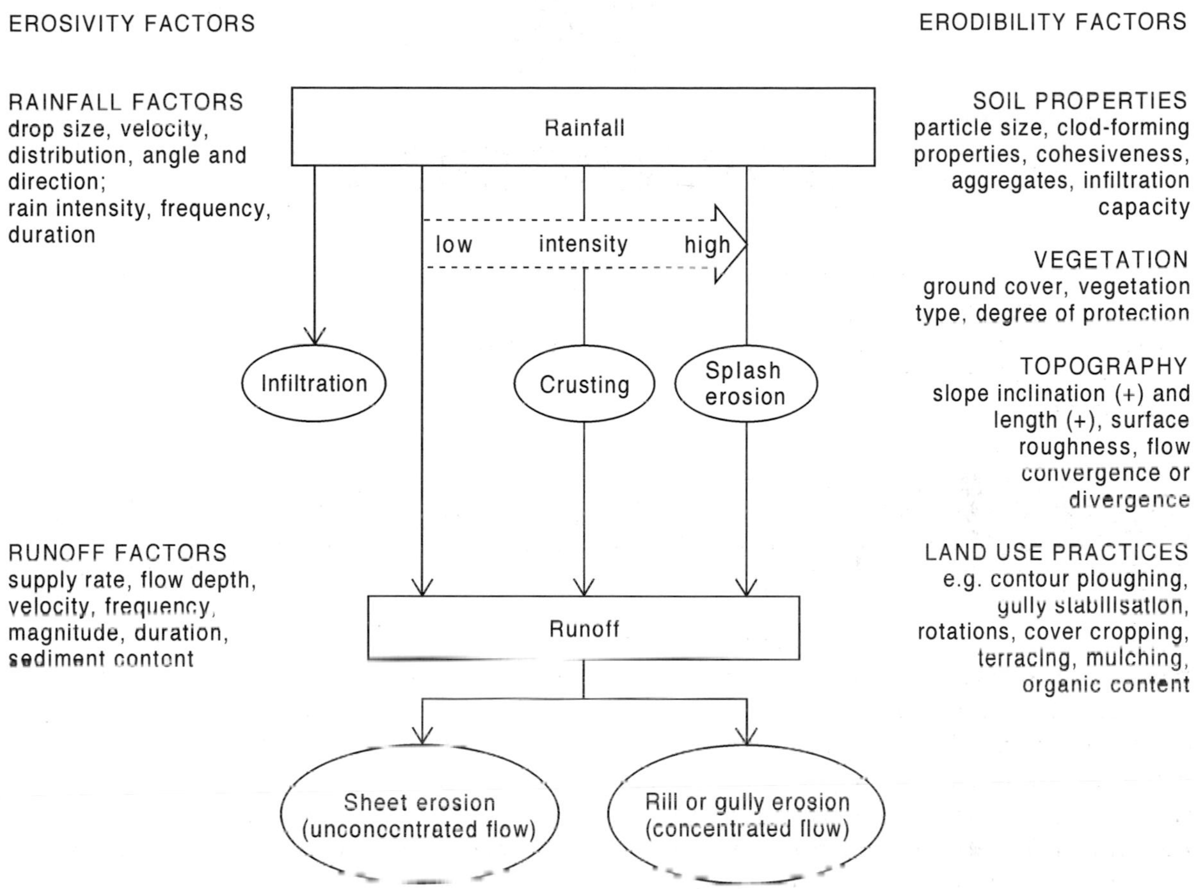

EROSIVITY FACTORS

ERODIBILITY FACTORS

RAINFALL FACTORS
drop size, velocity, distribution, angle and direction;
rain intensity, frequency, duration

Rainfall

low intensity high

Infiltration

Crusting

Splash erosion

SOIL PROPERTIES
particle size, clod-forming properties, cohesiveness, aggregates, infiltration capacity

VEGETATION
ground cover, vegetation type, degree of protection

TOPOGRAPHY
slope inclination (+) and length (+), surface roughness, flow convergence or divergence

RUNOFF FACTORS
supply rate, flow depth, velocity, frequency, magnitude, duration, sediment content

Runoff

LAND USE PRACTICES
e.g. contour ploughing, gully stabilisation, rotations, cover cropping, terracing, mulching, organic content

Sheet erosion (unconcentrated flow)

Rill or gully erosion (concentrated flow)

Figure 14.1 Main factors affecting types of soil erosion by water. Generally, erosion will be reduced if the value of an erosivity factor is reduced and/or the value of an erodibility factor is increased, with the exception of factors shown with a (+), where the reverse is the case (after Cooke and Doornkamp, 1990).

by runoff, which transports material either in sheet flow or in concentrated flows, which form rills or gullies. Wind moves soil particles by one of three processes depending on the size and mass of the particles: the largest particles move along the surface by the process of surface creep; sand-sized particles usually bounce along within a few metres of the surface by saltation, and finer dust particles are transported high above the surface in suspension. Mass movement occurs principally by landsliding and various forms of flow, depending upon the amount of moisture in the soil profile.

When and where soil erosion occurs is determined by the mutual interaction of the erosivity of the eroding agent and the erodibility of the soil surface. These variables of erosivity and erodibility change through time and space, at varying rates and differing scales, so that the relationship between the variables is in a constant state of flux. The various factors affecting erosivity and erodibility in relation to water erosion are shown in Fig. 14.1. Most of the numerous human activities that affect the erosion system do so by altering the erodibility of the soil surface (see below).

MEASURING SOIL EROSION

A number of techniques are commonly used to measure soil erosion and they do not all measure the same thing. Some are designed to monitor the redistribution of soil particles over distances of a few centimetres or metres on a field surface. Others measure soil loss from a field or the quantity of soil that leaves a catchment in a river. Measurements of the rate of soil loss from an area over long periods of time – up to thousands of years – are calculated from the depth of sediment accumulated in lakes or on the sea bed. As a general rule, the larger the area under the consideration, the lower will be the rate obtained.

Many older estimates of erosion are based on studies of experimental plots, usually of a standard size (22 x 1.8 m), on which controlled experiments are undertaken by varying such factors as slope angle, crop type and management practice. This approach suffers from the serious drawback that actual landscapes have much more complex topography than standard plots, and certain types of erosion form, such as rills and gullies, may not develop on such a small scale. Some measurements on real fields have concentrated on these important forms, with field workers repeatedly visiting an area to measure rills and gullies (e.g. Boardman, 1990a), an exercise that can be supplemented by aerial photography.

Measurement of suspended sediment loads in rivers is another commonly used approach, which gives a rough estimate of current erosion on slopes and fields in the catchment area. Not all eroded soil reaches the river, however, and there have been many attempts to estimate 'sediment delivery ratios'. The technique is further hampered by the fact that a proportion of a river's sediment load is eroded from the river's own banks. A recent technique used to estimate both erosion and deposition measures the radionuclide caesium-137 present in a soil profile, which can be compared to amounts in nearby undisturbed profiles. The caesium-137, most of which has been released by atmospheric bomb testing, has been deposited on soils in the last 35 years or so, and the technique can provide average rates over that period and patterns of erosion and deposition at particular sites.

Field measurements and experiments have also provided data from which general relationships between the factors affecting soil erosion have been derived, and these relationships have been incorporated into models used to predict erosion. The most widely used is the Universal Soil Loss Equation (USLE), developed in the USA to predict soil loss by runoff from US fields east of the Rocky Mountains under particular crops and management systems. The USLE is calculated as:

$$E = R x K x L x S x C x P$$

where E is mean annual soil loss, R is a measure of rainfall erosivity, K is a measure of soil erodibility, L is the slope length, S is the slope steepness, C is an index of crop type, and P is a measure of any conservation practices adopted on the field.

The USLE can be adapted for use outside the temperate plains of North America, although it has often been used inappropriately, without such adaptations (Wischmeier, 1976). An equivalent wind erosion equation has also been developed (Woodruff and Siddoway, 1965), and both of these relatively simple models have been further refined to produce more complex

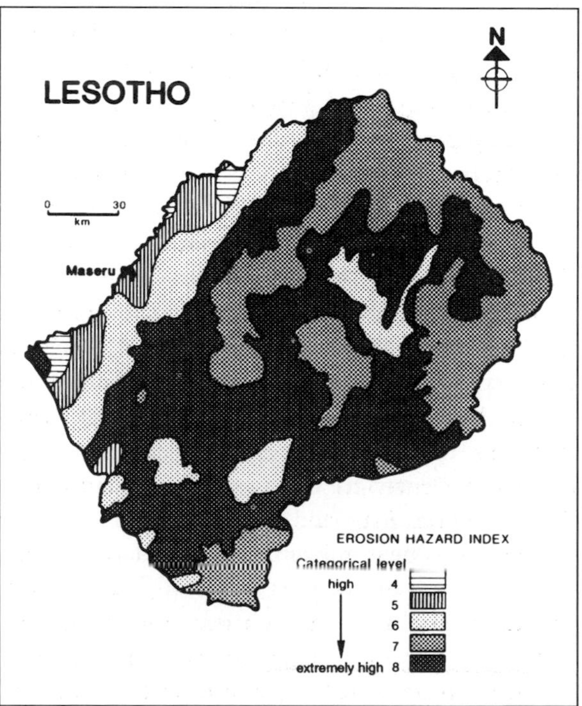

Figure 14.2 Erosion hazard map of Lesotho (after Chakela and Stocking, 1988).

TABLE 14.1 *GLASOD estimates of land area affected by human-induced soil erosion*

Continent	Area affected in million ha (% total continental area)	
	Water erosion	Wind erosion
Africa	227.4 (7.7)	186.5 (6.3)
Asia	439.6 (10.3)	222.1 (5.2)
Australasia	82.9 (9.4)	16.4 (1.9)
Europe	114.5 (12.1)	42.2 (4.4)
North America	106.1 (4.8)	39.2 (1.8)
South America	123.2 (7.0)	41.9 (2.4)

Source: Deichmann and Eklundh (1991)

computer-run models such as EPIC, one of the few erosion models that can be used to predict erosion by both water and wind. Models have also been used to produce soil erosion hazard maps such as that shown in Fig. 14.2 for Lesotho, the country with the highest erosion hazard in southern and central Africa thanks to its steep slopes, high rainfall totals, poor soils and average vegetation covers.

Despite the availability of numerous measurement and prediction techniques, the fact remains that there is virtually no soil erosion data for most of the world's land surface. An attempt to overcome this deficiency has been made with the Global Assessment of Human-Induced Soil Degradation (GLASOD), which employed more than 250 local soil experts around the world to give their opinion on soil degradation problems to supplement what data are available. The project, carried out by the International Soil Reference Center in conjunction with the UN Environment Programme, followed a strict methodology by dividing the world's land surface into mapping units corresponding to physiographic zones, and for each unit an assessment was made of water and wind erosion as well as chemical and physical degradation processes. The degree of degradation, its extent and causes were assessed to produce an overall estimate of degradation severity in each unit (Oldeman *et al.*, 1990). Table 14.1 shows the continental areas affected by water and wind erosion.

EFFECTS OF EROSION

The movement of soil and other sediments by erosive forces has a large number of environmental impacts that can affect farmers and many other sectors of society. Many of these effects are consequent upon natural erosion, but are exacerbated in areas where rates are accelerated by human activity. The environmental effects associated with erosion occur due to the three fundamental processes of entrainment, transport and deposition. This three-fold division is used in Table 14.2 to illustrate the hazards posed to human populations by wind erosion, while the following sections are divided into on-site and off-site effects.

TABLE 14.2 *Some environmental consequences and hazards to human populations caused by wind erosion and duststorms*

ENTRAINMENT
Soil loss
Nutrient loss
Crop root exposure
TRANSPORT
Sand-blasting of crops
Air pollution
Radio communication problems
Local climatic effects
Transport disruption
Disease transmission (human and plants)
DEPOSITION
Nutrient gain (soils, plants and oceans)
Salt deposition and groundwater salinisation
Burial of structures
Rainfall acid neutralisation
Machinery problems
Reduction of solar power potential
Electrical insulator failure

Source: after Goudie and Middleton (1992)

Figure 14.3 A 15 m-deep gully in central Tunisia.

On-site effects

The loss of topsoil changes the physical and chemical nature of an area. Deformation of the terrain due to the uneven displacement of soil can result in rills, gullies, mass movements, hollows, hummocks or dunes. For the farmer, such physical changes can present problems for the use of machinery, and in extreme cases such as gullying, the absolute loss of cultivable land (Fig. 14.3). Uneven displacement of soil may also result in burial of plants and seedlings; loss of soil may expose roots; and sand-blasting by wind-eroded material can both damage plants and break down soil clods, impoverishing soil structure and rendering soil more erodible. Splash erosion can cause compaction and crusting of the soil surface, both of which may hinder germination and the establishment of seedlings, while exposure of hardpans and duricrusts presents a barrier to root penetration.

Erosion has implications for long-term soil productivity through a number of processes.

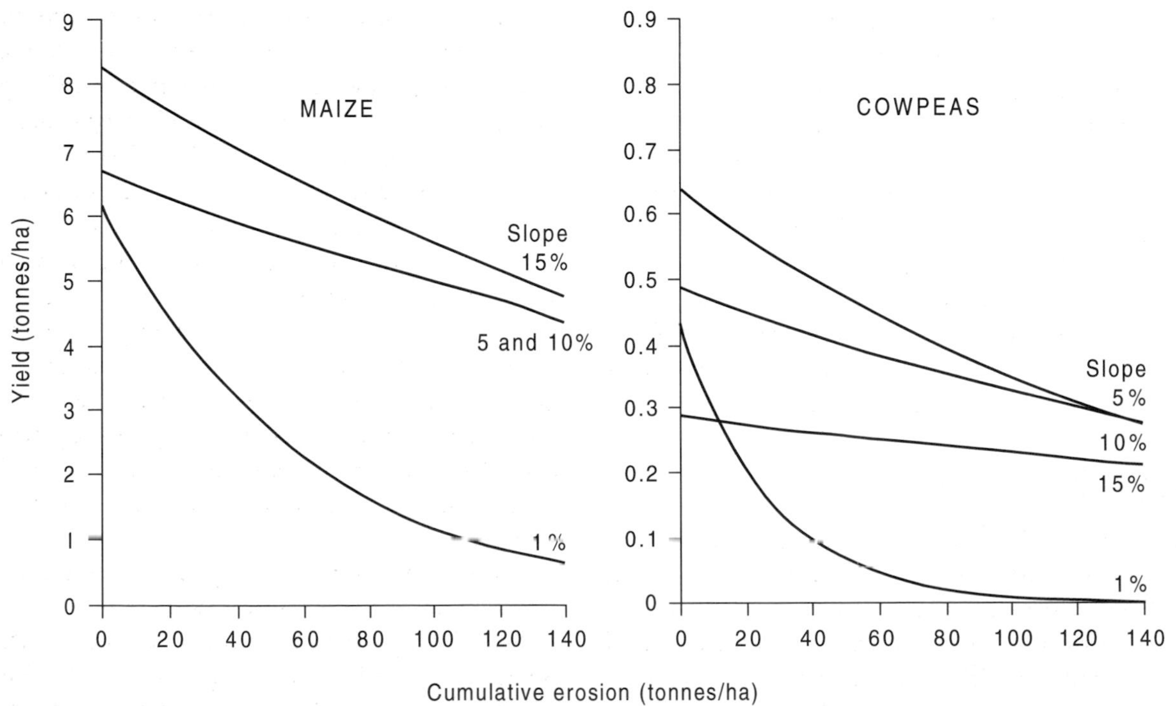

Figure 14.4 Decline in yields of maize and cowpea with cumulative loss of soil in south-west Nigeria (Lal, 1993).

TABLE 14.3 *Increase in fertiliser needs for corn as soil erodes in southern Iowa, USA*

Change in erosion phase	Additional fertiliser needed (kg/ha)		
	Nitrogen	Phosphate	Potash
Slight to moderate	11	2	7
Moderate to severe	34	1	8

Source: after Rosenberry *et al.* (1980)

Reduced soil thickness diminishes the depth through which roots can penetrate and the soil's capacity for holding water. The effects on soil structure, by preferential removal of organic material and fine particles, can also reduce the water-holding capacity of the soil. These problems are most serious on thin soils where plants root to depths greater than that of available topsoil. Erosion also carries away soil nutrients, including fertilisers, and even seeds.

Some researchers sound a note of caution in directly relating erosion rates to losses in productivity (e.g. Larson *et al.*, 1983), not least because the relationship is highly variable depending upon soil type and crop type, and because numerous other factors can affect crop yields. However, many experiments have shown that as erosion proceeds, crop yields do decline. Figure 14.4 illustrates this for two staple crops in south-west Nigeria for a range of slope angles on the same soil type. Yield declines can, of course, be compensated for by adding fertilisers, assuming the farmer can afford to do so. The additional fertiliser needed to maintain corn yields for an area of the USA are shown in Table 14.3.

Off-site effects

The off-site effects of eroded soil are caused by its transport and deposition. Material carried in strong winds can cause substantial damage to structures such as telegraph poles, fences and larger structures by abrasion. Material transported in duststorms, which can affect very large areas (Fig. 14.5), severely reduces visibility, causing a hazard to transport, and adversely affects radio and satellite communications. Inhalation of fine particles can aggravate human diseases such as bronchitis and emphysema, and the transport of soil pathogens spreads human and plant diseases. The nutrients attached to soil particles transported by water erosion can cause the eutrophication of water bodies (see Chapters 7 and 8).

The deposition of eroded material can also cause considerable problems for human society. The flood hazard can be increased due to riverbed infilling, and the siltation of reservoirs, harbours and lakes presents hazards to transport and loss of storage capacity in reservoirs. The economic costs thus imposed are often considerable. In Java, the off-site costs due to siltation of irrigation systems and reservoirs, and harbour dredging were estimated to be US$58 million in 1987 (Magrath and Arens, 1989). About 25 per cent of the sediment deposited in lakes and reservoirs in the USA is thought to originate from cropland. The resulting damage, which contributes to a 0.22 per cent annual loss in national water storage capacity, has been valued at from US$144 million to US$197 million per year (Crowder, 1987). Deposition of sediment reaching the coastline can adversely affect marine environments used by local populations, including coral reefs and shellfish beds.

Despite the numerous negative aspects of sediment deposition for human society, the positive side of the equation should also be highlighted. Relatively flat, often low-lying areas of deposition provide good, fertile land for agriculture and many other human activities. Floodplains are an

Figure 14.5 Severe duststorm at Melbourne, Australia, in February 1983 (Photo: Australian Bureau of Meteorology).

obvious example, as are valley bottoms that receive material transported from valley slopes. Wind-deposited dust, or loess, also provides fertile agricultural land – the loess plateau of northern China is the country's most productive wheat-growing area, for example – although it requires careful management because it remains highly erodible. Similarly, inputs of soil nutrients to the oceans enhance productivity in marine ecosystems.

ACCELERATED EROSION

Although many of the adverse effects of soil erosion for human society outlined above may occur due to natural rates, the most pressing problems are found in areas where accelerated rates occur. Natural rates vary enormously depending on such factors as climate, vegetation, soils, bedrock and landforms. Information on natural erosion rates, combined with our limited knowledge of the rates of soil formation, are used as yardsticks against which to measure the degree to which human action exacerbates natural processes and the sort of rates that soil conservation measures might aim to achieve (see below). Natural rates of soil formation depend on the weathering of rocks, the deposition of sediment and the accumulation of organic material from plants and animals. They are generally taken to be less than 1 t/ha/year, and for many practical purposes acceptable target rates for soil conservation are commonly set between 2.5 and 12.5 t/ha/year, depending on local conditions (Cooke and Doornkamp, 1990). In some countries, however, much lower limits have been established. The Swedish Environmental Protection Board, for example, considers a soil loss of 0.1–0.2 t/ha/year as a recommended limit for preventative measures to be applied on arable land (Alström and Åkerman, 1992).

The two most common effects of human activity that lead to accelerated erosion are modi-

TABLE 14.4 *Effects of human activities on erosion rates in Oahu, Hawaii, USA*

Initial land cover	Disturbance	Increase in erosion rate
Forest	Planting of row crops	×100–1000
Grass	Planting of row crops	×120–100
Forest	Building logging road	×220
Forest	Woodcutting and skidding	×1.6
Forest	Fire	×7–1500
Forest	Mining	×1000
Forest	Construction	×2000
Pasture	Construction	×200
Row crops	Construction	×10

Source: El-Swaify *et al.* (1982)

fications to, or removal of, vegetation, and destabilisation of natural surfaces. Such actions have a variety of motives: vegetation may be cleared for agriculture, fuel, fodder or construction; vegetation may be modified by cropping practices or deforestation for timber; land may be disturbed by ploughing, off-road vehicle use, military manoeuvres, construction, mining or trampling by animals. The effects of several of these disturbances on soil erosion by water in central Oahu, Hawaii, are shown in Table 14.4. Other processes of erosion also follow such disturbances: an example of the effects on slope stability and resulting mass movement problems caused by highway construction is given in Table 16.3. Transport routes can cause accelerated landsliding by increasing disturbing forces acting on a slope, both during construction when cuts and excavations remove lateral or underlying support (Fig. 14.6), and through earth stresses caused by passing vehicles. Other human activities that can increase the chance of slope failure do so by decreasing the resistance of materials that make up slopes. This can occur if the water content is increased, as happens when local water tables are artificially increased by reservoir impoundment, for example.

Figure 14.6 A landslide blocking a road in the Khasi Hills in north-east India. Very heavy rainfall in this area, where average annual totals approach 12 000 mm in some parts, causes frequent landslides during the monsoon.

The initial impact of certain activities may be reduced when a new land use is established, however. Many construction activities initially cause marked increases in soil loss, but erosion rates can be reduced to below those recorded under natural conditions when a soil surface is covered by concrete or tarmac. Conversely, soil erosion problems may become displaced by the effects of construction, with water flows from drainage systems causing accelerated soil loss where they enter the natural environment.

Another human impact that may result in temporarily accelerated erosion rates occurs through the use of fire as a management tool. The burning of bush and grasses is a long-established and widely used management technique in the savannas of Africa, used to encourage tender green shoots from perennial grasses for livestock to feed on and to release phosphorus and other nutrients for use by crops. The exposed soil is particularly susceptible to erosion if a new vegetation cover does not become established before an erosive event. Controlled studies on the effects of burning on semi-arid gorse scrubland in southeastern Spain indicate that when the first significant rains fall on the study plots, the loss of nutrients by water erosion can be an order of magnitude higher in the burned areas compared to the undisturbed control plot (Carreira and Neill, 1995).

There is widespread agreement that the prime causes of accelerated soil erosion are deforestation and agriculture. Deforestation removes the protection from raindrop impact offered to soil by the tree canopy, and reduces the high permeability humus cover of forest floors, a permeability that is enhanced by the many macropores produced by tree roots. Cultivation also removes the natural vegetation cover from the soil, which is particularly susceptible to erosion when bare after harvests and during the planting stage. Some crops, such as maize and vines, usually leave large portions of the ground unprotected by vegetation even when the plants mature. Furthermore, mechanical disturbance and compaction of the soil by ploughing and tilling can enhance its erodibility.

An illustration of the dramatic effects on soil loss instigated by deforestation of a tropical forest area is given by experimental work on plots in French Guyana (Fritsch and Sarrailh, 1986). Suspended sediment yields from undisturbed rain forest plots of less than 0.7 t/ha/year increased by up to 50 times after mechanised clear-cutting. When the plots were planted with grasses, erosion rates fell, but were still two to three times higher than under forest cover. A similar pattern of events over a longer period, derived from analysis of sediment and pollen in a lake bed, is shown for an area cleared of woodland in the mid-nineteenth century in Michigan, USA (Fig. 14.7). The initial response to clearance was a sharp rise in erosion by 30 to 80 times the rates derived for the pre-settlement period. Fluctuating rates characterised a period of about 30 years as a new steady state was reached under farmland, which is none the less some ten times greater than under undisturbed woodland.

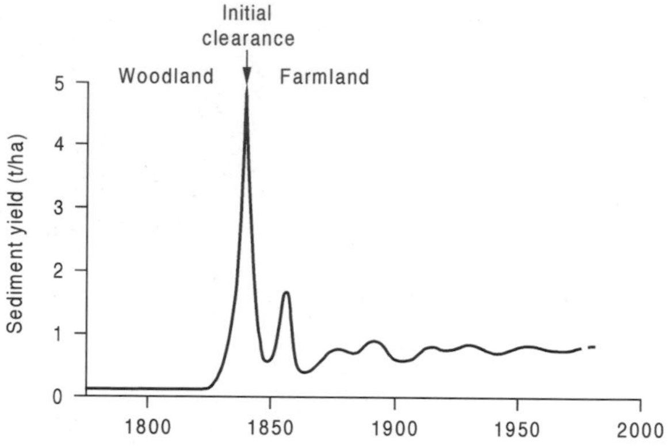

Figure 14.7 Historical reconstruction of sediment yield at Frains Lake, Michigan, USA (Davis, 1976).

These types of effect caused by deforestation are widespread, as Fig. 14.8 indicates for most of the countries in South and South-East Asia. The map depicts areas where deforestation and the removal of natural vegetation in recent times has resulted in some form of soil degradation (the same classes used for the GLASOD assessment – see above). For the most part, the areas shown on Fig. 14.8 are suffering from soil erosion. The driving forces behind deforestation in these areas are diverse. Some analysis of the situations in Viet

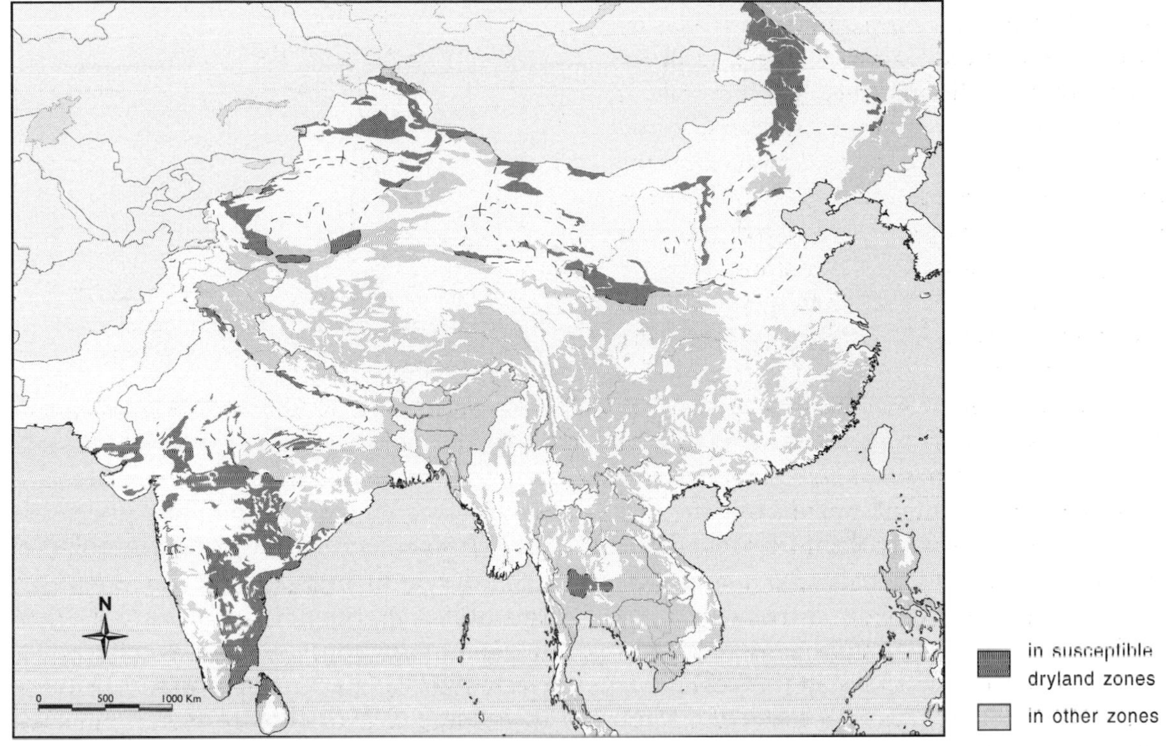

Figure 14.8 Areas where deforestation is a cause of soil degradation in south and south-east Asia (after Middleton and Thomas, 1997; data from the International Soil Reference and Information Centre).

Nam and the Philippines are given in Chapter 4. In lowland Nepal, locally serious water erosion has resulted from land clearance for subsistence agriculture. A large influx of population from the hills and mountains has felled vast tracts of cool tropical forest on the Terai alluvial plains since the 1960s when malaria was eradicated from the area (Tamang, 1995). Water erosion has also been severe in the Aravalli Hills in eastern India where the rates of vegetation clearance have been amongst the highest in the country in recent times. The major portion of these hills in Haryana state is village common land, and the situation has become so critical that the state government has begun a village-based plan to combat the erosion problem (Srivastava and Kaul, 1995).

Some of the most significant cases of accelerated erosion by wind have occurred in dry grassland areas used for grain cultivation. In the Maghreb countries of North Africa in the nineteenth century, French settlers brought European agricultural machinery that turned the soil to twice the depth of the traditional hoes, and removed shrubs and weeds, baring the soil to erosive forces. By the early twentieth century, the cropland area of the 'telle' zone had quadrupled, and traditional cultivators had been pushed on to steeper slopes and to the climatic limits of dry cereal crops (Dresch, 1986). The situation has changed little since the independence of Algeria and Tunisia, with the desire to enlarge cropland pushing the tractor and multidisc plough further into the steppe. In 1983, the Algerian Government passed legislation positively encouraging the cultivation of marginal lands in the Sahara and in the country's high plateau region in an effort to expand the agricultural resource base, to increase food supply, to combat the exodus of peasants to urban areas and to counterbalance coastal urban development. This

TABLE 14.5 *Effect of the Virgin Lands Scheme on the frequency of duststorms in the Omsk region of western Siberia*

Station	Mean annual number of duststorm days		Increase
	1936–50	1951–62	
Omsk, steppe	7	16	×2.3
Isil'-Kul'	8	15	×1.9
Pokrov-Irtyshsk	4	22	×5.5
Poltavka	9	12	×1.3
Cherlak	6	19	×3.2
Mean	6.8	16.8	×2.8

Source: Sapozhnikova (1973)

homesteading programme was part of the 'new lands' scheme, a plan that also involved the reduction of fallow in traditional crop rotation systems. The extensification of cropland into the desert margins of neighbouring Morocco also proceeded apace in the 1980s, in this case driven by a doubling of prices paid to barley and wheat producers, and relatively good rainfall totals. The average annual area under cereals grew from just over 4.4m ha in 1980–84 to 5.5m ha by 1990 (Swearingen, 1994). In addition to the resulting degradation in these regions as evidenced by the high frequency of wind erosion events, such marginal cropland is almost by definition also severely prone to the effects of drought.

Probably the most infamous case of wind erosion in the western world came after widespread ploughing of the grasslands of the US Great Plains, which created the Dust Bowl of the 1930s (see page 82), but similar environmental mishaps occurred in the former USSR in the 1950s and in a copycat exercise in the Mongolian steppes in the 1960s. Some 40 million hectares of Virgin Lands were put to the plough in northern Kazakhstan, western Siberia and eastern Russia between 1954 and 1960. Deep ploughing was employed, removing the stubble from the previous year's crop to allow planting earlier in the year, so reducing the chance of losses to early snows at harvest time. Land was also used more intensively, doing away with alternate years when land was traditionally left fallow under grass. Wind erosion soon began to take its toll (Table 14.5), coming to a head in the early 1960s when drought hit the region.

The introduction, from outside, of agricultural techniques that are inappropriate to local conditions is a widely cited cause of erosion. In many parts of Latin America, management techniques brought by Spanish and Portuguese colonists have been blamed for widespread soil degradation. In Ecuador, for example, farmers have neglected the main principles of pre-colonial agriculture, which were more suited to the mountainous areas and, today, accelerated soil erosion is estimated to affect 50 per cent of the national territory at rates that reach 200–500 t/ha/year in the great basin of Quito (De Noni *et al.*, 1986). In addition to the Spanish conquest, poor agricultural reform and the population explosion at the beginning of the twentieth century are highlighted as significant factors behind the soil erosion situation.

However, evidence from lake sediment cores in central Mexico suggests caution over the ease with which invading Iberians are blamed for unbalancing traditional, supposedly harmonious, systems of soil use. Analysis of the cores indicates that pre-Hispanic agriculture in the Lake Patzcuaro Basin was not as conservationist in practice as was previously thought. Several periods of accelerated erosion have been identified

TABLE 14.6 *Estimated rates of soil loss on slopes in Ethiopia by type of land cover*

Land cover/use	Proportion of national land area (%)	Soil loss (t/ha/year)
Cropland	13.1	42
Perennial crops	1.7	8
Grazing and browsing	51.0	5
Totally degraded	3.8	70
Currently uncultivable	18.7	5
Forest	3.6	1
Wood and bushland	8.1	5

Source: after Hurni (1993)

that occurred before the arrival of the Spanish and were of comparative magnitude to those during colonial times. This suggests that the introduction of the plough did not have a greater impact on soil erosion than traditional methods (O'Hara *et al.*, 1993).

Despite questions over where exactly to lay the blame, a long history of accelerated erosion in some regions has meant that a state of system collapse has been reached, as in parts of the Caribbean where erosion has been significant since plantation monoculture was introduced in the early eighteenth century. Large-scale deforestation, the use of fire for land clearance and clean weeding of cropland, which continued after the emancipation of plantation slaves, enhanced rates of degradation. Population increase became a significant factor from the latter years of the nineteenth century, the pressure on resources being exacerbated by the fragmentation of land holdings. Some of the worst-affected areas are on the island of Hispaniola (Haiti and Dominican Republic). In Haiti, large areas of marginal land are thought to be irreversibly degraded and an estimated 6000 ha of land are abandoned to erosion each year (Paskett and Philoctete, 1990).

Another country where accelerated erosion is widely accepted to have reached crisis proportions is Ethiopia. Erosion rates that average 42 t/ha/year on Ethiopian cropland (Table 14.6) reach as high as 300 t/ha/year on some fields in western parts of the highlands where rainfall erosivity is highest. Hurni (1993) suggests that this level of soil loss reduces soil productivity in Ethiopia by 1–2 per cent per year and that at current erosion rates most of the country's cropland soils will be completely lost within 150 years. The crucial factor behind the degradation of Ethiopia's soils and other land resources is the country's large population, and Hurni believes that whatever scenario for soil conservation is proposed, sustainable use of land resources can only be achieved if population growth rates are reduced to zero within 50 years.

Although serious soil erosion is often associated with the humid tropics and semi-arid areas, it does give cause for concern in more temperate regions. The intensification of agriculture in the UK, for example, has brought erosion problems to the fore, with about 36 per cent of arable land in England and Wales classified as being at moderate to high risk of erosion (Evans, 1990). Sandy and peaty soils in parts of the Vale of York, the Fens, Breckland and the Midlands suffer from deflation during dry periods, but water is the most widespread agent of soil loss. The central reason for the increase in erosion in the past 20 years or so has been a shift in the sowing season, from spring to autumn, for the main cereal crops of wheat and barley. The change has come about because autumn-sown 'winter cereals' produce higher yields, but the shift in sowing season leaves winter cereal fields exposed in the wettest months – October and November – in many parts of the country, since arable fields are at risk from erosion until about 30 per cent of the ground is covered by the growing crop.

Several other aspects of more intensive arable farming have also contributed to the enhanced erosion rates (Boardman, 1990b):

- the expansion of arable crops on to steeper slopes made possible by more powerful tractors
- the creation of larger fields by removal of walls, hedges, grass banks and strips has produced longer slopes and larger catchment areas, which generate greater volumes of water on slopes and in valley bottoms
- more powerful machinery is able to work the land even when damp, thus compacting soils along wheel tracks, which tend to channel runoff
- the tendency to break up soil into a fine tilth using powered harrows to aid seed germination – a technique that also increases erodibility
- decreasing aggregate stability of the soil, due to reduced organic matter content, renders soils more likely to form crusts under raindrop impact.

SOIL CONSERVATION

Since soil is such a vital natural resource, using it sustainably involves employing methods to reduce accelerated erosion. Numerous techniques are available for protecting cultivated soils from erosion and these can be classified into three groups:

1 agronomic measures, which manipulate vegetation to minimise erosion by protection of the soil surface
2 soil management techniques, which focus on ways of preparing the soil to promote good vegetative growth and improve soil structure in order to increase resistance to erosion
3 mechanical methods, which manipulate the surface topography to control the flow of water or wind.

Maintaining a sufficient vegetative cover on a soil is sometimes referred to as the 'cardinal rule' for erosion control. One of the oldest of agronomic measures designed to reduce soil erosion is to rotate the location of crops, so-called shifting cultivation, in which an area of forest is cleared and cultivated for a year or two and then allowed to revert to scrub or secondary forest. An essentially similar method involves rotating crops grown in rows with cover crops such as grasses or legumes grown on the same field every other year. Mulching – the practice of leaving some residual crop material, such as leaves, stalks and roots, on or near the surface – is another widely used agronomic technique. It is successful in reducing erosion and in reducing the loss of water from fields by decreasing evaporation. But crop residues also provide a good habitat for insects and weeds so that the method often requires higher chemical inputs of pesticides and herbicides where available, which is expensive and increases hazards of off-field pollution and the killing of non-target species. Mulching can also be conducted using vegetable matter from elsewhere. Grass is widely used on the central plateau of Burkina Faso just before the rainy season from February to May where soils are liable to crust formation, which produces considerable runoff during heavy rainstorms (Slingerland and Masdewel, 1996). The mulch is not only an effective method for soil and water conservation; the grass helps to fertilise the soil by decomposition, and attracts termites that burrow into the soil, breaking up the crust and increasing soil porosity and permeability. Although the termites eat the grass, if sufficient is used the mulch still protects the soil from rainfall and reduces runoff. The effects on soil and water loss of various cultivation practices for a range of crop types in the humid zone in Brazil are shown in Table 14.7.

TABLE 14.7 *Soil and water losses for different crops and cultivation practices in the humid zone of north-east Brazil*

Crop and cultivation practice	Soil loss (t/ha/year)	Water loss (% of rainfall)
Cotton planted downslope	32	15
Cotton planted across-slope	13	11
Mixed cotton, maize and beans planted across-slope	24	13
Mixed cotton, maize and beans planted across-slope with herbaceous anti-erosion strips	15	12
Mixed maize and beans planted across-slope	13	12
Beans planted across-slope	12	13
Maize planted downslope	10	9
Maize planted across-slope	11	10
Permanent planted pasture	1	3
Bare soil	94	26

Source: Leprun *et al.* (1986)

Most soil management techniques are concerned with different methods of soil tillage, an essential management technique that provides a suitable seed bed for plant growth and helps to control weeds. However, different types of soil respond in different ways to tillage operations and a range of methods has been developed to reduce erosion effects. Strip or zone tillage leaves protective strips of untilled land between seed rows, requiring weed control on the protective strips. Other tillage methods are designed to leave varying degrees of vegetative matter from the previous crop still on the soil surface to provide protection against erosive forces. Minimum tillage, which incorporates the idea of stubble mulching mentioned above, is widely practised in maize/soybean agriculture in the Corn Belt of the USA, while no tillage, which leaves most of the soil covered with plant residues, is a widely used anti-erosion method on the soils of the USA's Southern Piedmont (Langdale *et al.*, 1992).

Mechanical methods, which are normally used in conjunction with agronomic measures, include such techniques as the building of terraces and the creation of protective barriers against wind, such as fences, windbreaks and shelter belts. Terraces are commonly built on steep slopes and effectively transform them by creating a series of horizontal soil strips along the slope contours. They intercept runoff, reducing its flow to a non-erosive velocity. If well maintained terraces are very effective but they are costly to construct and their physical dimensions act as a constraint on the use of mechanised agriculture. Windbreaks reduce wind velocity, thereby lowering its erodibility, and encourage the deposition of material that is already entrained. Fences or walls placed at right angles to erosive winds serve this purpose, or windbreaks may be created from living plants such as trees or bushes, in which case they are known as shelter belts.

The beneficial effects of some of these techniques for erosion control and crop yields are shown in Table 14.8, in which the range of percentage figures is derived from a review of more than 200 case studies.

TABLE 14.8 *Effect of low-cost soil conservation techniques on erosion and crop yields*

Method	Decrease in erosion (%)	Increase in yield (%)
Mulching	73–98	7–188
Contour cultivation	50–86	6–66
Grass contour hedges	40–70	38–73

Source: Doolette and Smyle (1990)

Implementation of soil conservation measures

Since we have a good understanding of the factors affecting soil erosion, the problems caused by the process, and the methods for its control, the question 'Why is soil erosion still a problem?' is an obvious one to ask. Innumerable soil conservation projects have been implemented, particularly in the poorer areas of the world, but often with disappointing results. An analysis of the reasons for the failure of such projects has been made by Hudson (1991), who found that the poor design of projects was a fundamental flaw. He concluded that errors are mainly a function of incorrect assumptions made at the design stage, by both donor agencies and host governments. The main donor errors are:

- overoptimism, including overestimating the effect of new practices
- overestimating the rate of adoption of new practices
- overestimation of the ability of the host country to provide back-up facilities
- underestimation of the time required to mobilise staff and materials for the project
- frequently, a quite unreal estimate of the economic benefits.

Some of the main problems arising from the assumptions of the host governments are:

- overestimating their capacity to provide counterpart staff and the funds for the recurrent costs arising from the project
- a tendency to underestimate the problems of coordination among different ministries or departments
- a tendency to overestimate the strength of the national research base and its ability to contribute to the project.

Such an analysis provides useful guidelines for future soil conservation projects, but other observers of the soil erosion problem believe that attention must also be focused on the central reasons for people misusing soils, which lie at a deeper level. The Haitian and Ethiopian examples cited earlier illustrate some of the underlying driving forces behind accelerated erosion rates, which are emphasised by social scientists (e.g. Blaikie, 1985). These driving forces are social, economic and political in nature, and include such factors as population growth, unequal distribution of resources, land tenure, terms of trade and subsidies, colonial attitudes and legacies, and class struggles. These factors limit the options available to poorer sectors of society who may be forced into degrading their soil resource simply in order to survive.

In less critical situations, erosion may result because it is not perceived as a serious problem since it often proceeds at relatively imperceptible rates, or because its effects are masked by fertilisers. Allied to this problem, the benefits of soil conservation on the farm may be long term, whereas farmers may have a short-term view of the relative importance of maintaining their income or repaying their debts. The costs involved in implementing soil conservation measures may also be high and there is an ethical and practical question of who should pay: the individual farmer or society as a whole? Often it is the off-site impacts of soil erosion that cause more immediate concern if a farmer is found liable for damage caused by sedimentation of material eroded from his fields.

FURTHER READING

Blaikie, P. 1985 *The political economy of soil erosion*. London, Longman. This book looks at soil erosion from a social perspective, emphasising the political and economic reasons behind accelerated rates of soil loss.

Boardman, J. and Favis-Mortlock, D.T. 1993 Climate change and soil erosion in Britain. *The Geographical Journal* 159: 179–83. A look at some of the ways in which climatic change may affect erosion patterns and processes in Britain.

Chappell, A. 1998 Using remote sensing and geostatistics to map 137Cs-derived net soil flux in south-west Niger. *Journal of Arid Environments* 39: 441–55. A regional-scale assessment of erosion using caesium-137.

Hallsworth, E.G. 1987 *Anatomy, physiology and psychology of soil erosion*. Chichester, Wiley. The author highlights the similarities between traditional and modern methods of soil conservation, proposes reasons why modern techniques are not always adopted and suggests how these problems can be overcome.

Mermuta, A.R. and Eswaran, H. 2001 Some major developments in soil science since the mid-1960s. *Geoderma* 100: 403–26. A wide-ranging review.

Morgan, R.P.C. 1995 *Soil erosion and conservation*, 2nd edn. London, Longman. A good basic text detailing the processes of soil erosion, methods of measurement and modelling, and the techniques developed for its control.

Pimental, D. (ed.) 1993 *World soil erosion and conservation*. Cambridge, Cambridge University Press. A collection of papers on numerous aspects of soil erosion, its measurement, the problems it causes, and its control.

Reij, C., Scoones, I. and Toulmin, C. 1996 *Sustaining the soil: indigenous soil and water conservation in Africa*. London, Earthscan. Numerous articles on farmers' practices in Africa.

Stocking, M.A. 1995 Soil erosion in developing countries: where geomorphologists fear to tread. *Catena* 25: 253–67. A thought-provoking analysis of the supposedly objective science of soil erosion study.

Websites

http://www.isric.nl/ the International Soil Reference and Information Centre contains information on international research programmes on all aspects of soil degradation.

http://www.swcs.org/ the Soil and Water Conservation Society site includes news, conferences and publications.

http://www.weru.ksu.edu/ the Wind Erosion Research Unit has much technical information on research programmes as well as a multimedia archive section.

http://topsoil.nserl.purdue.edu/nserlweb/ US research on soil erosion by water.

http://www.soil-convention.org/ information about a proposed Convention on the Sustainable Use of Soils.

http://www.wocat.net/ the World Overview of Conservation Technologies and Approaches (WOCAT) provides a forum for the exchange of information on soil and water conservation.

POINTS FOR DISCUSSION

- Should we be concerned about soil erosion? If so, why?
- Design a research programme to determine whether the implications of soil erosion are more serious on-site or off-site.
- Outline the main ways in which people influence erosion rates.
- There are many techniques to prevent soil erosion, so why are there still areas where soil erosion is a problem?

15

BIODIVERSITY LOSS

TOPICS COVERED
Understanding biodiversity
Threats to biodiversity
Threatened species
Threats to flora and fauna
Conservation efforts
Convention on Biological Diversity

Key words: extinction, endemism, *K*-strategist, *r*-strategist, keystone species, *ex situ* conservation, hotspot

Biodiversity, a term that refers to the number, variety and variability of living organisms, has become a much-debated environmental issue in recent times. Although we live on a naturally dynamic planet, where species are prone to extinction and ecosystems are always subject to change, biodiversity has become an issue of major concern because of the unprecedented rate at which human action is causing its loss. The very large and rapid increase in human population over recent centuries has been accompanied by an unprecedented scale of modification and conversion of ecosystems for agriculture and other human activities. At the same time, we have documented increasing numbers of cases where species have been driven to extinction by human activities.

Although the exact definition of biodiversity (or biological biodiversity) is the subject of considerable discussion, it is commonly defined in terms of genes, species and ecosystems, corresponding to three fundamental levels of biological organisation. Some authorities also include a separate human element in their definitions: cultural diversity. The Convention on Biological Diversity defines biodiversity as 'the variability among living organisms from all sources including, *inter alia*, terrestrial, marine and other aquatic ecosystems and the ecological complexes of which they are a part; this includes diversity within species, between species and of ecosystems'. Genetic diversity includes the variation between individuals and between populations within a species. Species diversity refers to the different types of animals, plants and other life-forms within a region. Ecosystem diversity means the variety of habitats found in an area. Much of this chapter is focused at the species level. Some aspects of genetic and ecosystem diversity are dealt with in more detail elsewhere (see Chapters 4 and 13).

UNDERSTANDING BIODIVERSITY

The 158 states that signed the Convention on Biological Diversity at the UN Conference on Environment and Development in Rio de Janeiro in 1992 agreed that there was a general lack of information on and knowledge of biodiversity, and that there was an urgent need to develop scientific, technical and institutional capacities to provide the basic understanding on which to plan and implement appropriate measures. Although scientists have been systematically counting and classifying other living organisms for at least two centuries, we remain remarkably ignorant of the most basic of information concerning the living things with which we share the Earth. While the numbers of species in some groups of organisms are relatively well known, our knowledge of others is extremely imprecise. The number of bird species, for example, is close to 10 000 (Sibley and Monroe, 1990), but estimates of the number of insect species vary widely (from 2 million to 100 million) with about 1 million having been described (UNEP, 1995). In total, about 1.75 million species have been scientifically described to date, and a reasonable estimate of the total number of species on the planet is 14 million (UNEP, 1995). Geographically, most estimates agree that more than half of all species live in the tropical moist forests that cover just 6 per cent of the world's land surface.

Our knowledge of the world's biomes and individual ecosystems is also unsatisfactory. Indeed, although several systems of classification have been developed no single measure of ecological community diversity can be uniformly applied to all ecosystems. We are also still trying to understand the ways in which biological diversity affects the functioning of ecosystems. Although genetic diversity is the ultimate basis for evolution and for the adaptation of populations to their environment, it is even less well understood. We know very little about the genetics of most living organisms. The few exceptions are for a handful of species identified as having direct importance to certain forms of economic activity, such as agriculture and human health.

Given the poor state of our knowledge, it is likely that many species that we have never known about have become extinct. Even among those organisms that have been described, problems emerge in documenting their disappearance. Just because no member of a species has been documented for some time does not necessarily mean that it has become extinct: absence of evidence is not evidence of absence.

Despite these and other difficulties, estimates of the rate at which species are becoming extinct, mostly due to human action, indicate that the rate has been growing exponentially since about the seventeenth century. Current and projected estimates of species loss are based upon the rate at which habitats are being destroyed, modified and fragmented – the most serious threats to species diversity – coupled with biogeographical assumptions relating to numbers of species and area of habitat. We should be aware, however, that estimates for habitat loss in tropical forest areas, the most diverse ecosystems, are themselves subject to wide variations (see page 49). While some earlier projections suggested that 20–50 per cent of species would be lost by the end of the twentieth century (Myers, 1979; Ehrlich and Ehrlich, 1981), these now seem exaggerated. Reid (1992) estimates a 1–5 per cent loss per century, and figures of 100 000 species lost per year are frequently quoted (WCMC, 1992).

THREATS TO BIODIVERSITY

Ecosystems change and species can become extinct under natural circumstances. We know that the Earth's climate is dynamic over a variety of timescales, and plant and animal communities have to adapt to these changes or run the risk of extinction. Species may also become extinct due to a range of other natural circumstances, such as random catastrophic events, or through competition with other species, by disease or predation.

Studies of the fossil record show us that long geological periods when the rate of species extinction was fairly uniform have apparently been punctuated by catastrophic episodes of mass extinction. In the last 570 million years of Earth's history five 'mass extinction events' have occurred. The most severe was during the late Permian period some 245 million years ago, and the most recent was at the end of the Cretaceous period 65 million years ago when the dinosaurs and several other families of species were wiped out.

Currently, however, there is widespread fear that another mass extinction event is occurring, one in which the Earth's human population is playing the key role. Although we are not sure of the current rate of species extinction, six fundamental causes of biodiversity loss have been identified (WRI, 1992):

1 the unsustainably high rate of human population growth and natural resource consumption
2 a narrowing spectrum of products from agriculture, forestry and fishing
3 economic systems that fail to value the environment and its resources
4 inequity in the ownership, management and flow of benefits from both the use and conservation of biological resources
5 deficiencies in knowledge and its application
6 legal and institutional systems that promote unsustainable exploitation.

THREATENED SPECIES

Some species are particularly at risk from the threat of extinction simply because they are only found in a narrow geographical range, or they occupy only one or a few specialised habitats, or they are only found in small populations. Species that occur in only one location are known as 'endemic'. Other factors may also affect the degree of risk faced by certain species. These include:

- low rates of population increase
- large body size (hence requiring a large range, more food and making the species more easily hunted by humans)
- poor dispersal ability
- need for a stable environment
- need to migrate between different habitats
- perceived to be dangerous by humans.

Combinations of some of these characteristics are found in species known as *K*-strategists, and it is these species that are generally more likely to become extinct because they tend to live in stable habitats, delay reproduction to an advanced age

TABLE 15.1 *IUCN Red List categories**

Extinct	No reasonable doubt that the last individual has died
Extinct in the wild	Only known to survive in cultivation, in captivity or naturalised well outside past range
Critically endangered	Extremely high risk of extinction in the wild in immediate future
Endangered	High risk of extinction in the wild in near future
Vulnerable	High risk of extinction in the wild in medium-term future
Lower risk	*Conservation dependent:* reliant on conservation programme, which if stopped would put taxon in one of above categories within five years *Near threatened:* not conservation dependent, but close to qualifying for vulnerable category *Least concern:* not conservation dependent or near threatened
Data deficient	Data insufficient to categorise, but listing highlights need for research, perhaps acknowledging the need for classification
Not evaluated	Not assessed against the criteria

*This ten-fold classification replaced the previous six-category system in 1994

and produce only a few, large offspring. By contrast, species that produce many offspring at an earlier age and have the ability to react quickly to changes in their environment, are known as *r*-strategists, and it is their speedier turnover and flexibility that make *r*-strategists less likely to experience extinction.

The wider ecological implications of the loss of a certain species vary between species. Another important aspect of the extinction issue is the fact that certain keystone species may be important in determining the ability of a large number of other species to persist. Hence, the loss of a certain keystone species could potentially result in a cascade of extinctions. Such fears have been expressed over tropical insects, many of which have highly specialised feeding requirements.

Species known to be at risk are documented, according to the severity of threat they face and the imminence of their extinction, by the International Union for Conservation of Nature and Natural Resources (IUCN, also known as the World Conservation Union) on so-called Red Lists (Table 15.1). The general term 'threatened' is used to refer to a species of fauna or flora considered to belong to any of the categories shown in Table 15.1.

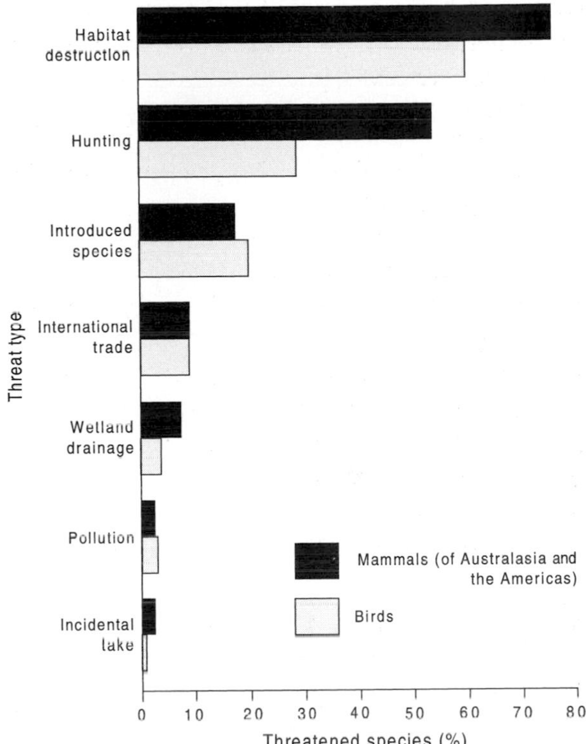

Figure 15.1 Threats to mammals and birds (after WCMC, 1992).

THREATS TO FLORA AND FAUNA

Most of the factors currently threatening species of both fauna and flora are induced or influenced by human action. Such actions may be deliberate, as in the case of destruction by hunting, or inadvertent, as in the case of destruction or modification of habitats in order to use the land for other purposes. In practice, many species are at risk from more than one threat and some threats tend to combine: the clearance of forests, for example, makes the hunting of large mammals easier. The threats that a particular species faces may also vary through time. The decline of the New Zealand mistletoe (*Trilepidea adamsii*) began as its habitat was reduced by deforestation, first by the Maoris and at an accelerating rate by British settlers in the late nineteenth century. The population was further reduced by collectors, and the decline of bird populations that were responsible for seed dispersal, due to forest clearance. The final specimens, which disappeared from North Island in 1954, may have been eaten by brush-tailed possums deliberately introduced from Australia during the 1860s to establish a fur trade (Norton, 1991). Although the detailed nature of threats may be complex and variable, some indication of the relative importance of the different threats is given in Fig. 15.1, which was compiled for the birds of the world and mammals in Australasia and the Americas.

All of these human-induced reasons for the demise of species have operated in the past. In Britain, for example, the combination of hunting and habitat destruction by deforestation has put paid to many original animal species over the centuries (Table 15.2).

TABLE 15.2 *Dates when the last wild member of selected animal species was killed in Britain*

Animal	Year
Beaver (*Castor fiber*)	Late 1100s
Wild swine (*Sus scrofa*)	1260
Wolf (*Canis lupus*)	1743
Goshawk (*Accipiter gentilis*)	1850
White-tailed sea eagle (*Haliaetus albicilla*)	1918

Sources: from information in Rackham (1986); Peters and Lovejoy (1990)

Habitat loss and modification

The destruction of habitats is widely regarded to be the most severe threat to biological diversity, while fragmentation and degradation, which are often precursors of outright destruction, also present significant cause for concern. Habitat loss and degradation was the most pervasive threat to birds, mammals and plants, according to the 2000 IUCN Red List, affecting 89 per cent of all threatened birds, 83 per cent of the threatened mammals assessed and 91 per cent of the threatened plants (IUCN, 2000).

In many countries, particularly on islands (see below) and where human population densities are high, most of the natural habitat has been destroyed to provide farmland, rangeland, and land for settlement and industry. No fewer than 49 out of 61 countries surveyed in the African and Asian tropics are thought to have lost more than 50 per cent of their wildlife habitats (IUCN/UNEP, 1986a; 1986b). In the tropical African countries, 65 per cent of the original wildlife habitat has been lost, with particularly high rates of destruction reported from Gambia (89 per cent), Liberia (87 per cent), Rwanda (87 per cent), Burundi (86 per cent) and Sierra Leone (85 per cent). In the Indomalayan countries the overall loss is 68 per cent, with particularly severe losses reported from Hong Kong (97 per cent), Bangladesh (94 per cent), Sri Lanka (83 per cent), Viet Nam (80 per cent) and India (80 per cent).

The destruction of tropical rain forests, coral reefs, wetlands and mangroves, documented elsewhere in this book, is particularly serious given

TABLE 15.3 *Habitat extent (1980s) and loss since pre-agricultural times in selected countries*

Country	Forests		Savanna/grassland		Wetlands/marsh	
	1980s extent (thousand ha)	% lost	1980s extent (thousand ha)	% lost	1980s extent (thousand ha)	% lost
Australia	13 000*	95	75 900	nd	17 000	*c.* 95
Canada	274 000*	48	27 663	nd	127 000	nd
Burundi	117	91	246	80	14	nd
Ethiopia	5 570	86	27 469	61	0	0
France	131	*c.* 99	250	nd	1 171	nd
Malaysia	18 008	42	0	0	2 214	35
Namibia	15 020	52	14 741	59	225	10
Peru	74 270	12	13 900	41	1 303	nd
Thailand	13 107	73	0	0	83	96

nd: no data
*Primary forest extent only
Source: after WRI (1994: 320–321, Table 20.3)

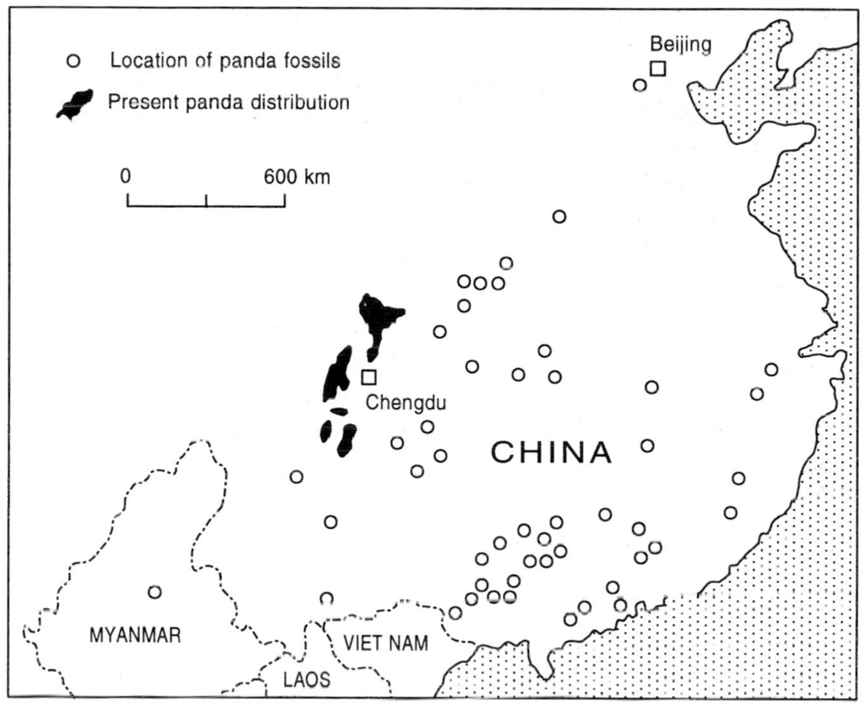

Figure 15.2 Past and present distribution of pandas (reprinted with permission from Roberts, 1988. Copyright, 1988, American Association for the Advancement of Science).

the high biodiversity of these habitats. Hence, the large majority of current human-induced extinctions are occurring in the world's tropical rain forest areas, and insects are the order of species most at risk. Nevertheless, other habitats have suffered equally severe destruction; Table 15.3 shows the percentage of three types lost since pre-agricultural times in selected countries.

There are numerous examples of species that have been driven to the edge of extinction because of the loss of their habitat. In the UK, for example, the distribution of the threatened green winged orchid (*Orchis morio*) has been severely reduced by the loss of 40 per cent of unintensified lowland grassland between 1932 and 1992 (DoE, 1992). Internationally, one of the best-known threatened species, the giant panda (*Ailuropoda melanoleuca*), has also suffered from progressive human encroachment into its habitat. Once found throughout much of China's high-altitude regions and beyond, the species is now confined to a few sites near Chengdu (Fig. 15.2). The panda's heavy reliance upon its specialised bamboo diet, which occasionally requires forays into lowland regions during natural bamboo die-offs, puts the normally shy creature in direct conflict with encroaching human populations.

These factors, combined with the extreme difficulties encountered by attempts at captive breeding, look set to add the panda to a long list of species whose ultimate fate can be traced to destruction of their habitat by humans; this is despite huge efforts to conserve the species. A network of panda reserves, designed to protect 60 per cent of the animal's range, has been established in China, but not all of these protected areas have been successful in their prime objective. Research on the 200 000-ha Wolong Nature Reserve has shown that panda habitat is still being destroyed faster inside one of the world's most high-profile protected nature reserves than in adjacent areas that are not protected. Further, the rates of destruction were higher after the reserve was established in 1975 than before (Liu *et al.*, 2001).

It appears that the Wolong Nature Reserve has been a victim of its own success. Towns and settlements have thrived in and around Wolong since the reserve was established and began attracting tourists to the area. The upturn in the local economy, fuelled by tourism, has been a major factor behind an increase in resident human population in the area of 70 per cent since the reserve was created. The people are cutting more trees for fuel and other uses, destroying prime panda habitat.

Fragmentation of habitats not only reduces the area of habitat but also affects normal dispersion and colonisation processes, and reduces areas for foraging. Fragmentation of the panda's habitat is not just a threat to the number of bamboo species remaining. Small, isolated populations of giant pandas also face a risk of inbreeding. This could lead to reduced resistance to disease, less adaptability to environmental change, and a decrease in reproductive rates. Fragmentation also increases the ratio of edge to total habitat area, hence increasing dangers from outside disturbances such as the effects of hazards (e.g. fire and diseases) and invasion by competitors and/or predators. The recent decline in numbers of songbirds throughout North America, for example, is at least partly explained by increased predation and parasitism of nests in fragmented nesting tracts (Terborgh, 1992). Species–area relationships derived from the study of oceanic islands suggest that the number of species a fragment is able to support is directly related to the fragment area (MacArthur and Wilson, 1967).

Research in the Brazilian Amazon on dung beetles shows that they are significantly affected by fragmentation, with small forest fragments having fewer species of beetles, lower population densities for each species and smaller-sized beetles than undisturbed forest areas. Since dung beetles are a keystone species in the forest ecosystem, through their role in burying dung and carrion as a food source for their larvae – thus facilitating the rapid recycling of nutrients and the germination of seeds defecated by fruit-eating animals, and reducing vertebrate disease levels by killing parasites that live in the dung – the effects on community interactions and ecosystem processes are probably very widespread (Klein, 1989).

Human-induced modifications to habitats can also lead to species loss, either directly or in a more pervasive manner through stress or subtle effects on ecological processes. Pollution in its various forms is probably the most significant problem in this respect. Examples of such effects are referred to in many of the chapters in this book; hence, just a few illustrations will be made here.

The detrimental effects of agrochemicals have been widely documented (see Chapter 13), with some of the most deleterious being observed in creatures at the top of the food chain where some materials bioaccumulate. Raptors are typical in this respect, suffering a steep decline in population numbers in the post-war years with the rise in use of organochlorine pesticides – particularly DDT, which had the effect of thinning egg shells, leading to breakage in the nest, and alterations in breeding behaviour. Use of the highly toxic dieldrin and aldrin in cereal seed dressings and sheep dip, which occurred from 1956 in the UK, also led to widespread declines through increased adult mortality.

However, the phasing out of organochlorine pesticides in many western countries has resulted in bird of prey populations rising once more. In the UK, the population of peregrine falcons (*Falco peregrinus*) numbering around 850 pairs in the 1930s, had fallen to 360 pairs by 1962, but had risen to 1050 pairs in 1991 (DoE, 1992).

Numerous forms of water pollution can have detrimental effects on aquatic flora and fauna. Acidification of lakes and streams, for example, has affected waterbirds and invertebrates (see Chapter 12). An example of a species extinction due to pollution is the splitmouth (*Lagochila lacera*), discovered in the Chickamauga River in Tennessee in the late nineteenth century and named as a new genus and species. The splitmouth was very common in the region and found in the

Tennessee, Cumberland, White and Ohio river drainages to Lake Erie. The species was last seen in the Auglaize River, Ohio, in 1893. Its extinction – the first North American freshwater fish species known to die out entirely due to human action – appears to have been caused by continuous silting and pollution of its habitat (Maitland, 1991).

Air pollution can also have insidious effects on plants and animals. Atmospheric acidification has been linked to numerous impacts on trees and other floral species (see Chapter 12); the widespread impoverishment of lichens, which are very sensitive to air pollution, across large parts of Europe and North America has been linked to sulphur dioxide and, to a lesser extent, to photochemical smogs (Hawksworth, 1990). Atmospheric pollutants from heavy industries such as smelting can be devastating to surrounding ecology (see Chapter 19).

Overexploitation

The 2000 IUCN Red List concluded that exploitation, including hunting, collecting, fisheries and fisheries by-catch, and the impacts of trade in species and species' parts, constitutes a major threat for birds (37 per cent of all), mammals (34 per cent of all), plants (8 per cent of those assessed), reptiles and marine fishes (IUCN, 2000). Humans have long been implicated in the extinction of species through overexploitation, particularly by hunting for food. Some believe that such effects can be traced back to the Stone Age and to the late Pleistocene, with the loss of large mammals such as the mammoth and sabre toothed tiger being intimately linked to overhunting, although the role of abrupt and substantial climatic change has also been suggested as being instrumental in the extinction of such creatures (Martin and Klein, 1984).

Larger creatures have been suffering from overexploitation ever since. In Roman times, hunters and trappers extirpated the elephant, rhino and zebra in Africa north of the Sahara, and lions from Thessaly, Asia Minor and parts of Syria. A similar fate befell tigers in Hercynia and northern Persia. In many instances, the loss of a species has knock-on effects for other organisms. The death of the last dodo (*Raphus cucullatus*), killed by seafarers on Mauritius in 1681, has meant that another of the island's endemic species, the tambalocque tree (*Sideroxylon sessilisiorum*), has been unable to reproduce for the past 300 years because the fruit-eating bird prepared the fruit in its gizzard for germination (Temple, 1977).

The advent of the firearm greatly enhanced humans' ability to exterminate creatures in large numbers, resulting in the decimation of the buffalo and the extinction of the passenger pigeon in North America, for example. The North American passenger pigeon (*Ectopistes migratorius*) is widely believed to have been the most abundant bird ever to have inhabited the planet. Conservative estimates suggest that its numbers may have been around 10 000 million in the first half of the nineteenth century. They lived in huge, very dense flocks, some containing more than 2000 million birds, which darkened the skies for up to three days on flying past. They were killed in very large numbers for their meat, and this intense exploitation, coupled with destruction of breeding habitats, led to a precipitous decline in numbers. The last passenger pigeon seen in the wild was shot in 1900 and the last known individual died a lonely death in Cincinnati Zoo in 1914. In just a few decades, the planet's most abundant bird species had been rendered extinct (Bucher, 1992).

In some parts of the world, migrating birds continue to be killed in huge numbers by hunters. Some of the worst culprits are the European Mediterranean countries such as Italy and Malta (Fig. 15.3). Overzealous hunting is also still a very serious threat for numerous large mammals, particularly for those whose products fetch high prices in local and international markets. The African elephant population, for example, was cut by half during the 1980s, falling from 1.3 million to under 600 000 (Cohn, 1990) before a ban on international trade in ivory was introduced in 1989 (although even this ban has not been 100 per

Figure 15.3 Hide for shooting migratory birds in rural Umbria, Italy. Italian hunters are thought to kill as many as 3 million birds a year.

cent effective – see below). Similarly, the market for whale products fuelled the hunting of oceanic cetacean species, many of which currently have indeterminate chances of survival (see page 91).

The market for certain floral species is also an underlying factor behind their overexploitation. Cacti and orchids are particularly at risk from collectors and many species of tree have been dramatically reduced in their range by selective logging. Prance (1990) notes the complete loss of the scented sandalo tree (*Santalum fernandezianum*) from Juan Fernández islands, all of which had been felled and shipped to Peru and Chile by 1908, and the bois de prune blanc (*Drypetes caustica*) of Mauritius and Réunion, renowned for its excellent hard timber, of which just 12 trees remain.

Island species

Although islands make up only a minor proportion of the world's land surface area, they feature prominently in any account of threatened flora and fauna. Island species are considered to be especially vulnerable for a number of reasons. Many are endemic species that have evolved in isolation and are thus also particularly susceptible to introduced competitors, diseases and predators. The physical size of islands also means that human activities can rapidly degrade large portions of their ecosystems and can have a great impact on relatively small island populations.

The susceptibility of island species to human-induced change is indicated by the fact that at least 75 per cent of vertebrate extinctions to date have occurred among island fauna (Diamond, 1984). For obvious reasons, it is often the largest species on islands that first suffer from direct human action. On Madagascar, six genera and at least 14 species of lemurs have become extinct, along with several species of flightless bird and land tortoises, since humans arrived there less than 2000 years ago. There is good evidence to suggest that people were responsible, and hunting was probably the primary cause (Mittermeir *et al.*, 1992).

Severe ecological damage on many islands dates from the arrival of explorers and colonisers from the European maritime powers, for whom oceanic islands had strategic importance. Goats were introduced to the South Atlantic island of St Helena in 1513, and within 75 years, vast herds roamed the island. St Helena's endemic plants had evolved without large grazing animals to contend with and so had few defences against them. Botanists first reached St Helena in the early 1800s, so we can only guess at the damage done by then. Today, 46 endemic species are known, of which seven are extinct, but some estimates put the original number of endemics at more than a hundred. St Helena remains an island with severely threatened endemic flora (Table 15.4). The most serious threat on the other islands shown in Table 15.4 is introduced plants, which have outcompeted endemic species.

TABLE 15.4 *Selected examples of islands known to have severely threatened endemic flora*

Island (ocean)	Status of endemic flora
St Helena (Atlantic)	7 species already extinct; all of 39 remaining are also threatened
Mauritius (Indian)	21 species already extinct; 89 per cent of 215 remaining are also threatened
Juan Fernández (Pacific)	1 species already extinct; 81 per cent of 122 remaining are also threatened
Hawaii (Pacific)	108 taxa already extinct; 39 per cent of 877 remaining are also threatened

Source: modified after WCMC (1992: 246, Table 17.6)

European colonisers are by no means the only culprits, however. Damage to the wildlife of the Hawaiian islands, where almost 100 per cent of the native insects are endemic, as well as 98 per cent of the birds, 93 per cent of the flowering plants, and 65 per cent of the ferns, began long before the arrival of Captain Cook, the first European, in 1778. Polynesians, who had settled the island by AD 750, cleared extensive lowland areas for cultivation and introduced rats, dogs, pigs and jungle fowl. Palaeontological evidence suggests that at least half of the known total of 83 species of Hawaiian birds became extinct in the pre-European period (Olson and James, 1984). However, further introductions and disturbance of the natural habitat since 1778 has resulted in the loss of at least 23 bird species and 177 species of native plants.

In some cases, the introduction of one exotic species to an island may result in the local extinction of numerous native species. An example here is the brown tree snake (*Boiga irregularis*), which has been introduced to a number of Pacific islands with devastating effects on endemic birds. The snake has caused the extinction of nine of the eleven endemic forest bird species on the island of Guam where the snake was accidentally introduced from New Guinea in the 1940s (Gillis, 1992). Similarly, the deliberate introduction of the predaceous Ferussac snail (*Euglandina rosea*) to several Pacific islands in an attempt at biological control of a previously introduced snail that had attained pest status, the giant African (*Achatina fulica*), has had severe repercussions for endemic snail species on Moorea (see page 225).

In other cases the exotic species introduced may represent a threat even after the people responsible for their introduction have gone. This is the case with the endangered short-tailed albatross (*Diomedea albatrus*), one of many albatross species facing extinction, whose chicks and eggs are easy prey for feral cats and rats on Torishima island, one of the Japanese Izu Islands, its main nesting site (Collar *et al.*, 1994). Once seen throughout the North Pacific Ocean, the short-tailed albatross has suffered a long history of decline. In the late nineteenth century its numbers began to fall because its long white wing and tail feathers became popular in the manufacture of pen plumes and its downy body feathers to stuff feather beds. Although a ban on the collection of short-tailed albatross feathers was instituted in 1906, it was not very effective. Indeed, illegal feather collection continued until the 1930s. Collection only declined then because the species was no longer economically significant, its numbers having been reduced so drastically. Once thought to number around five million individuals, the species was mistakenly declared extinct in 1949.

Introduced species on Torishima are not the only current threat the short-tailed albatross faces. People no longer live on Torishima but this is because it is an active volcano. In 1902, an eruption killed all 125 of the island's human occupants. A repeat eruption, if it occurred during the breeding season, would wipe out most of the world population that nests on the ash-covered slopes. The short-tailed albatross is also threatened by longline fishing while at sea.

CONSERVATION EFFORTS

The two ends of the spectrum of arguments in favour of conserving biodiversity are rooted in morals and pragmatism. Some believe that the destruction of any living organism is morally unacceptable, and that people who share the planet with plants and animals have no right to exterminate other species. More technocentric arguments point out that the extermination of other species is not in the interests of humankind. Biodiversity is useful to us in the wide perspective, in maintaining the biosphere as a functioning system of which we are a part (so-called 'ecosystem services'), and at a more functional level in providing resources for agriculture, industry, medicine and other utilitarian needs. These values are both direct (using animals for food and cutting timber for export, for example) and indirect (such as for aesthetic and recreational purposes). There are also potential uses of biodiversity not yet realised. The importance of maintaining genetic diversity in the wild because it might one day prove useful for agricultural crops is outlined on page 227.

Conversely, however, certain species are considered detrimental to human activities and are therefore the subject of deliberate attempts to control or exterminate their populations. Species regarded as pests are obvious examples here. While deliberate efforts to make an organism extinct are not common, the smallpox virus (although not strictly a species) is a case in point. Subject to a successful international effort to eradicate the disease by vaccination, the virus remains only in laboratory stocks in Russia and the USA. Periodic recommendations to destroy these stocks have been countered by preservation arguments because they can be used for research purposes.

Human perceptions of biodiversity are always subject to change. For a long time many societies considered wetlands to be worthless and so these habitats have been altered in many parts of the world to make the land useful for human activities. It is only relatively recently that the inherent values of undisturbed wetlands have been widely recognised (see pages 134–5).

Habitat protection

Since the destruction, modification and fragmentation of habitats is the primary threat to biodiversity, the protection of habitats from the causes of their destruction is the most obvious conservation method. Currently, protected areas cover just over 6 per cent of the Earth's land area, although the proportion varies widely at the national level (Table 15.5). While some countries give some form of protection to more than one-fifth of their land area (e.g. Austria, Bhutan, UK, New Zealand), others protect less than 1 per cent of their national space (e.g. Equatorial Guinea, Jamaica, Uruguay). In practice, some types of habitat are better protected than others, largely due to the fact that the actual designation of sites is often based more upon socio-economic and political factors than on conservation ideals. Biomes such as mixed mountain systems and island systems, which are typically not intensively used by humankind, are better represented than systems such as temperate grasslands and lakes. It comes as no surprise to learn, therefore, that the world's largest protected area is the Greenland National Park, which covers 972 000 km^2, and that the area proposed for the first World Park is Antarctica.

The types and quality of protection and management in these areas also vary greatly. A lack of political and financial support limits the success of many protected areas. The need to maintain levels of protection is well illustrated by the Operation Tiger reserves set up in several Asian

TABLE 15.5 *Protected areas by continent and in selected countries*

Continent/country	Area protected (thousand ha)	Area protected (% land area)
WORLD	**851 511**	**6.4**
ASIA (exc. Middle East)	**148 692**	**6.0**
Bhutan	998	21.2
Japan	2 561	6.8
Uzbekistan	818	1.8
EUROPE	**109 297**	**4.7**
Austria	2 451	29.2
Greece	469	3.6
UK	5 000	20.4
MIDDLE EAST AND NORTH AFRICA	**25 863**	**2.1**
Algeria	5 891	2.5
Israel	326	15.5
Yemen	0	0
SUB-SAHARAN AFRICA	**146 904**	**6.0**
Botswana	10 499	18.0
Equatorial Guinea	0	0
Zambia	6 366	8.5
NORTH AMERICA	**213 822**	**11.1**
Canada	90 702	9.1
USA	123 120	13.1
CENTRAL AMERICA AND CARIBBEAN	**16 450**	**6.1**
Costa Rica	723	14.2
Cuba	1 909	17.2
Jamaica	1	0.1
SOUTH AMERICA	**131 663**	**7.4**
Chile	14 142	18.7
Surinam	736	4.5
Uruguay	48	0.3
OCEANIA	**60 784**	**7.1**
Australia	54 250	7.0
Fiji	20	1.1
New Zealand	6 334	23.4

Note: The disparity between summed continental totals and world total is due to rounding.
Source: after WRI (2000: 244–5, Table BI.1)

Figure 15.4 Tiger (*Panthera tigris*) in Rhanthambore National Park, an Operation Tiger reserve in northern India (Photo: Mark Carwardine).

countries in the early 1970s to protect the Asian tiger from local people faced with the loss of their domesticated livestock, and in some cases fellow villagers. Threats from the tiger have been enhanced in many areas due to the loss of its forest habitat to agricultural lands.

While Operation Tiger has long been hailed as a conservation success story, the story's most recent chapter has not been a happy one. Indonesia's tiger population, currently thought to be 400–650, has been declining progressively in the last few years due to habitat destruction, poaching and the removal of individuals in conflict with farmers (Plowden and Bowles, 1997). Poor management and continued human encroachment into protected areas threatens Thailand's remaining tiger population, estimated to be around 250 (Rabinowitz, 1993). Russian experts think that between 50 and 100 Siberian tigers were poached in 1992 in the Russian Far East, where poaching of all wildlife is rampant due to the breakdown in law and order following the demise of the USSR. The Siberian tiger population had fallen to fewer than 100 in the early twentieth century, and a few years of poaching are rapidly negating a half-century of conservation efforts, which had allowed the Russian population to approach an estimated 350. Poaching for skins and bone smuggling also rapidly increased in the 1990s in the Indian subcontinent where over half of the remaining world's wild tiger population of fewer than 6000 is located (Fig. 15.4). Many known tigers disappeared in India and Nepal in the 1990s and guards in Indian Operation Tiger reserves have been killed in clashes with poachers (IUCN, 1993). The South China tiger is probably closest to extinction since only 20 or 30 are thought to remain in the wild,

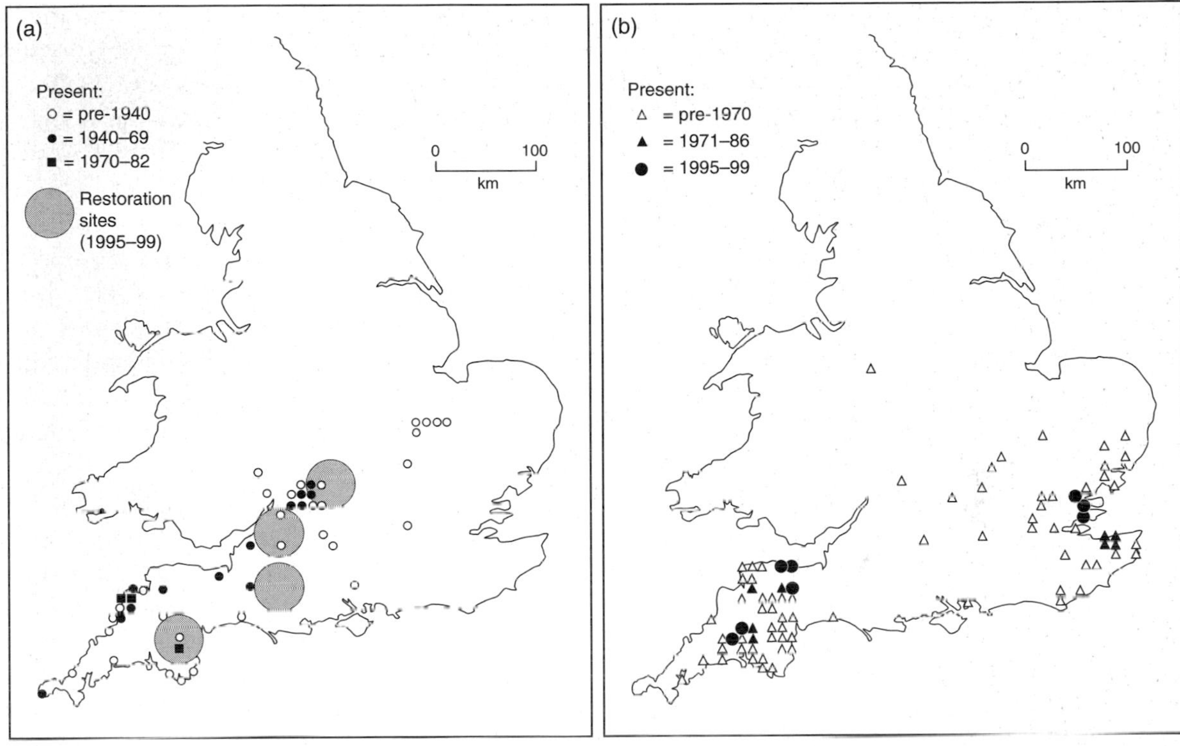

Figure 15.5 Decline in the distribution of butterflies in southern Britain: (a) the large blue (*Maculinea arion*); (b) the heath fritillary (*Mellicta athalia*) (after Heath *et al.*, 1984; Asher *et al.*, 2001).

down from an estimated 4000 in the 1950s (WWF, 1998).

Conflict between the aims of conservation and those of local people is a key issue in the threatened species debate since without local involvement in both the design and management of protected areas, adequate protection can only be achieved if the park agency has the authority and ability to enforce regulations. This is both undesirable and often unattainable, and this realisation represents a major recent shift in practical conservation philosophy.

Even in those areas that receive maximum protection, however, active conservation management is usually needed because often it is not sufficient simply to cordon off an area and leave it be. Such management is not always an easy task, however. A ban on hunting, for example, is theoretically easy to introduce, and grazing goats, sheep, pigs and even rabbits are relatively easy to control and even eliminate, but a threat from an introduced plant is much more intractable.

Effective management is also dependent upon a sufficient understanding of the ecology of the organisms in question, which is not always the case. Although the gradual decline in British colonies of the large blue butterfly (*Maculinea arion*), a dry grassland species, was largely due to the fact that 50 of its former 91 sites were ploughed or otherwise fundamentally changed in the period 1800–1970 (Fig. 15.5a), its eventual extinction from Britain in 1979 was due to ignorance. Conservationists were aware of the large blue's highly specialised lifecycle: eggs are laid on *Thymus praecox*, on which the larvae feed briefly before being adopted and raised by *Myrmica* spp. of ant. But the disappearance of the large blue from another 41 sites, including four nature reserves where *Thymus* and *Myrmica* remained abundant, proved enigmatic. The realisation that only one species of *Myrmica* ant could act as host, and that regular heavy grazing was necessary to maintain appropriate soil surface temperatures came too late to save the large blue.

A similar pattern of events occurred with the heath fritillary (*Mellicta athalia*), a species that has declined as its woodland clearing habitat has disappeared and because specially designated nature reserves were not adequately managed (Fig. 15.5b). The heath fritillary is still one of the rarest butterflies in Britain, but a clearer understanding of the make-up of its narrow niche may yet save it from extinction (Thomas, 1991). Likewise, the lessons learned over the loss of the large blue have been put to good use in recent attempts to re-introduce a similar subspecies from Sweden. Following a re-establishment trial on the edge of Dartmoor in 1983, it has since been re-introduced or has spread to seven other sites in the Cotswolds in Gloucestershire, and the Polden Hills and Mendip Hills in Somerset (Asher *et al.*, 2001).

Bans on hunting and trade

Legislative bans on threatening activities have been widely implemented to safeguard certain species. One example is the ban on the catching and selling of coelacanths introduced in the Indian Ocean island state of the Comoros as part of a conservation plan to protect the small population of this fish, first found in 1938, which previously had been thought to have become extinct 65 million years ago (Fricke and Hissman, 1990).

However, a ban on hunting and trade is, in practice, like any legislation, only as effective as the ability of states to uphold it. If enforcement is weak, and incentives are large enough, poachers will continue to operate in spite of the law. The seemingly intractable problem of African elephant poaching during the 1980s inspired drastic conservation measures, with some countries implementing shoot-on-sight policies to deal with poachers in protected areas. But the placement of the African elephant on Appendix I of the Convention on International Trade in Endangered Species (CITES) in 1989 has dramatically reduced the world price of ivory, resulting in a downturn in poaching of the African elephant. None the less, the listing on Appendix I was not welcomed by all conservationists. In the southern African countries of Botswana, Malawi, Namibia, South Africa and Zimbabwe, well-managed elephant herds are large and expanding, to the extent where they have to be culled to offset the risk of degradation in their ranges. The ban meant that ivory from the culled animals could not be sold to maximum profit. As a result, a decision was taken in 1998 to allow the resumption of some limited trade in African elephant ivory.

Problems still occur with trade bans, and the case of the Old World fruit bats (or flying foxes) is an interesting one in this respect. The flying foxes constitute a single family (*Pteropodidae*) with 41 genera and 161 species, which play an important ecological role among Pacific islands as pollinators and seed dispensers for hundreds of plant species, many of which have important economic and subsistence value to humans (Fujita and Tuttle, 1991). One of the central threats to the fruit bats is from hunters. Indigenous people on Guam have eaten them for at least 2500 years, but Guam's bat population was decimated in the years after the Second World War with the proliferation of firearms used in hunting. Many other south Pacific islands' fruit bat populations have subsequently suffered declines as exporters have supplied the Guam market. Listing on CITES Appendix I has had some success in curbing the threat to some of the species, but problems have arisen because of the lack of wildlife inspectors to uphold the law and the fact that several other islands in the region are, like Guam, effectively part of the USA, and therefore the trade is considered domestic rather than international (Sheeline, 1993).

Some conservationists argue that bans can only be a short-term measure to protect certain species. Long-lasting protection can only be envisaged if the people who demand certain products can be persuaded against them. Concerted publicity campaigns by animal rights groups have had

Figure 15.6 Saiga antelopes (*Saiga tatarica*) fleeing a hunter in Siberia. This engraving was made in the late 1800s when the saiga was widely hunted for its meat, hide and horn. Having recovered from the brink of extinction thanks to a hunting ban introduced in the 1920s, the saiga became threatened again by a resumption in illegal trade in the 1990s.

considerable effect in this respect in certain European countries where ornamental furs are concerned. The fact that if a market for products persists, the threat it represents to species does not disappear can be illustrated by the case of the saiga antelope. Hunting of the saiga during the nineteenth century in Russia (Fig. 15.6) pushed the species to the brink of extinction before being banned in the 1920s. Subsequently, controlled cropping allowed the saiga to recolonise most of its original range, but deterioration in law enforcement since the collapse of the Soviet Union in the early 1990s has resulted in an upsurge in the illegal trade in saiga horn destined for medicinal markets, particularly in China. The saiga population in the Russian Autonomous Republic of Kalmykia exceeded 700 000 in the 1970s, but had declined to between 120 000 and 150 000 animals in the mid-1990s. The plight of the saiga in Kalmykia has also been exacerbated by competition for pastures from other grazing animals (Zonn, 1995).

Off-site conservation practices

While the conservation of species is best achieved by their maintenance in the wild, through protected area programmes and legislative measures, other practices may be necessary for species whose populations are too small to be viable in the wild or are not located in protected areas. Maintenance of species in artificial conditions under human supervision is a strategy known as off-site, or *ex situ*, conservation. The off-site approach includes game farms, zoos, aquaria and captive breeding programmes for animals – although in reality few of these are actively involved in the conservation of endangered species – while plants are

Figure 15.7 Feeding an Aldabra giant tortoise (*Geochelone gigantea*) on the Indian Ocean island of Changuu. This rare species, endangered by poachers supplying the illegal exotic pet trade, is protected here in a sanctuary that is open to visitors. Income from tourists helps to finance the programme, which includes educational visits from local schoolchildren.

maintained in botanical gardens, arboreta and seed banks.

In practice, off-site techniques complement on-site approaches in a number of ways, such as by providing individuals for research, the results of which can be fed back into management techniques in the wild. Perhaps the ideal way in which the maintenance of captive individuals of species can help in the biodiversity issue is by providing individuals that can be re-introduced into the wild.

Such programmes are expensive, logistically demanding, and require long-term management and monitoring to assure and assess their success. Plant re-introductions are generally viewed as a high-risk strategy with uncertain indications of their long-term success (WCMC, 1992). Re-introduction of animals is also a difficult task, but one notable success has been the re-introduction of the Arabian oryx (*Oryx leucoryx*) to Oman, a species that is thought to have become extinct in the wild in 1972. Individuals kept in captivity in the Middle East and elsewhere were released to the wild in Oman in batches throughout the 1980s and now seem to be established. An equivalent re-introduction programme has also been established in neighbouring Saudi Arabia. A similar programme was initiated in Mongolia in 1992 when some Przewalski's horses (*Equus caballus przewalski*) were taken to Mongolia from zoos and game parks in Canada, the Netherlands and Russia. The horses, which are extinct in the wild, are kept under close observation in corrals for a period before being released. Given the long-term commitment and substantial funding that such programmes require, their contribution to the maintenance of species diversity can only be limited. Perhaps the greatest value of such programmes is in their symbolic and educational importance (Fig. 15.7).

Hotspots

Given the large numbers of species thought to be at risk of extinction it is unlikely that the conservation efforts outlined above can be successfully applied to save them all, if only for lack of sufficient funding. This being the case, some conservationists have used the fact that species diversity is geographically uneven to concentrate efforts on particularly endangered areas with high biodiversity. Myers *et al.* (2000) have identified 25 biodiversity 'hotspots', where exceptional concentrations of endemic species are undergoing exceptional loss of habitat, as priority areas for conservation efforts. As many as 44 per cent of all species of vascular plants and 35 per cent of all species in four vertebrate groups are confined to these 25 hotspots, which comprise just 1.4 per cent of the Earth's land surface. The risks of continued human impact in these hotspots is emphasised by the fact that human population growth rates in these areas are higher than the global average (Cincotta *et al.*, 2000).

For a long time it was thought that most marine species were less likely than terrestrial species to become extinct as a consequence of human activities because of their vast geographical ranges in the oceans. But similar hotspots have also been identified for coral reefs (Roberts *et al.*, 2002). The ten richest centres of coral reef endemism cover 15.8 per cent of the world's coral reefs (0.012 per cent of the oceans) but include about half of the restricted-range species. Eight of the ten coral reef hotspots are also adjacent to a terrestrial hotspot.

CONVENTION ON BIOLOGICAL DIVERSITY

Increasing realisation of the adverse human impact on biodiversity in recent decades has given rise to a number of international efforts to curb the worst effects. Some of the international agreements have focused on specific habitats, such as the Ramsar Convention for wetlands (see page 138), others on particular threats such as international trade (e.g. CITES). The culmination of these efforts was the Convention on Biological Diversity, which came into force in 1993. It is designed to ensure the conservation of biological resources and their sustainable use, and to promote a fair and equitable sharing of the benefits arising from genetic resources. In the latter respect, this means regulating biotechnology firms, access to and ownership of genetic material, and ensuring that compensation is paid to developing countries for extraction of their genetic materials. The Convention requires countries to develop and implement strategies for the sustainable use and protection of biodiversity. These strategies, now being developed in the countries that signed the Convention, will include the traditional approaches to maintaining biodiversity outlined above. However, scientists agree that it is impossible to shield all genes, species and ecosystems from human influence. The aim must be towards adapting all human activities so that they can take place in ways that minimise adverse impacts on the planet's biodiversity.

FURTHER READING

Groombridge B. and Jenkins M.D. 2002 *World atlas of biodiversity: Earth's living resources in the 21st century*. Berkeley, University of California Press. A comprehensive overview of the state of biodiversity.

McNeely, J. 1988 *Economics and biological diversity*. Gland, IUCN. This book highlights the pragmatic reasons for conserving biological resources and focuses on the economic incentives for so doing.

Norton, D.A. 1991 *Trilepidea adamsii*: an obituary for a species. *Conservation Biology* 5: 52–7. Case study of the multiple threats that eventually led to the extinction of the New Zealand mistletoe.

Pimm, S.L., Russell, G.R., Gittleman, J.L. and Brooks, T.M. 1995 The future of biodiversity. *Science* 269: 347–50. A discussion of the uncertainties in estimating the global total number of species and extinction rates.

Primack, R.B. 1993 *Essentials of conservation biology*. Sunderland, MA, Sinauer Associates. A comprehensive review of biodiversity, the threats it faces, its value and strategies for its conservation.

Spellerberg, I.F. (ed.) 1996 *Conservation biology*. London, Longman. A collection of articles on both the theoretical and practical sides of biodiversity conservation and management.

Wilson, E.O. 1992 *The diversity of life*. Cambridge, MA, Harvard University Press. This book covers the evolution and creation of biodiversity and human impact.

Wood, A., Stedman-Edwards, P. and Mang, J. (eds) 2000 *The root causes of biodiversity loss*. London, Earthscan. A close look, with good case studies, at the complex reasons behind biodiversity loss.

Websites

http://www.redlist.org/ IUCN's Red List of threatened species.

http://www.biodiv.org/ official site for the Convention on Biological Diversity.

http://www.conservation.org/ Conservation International is a major NGO that promotes biodiversity conservation in rain forests and other biomes.

http://www.unep-wcmc.org/ the World Conservation Monitoring Centre site has reports, databases and information on biodiversity, and hosts the Convention on Migratory Species homepage.

http://www.caff.is/ the Conservation of Arctic Flora and Fauna site with information on biodiversity and conservation.

http://www.cites.org/ official site for the Convention on International Trade in Endangered Species of Wild Fauna and Flora.

http://wcpa.iucn.org/ the World Commission on Protected Areas site.

POINTS FOR DISCUSSION

- Is the loss of biodiversity simply the price we pay for progress?
- Are there any circumstances in which you would feel the human-induced extinction of a species is morally justified?
- Since the destruction, modification and fragmentation of habitats is the primary threat to biodiversity, why don't we protect more habitats?
- Is it fair to say that we cannot preserve all biological diversity, so all we have to do is decide which bits are most important?

16

TRANSPORT

TOPICS COVERED

- Impacts on land
- Impacts on the biosphere
- Atmospheric impacts
- Noise
- Transport policy

Key words: biological invasions, smog, catalytic converter, *Nefos*

Movement is a basic element of the daily rhythm of life in every society, a fundamental human activity and need. Transport modes make such movement easier, whether it be a trip to the shop, or the movement of raw materials or goods from one place to another. From the earliest times, transport has been an integral part of the evolution of civilisations, and rapid transport and good communications using modern transport networks of road, rail and air are essential elements of all advanced economies, while global economic integration relies upon efficient maritime transport. Development of most parts of the less prosperous economies is likewise intimately linked to transport facilities. Transport is becoming faster and more efficient, and there is no better indicator of this trend than the relentless growth of car ownership and travel that has occurred in recent decades: in 1950 there was one car for every 46 people worldwide; by 1970, this number had risen to one per 18 people, and by the early 1990s there was one car per 12 people (Lowe, 1994).

Such mobility comes with an environmental price tag, and the effects of transport on the environment are wide-ranging. Some of the major issues are summarised in Table 16.1; they include air pollution and noise from road and air traffic, and marine pollution from shipping (see Chapter 6). The transport sector of economies is also a

TABLE 16.1 *Major environmental effects of transport*

LAND
- Use and wastage of land and its associated ecosystems
- Excavation and use of minerals (e.g. gravels) for road construction
- Generation of solid waste as vehicles are withdrawn from use

BIOSPHERIC
- Introduction of immigrant species to new environments
- Barriers to migration of species

ATMOSPHERIC
- Emissions of greenhouse gases, particulates, fuel and fuel additives
- Noise and vibrations

HYDROLOGICAL
- Contamination of surface and groundwater from surface runoff, and spillages of petrol and oil and transported substances
- Modifications of hydrological regimes during construction of roads, ports, canals and airports

Source: modified after OECD (1991a)

TABLE 16.2 *Differential impact of transport modes on a number of environmental variables*

Environmental variable	Unit	Air	Car	Car with three-way catalytic converter	Rail	Bus	Bicycle	Pedestrian
Land use	m^2/person	1.5	120	120	7	12	9	2
Primary energy use	g coal equivalent units/pkm	365	90	90	31	27	0	0
Carbon dioxide emissions	g/pkm	839.5	200	200	60	59	0	0
Nitrogen dioxide emissions	g/pkm	6.4	2.2	0.34	0.08	0.02	0	0
Hydrocarbons	g/pkm	1.4	1	0.15	0.02	0.08	0	0
Carbon monoxide emissions	g/pkm	8.1	8.7	1.3	0.05	0.15	0	0
Air pollution	Polluted air m^3/pkm	95 000	38 000	5900	1200	3300	0	0
Accident risks	Hours of life lost/1000 pkm	1.4	11.5	11.5	0.4	1	0.2	0.01

Note: pkm = passenger kilometre. One passenger travelling 1 km is 1 pkm, 10 passengers travelling 50 km equals 500 pkm. The same logic is used in freight transport where the unit of work done is the tonne kilometre (tkm)
Source: Teufel (1989)

major consumer of resources, including energy, minerals and land. The burdens imposed vary greatly between different types of transport, however, and some indication of the differential impacts of transport modes is given in Table 16.2.

IMPACTS ON LAND

All forms of transport consume land resources, whether for nodes such as airports, railway stations and ports, or for the route corridors of roads, railways or canals. New transport developments are often opposed by certain groups for these very reasons, because they alter the nature of the landscape in which they are placed. Concern for the impacts of new transport routes on the countryside is not a recent phenomenon. Canal and railway companies in eighteenth- and nineteenth-century Britain faced opposition from landowners, artists and writers on environmental grounds in the same way that motorway construction is opposed today by individuals and pressure groups intent on preserving the countryside. In many countries, however, desire by governments to improve communications often overrides other concerns, even to the detriment of areas supposedly protected from damage by developments. In the UK, the most recent road programme affects no less than 161 Sites of Special Scientific Interest (SSSIs), one national nature reserve, three local nature reserves, five county sites of special wildlife interest and two trust reserves (Whitelegg, 1992). New motorways are often built in response to congestion problems as more cars are used, but unfortunately for those who decry the loss of landscapes to motorways, more roads can also create more traffic, spurring the demand for still further expansion of road networks.

Transport developments not only despoil landscapes, but also consume resources in their construction and operation. There are numerous historical examples of the impacts new routes have upon local resources. The introduction of wood-burning steamships to West African rivers, for example, which occurred in the 1830s on the River Niger, heralded a significant pulse of deforestation along river banks. Similarly, when rail travel began with the line from Dakar to Saint Louis in Senegal in 1885, timber was felled on a large scale to clear the way for tracks, to make sleepers, to build bridges and to fuel steam engines. In many countries today, road building is the largest consumer of aggregates from quarries, which leave permanent scars on the landscape. About 125 000 t of crushed rock are needed to construct 1 km of motorway, for example.

Impacts upon the landscape can, in turn, present hazards to the transport routes themselves, as well as other land uses in the vicinity. In permafrost areas, the importance of maintaining the thermal regime of ground ice during and after construction of routeways was learnt the hard way during the construction of the Alaska Highway in the early 1940s. Even minor disturbances to the thermal equilibrium caused

TABLE 16.3 *Gradient and proportion of slopes (%) with major slope stability problems on the Kuala Lumpur–Ipoh highway, Malaysia*

Stability problem	Slope gradient		
	33–45 degrees	63 degrees	83 degrees
Slope failure	8	19	0
Slumps	21	31	54
Gullies	11	8	0

Source: modified after Bayfield *et al.* (1992: 80, Table 2)

Figure 16.1 Heavy industry on the Trans-Siberian railway just outside Irkutsk, Russia.

permafrost to thaw, presenting problems of heaving and subsidence due to frost action, and the creation of impassable mires. Soil erosion and problems of slope failure are other hazards the highway engineer has to confront. Areas receiving high rainfalls are among the most susceptible, and slopes that are lacking in vegetation are often the most likely to fail. The major stability problems faced on steep slopes along the Kuala Lumpur–Ipoh highway in peninsular Malaysia are shown in Table 16.3. Serious soil erosion problems can also occur when vehicles are used off roads. Such off-road driving was found by Jones *et al.* (1986) to be a prime cause of surface destabilisation and consequent soil erosion by wind in a Middle Eastern town, the blows of dust causing a considerable problem to the occupants of the town, particularly for the operation of machinery.

New transport routes also spawn environmental impacts indirectly through their raison d'être of improving access to places and resources, thereby encouraging industrial and other land uses. In Russia, for example, construction of the Trans-Siberian Railway in the late nineteenth century opened up the vast mineral resources of Siberia, resulting in numerous sites of environmental degradation (Fig. 16.1). Agriculturalists too often follow in the wake of new transport routes, to convert new land to food production. Cultivation only began in Australia, for example, with the arrival of European settlers, and in modern times it is generally agreed that agriculturalists are the most serious agent of tropical forest loss, their movement often closely following transport routes. The deforestation of Rondônia State in Brazil, following the arrival of the BR-364 highway in the 1970s is a well-known example (see Chapter 4).

IMPACTS ON THE BIOSPHERE

Transport both facilitates and hinders movements of flora and fauna, and opens up new routes for species dispersal. The movement of people has assisted numerous 'biological invasions' of great ecological significance, both deliberate and inadvertent (e.g. Drake *et al.*, 1989; Hengeveld, 1989), and competition from introduced species remains a major threat to the continued existence of native species on the global scale (see Chapter 15). Recognition of the great disturbances that can be caused by this form of biological pollution prompted the inclusion of a call to prevent the introduction of alien species in the Convention on Biological Diversity, but implementing such an objective will be no easy task.

History is punctuated with many examples of plants and animals deliberately transferred between regions for commercial purposes. Western Europe, for example, imported new crops such as potatoes, maize, tomatoes and tobacco from the Americas, and several of today's staple African food crops (e.g. maize and cassava) have been brought to the continent from Latin America by outsiders. The introduction of alien species can bring great benefits if managed properly (Fig. 16.2), but there are many cases where introduced species have become such successful colonisers that they have given cause for concern. The golden apple snail is one such example, introduced into Asia from South America in 1980 to be cultivated as a high-protein food source. It has dispersed into rice paddies, feeding on rice seedlings and causing significant crop damage.

Figure 16.2 A carefully managed facility outside Riyadh in Saudi Arabia where plants brought from desert areas all over the world are cultivated to ascertain their suitability for use on the Arabian peninsula.

One estimate of the damage caused in the Philippines put the cost in economic terms at up to US$1.2 billion in 1990 (Naylor, 1996). A successful attempt to reverse the rapid spread of an introduced species that had reached pest status can be seen in the case of the prickly pear in Australia (see page 224).

There are also numerous cases of unintentional dispersal of, and subsequent colonisation by, species that have 'hitched a ride' on various forms of transport. More than 120 aquatic species have been introduced into marine and estuarine systems and inland seas in this way, most as a result of canal building or through transport in ships' ballast water (Baltz, 1991). Following most of these introductions, native fish species disappear or are greatly reduced due to competition or predation.

A classic example is the invasion of the North American Great Lakes by two species from the Atlantic, the alewife and sea lamprey, following completion of the Erie Canal (now the New York State Canal) and the Welland Canal during the nineteenth century (Fig. 16.3). As a result, the once common Atlantic salmon, lake charr, lake trout and lake herring have been severely depleted by competition for food from the alewife and predation by the sea lamprey. Since several of these greatly reduced native species are of commercial importance, the Great Lakes Fishery Commission currently spends about US$12 million per year in lamprey control, mainly through the use of chemical lampricides in nursery streams. Ironically, in Europe the sea lamprey has declined over most of its range because of river pollution and obstructions, is now extinct in many large rivers where it was formerly abundant, and is currently regarded as a threatened species to be conserved (Maitland, 1991).

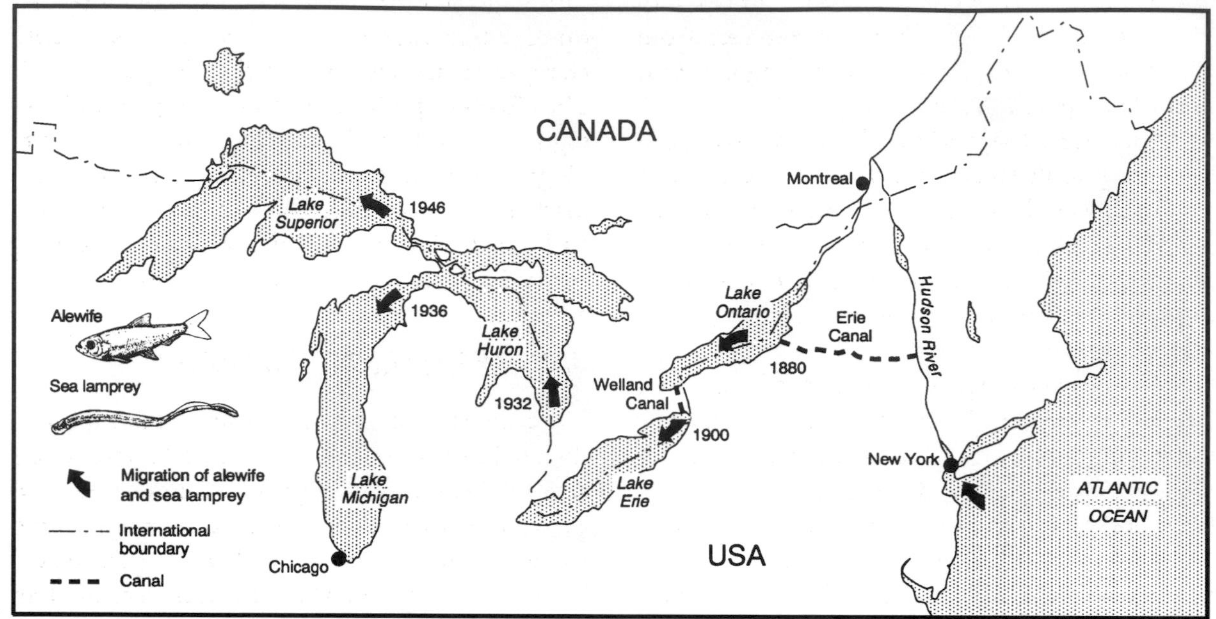

Figure 16.3 Migration of alewife and sea lamprey into the North American Great Lakes.

More recently, the Great Lakes have been invaded by another species, the zebra mussel: a small, striped native of the Caspian Sea, which probably arrived in the ballast tanks of a European tanker. Within two years of their first appearance in 1988, zebra mussel densities had reached 70 000 individuals per square metre in parts of Lake Erie, choking out native mussels in the process. There has been a subsequent spread southwards into the Detroit, Cumberland and Tennessee rivers with potential to cause enormous damage to fisheries, dams and boats, as well as devastation to the rich endemic aquatic communities (Stolzenburg, 1992). One reaction to the zebra mussel invasion was the introduction in 1990 of mandatory controls on the release of ballast water from commercial ships in the Great Lakes in an effort to prevent any future biological invasions.

Similar examples can be quoted for land transport. In Brazil, the house sparrow had spread 1500 km to Belém within a short time of the Belém–Brasilia highway being opened, and within six years had spread on to reach Marabá in 1973 (Goodland and Irwin, 1974). The sparrow is a carrier of the bugs that transmit Chagas' disease (American trypanosomiasis), which is debilitating and usually fatal.

The dispersion of diseases is a particularly notable aspect of the transport revolution, with rapid journey times sharply reducing the natural checks upon the spread of diseases. The point can be illustrated by the case of smallpox:

> *In Columbus's time, crossing the Atlantic was slow compared to the progression of smallpox. Since all carriers of the virus manifest symptoms of the disease, most of the infected travellers would have either become sick and died or recovered before reaching the New World. As a result, smallpox probably did not reach the Americas until several decades after Columbus's voyage.*
>
> (*Levins* et al., *1994: 57*)

Today the situation is very different; travel to almost anywhere in the world can be accomplished in a few days at most, less than the average incubation time of many disease pathogens. Modern rapid transportation can turn a local disease into a worldwide pandemic. A good example is the outbreak of cholera in South America in 1991, thought to have been introduced from China in the ballast water of a freighter that docked at a port in Peru (see page 365).

Many plant diseases have also been introduced to native populations through modern transport. An example is the devastating Dutch elm disease fungus, which killed many elm trees in Britain in the 1970s. The fungus is thought to have been brought to the island on timber imported through southern ports.

In contrast to these examples of transport facilitating dispersal, the construction of new routes can also have the opposite effect by introducing a barrier to the movement of organisms. Klein (1971) describes the impact upon wild reindeer of the construction of a railway line in northern Norway, which disrupted normal herd movements. The track effectively cut off access to one part of the reindeers' range, fragmenting their habitat and leaving a smaller area of their former range to be heavily grazed. In another example involving railways, rock tipping and sliding during construction of a line through Hell's Gate, a narrow gorge in the Fraser River canyon in western Canada, blocked the upriver spawning migration path of Pacific salmon in 1913 and 1914. Pacific salmon were a vitally important economic resource on the Canadian west coast at the time and their unusual biology was instrumental in causing a dramatic and long-lasting drop in the numbers spawning. The effect on the sockeye salmon catch from the Fraser River is estimated to be as high as US$2.2 billion over the period 1914–45 (Ellis, 1989).

The numerous forms of pollution derived from transport sources (see below) also have impacts on the surrounding biosphere. Noise has been shown to affect some wildlife. Disturbance from noise was thought to be a possible cause of the difficulty male willow warblers had in attracting and keeping a mate along roadsides in the Netherlands due to the distortion of their song by traffic (Reijnen and Foppen, 1994). Emissions of nitrogen oxides from passing motor vehicles on roads in southern England was found to have caused greater growth in heathland plant species up to 200 m away from the road and consequent changes in species composition (Angold, 1997). The effects of pollutants can be complex and synergistic. A study in Switzerland found that physiological stress in plants caused by road pollutants made them more susceptible to attack by pests, with greater infestations of aphids on trees along roadsides (Braun and Fluckiger, 1984).

ATMOSPHERIC IMPACTS

The atmospheric impacts of transport routes and their vehicles include microclimatological effects and the effects of pollution. Microclimate and vegetation-edge effects were studied by Young and Mitchell (1994) in a broadleaf forest in North Island New Zealand. They found that more than 100 years after road construction microclimatic edge effects were discernible about 50 m into the forests, influencing processes such as germination and the early establishment of young trees. However, the pollution problems associated with transport are better known. Transport routes act as corridors for the dispersion of pollutants from vehicles with numerous consequent effects. These pollutants include noise (see below), light, particulate matter such as sand and dust, and a number of gases. Transport is a major source of air pollution due to its heavy dependence upon the combustion of fossil fuels, either in vehicles or at power stations. The major pollutants involved are:

- carbon dioxide
- carbon monoxide
- nitrogen oxides
- hydrocarbons
- sulphur oxides
- lead
- suspended particulate matter.

These pollutants have a wide variety of environmental impacts, being detrimental to human health (see Chapter 10) and a significant disruption to flora and fauna (see above). In the wider perspective, carbon dioxide is a greenhouse gas, and nitrogen and sulphur oxides contribute to acid rain.

Pollution emissions per passenger kilometre are greatest for air travel (Table 16.2), and since aircraft emissions are largely released into the sensitive upper atmosphere, where gases are longer-lived than at ground level, they give rise to particular concern (Archer, 1993). The emission of gases and particles from aircraft directly into the upper troposphere and lower stratosphere have an impact on atmospheric composition. These gases and particles alter the concentration of atmospheric greenhouse gases, including carbon dioxide (CO_2), ozone (O_3), and methane (CH_4); trigger formation of condensation trails (contrails); and may increase cirrus cloudiness, all of which contribute to climate change (Penner *et al.*, 1999).

Emissions of nitrogen oxides and water vapour during cruise mode are perhaps the most critical in this respect. In the troposphere, nitrogen oxides contribute to ozone formation, in turn contributing to smogs at ground level, while as a greenhouse gas, nitrogen oxides emitted directly to the troposphere may remain resident there as much as a hundred times longer than those released from terrestrial sources. Although estimates are difficult, some suggest that up to 30 per cent of aircraft emissions occur while cruising in the stratosphere, and in this region nitrogen oxides deplete ozone at a level that some researchers regard to be equivalent to the depletion caused by CFCs. Sulphur dioxide emitted by air traffic may also contribute to stratospheric ozone destruction.

Water vapour released during stratospheric flight is another area of some concern. Since the natural water vapour content of the stratosphere is low, the effect of additions could be marked, leading to increased frequency of cirrus clouds, with a possible increase in atmospheric temperature. Long residence times, thin air and low temperatures mean that the global warming effect of water vapour emissions at flight altitudes could be up to 200 times more effective than a CO_2 molecule (Grasl, 1990). Since the total aviation fuel use from passenger transport, freight and military uses is projected to increase by 3 per cent per year between 1990 and 2015 (Penner *et al.*, 1999), these environmental impacts will be the subject of increasing concern in the years to come.

Despite the proportionally large impact of air travel, the heavy reliance on motor vehicles in most countries makes road transport the largest polluting source in the transport industry. In the UK, for example, road transport accounted for 93 per cent of passenger travel and 81 per cent of freight moved in 1990 (DoE, 1992). Concern over the environmental effects of motor vehicle pollution has prompted some concerted action to reduce certain pollutants. Lead is a particular example, due to its association with reduced mental development in infants and children, inhibition of haemoglobin synthesis in red blood cells in bone marrow, and impairment of liver and kidney function. Indeed, lead is toxic even at very low levels: no threshold blood concentration has yet been identified below which no adverse health effects occur (Schwartz, 1994). Besides its direct health risk through inhalation, lead can also enter the food chain via accumulation in soil and contamination of drinking water.

Lead's use in petrol as an anti-knock agent has been reduced considerably in many countries in the past 20 years or so, with consequent reductions in concentrations in human blood. The reduced lead content of petrol was also partly a response to the realisation that pollution by this highly toxic metal had dramatically increased at great distances from areas where motor vehicles are used. The concentration of lead in Greenland ice and snow increased by about 200 times between 5500 years BP and the mid-1960s, but has decreased significantly since then (Boutron *et al.*, 1991). In many countries, lead-free petrol has been widely introduced, with its adoption by motorists encouraged by lower taxes, although only a handful of countries have completely phased out the use of leaded petrol. The lead content in petrol in 20 megacities of the world is shown in Table 16.4.

TABLE 16.4 *Lead content of petrol used in 20 megacities*

City	Lead content of petrol (g/l)	Comments
Bangkok	0.15	
Beijing	0.4–0.8	80 per cent unleaded
Bombay	0.15	
Buenos Aires	0.6–1.0	
Cairo	0.8	
Calcutta	0.1	
Delhi	0.18	
Jakarta	0.6–0.73	
Karachi	1.5–2.0	
London	0.15	>33 per cent unleaded
Los Angeles	0.026	>95 per cent unleaded
Manila	1.16	
Mexico City	0.54	
Moscow	0	No leaded fuel sold
New York	0.026	95 per cent unleaded
Rio de Janeiro	0.45	Ethanol and gasohol used
São Paulo	0.45	40 per cent ethanol, 60 per cent petrol/gasohol
Seoul	0.15	
Shanghai	0.4	
Tokyo	0.15	>95 per cent unleaded

Source: UNEP/WHO (1992: 41, Table 4.2)

Karachi remains one of the worst cities for lead pollution due to Pakistan's high lead concentrations in petrol and the high concentration of traffic in the city. The number of registered vehicles more than doubled between 1980 and 1989 from 300 000 to about 650 000 at an annual growth rate of 12.5 per cent (2.5 times higher than the population growth rate). Vehicle numbers were projected to continue climbing and to reach 1.1 million by 2000 (Beg, 1990). Mean concentrations of inorganic lead in Karachi were about 1–3 g/m^3 with maximum values of 7–9 g/m^3 in areas of heavy traffic. These values are far above the WHO annual mean guideline range of 0.5–1.0 g/m^3. Beg's study showed that more than four-fifths of the lead is in respirable particles, which underlines the threat to human health.

At the continental scale, Africa has the highest average levels of lead in petrol (0.5–0.8 g/l) and although data are sparse, childhood lead poisoning is thought to be a widespread and growing urban health problem throughout the continent (Nriagu *et al.*, 1996). Although national vehicle fleets are smaller than for many more developed countries, the vehicles tend to be older, in worse repair and more polluting on average. Certain sectors of the population are most at risk, principally those with a roadside, open-air lifestyle, but vehicle emissions are not the only source of lead. Other sources include emissions from both large-scale and cottage industries, battery casings and lead paint.

The dramatic effect on air quality of lowering the lead content of petrol is indicated in Los Angeles, where in the mid-1970s atmospheric lead concentrations exceeded Federal standards virtually every day of the year but had been reduced to zero within ten years (UNEP/WHO, 1992). The importance of legislation to reduce vehicle pollution is paramount in Los Angeles, where the public transport network is poor, so that residents have to rely on personal vehicles for virtually all transportation. This situation makes the Los Angeles Basin the area with the worst air quality in the USA. Ten million vehicles in the urban area of 14 million people (in 1990) represent probably the greatest number of vehicles per person in the world. However, Los Angeles' long history of pollution problems has led to intensive efforts to combat the problem. In the mid-1950s, for example, California established the first state agency to control motor vehicle emissions in the USA. An Air Quality Management Plan for southern California, approved in 1989 and updated regularly, aimed to reduce emissions of NO_x and VOCs by 80 per cent in 20 years, sulphur oxides by 62 per cent and particulates by 20 per cent.

Part of the plan that targets transport is the promotion of ride-sharing, use of alternative forms of transportation (e.g. buses, trains), and the use of alternative-fuel vehicles or other clean-engine technologies. Introduction of the use of emission-free vehicles in southern California has been subject to delays (see below), but improvements in air quality have been achieved. The widespread use of lead-free petrol has brought atmospheric lead levels under control in recent years, and ozone concentrations have also fallen (Fig. 16.4), although the number of days per year when the state health standard for ozone (which is more stringent than the federal government's standards) was exceeded was still more than 120 in the early 2000s. Motor vehicles in Los Angeles have been running on a specially reformulated petrol, designed to reduce the emission of VOCs, since the beginning of the 1990s. Reformulated petrol was introduced to southern California

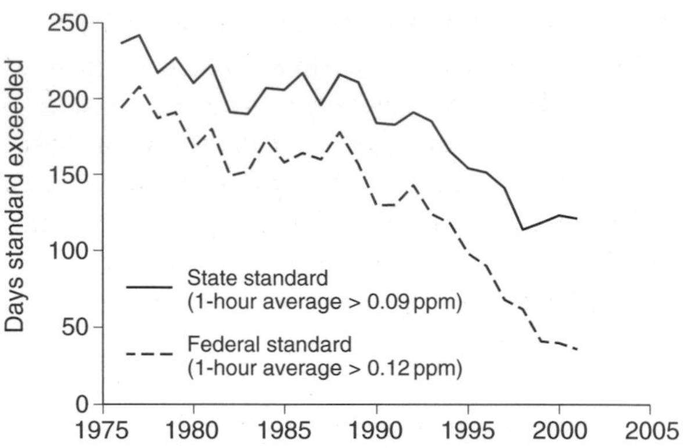

Figure 16.4 Number of days when atmospheric levels of ozone exceeded state and Federal standards in southern California, USA, 1976–2001 (data from http://www.aqmd.gov/smog/, accessed April 2002).

under the 1990 Clean Air Act and it is part of the reason for the reduction in ozone pollution over Los Angeles.

Much of the success in reducing pollutants from motor vehicles in Los Angeles and elsewhere in the western world is attributable to the increasingly widespread use of catalytic converters, which were first developed in the USA in the 1950s and 1960s, and fitted to US cars from the mid-1970s. The catalytic converter is a device fitted to vehicles, which removes certain pollutants by a chemical reaction. The pollutants targeted in vehicle exhausts are unburnt hydrocarbons, carbon monoxide and nitrogen oxides, which are converted by platinum, palladium and rhodium into carbon dioxide, water vapour and nitrogen. The ability of the three-way catalytic converter to reduce by more than 80 per cent the output from the internal combustion engine of these three pollutants, has led to its widespread adoption. From 1993, for example, all new cars sold in the European Union have had to be fitted with catalytic converters, and since these systems are only effective with lead-free petrol, their introduction will also continue to reduce the amount of lead pollution from motor vehicles.

However, even with catalytic converters fitted, cars still produce atmospheric pollution. Carbon dioxide, a greenhouse gas, is the most significant pollutant, produced in large quantities because cars fitted with catalytic converters tend to be less fuel efficient than those without. Other pollutants, which are also still produced, are emitted before the engine has warmed up, a particular problem where short journeys are concerned, as with a high proportion of city car travel. Table 16.5 indicates the pollution produced by a car during its ten-year lifespan, assuming it is driven 13 000 km per year, consuming ten litres of petrol per 100 km. The high carbon dioxide output is greater still if the manufacturing process and eventual scrapping of the car is included, bringing the total carbon dioxide produced to 59.7 t (Whitelegg, 1994).

The use of catalytic converters, and other emission-reducing technologies, such as lean-burn engines, electronic fuel injection, computer-controlled spark ignition and exhaust recirculation systems, can only go so far in reducing polluting emissions. A technical fix to the vehicle engine system or an additive to fuel cannot solve the global emission problem. Indeed, recent experience indicates that these efforts do not keep pace with increasing traffic. It seems that we may have to abandon crude oil as a primary energy for transportation within three or four decades, due to dwindling supplies (Svidén, 1993). Even if this estimate is pessimistic, the end of crude oil is undoubtedly a foreseeable event, and thus provides humankind with the opportunity 'to redesign the complete energy supply and conversion chain from mining and refining to the combustion in the vehicle engines in such a way that it can meet the global ecological criteria' (Svidén, 1993: 145).

TABLE 16.5 *Estimated emissions of air pollutants over ten years from a car fitted with a three-way catalytic converter*

Pollutant	Total emissions
Carbon dioxide	44.3 t
Carbon monoxide	325 kg
Nitrogen dioxide	46.8 kg
Hydrocarbons	36 kg
Sulphur dioxide	4.8 kg

Source: after Whitelegg (1994)

Some examples of switches to less-polluting fuels have already occurred. Diesel has some advantages over petrol. Diesel engines use around 30 per cent less fuel, hence they emit less carbon dioxide; they also emit fewer hydrocarbons and fewer nitrous oxides. Conversely, diesel engines produce larger quantities of sulphur dioxide, particulate matter, noise and smoke. Other alternative fuels are also in use in various parts of the world. In Tokyo, for example, taxis are obliged to use liquefied petroleum gas (LPG) rather than petrol because LPG, like all natural gas derivatives, contains less carbon than gasoline and hence produces less carbon dioxide, carbon monoxide and hydrocarbons, as well as less nitrous oxides, when burnt.

Other alternatives already in use include methanol and ethanol, which when burnt produce lower quantities of carbon dioxide and nitrous oxides than petrol. Methanol and ethanol also have the advantage that they can be produced renewably from biomass. At one time in the 1980s, virtually all cars in Brazil were run on either pure ethanol or a mix of ethanol and petrol, the ethanol being fermented from sugar-cane juice, although, more recently, lower international oil prices and supply problems have stalled the country's ethanol programme. A significant problem with these fuels derived from biomass is the large areas necessary for cultivation. For example, replacement of all the petrol used by cars and taxis in the UK with ethanol derived from biomass would require an area three times the size of the UK to grow the biomass (Transnet, 1990).

Overall, alternative fuels, such as methanol, diesel and gas have the potential as short-term solutions to the growing problem of pollution from transport, but unless cleaner ways of creating fuels are developed, the growing flow of traffic will wipe out these gains. One of the fuels that

Figure 16.5 Severe traffic congestion in Kampala, Uganda.

is a strong candidate for a cleaner replacement to those derived from crude oil is hydrogen, sometimes referred to as the ultimate clean combustible fuel since its combustion emits only water vapour and small quantities of nitrogen oxides. Electrically driven vehicles also present some hope for future development when the electricity is generated from renewables, although at present their range is limited by the need for frequent recharging, a function of battery storage capacity. Vehicle manufacturers are actively involved in developing such alternatives, in some cases motivated by legislation. A leader in this respect is the US state of California, where a statewide mandate was introduced in 1990 requiring that 2 per cent of all passenger vehicles sold in the state by 1998 be emission free, increasing to 5 per cent in 2001 and 10 per cent in 2003. However, in 1995 the mandate was eased, and the requirements for zero-emission vehicles were suspended until 2003 because the technology has not been developed fast enough. Critics suggest that the relaxation was motivated in part by political pressure from the auto and oil industries.

Nevertheless, such technological developments to reduce pollution would still leave many of the undesirable aspects of society's heavy dependence on cars, such as congestion and sprawl, unresolved (Fig. 16.5). Hence, a wider perspective on the problem from the policy viewpoint is needed (see below).

NOISE

Road, air and rail transport is a major source of noise in both urban and rural areas. It has been estimated that about 17 per cent of the population of all OECD countries is exposed to transport noise, which is regarded by most authorities to be unacceptably high, and the proportions of urban populations exposed to unacceptably high transport noise is often 50 per cent, although overall transport noise levels have tended to stabilise since the late 1970s (Farrington, 1992). Transport noise affects people to varying degrees, ranging from mild annoyance and minor interruptions to everyday activities, to mental and physical damage, as well as an adverse effect on property prices in some cases. As such, noise is a significant aspect of the transport sector's environmental impact, and the clear importance of major transport corridors as sources of noise is shown, in Fig. 16.6, in the urban environment for the Chinese city of Nanjing. On the national level, one estimate of the economic cost of traffic noise, measured by the depreciation of house prices in France, put the cost to the nation at US$0.27–0.45 billion a year (Pearce *et al.*, 1984).

Although it is difficult to arrive at a definition of what level of sound constitutes a problem or a nuisance, and hence when a certain sound becomes unacceptable, guidelines are available. Transport noise is usually measured using the logarithmic dB(A) scale, a variation on the decibel

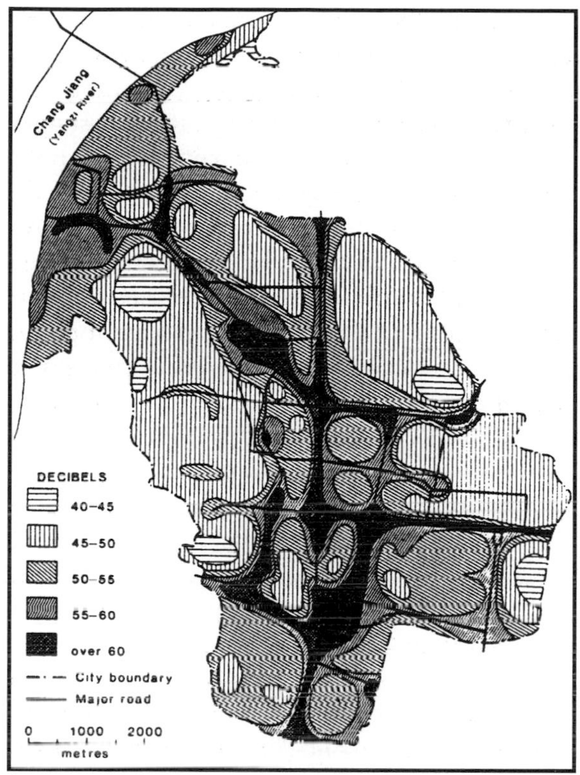

Figure 16.6 Spatial distribution of daytime environmental noise levels in Nanjing, China (Noble, 1980).

scale that is weighted towards the frequencies most affecting the human ear. At 80 dB(A), people standing next to each other would need to shout to be heard, and at 90 dB(A) – a typical level for a heavy lorry passing at 7 m – temporary loss in acuteness of hearing would be felt for a few minutes. The 120 dB(A) noise typically heard at 200 m from a jet aircraft on take-off is approaching the level at which physical pain, and ultimately deafness, is experienced. Annoying or unacceptable noise is often taken to be 65 dB(A), although a continuous noise tends to be regarded as less of a nuisance than individual events, and the time of day when a noise is heard is also a factor – night-time noise being less acceptable than that during daylight hours.

Mitigation of noise problems can be approached on four fronts (Nelson, 1987):

1. reduction of noise at source – through vehicle design, traffic management, tunnelling and noise-abatement procedures for aircraft
2. measures to control noise along its transmission path – mainly by barriers such as fences and embankments, and the use of buildings as noise barriers
3. measures to protect the observer from noise at the point of hearing – such as through building design, ensuring smaller windows on the noisiest façades and incorporating double glazing and acoustic insulation
4. land use planning and zoning – effective on the larger scale by zoning noisy traffic paths away from residential and working areas.

The best results for noise abatement are derived from a combination of all these approaches, and the success of such strategies in reducing the numbers of people affected is well illustrated for US airports in Table 16.6.

TABLE 16.6 *Numbers of people exposed to daytime noise of 62 dB(A) around US airports*

Airport	1970	1980	1985	2000 (estimate)
New York				
JFK	–	–	606 000	571 000
La Guardia	–	–	462 000	283 000
All airports	6.0 million	4.2 million	3.2 million	2.4 million

Source: modified after OECD (1993)

TRANSPORT POLICY

The future environmental impacts of transport, as with many other environmental issues, will depend on policy-makers. Passenger travel by public transport and bicycles, all more environmentally sound methods of travel than the car (Table 16.2), depend upon appropriate emphasis being given to such modes. In Britain, the 1990–91 National Transport Survey showed a 22 per cent increase in the average distance people travelled per week since 1985–86, with little change in the number of journeys made. Most of these journeys were made by car. Such a trend reflects policy and planning decisions that feed back on society and change lifestyles: decisions that reinforce dependence upon cars increase car use. A similar attitude has led to the great increase in road haulage of freight at the expense of rail, for example.

Technological innovations can only go so far in reducing the environmentally damaging aspects of transport, and will need to be complemented with better planning that can reduce the need for travel that is costly in environmental terms. More practical urban and suburban land use patterns can help to reduce car trips, both by shortening them and by reducing the need to travel by car. Public transport, cycling and walking need to be encouraged and provided for as alternatives (Figs 16.7 and 16.8).

Figure 16.7 The use of electrified trams like this one in Riga, Latvia, reduces the number of car journeys, relieving congestion and cutting down on urban pollution.

The current obsession with the car, which characterises many governments' attitude to transport, might change if the true costs of the car and other transport modes is calculated. Lowe (1994: 95) suggests that 'the single most important goal for policy reform may be to ensure that a more accurate tally is used to guide all public and private decisions in the future'. Important, in this respect, is the need to clarify and quantify the true costs of an increasingly mobile society. External costs, which a particular mode of transport imposes on society as a whole, are difficult to quantify. They include traffic congestion, road accidents, pollution, dependence on imported oil, solid waste, loss of cropland and natural habitats, and climatic change.

Even in cases where the costs of personal mobility are well known to be high, changing people's behaviour is no simple matter. The experience of Athens, capital city of Greece, illustrates this point well. The city's air quality has deteriorated significantly since the 1970s to become one of the worst in Europe, and road traffic is the principal source of pollutants that include ozone and nitrogen dioxide from petrol-driven vehicles, and black smoke particles emitted by diesel engines. Other common atmospheric pollutants produced by burning fossil fuels are also present, including carbon monoxide and sulphur dioxide. Athenians call their pollution the *Nefos* (literally meaning cloud); *Nefos* pollutants have negative effects on pubic health (Touloumi *et al.*, 1994), damage the city's numerous ancient monuments (Kirkitsos and Sikiotis, 1996), and probably retard the growth of fir trees on nearby Mount Parnis (Heliotis *et al.*, 1988).

Motor vehicles have been one of the main targets of local authority action to combat the *Nefos*. In 1982, a scheme designed to reduce the number of private cars in the city centre was introduced: vehicles with licence plates ending in an even number were only allowed access on even-numbered dates, while odd-numbered dates were

Figure 16.8 More than 6 million bicycles are used in the Chinese capital of Beijing, a city of 11 million people.

reserved for cars with odd-numbered licence plates. Many Athenians reacted by buying a second vehicle, with a licence plate allowing them in on the days when their first car was banned. These second cars tended to be older and more polluting, so air quality actually got worse. A year later, the authorities modified the scheme, and warned that the regulations would be changed again if people defied the spirit of the programme. In 1985, the impact on vehicle numbers was assessed. Although there were 22 per cent fewer private cars in the city centre, atmospheric pollution showed little change. It transpired that many people were travelling into town by taxi instead. Their numbers had increased by 26 per cent. It was decided to include taxis in the ban but the taxi drivers went on strike and deliberately staged traffic jams until their exemption was restored (Elsom, 1996).

In April 1995, all motor vehicles were banned from 2.5 km^2 of the city centre for a three-month experimental period, but the experiment was not made permanent. During smog alerts, various levels of restrictions on vehicle access are still implemented, while factories are required to reduce or stop production. Vehicle emission tests have also been introduced, but with limited success.

Some improvements in air quality in Athens have been observed thanks to these and other measures. There has been a definite decline in carbon monoxide concentrations since 1990, probably a result of the increasing number of motor vehicles fitted with three-way catalytic converters (Viras *et al.*, 1996). But ozone concentrations have shown only very gradual change since peaking at the end of the 1980s (Fig. 16.9). After 1987, the monthly mean ozone concentration in the northern suburb of Liosia has repeatedly exceeded

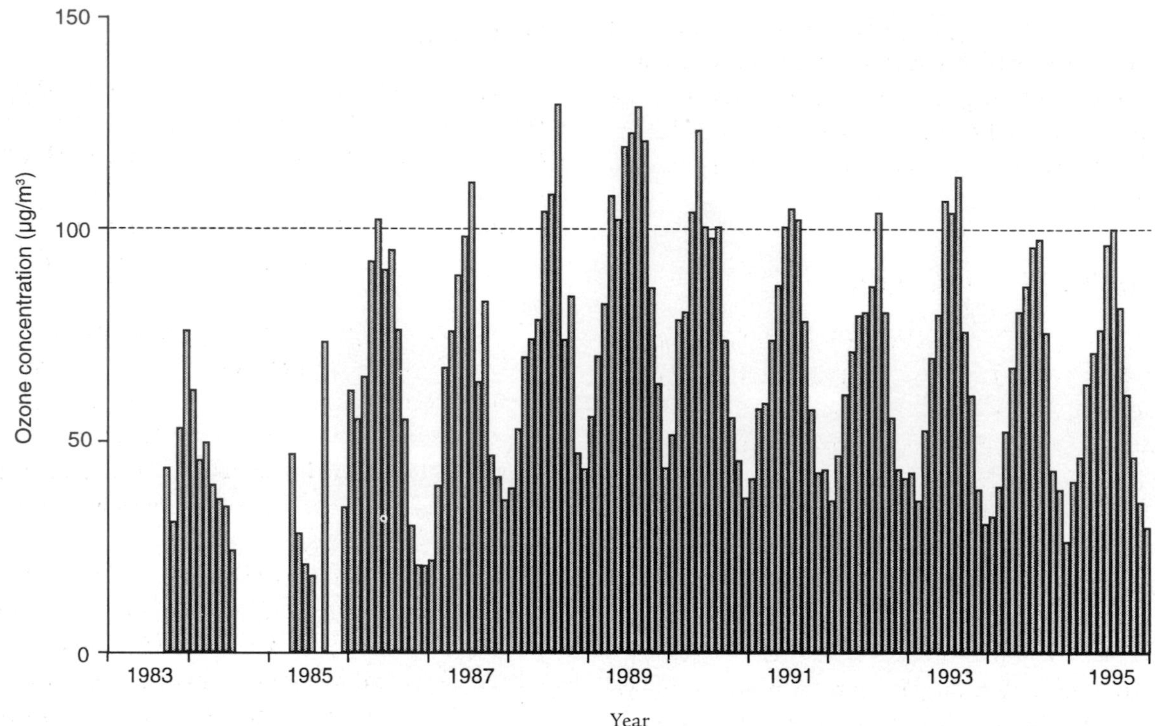

Figure 16.9 Monthly mean ozone concentrations at Liosia, northern Athens (EEA, 1997).

110 μg/m^3, the health protection threshold defined by the EU Ozone Directive, computed as a mean over eight hours. The moving eight-hour average of ozone concentration at Liosia exceeded 110 μg/m^3 on 140 days in 1988 (EEA, 1997).

Many observers believe that most of the environmental issues that make transport unsustainable can only be properly addressed by focusing on ways in which behaviour can be changed, to reduce the demand for the activity that produces the problem in the first place (Whitelegg, 1993). As the Athens example shows, such changes can be very difficult to make, but moderation of the environmental impacts of transport will come about ultimately from

TABLE 16.7 *Suggested complementary policy solutions to transport problems for different levels of government in Europe*

Government level	Policies
Local	Improve land-use planning and traffic management
Central	Change pricing and tax policies to assist development of alternative fuels, to improve transport efficiency, to encourage shifts to 'greener' modes of travel, and to reduce unnecessary travel
European	Impose 'environmentally optimal' community-wide policies involving fuel and vehicle tax harmonisation, vehicle speed limits and heavy vehicle weights

Source: after Transnet (1990)

government, through regulation, legislation, and other ways in which transport modes are encouraged and discouraged. Some suggestions as to the types of approaches that can be taken at different levels of government to ameliorate the problems are suggested in Table 16.7.

FURTHER READING

Bannister, D. and Button, K. 1993 *Transport, the environment and sustainable development*. London, E & FN Spon. A collection of papers on transport policy in various countries, and views of the issue from a number of different perspectives.

Carpenter, T.G. 1994 *The environmental impact of railways*. Chichester, Wiley. This book details impacts on people, such as noise and pollution, and resources.

Drake, J.A., Mooney, H.A., di Castri, F., Groves, R.H., Kruger, F.J. and Williamson, M. (eds) 1989 *Biological invasions: a global perspective*. SCOPE Report 37, Chichester, Wiley. A comprehensive review of the invasion of natural ecosystems by animals, plants and micro-organisms, the environmental problems caused and attempts to control the process.

Greene, D.L. and Wegener, M. 1997 Sustainable transport. *Journal of Transport Geography* 5: 177–90. A useful overview of the environmental impacts of transport, and suggestions for research and policy.

Renner, M. 1988 *Rethinking the role of the automobile*. Worldwatch Paper 84. Worldwatch Institute, Washington, DC. A discussion of the environmental impacts of motor vehicles and the need to reduce our reliance upon them.

Spellerberg, I.F. 1998 Ecological effects of roads and traffic: a literature review. *Global Ecology and Biogeography* 7: 317–33. A good summary of studies on impacts and methods for mitigation.

Whitelegg, J. 1993 *Transport for a sustainable future: the case for Europe*. London, Belhaven. An overview of transport's pollution problems and costs, which emphasises the need for fundamental changes in our attitude to personal mobility before transport in Europe becomes sustainable.

Websites

http://www.cities21.com/ the International Council for Local Environmental Initiatives site has information on programmes for sustainable urban transport.

http://www.dot.gov/ the US Department of Transportation covers issues related to all forms of transport.

http://www.geocities.com/sustrannet/ the SUSTRAN network promotes and popularises people-centred, equitable and sustainable transport with a focus on Asia and the Pacific.

http://invasions.bio.utk.edu/ the Institute for Biological Invasions site contains a wide range of information.

http://www.vantaa.fi/mipim/aircity.htm details of the international Helsinki-Vantaa Airport with links to its environmental implications.

http://www.aqmd.gov/ southern California's Air Quality Management Plan site has news, pollution data and information on transport programmes.

http://www.sima.com.mx/ information on air quality and transport programmes in Mexico City.

http://greenercars.com/ a guide to 'green' cars and trucks.

POINTS FOR DISCUSSION

- Is it possible to prevent biological invasions? Would it be desirable?
- Is air pollution just an unfortunate side-effect of our need to move?
- If you were asked to propose measures to make transport more sustainable in your area, what would you suggest?
- Assess the ways in which transport is related to other environmental issues covered in this book.

17

WASTE MANAGEMENT

TOPICS COVERED
Types of waste
Disposal of waste
Reuse, recovery, recycling and prevention

Key words: landfill, prior informed consent, polluter pays principle, Superfund programme, lifecycle analysis, cleaner production, closed-loop processing cycle

Wastes are produced by all living things – excreted by an organism or thrown away by society because they are no longer useful and if kept may be detrimental. But the make-up of waste means that once discarded by one body its constituents may become useful to another. In the natural world, a waste produced by an animal, for example, is just one stage in the continual cycle of matter and energy that characterises the workings of the planet: an animal that urinates is disposing of waste products its body does not need, but the water and nutrients can become resources for other organisms. Similarly, sewage from a town can be used by bacteria that break it down in a river, for example, or an old shirt discarded by one person might be worn by another, often poorer, individual. The term 'waste' is therefore a label determined by ecology, economics and/or culture.

Much of the waste produced by human society consists of natural material and energy, although we have also created products not found in the natural world, such as CFCs and plastics, which can become wastes. The issues surrounding wastes stem from problems of disposal. People's use of resources, and hence production of wastes, has accelerated in the period since the Industrial Revolution, and this fact, combined with the growing number of people on the planet, has created increasing volumes of waste that need to be disposed of. This disposal requires careful management since too much waste disposed of in a certain place at a particular time can contaminate or pollute the environment, creating a hazard to the health, safety or welfare of living things. Examples of waste products and their impacts on the environment are found throughout this book. They include wastes produced by agriculture (Chapter 13), energy production (Chapter 18) and mining (Chapter 19), and pollution of the atmosphere (Chapters 11, 12 and 16), aquatic environments (Chapters 6, 7 and 8) and urban environments (Chapter 10).

TYPES OF WASTE

Waste can take many different forms: solid, liquid, gas, or energy in the form of heat or noise. The problems associated with its disposal can stem from a range of other properties, including its chemical make up, whether it is organic or inorganic, the length of time taken for it to be broken down into less hazardous constituents, and its reactivity with other substances – so-called synergy. The sources of waste are also diverse, stemming from virtually every human activity, including:

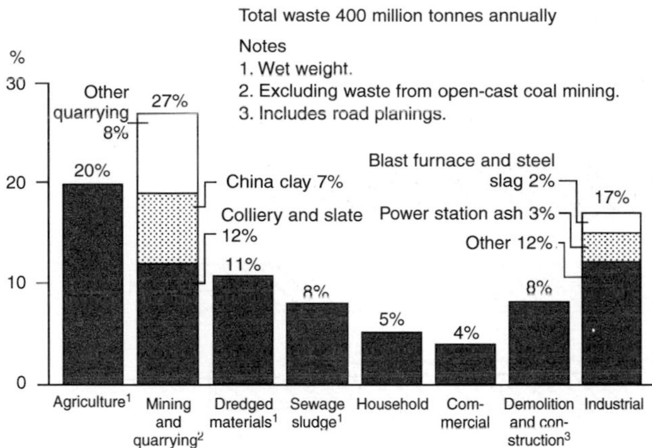

Figure 17.1 Waste arisings by sector in the UK (after DoE, 1992).

- mining and construction
- fuel combustion
- industrial processes
- domestic and institutional activities
- agriculture
- military activities.

Collection of data on waste has had a relatively low priority in many countries, and the methods for their compilation vary widely. Statistics that are available are often given as weights, but the large differences between quantity and quality mean that these figures can only give a general indication of the nature of the waste disposal problem. The waste arisings for the UK, which totalled about 400 million tonnes annually in the early 1990s, are shown by source in Fig. 17.1. The general absence of annually updated data makes the recognition of trends difficult.

Since some wastes are inherently more dangerous than others, categories such as 'special', 'controlled' and 'hazardous' are often identified and such wastes dealt with in a more careful manner. In the UK, where the disposal of different types of waste is regulated by different laws, so-called special wastes, for example, are those deemed dangerous to life and are subject to regulations designed to track their movement from production to safe disposal, or from 'cradle to grave'. Although there is no universal agreement over what constitutes hazardous wastes, they include substances that are toxic to humans, plants or animals, are flammable, corrosive, or explosive, or have high chemical reactivity. Such hazardous substances include acids and alkalis, heavy metals, oils, solvents, pesticides, PCBs (polychlorinated biphenyls) and various hospital wastes.

DISPOSAL OF WASTE

All wastes are disposed of into the environment, but some enter the environment in a more controlled manner than others. Some wastes are emitted directly from the source without treatment, others are collected and sometimes treated

TABLE 17.1 *Treatment and disposal technologies for hazardous wastes*

General approach	Specific technology
Physical/chemical	Neutralisation Precipitation/separation Detoxification (chemical)
Biological	Aerobic reactor Anaerobic reactor Soil culture
Incineration	High temperature Medium temperature Co-incineration
Immobilisation	Chemical fixation Encapsulation Stabilisation Solidification
Dumping	Landfill Deep underground Marine
Recycling	Gravity separation Filtration Distillation Solvent extraction Chemical regeneration

Source: after Tolba and El-Kholy (1992)

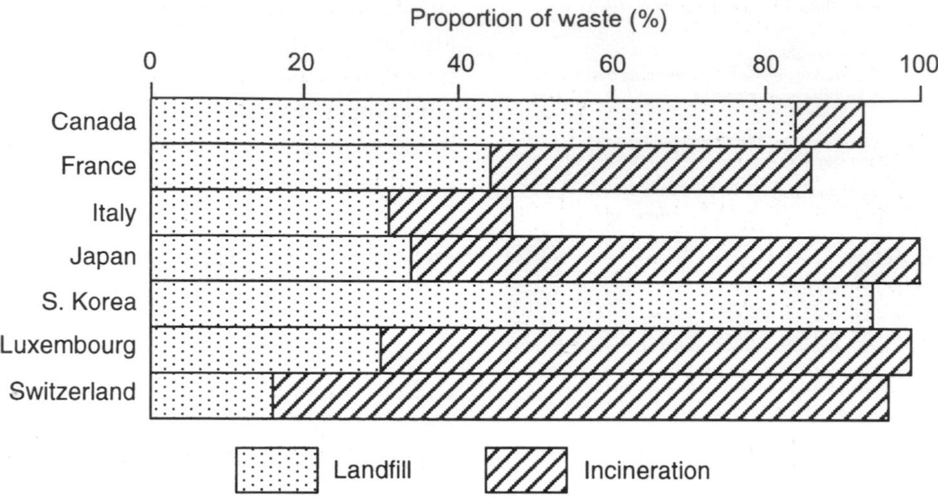

Figure 17.2 Main forms of treatment and disposal of municipal wastes in selected countries (from data in UNEP, 1993).

before disposal. Wastes produced from the combustion of fuel by motor vehicles, for example, are emitted directly into the atmosphere, while domestic sewage wastes are often collected by municipal authorities and disposed into specific locations such as a river or ocean. Three of the compartments into which wastes are emitted – air, rivers and oceans – are usually publicly owned and this common ownership has facilitated unregulated emissions of wastes. In many countries, however, the degradation of these common property resources by wastes has spawned numerous controls on their emission.

This chapter is concerned with wastes that are actively managed and where a wide range of management options is available, as Table 17.1 indicates for hazardous wastes. The following sections will look more closely at two of the most commonly used options: landfill and incineration.

Landfill

Dumping of waste is by far the most commonly used method of waste disposal in most countries. For example, much of the mining and quarrying wastes arising in the UK, shown in Fig. 17.1, remains as tailings and spoil heaps, while most of the dredged material is dumped at sea. A significant quantity of other wastes is buried in landfill sites, and in many countries it is the most commonly used management method for disposing of municipal wastes (Fig. 17.2) as well as being a frequently used method for hazardous waste disposal. Of the estimated 24 million tonnes of hazardous wastes generated in the OECD member countries of Europe, 70–75 per cent is deposited in landfill (Yakowitz, 1993).

If properly designed and managed, landfill sites may do little harm to the environment and can eventually provide a surface for such land uses as playing fields or reforestation, though not usually for heavy structures such as housing. The two main hazards associated with landfills are leakage of toxic leachates, which can contaminate surface and groundwater, and leaving refuse open to the air, which can allow infestation by rats or fermentation bacteria, which can generate methane and cause a fire hazard. Each day's addition must be covered with soil to prevent infestation and fermentation, although in some cases, methane generation is actively encouraged and collected for use as a fuel (Lisk, 1991). Four typical landfill designs are shown in Fig. 17.3. If the site is located on impermeable strata, problems from leaching to groundwater are not faced (Fig. 17.3a), but when sited on an aquifer, various

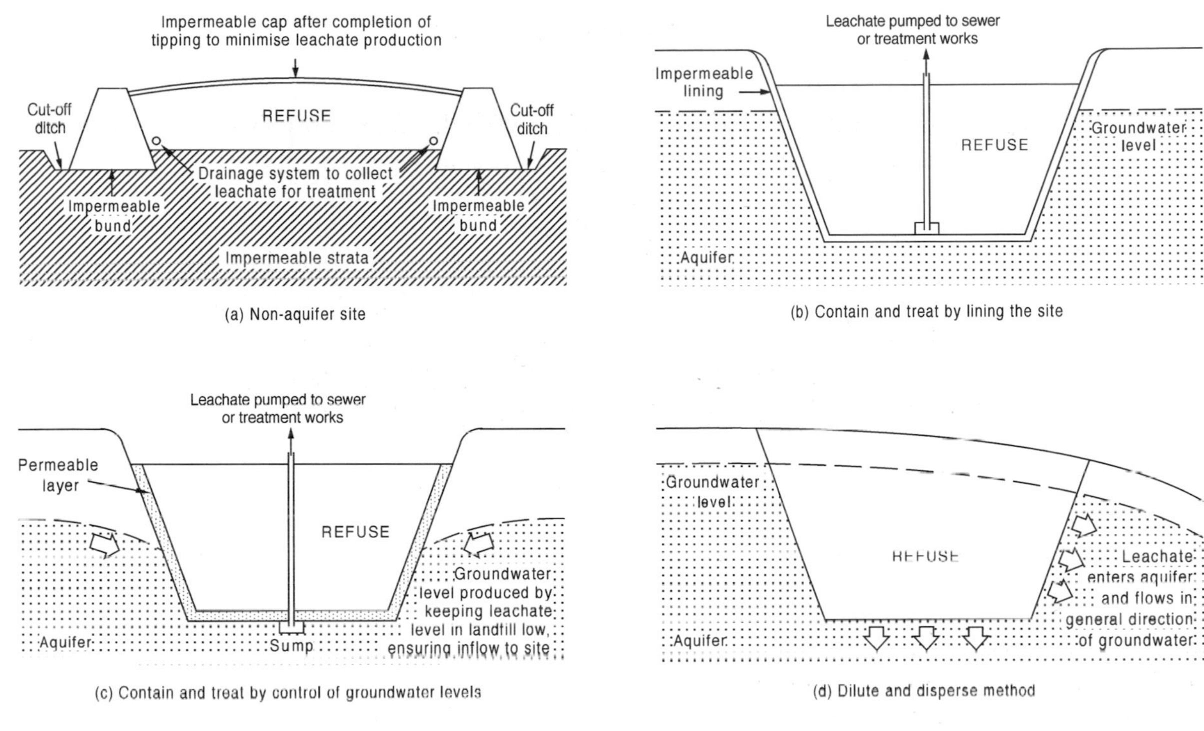

Figure 17.3 Four landfill designs (after Swinnerton, 1984).

approaches can be taken to prevent leaching, such as an impermeable landfill lining of plastic or rubber (Fig. 17.3b) or by careful control of the local groundwater table (Fig. 17.3c). In these designs the aim of the landfill is to concentrate, isolate and contain wastes to minimise the hazard they represent, but in areas where leachate is not expected to be produced in harmful quantities a dilute and disperse philosophy may be adopted, allowing seepage into the aquifer (Fig. 17.3d).

Designs like that shown in Fig. 17.3d have been used all too commonly in the past for disposal of hazardous wastes, and leakage from such old sites is causing increasing concern in many countries. A classic example of the dangers and costs of such sites is at Love Canal, a disposal site in the city of Niagara Falls, north-eastern USA, where between 1942 and 1953 a chemical company disposed of more than 20 000 t of chemical wastes. Not long after the landfill was sealed, the site was redeveloped. A school was constructed right at the edge of the filled canal and many houses were built in the surrounding area. Problems with odours and residues, first reported at the site during the 1960s, came to a head more than 20 years after the redevelopment. Heavy rains in the winter of 1975 and spring 1976 caused land subsidence and created pooling of surface waters, which were heavily contaminated with numerous toxic chemicals from the dump. Infiltration of these waters into nearby buildings caused a public outcry over the possible health hazards, and after a startling increase in skin rashes, miscarriages and birth defects, a state of emergency was declared at Love Canal by President Carter in 1978, the first executive order to address a hazardous waste problem in the USA (Deegan, 1987). Residents from 239 houses were relocated in that year, followed by another 710 families in 1980. Subsequently, the school and many houses were razed to the ground. Clean-up of the site and relocation costs amounted to about US$250 million (Lundgren, 1999).

The Love Canal incident played a major role in inducing the US Congress to create in 1980 the so-called Superfund, a programme run by the US

Environmental Protection Agency (USEPA) to provide financial support for clean-ups of such environmental disasters. The USEPA also created a National Priorities List of hazardous waste sites that needed remediation. Subsequent to the Love Canal incident, many of the world's industrialised countries have also begun a slow and costly process of identifying and cleaning up potentially dangerous old landfill sites. About 50 000 contaminated sites have been identified in Germany and 4000 in the Netherlands. In the USA, more than 40 000 potential hazardous waste sites have been identified, of which about 1200 are on the National Priorities List and thus the subject of immediate attention. The costs of clean-ups in the USA are shared between the current and previous landowners, those who have dumped waste at the site and those who have transported it there, with some money also provided by the Superfund. Legal arguments over who should pay are proving to be a considerable delaying factor in the process, however.

Further evidence of the need to tackle sites containing hazardous wastes comes from continuing worries over the possible health effects among people living near them. One study of landfills containing hazardous industrial waste in five European countries showed a strong association between mothers living within 3 km of a site and a significantly raised risk that their babies would be born with birth defects (Dolk *et al.*, 1998). Deformities and illnesses such as spina bifida, cardiac septa (hole in the heart), and malformations of the arteries and veins were noted in the analysis of children born to mothers who lived near 21 landfills in Belgium, Denmark, France, Italy and the UK. Investigation of the pathways by which people living near such sites may be exposed to toxic waste will help to establish one way or the other whether the link between hazardous waste and birth deformities is a real one, but the study also showed a fairly consistent decrease in the risk of birth defects with increasing distance away from the sites, suggesting the strong possibility of a causal relationship.

Although today's landfill sites are subject to much stricter controls in many countries than in the past, the future of burial in landfills is uncertain, given the pressure on available space and opposition from local residents to the creation of new sites. However, the decline of waste disposal at sea, which has been banned for many waste forms (see Chapter 6), is likely to increase the pressure for more sites, although some of the additional burden will undoubtedly be absorbed into the incineration option. One possible way of making landfill a sustainable option for waste management is to view it as a treatment process that can allow eventual re-assimilation of the site into the surrounding environment (Westlake, 1997). The location, design and operation of landfills must be appropriate to local conditions, and reduce risks associated with such sites to acceptable levels. The dangers of not so doing were illustrated at Love Canal.

Incineration

Incineration is increasingly being used by many countries to reduce the bulk of wastes and to break down hazardous compounds, so rendering them less dangerous. Incineration can also be combined with energy recovery. In Japan, Luxembourg and Switzerland, for example, more than 65 per cent of all municipal waste is incinerated (Fig. 17.2) and the proportion of this waste that also generates energy in these countries is 27 per cent, 100 per cent and 80 per cent, respectively. The rise in use of this form of waste disposal in the Netherlands is shown in Fig. 17.4.

Incineration is proving to be a good disposal method for tyres, which are otherwise problematic to get rid of. Landfill sites do not welcome them since tyres represent a fire hazard and tend not to stay buried. In practice, they are simply stockpiled, at the rate of 230 million per year in the USA, for example. The high energy content of tyres makes them a good fuel, however: an incineration plant at the largest tyre dump in the USA, near Modesto in California, produces electricity for 15 000 homes, and tyres could be more widely used as supplementary fuel in a number of

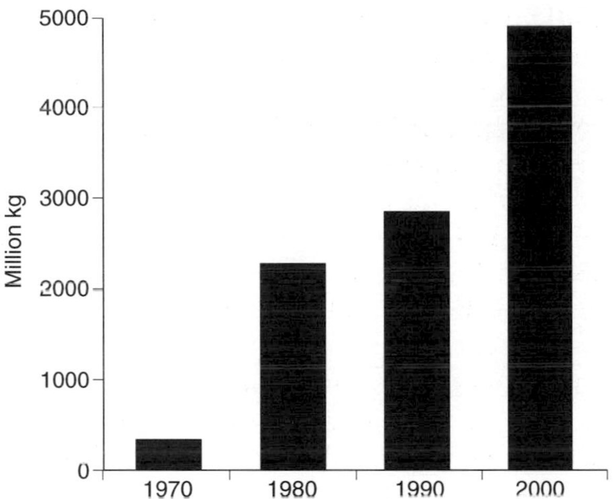

Figure 17.4 Waste incinerated in the Netherlands, 1970–2000 (from data at http://www.rivm.nl/environmentaldata/, accessed April 2002).

industries such as cement and paper mills (Barlaz *et al.*, 1993).

Incinerators do, however, release harmful emissions into the atmosphere, such as particulate matter, heavy metals and trace organics (in particular, highly toxic dioxins). Such emissions can be limited with a range of approaches, such as improved combustion techniques, sorting of wastes prior to incineration, and by fitting pollution control devices. Ash produced in incinerators must still be disposed of, and this material is commonly buried in landfills, although some attempts have been made to use the ash as a component of building materials. Site selection for incinerators, as for landfills, is another controversial issue for the waste disposal industry. While people may recognise that such facilities are necessary, the 'not in my backyard' syndrome is a powerful force in determining where such facilities are located. In practice, many such locally undesirable land uses, or LULUs, are sited in poor and minority communities. While, at first, this pattern may appear to reflect discrimination in siting procedures, research in the USA suggests that the situation may not always be so clear-cut. In some examples, LULUs appear to change community dynamics by driving down property values, resulting in a higher proportion of African-Americans and the poor around such sites, so that the dynamics of the housing market are more to blame than any discrimination in the choice of LULU site (Been, 1994).

Strong public feelings have often been foremost in pushing for more stringent emission standards for incinerators. In many European countries, these standards are now stricter than those applied to other energy sources, and some observers fear that the use of incinerators may be discouraged as a consequence. In practice, the total environmental impact of incinerators should be assessed and compared to similar measures for both alternative energy production and alternative waste treatment facilities (Nilsson, 1991).

Part of this total environmental impact stems from the need to dispose of the ash generated by incineration, a final waste product that is usually buried in landfill sites. Pressure on available space for landfills, combined with the pollution dangers of heavy metals and dioxins remaining in the ash, have prompted research into the possible reuse of incinerator ash. One form of treatment developed in Japan is to convert incinerated waste ash into melted slag, which is then artificially crystallised to form stones (Nishida *et al.*, 2001). Crystallisation gives the stones a hardness equal to natural stones, decomposes most of the dioxins in the process, and successfully contains hazardous heavy metals. The stones can then be used in civil engineering works as aggregates and building materials.

International movement of hazardous waste

As controls on the disposal of hazardous wastes have tightened in developed countries, there has been movement of both operations and wastes themselves to areas where legislation is less stringent or poorly enforced. During the 1980s, it became increasingly apparent that such trade in hazardous wastes was on the increase, both between industrialised and less-developed countries, and also between western and eastern Europe (Yakowitz, 1993). International concern over this trend led ultimately to the Basel Convention on the Control of Transboundary Movements of Hazardous Wastes and their Disposal, which came into force in 1992. The convention is based on a series of guiding principles originally adopted by the OECD countries, three of which are particularly important (OECD, 1991a).

1 The principle of non-discrimination – OECD members will apply the same controls on transfrontier movements of hazardous wastes involving non-member states as those applied to movements between member states.

2 The principle of prior informed consent – movements of waste will not be allowed without the consent of the appropriate authorities in the importing country.

3 The principle of adequacy of disposal facilities – movements of waste will only be permitted when wastes are directed to adequate disposal facilities in the importing country.

The Basel Convention also bans exports to countries that have not signed and ratified the treaty, and incorporates requirements to control the generation of hazardous wastes and obligations to manage them within the country of origin unless there is no capacity to do so (Hilz and Radka, 1991).

TABLE 17.2 *Priorities for waste treatment in the EU*

CLEANER TECHNOLOGY	1. Prevention
RECOVERY	2. Reuse 3. Recycling 4. Main use as fuel or other other means of generating energy
DISPOSAL	5. Incineration or composting 6. Landfill

Source: European Environment Agency

REUSE, RECOVERY, RECYCLING AND PREVENTION

Although landfill and incineration are currently the most frequently used methods of waste disposal, they are generally considered to be low down on the list of possible techniques devised from an environmental impact perspective, as the EU's hierarchy of treatment techniques for solid wastes indicates (Table 17.2). Reusing a product, as opposed to discarding it, obviously makes environmental sense, and the advantages of waste recovery and recycling have also long been recognised. As for reuse, these advantages include a reduction of resource consumption, and a curb to the cost and inadequacies of disposal. In practice, waste products are reused when it is economically viable to do so and viability is assessed for a variety of motives. The history of industrial development is punctuated with examples of waste products being reappraised and turned into valuable resources. In the early nineteenth century, for example, Britain's fledgling chemical industry on Merseyside was using the Leblanc soda process to produce alkalis by treating salt with sulphuric acid, producing highly corrosive hydrogen chloride as a waste product. Realisation that this pollutant was a lost resource, combined with fears of legal action from local landowners over the effects on surrounding vegetation, spurred industrialists to convert the hydrogen chloride to chlorine, which was used to make bleaching powder (Elkington and Burke, 1987). Another example from the nineteenth century concerns a resource that, today, most people take for granted. In the 1870s, when Nicolaus Otto found that a mix of flammable gas and air could be spark-ignited within the cylinder of a piston machine to generate movement, the flammable gas used was obtained from gasification of a waste product that came from the petroleum refinery process that produced paraffin for lamps.

Figure 17.5 A novel reuse for soft drinks cans in Namibia.

Figure 17.6 Recycling collection point in a suburb of Berlin. Germany has been a European trendsetter in recycling.

The waste product was called petrol (Svidén, 1993).

The perception shift that underlies these and many other examples highlights the fact that something that is considered a waste today can become a resource tomorrow. As Dijkema *et al.* (2000) put it, a substance or object is qualified as waste when it is not used to its full potential. This type of thinking can be summarised in a formula adopted by the multinational Minnesota Mining and Manufacturing (3M), a company distinguished by its commitment to a constant succession of innovative products:

pollutants (waste materials) +
knowledge (technology) =
potential resources

Reuse, recovery and recycling

Realisation of the economic value of certain wastes promotes reuse, recovery and recycling. In many developing countries, informal scavenging of municipal wastes from city dumps is widely practised and the materials recovered are put to a variety of uses (Fig. 17.5). It provides employment to individuals who otherwise would have none, particularly important in those countries where social security systems are inadequate or non-existent. Studies of this informal sector have also highlighted the important role of such scavenged resources as inputs to small-scale manufacturing industries. In the Tanzanian capital of Dar es Salaam, these industries provide consumer goods that are much cheaper than if they were made from imported raw materials. Low-cost buckets, charcoal stoves and lamps are all made from scavenged metals from city dumps and sold to people who live in squatter settlements in the city (Yhdego, 1991).

In today's more developed countries such informal waste recovery was also common more than 100 years ago, but more formalised recovery and recycling schemes have now sprung up. These schemes are also most advanced for metals and some other materials such as paper and glass, and collection points for such solid wastes have become a familiar sight in many cities in recent years (Fig. 17.6). Some of the benefits of these schemes in environmental terms are shown in Table 17.3.

Among the critical factors affecting the quantities of material that are recycled are the capacity of the recycling plant and equipment, and the size of the market for recycled materials. In some cases, these factors have been influenced by legislation. In extreme cases, some types of packaging are banned altogether. In Denmark, for example, beer and soft drinks were simply not allowed to be sold in cans for many years (although the ban was

TABLE 17.3 *Environmental benefits of substituting secondary materials for virgin resources*

Environmental benefits	Aluminium	Steel	Paper	Glass
Reduction (%) of				
energy use	90–97	47–74	23–74	4–32
air pollution	95	85	74	20
water pollution	97	76	35	—
mining wastes	—	97	—	80
water use	—	40	58	50

Source: Bartone (1990)

repealed and replaced with a deposit and return system in 2002). More commonly used measures include laws designating a certain level of use of recycled materials and the imposition of a tax on products that do not incorporate a certain amount of recycled material. The Canadian city of Toronto, for example, has a local law which states that daily newspapers must contain at least 50 per cent recycled fibre or the publishers will not be allowed to have vending boxes on the city's streets. Taxes on waste products are in line with the 'polluter pays' principle and can encourage both the reduction of waste and the increase of recycling rates. Taxation has been suggested as a possible solution to the widespread problem of packaging waste, a priority waste problem in Europe and elsewhere. Imposition of a tax on packaging would effectively incorporate the full social cost of a product into the retail price by including the cost of disposal of its wrapping (Pearce and Turner, 1992). The EU Packaging Directive imposed an obligation on industry to recover or recycle at least half of all its packaging by 2001.

Denmark was one of the first countries to introduce a comprehensive waste taxation scheme to promote reuse and recycling as part of a waste management policy developed in the mid-

TABLE 17.4 *Metal consumption and recycling in the USA, 1990*

Metal	Consumption (thousand tonnes)	Share of consumption provided by recycling (%)
Lead	1297	73
Copper	2168	60
Gold	0.2	47
Aluminium	5263	45
Tin	45	38
Tungsten	8	29
Chromium	423	21
Molybdenum	21	5

Source: Young (1992)

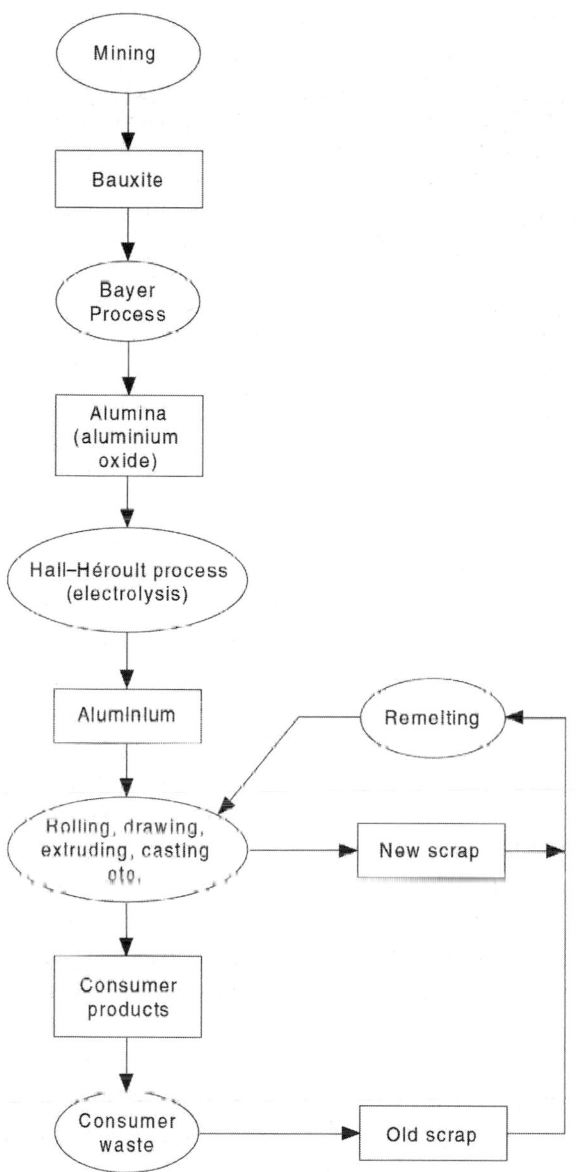

Figure 17.7 Stages in the production of aluminium goods.

1980s (Andersen, 1998). The policy was in response to the country's serious waste disposal problem: per capita generation of waste was among the highest in Europe, Denmark was running out of landfill space and there was widespread public concern about air pollution from incinerators. The tax is levied on most of the household and industrial waste delivered to Danish landfills and incinerators, designed to reduce the overall quantity of waste, to ease the burden on the country's limited landfills and to lower dioxin emissions from incineration. It was also intended as an integral part of Denmark's Action Plan for Waste and Recycling, introduced in 1989 with the aim of achieving a 54 per cent recycling rate for all waste by 1996. No taxes are levied on waste that is reused or recycled.

The Danish waste-reduction policy, in which the taxation system has played an important role, has registered considerable success. In the ten years from 1987 to 1996, Denmark achieved a reduction of 26 per cent in the quantity of waste delivered to landfills and incinerators, and reached an overall recycling rate of 61 per cent.

Taxes on waste are used to increase the economic incentives to recycle, but such intervention is not always necessary. In economic terms, the large energy savings gained from recycling of most metals make the practice worthwhile anyway. In the case of aluminium, the conversion of alumina to aluminium by electrolysis – the Hall–Héroult process – accounts for 80 per cent of the energy used in the entire production process, a stage that is bypassed when aluminium is recycled (Fig. 17.7). These and other benefits have encouraged the increasing use of recycling in those industries that use metals (Table 17.4).

Manipulation of the retail price of a consumer product is an economic instrument that has long been used to encourage reuse of certain articles. The charging of a returnable deposit on glass bottles is a good example, providing an economic incentive to return the bottle for reuse. It is a practice that is still widely used in many developing countries, but one that has declined in some more developed economies as returnable glass bottles have largely been replaced with throw-away plastic ones. In some countries, such as Germany and Switzerland, recent legislation has been passed to encourage moves back to the refilling of drinks containers, but in some other countries similar measures could significantly increase the price to the consumer, a politically sensitive step to take. In the USA, for example, reusable bottles were

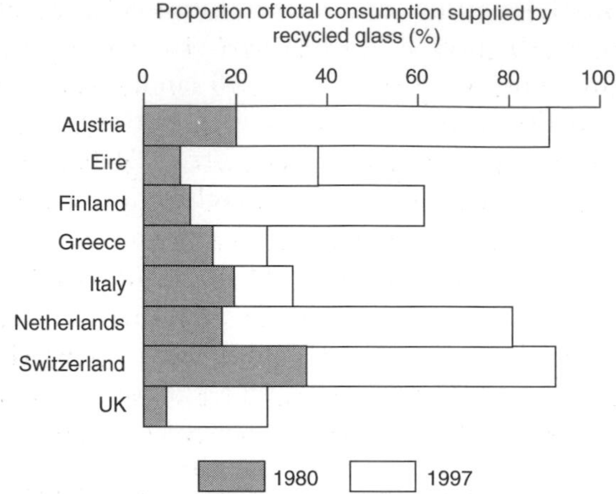

Figure 17.8 Glass recycling in selected European countries (from data in *Eurostat*, 1997).

common when bottling plants were small and widely distributed throughout the country, but recent decades have seen a centralisation of bottling facilities, in part as a response to the lower transport costs imposed by lighter plastic bottles. Hence, an increased emphasis on returnable glass bottles would entail considerably higher transport costs. Such economic factors have been partly responsible for the increasing use of the recycling option, which has become widespread practice in many developed countries in recent years. For the OECD countries, the average recovery rate for glass rose from 22 per cent in 1980 to 32 per cent in the late 1980s (OECD, 1991a), and in a number of European countries recycled material makes up more than 70 per cent of all glass used (Fig. 17.8).

An industrial process that has a long history of waste product recycling is the steel industry. Steel slags from the Basic-Bessemer or Thomas process have been used as a phosphate fertiliser since 1880. The hardness and structural stability of blast furnace and steel slags have also enabled their use as aggregates for road construction and as armourstones in hydraulic engineering works, to stabilise and refill river beds and banks scoured by erosion (Motz and Geiseler, 2001). These uses help to relieve the pressure on natural aggregates like gravel, sand and processed rocks, the excavation of which leaves scars on the landscape. Steel converter slag has also been used more recently to treat wastewater. The process utilises the slag's iron oxide, or magnetite component, which can adsorb nickel, a heavy metal, from wastewater (Ortiz *et al.*, 2001).

Wastewater is commonly reused or recycled in many countries, and is widely recognised as a significant, growing and reliable water source that is particularly important in drylands where water is a scarce resource. Indeed, as Bakir (2001) points out, wastewater production is the only potential water source that will increase as the population grows and the demand on freshwater increases. The use of treated or untreated wastewater in landscaping and agriculture is common in many countries of the Middle East and North Africa, including the United Arab Emirates, Oman, Bahrain, Egypt, Yemen, Syria and Tunisia. Water resources management strategies in several countries, such as Jordan, sensibly consider wastewater as a part of the water budget.

In many cases domestic wastewater is diverted to cropland for irrigation. The water itself is a valuable resource and sewage is typically high in key plant nutrients (phosphorus and nitrogen). In Mexico, wastewater from almost all cities that have a sewerage system is used in this way to irrigate about 150 000 ha of crops nationally. The largest such zone is in the Mezquital Valley where wastewater and storm runoff channelled out of Mexico City considerably increases yields (Table 17.5). However, these increased yields come at an environmental price, since none of Mexico City's wastewater receives any conventional sewage treatment before it is used in this way. One characteristic of sewage is its high heavy metal content, a function of sewage systems mixing industrial and household wastes, and long-term applications of wastewater in the Mezquital Valley have led to an accumulation of heavy metals in soils and their uptake into the crops themselves. There is also evidence of an increase in parasitic infections among agricultural workers and their families due to their exposure to raw sewage (Siebe and Cifuentes, 1995). In some cases,

TABLE 17.5 *Crop yields (t/ha) with different agricultural practices in the Mezquital Valley, central Mexico*

Crop	Wastewater irrigation	Fluvial irrigation	Rainfed agriculture
Maize	4.8	2.9–4.0	0.37
Bean	1.4	0.28–1.2	0.18
Barley	3.1	2.0	1.4

Source: after Siebe and Cifuentes (1995)

infectious diseases can be transmitted from sewage-irrigated crops to the general public. An outbreak of cholera in Jerusalem in 1970 was thought to be caused by consumption of vegetables irrigated with wastewater. The risks of this kind of disease transmission can be limited by simple measures, however. Irrigation with wastewater during planting is less hazardous than during the growing cycle, and certain crops (e.g. fruit and vegetables) are more likely to carry disease than others.

In countries where sewage is treated, the sludge that remains after treatment is also reused as a fertiliser due to its high nutrient content. About half the sewage sludge produced in the UK and France, and one-third of that in Germany, is reused in this way, but in these and other countries, further use is limited by the high heavy metal content. Hence, sewage sludge not used as fertiliser is disposed of by ocean dumping, dumping in landfill and municipal garbage dumps, or by incineration. Indeed, continued concern over the pollution content of sewage sludge, combined with a rising demand for organic and quality-assured food products, has prompted Switzerland to phase out its use as a fertiliser by 2005.

As population and industrialisation continue to increase, however, and the treatment of water and sewage becomes more widely used in order to reduce the pollution impacts from untreated sewage outlets, so more sludge will need to be disposed of. Switzerland intends to incinerate all of its sewage sludge by 2005, but a more innovative approach to this increasing dilemma of disposal can turn a problem into a benefit by realising the value of the heavy metals themselves. One estimate of the potential yield of some of the high-value metals in sewage sludge suggests that global production of palladium from sewage sludge could be of the same order of magnitude as that from mining production (Table 17.6). The future disposal of sewage sludge should concentrate on the control of pollutants at source and the extraction of metals (Lottermoser and Morteani, 1993), an approach that would yield numerous benefits, including:

- more widespread use of sludge as a fertiliser
- revenue from metals extracted
- conservation of geological metal resources
- saving on expensive waste repository space
- prevention of environmental impacts on terrestrial and marine ecosystems.

TABLE 17.6 *Estimated worldwide annual production of noble metals from sewage sludge compared with current production from geological resources*

Metal	Current production (t/year)	Production from sewage sludge (t/year)
Gold	1600	100
Platinum	100	8
Palladium	100	80

Source: after Lottermoser and Morteani (1993)

Waste prevention: cleaner production

There is no doubt that the best way to manage waste is to prevent it at source wherever this is possible. The argument that prevention is better than cure is put by UNEP's Industry and Environment Programme Activity Centre:

> ***When end-of-pipe pollution controls are added to industrial systems, less immediate damage occurs. But these solutions come at increasing***

TABLE 17.7 *Examples of the application of cleaner production*

Country	Industry/process	Cleaner production	Advantages	Payback time of capital costs incurred
France	Metal processing – steel galvanising	Development of a new process for zinc coating of steel	Total suppression of conventional plating waste; lower operating costs and improved quality control	3 years
India	Textile – sulphur black colour dyeing using sodium sulphide to convert sulphur dyes	Substitution of sodium sulphide with hydrol, a by-product of the maize starch industry	Reduction of sulphide in the effluent; less corrosion in the treatment plant and reduction of odour problems	No capital expenditure
Indonesia	Cement production	Optimisation of kiln operation and improved process control through the installation of expert control systems	Some reduction in NO_x and SO_2 emissions, energy savings and reduction in the quantities of off-specification material	<1 year
Poland	Automobile component manufacture – Cu/Ni/Cr and Zn electroplating of aluminium alloy and steel components	Modification of the rinsing systems and addition of facilities that permit water recycling and raw materials recovery	A decrease in water and raw materials consumption; reductions in contaminant levels in wastewater streams of 80–98 per cent	4 months
Sweden	Metal fabrication – de-greasing of metal sections using trichloroethylene	A switch to an alkaline de-greasing procedure that utilises biodegradable cutting oils	Zero emission of trichloroethylene and reduction in the quantity of hazardous waste sludge produced	11 months
UK	Adhesives	Development of a new range of water-based adhesives in place of solvent-based ones	Compared with solvent-based adhesives, water-based ones are less toxic and require no special handling, yet still offer a wide range of applications	New product

Source: after UNEP IE/PAC (1993)

> *monetary costs to both society and industry and have not always proven to be optimal from an environmental aspect. End-of-pipe controls are also reactive and selective. Cleaner production, on the other hand, is a comprehensive, preventative approach to environmental protection.*
> *(UNEP IE/PAC, 1993: 1)*

Cleaner production is achieved by examining all phases of a product's lifecycle from raw material extraction to its ultimate disposal – so-called 'lifecycle analysis' – and reducing the wastefulness of any particular phase. Hence, cleaner production encompasses such aims as:

- conservation of energy and raw materials
- reduction in the use of toxic or environmentally harmful substances
- reduction of the quantity and toxicity of wastes and pollutant discharges
- extension of product durability.

Cleaner production embodies both the ideas of environmental sustainability and economic efficiency. In practice it involves improving the productivity of energy and material use to reduce consumption of resources and cut pollution per unit output. A range of industrial examples of such cleaner production is shown in Table 17.7. They range from low-tech changes in production like the example of a textile factory in India to more capitally intensive modifications, to large-scale production facilities such as cement and automobile manufacture, and new product development. Nevertheless, for all the examples shown, the payback time for the capital expenditure incurred is three years or less, indicating the immediate economic feasibility of such actions.

Such economic incentives are important in encouraging companies to adopt cleaner production methods, although many such initiatives to date have been driven by legislated pollution controls, both actual and anticipated. A number of companies go beyond legal requirements, however, and attempt to develop 'closed-loop' processing cycles in which waste products are completely recycled and never enter the environment to become pollutants. One such example is the programme established by SC Johnson Wax in 1990, which reduced the company's manufacturing waste by half, cut virgin packaging waste by 25 per cent and reduced the use of volatile organic compounds by 16 per cent while increasing production by more than 50 per cent (Schmidheiny *et al.*, 1997). One-third of the energy used in the company's largest plant comes from methane extracted from a nearby landfill and the recycling of organic vapours from its process lines. Another plant has been modified to continuously reuse 95 per cent of its wastewater. These and other changes in the manufacturing process have saved the company more than US$20 million a year in production costs.

FURTHER READING

Bradshaw, A.D., Southwood, R. and Warner, F. (eds) 1992 *The treatment and handling of wastes.* London, Chapman & Hall. The papers in this book cover a range of topics on the various types of waste and the methods for disposal.

El-Fadel, M., Zeinati, M., El-Jisr, K. and Jamali, D. 2001 Industrial-waste management in developing countries: the case of Lebanon. *Journal of Environmental Management* 61: 281–300. An overview of wastes generated and the need for an environmental management plan.

Gandy, M. 1994 *Recycling and the politics of urban waste.* London, Earthscan. A look at the politics and economics of urban waste disposal with case studies from London, New York and Hamburg.

Heinen, J.T. 1995 A review of, and research suggestions for, solid-waste management issues. *Environmental Conservation* 22: 157–66. An interesting overview of the complexities of managing garbage disposal.

Hocking, M.B. 1991 Paper versus polystyrene: a complex choice. *Science* 251: 504–5. Using lifecycle analysis to compare two types of an apparently simple product: the throw-away cup. The polystyrene version may not deserve its poor reputation.

Rhyner, C.R., Schwartz, L.J., Wenger, R.B. and Kohrell, M.G. 1995 *Waste management and resource recovery*. Boca Raton, CRC Press. A useful broad-ranging textbook focusing almost exclusively on the issue in the USA.

Young, J.E. 1990 *Discarding the throwaway society*. Worldwatch Paper 101. Washington, DC, Worldwatch Institute. An overview of the rise of resource consumption and their wasteful use, focusing on the role of the individual consumer as the ultimate decider of how wasteful a society we live in.

Websites

http://www.awma.org/ the Air and Waste Management Association includes much information and news, and access to the association's online journal.

http://www.iswa.org/ the International Solid Waste Association covers all aspects of solid waste management.

http://www.wastewatch.org.uk/ much information on waste reduction, reuse and recycling, particularly in the UK.

http://www.epa.gov/swerosps/ the US Environmental Protection Agency's office of solid waste and emergency response.

http://www.epa.gov/superfund/ the USEPA's Superfund site with details of Love Canal and many other hazardous waste sites.

http://www.unep.ch/basel/ site for the Basel Convention on hazardous wastes.

http://www.unepie.org/ UNEP's Division of Technology, Industry and Economics site contains information on cleaner production and waste management, and on many other environmental issues related to industry (e.g. ozone-destroying chemicals and energy efficiency programmes).

http://www.financingcp.org/ a UNEP project aiming to increase investment in cleaner production in developing countries.

http://waste.eionet.eu.int/ the European Topic Centre on Waste and Material Flows has information about waste and waste management in the European Union.

POINTS FOR DISCUSSION

- Are all waste products simply resources we haven't discovered yet?
- How would you set about trying to prove a causal relationship between hazardous landfill sites and deteriorating human health?
- Is cleaner production the answer to environmental problems caused by industrial emissions?
- Dilute and disperse or containment?

18

ENERGY PRODUCTION

TOPICS COVERED

Energy sources
Energy efficiency and conservation
Renewable energy
Nuclear power

Key words: cogeneration of heat and power (CHP), photovoltaic (PV) power, biomass fuel, dilute and disperse philosophy, Chernobyl

Energy has long been a basic requirement of human societies. Archaeological evidence from caves in Africa suggests that fire was used by early humans, the hominids, as long ago as 1.5 million years BP, and fire has since fuelled the technologies on which civilised societies have been built. Today, regular supplies of energy drive the motors, appliances, cities, industries and transport on which the lifestyle of most of the Earth's human population depends. Through time, the ability of human cultures to access energy, and the amounts of energy used, are often seen as indicators of society's level of development and resource use. Energy use's attendant environmental problems have long been with us, however. One of the first examples of environmental legislation in England occurred during the reign of Edward I when coal burnt for industrial and domestic purposes caused so great a smoke nuisance that the nobility, strongly backed by London residents, were able to obtain a royal proclamation forbidding coal burning in 1306 (although it proved impossible to enforce). Since then, the generation of energy and the use of fuels in multifarious ways to drive the industries and machinery of modern societies have facilitated an unprecedented level of both deliberate and inadvertent impacts on the global environment. The production, transportation, conversion and use of energy, particularly that derived from fossil fuels, are responsible for some of the world's most serious environmental problems. These include global climatic change (see Chapter 11), acid rain (see Chapter 12) and pollution from motor vehicles (see Chapters 10 and 16), but human society's harnessing and use of energy lies behind virtually every one of the environmental issues found in this book. This chapter will focus on energy sources, specifically on the possible future of those not based on fossil fuels.

ENERGY SOURCES

Ultimately, most of the energy we use today comes from the sun. This radiant energy is converted by photosynthesis into chemical energy, which makes plant life and ultimately all animal life possible. It is also used by human society both as biomass and as fossil fuels, which were themselves once living plant matter. Solar radiation is also used directly by human society and indirectly by harnessing the movement of wind and rivers as part of the hydrological cycle. Three other ultimate energy sources are also commonly used: the processes of cosmic evolution preceding the origin of the solar system (nuclear power); the forces of lunar motion (tidal power), and energy from the Earth's core (geothermal power).

TABLE 18.1 *Effects of various energy-saving options for heating a three-bedroomed house in Britain*

Heating/ insulation packages	Heat loss (kW)	Average heating bill (£/week)	Area heated
Gas fire in living room	7.22	4.75	Living room only
Gas central heating and loft insulation	5.78	6.60	Whole house
		5.10	Downstairs only
Gas central heating, loft insulation, cavity wall insulation, draught-stripping and humidistat-controlled extractor fans	4.11	4.50	Whole house
		3.65	Downstairs only
Gas unit heaters and gas water heaters plus insulation and fans as above	4.11	3.75	Whole house
		3.00	Downstairs only

Source: after Musannif (1992)

These sources of energy are often categorised into renewable and non-renewable resources, based on a timescale of human lifetimes. Hence, although fossil fuels are renewable on a geological timescale, the rate at which we are using them effectively means that they are finite. Although estimates of just how much oil, gas and coal remain for society to use are constantly being revised – and are to an extent a function of technology, price and the rate at which we use them – they are ultimately non-renewable. Estimates made by the World Energy Council (WEC, 1992) suggest that we have about 40 years' worth of oil at current production rates, 60 years for gas and more than 200 years for coal. While the 1970s were characterised by fears over the imminent exhaustion of fossil energy resources, such fears have dissipated following the re-evaluation of existing reserves and the discovery of new ones. Indeed the WEC stated in 1992 that 'The concepts of exhaustion, or even scarcity, fail to appear anywhere in this survey' (WEC, 1992: 13). Nevertheless, the finite nature of fossil fuels, combined with increasing awareness of the environmental impacts of their use, has spurred interest in the conservation of energy and the development of more sustainable forms of production from renewable resources.

ENERGY EFFICIENCY AND CONSERVATION

Perhaps the most obvious approach to reducing the environmental impacts associated with energy use is to make energy production more efficient and to reduce the amounts of energy that are wasted during its use. Better energy conservation can be promoted in many sectors of society and the adoption of energy-saving technology and methods is usually encouraged by a fairly rapid return in terms of reduced energy bills.

In industry, for example, the conservation of energy is one of the basic aims of campaigns for cleaner production (see page 295), while in the domestic and commercial sectors, substantial savings can be made with relatively simple building design and technologies. In many countries of the developing world, energy lost by burning fuelwood on open fires can be reduced by using closed stoves, which is also safer, while converting wood to charcoal is more energy-efficient still. Heating and hot water often represent the largest proportion of energy use in the home and office of more developed settings, and Table 18.1 illustrates the scale of savings that can be made with readily available technology.

On the national level, the US Environmental Protection Agency (EPA) estimates that the energy used to run commercial and industrial buildings produces 19 per cent of US carbon dioxide emissions (Lupinacci, 2000). The potential savings from using energy more efficiently are summed up by Ruckelshaus (1989: 120), who states that: 'Right now more energy passes through the windows of buildings in the US than flows through the Alaska pipeline.' The use of energy for lighting, which consumes about 25 per cent of electricity in the USA, could also be substantially reduced with a more widespread adoption of energy-efficient lighting hardware such as compact fluorescent lamps. The lighting innovations that are commercially available in the USA today could potentially save up to 20 per cent of the country's electricity use (Fickett *et al.*, 1990). In other areas, energy is being used for little apparent purpose. The consumption of electricity by home entertainment electronic goods (e.g. televisions, hi-fi equipment and video cassette recorders) while in standby mode in the UK was about 5 terrawatt hours in 1994 (DECADE, 1995). This is the equivalent output of one large power station being used solely for the luxury of being able to activate the appliance with a remote control device.

In commercial office space, energy is the single largest operating cost, amounting to roughly one-third of operating expenses. Improving the energy performance of buildings using readily available techniques can result in average savings of 30 per cent (Lupinacci, 2000). This can be achieved by a cut in heating, properly sizing the heating and cooling equipment, and using equipment only when it is needed.

Another area with substantial room for improvement in efficiency is in the generation of electricity. In many developing countries, power plants that run on fossil fuels use five or six units of fuel to make one unit of electricity. In more developed countries, the ratio is about three to one but can reach two to one in combined-cycle gas turbines. This ratio can be improved further by the cogeneration of heat and power (CHP).

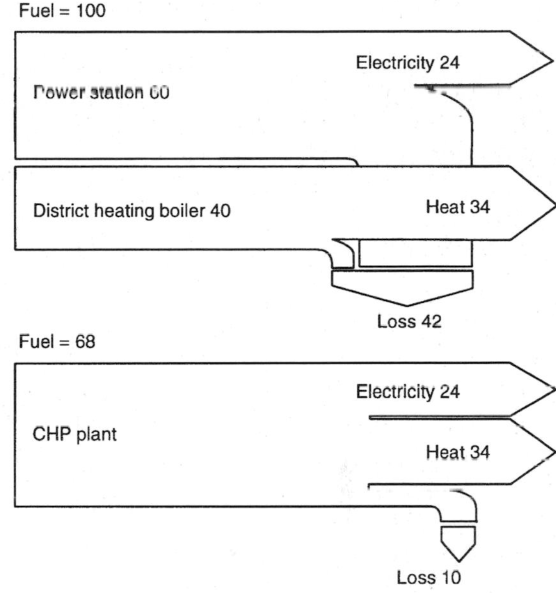

Figure 18.1 Energy efficiency with combined production of power and heat (distributed through a district heating system). With cogeneration, fuel consumption is reduced by about 30 per cent.

Much of the energy wasted in conventional electricity generation is in the form of hot water so using this water in some way makes economic and energy-efficient sense. In Denmark this hot water is increasingly being used in district heating systems for houses, factories and offices. While in a country like the UK most buildings are supplied with cold water that is heated by individual boilers, some Danish buildings have received hot water from district heating boilers since the 1920s. Providing this hot water from a CHP plant is more efficient still: reducing fuel consumption by about 30 per cent (Fig. 18.1). Production of electricity and district heating in Denmark is already among the most efficient in the world, since about one-third of heating energy used in Denmark is produced by CHP. Increasing the contribution made by CHP plants is a central aim of the Danish Energy 2000 programme, an energy efficiency plan that aims to reduce gross national energy consumption in 2005 by about 15 per cent relative to 1988 (DME, 1990). More buildings are being connected to CHP heating, and existing district heating plants are being converted to CHP plants run on natural gas and organic waste biomass (see below).

RENEWABLE ENERGY

The current use of renewable energy sources is most significant in the developing countries where biomass fuels represent a major energy source. Hydropower provides the main source of electricity for some countries, such as Brazil, Canada and Norway. Overall, of course, fossil fuels currently make the largest contributions to global energy use (Fig. 18.2).

Although the current cost of renewable energy supplies is not competitive with fossil fuels in many areas, this is largely a function of the lack of finance and effort put into their development, and the subsidies given to energy production from oil, coal and natural gas. Given more attention, renewable energy sources could become cost-effective more quickly, and with a number of other advantages in addition to the environmental benefits most have relative to the burning of fossil fuels. One such advantage is that most renewable energy equipment is small, allowing much faster construction than conventional technologies. While most large conventional facilities are constructed in the field, most renewable energy equipment is constructed in factories where manufacturing techniques can facilitate cost reduction. The small scale of equipment also makes for a short time between design and operation, so that modifications stemming from field tests can quickly be incorporated into new designs. Hence, many generations of technology can be introduced in a short period. There are, however, some difficulties with certain renewable energy sources. The intermittent nature of wind and solar power sources is one such problem. This can be resolved by appropriate use or storage of energy in many cases, but not all. It is unlikely that solar power will ever be a realistic option in high-latitude areas because of the lack of solar intensity and the long, dark winters. It is also important to note that not all renewable sources are appropriate for all types of energy need: wind and wave energy will not fuel a motor vehicle, for example, although electricity so generated may eventually supply battery-powered vehicles. The following sections will look at the main forms of renewable energy to assess their future potential to contribute to world energy needs.

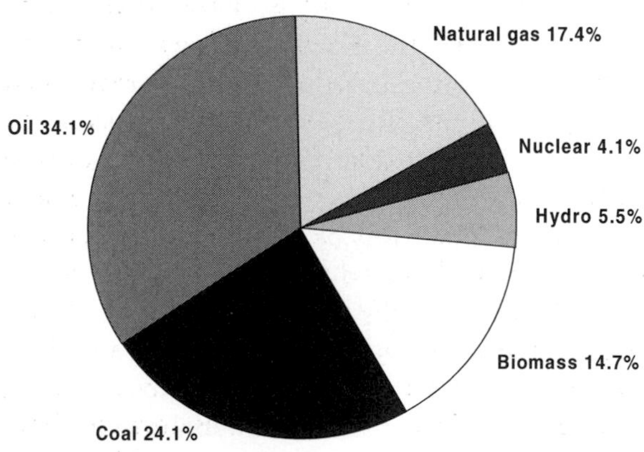

Figure 18.2 Global energy use (Hall *et al.*, 1993).

Hydropower

Hydropower is the only renewable resource used on a large scale today for electricity generation. At the turn of the current century, one-third of the countries in the world relied on hydropower for more than half their electricity supply, and large dams generate about 19 per cent of electricity overall (World Commission on Dams, 2000). The main constraints on further development are the social and environmental impacts associated with dams, which are outlined in Chapter 9, although many of these problems can be mitigated by improved planning with the inclusion of public participation at an early stage. It should also be noted that reservoirs used for hydroelectricity generation are likely to emit significant quantities of atmospheric pollutants (see page 149).

Although much uncertainty surrounds the estimates of potential hydroelectric power that could be generated, there is agreement that, to date, only a relatively small proportion of this resource has been developed, and the greatest potential lies in the developing world and the countries of the former USSR. The potential for small-scale hydroschemes, designed to serve local needs, is thought to be particularly great (Moreira and Poole, 1993).

Wind energy

The wind's energy has been used by sailing ships and windmills to power human activities for thousands of years, and wind pumps are still a common feature of rural landscapes today, particularly in dry regions where they are used to pump groundwater to the surface (Fig. 18.3). Modern wind turbines for electricity generation are a comparatively recent phenomenon, which some suggest have the potential to supply as much as 20 per cent of global electricity demand (Grubb and Meyer, 1993).

Much of the pioneering work on the large-scale generation of wind-derived electricity has been carried out in California where developments were encouraged by favourable tax policies. California now produces just over 1 per cent of its electricity from wind farms, much of it coming from the 7500 turbines at Altamount Pass. Denmark has also put a great deal of effort into developing its wind power potential, encouraged by investment incentives and enthusiastic public support. Some 5000 turbines in the Danish countryside generated 2.5 per cent of its electricity consumption in the mid-1990s. Denmark aims to provide half the country's electricity needs from renewables by 2030 and a large part of this will come from offshore wind farms to be built in the North Sea and the Baltic (EEA, 2001). The 2000 new turbines will be four times as big as existing models, with blades 60 m in diameter on 55-metre towers. Long-term development of the technology means that Danish offshore wind power has become as cheap to produce as building new coal-fired power stations and comparable in cost to gas-fired stations, without the disadvantage of CO_2 emissions.

There are several environmental impacts associated with wind farms. Noise was a serious concern surrounding earlier generations of turbines, but modern designs make little sound above the rush of the wind. Problems with bird kills persist, however, for some larger species such as eagles, and interference with TV reception and sensitive electronics is also experienced around turbines with steel blades. The land requirement of large wind farms has been another issue of concern, although land between individual machines can still be used for other activities such as farming and ranching. Land values at Altamount Pass, for example, have increased markedly following installation of the wind farms, as royalties have generated extra income for ranchers while ranching has continued. Perhaps the most serious

Figure 18.3 Wind power is widely used in rural areas, particularly to pump groundwater to the surface in regions where surface water is in short supply, as shown here on the Mediterranean island of Malta.

objection to wind power in many areas is on aesthetic grounds, since some people frown upon the idea of landscapes covered in windmills. In the UK, for example, more than 50 well-known literary figures wrote to *The Times Literary Supplement* in 1994 to protest at plans to increase the number of wind turbines on the moors above Haworth because of the area's associations with the novels of the Brontë sisters.

Solar power

In the very long term - beyond this century - solar power in its various forms is expected to be able to meet the bulk of the world's energy needs, although in the shorter term the contribution can be expected to be much more modest (WEC, 1993). Solar power can be used passively by designing buildings to optimise the sun's light and heat energy, or actively to heat water and building space, and to generate electricity. Photovoltaic (PV) power is one of the most direct methods of converting the sun's rays into electricity; it is produced when individual light particles - photons - absorbed in a semiconductor create an electric current. PV electricity is thus created with no pollution and no noise. PV systems also need minimal maintenance and no water, and have been particularly successful in remote areas such as deserts, although high temperatures can adversely affect their performance (Diarra and Akuffo, 2002). Another advantage of PV systems is that they can operate on any scale, from portable modules for remote communications and instrumentation to huge power plants covering millions of square metres, enabling them to be positioned near to the electricity users and reducing the need for transmission systems. They are potentially cost-effective even at high latitudes and in relatively cloudy regions (Fig. 18.4) and the land areas required are not great: if the entire electricity needs of the USA were generated by PV systems, for example, the area required would be about 34 000 km^2, less than 0.4 per cent of the US land area (Weinberg and Williams, 1990). The cost to the consumer of electricity from PV arrays

Figure 18.4 A parking meter run on photovoltaic power in Berlin, Germany.

has fallen rapidly in the last three decades. They are already cost-effective in remote areas and are likely to rival the cost-effectiveness of conventional energy sources within the next 20 years, as the cost of manufacturing the sheets of semiconductor materials falls.

Another way in which solar radiation is used directly to generate electricity is the solar thermal electric generator, which uses mirrors to track the sun and focus its rays to heat a fluid in a pipe. The heated fluid is then used to create steam that drives a turbine generator, or directly to produce steam and hot water for industrial and domestic purposes. Several designs have been developed, ranging in scale from an individual parabolic mirror that focuses the sun's rays on to its receiver, to a field of sun-tracking mirrors, which all reflect solar energy to a receiver mounted on a central

Figure 18.5 The central receiver solar thermal power plant at Solar One in California, USA (Photo: US Department of Energy).

tower (Fig. 18.5). The nine commercial solar thermal power plants now operating in the Mojave Desert in the USA indicate the viability of the technology, which like other renewable energy sources, has been hampered by a lack of interest and finance from governments.

Biomass

Plant matter, or biomass, is a form of energy that has been used by people for heating, lighting and cooking since the discovery of fire. In many developing countries, fuelwood is still the main source of energy (Fig. 18.6) and an estimated 2.5 billion people, nearly half the world's population, rely mainly or exclusively on biomass for their daily energy needs (Hall and Rosillo-Calle, 1991). Although in some parts of the developing world fuelwood collection is being carried on unsustainably, with severe environmental impacts (see page 76), biomass has a large potential to be used as a renewable fuel in many different applications.

Biomass fuels come in a variety of forms, both unprocessed (wood, straw, dung, vegetable matter and agricultural wastes) and processed (e.g. charcoal, methane from biogas plants and landfills, logging waste and sawdust, and alcohol produced by fermentation). These fuels were widely used in the more developed countries before the large-scale switch to fossil fuels that came with industrialisation, but biomass also has a more recent history of use in the industrialised countries. Industrial and agricultural residues have been used for many years to provide electricity in conventional steam-turbine power-generators. In some of the Scandinavian countries, where biomass contributes a fairly high percentage of national energy consumption (Table 18.2), woody waste from the pulp and paper industries is the most significant component. Some countries, notably Brazil, have made considerable progress in substituting fuel derived from sugar cane for petrol used in motor vehicles (see Chapter 16), while on the local scale, municipal wastes are

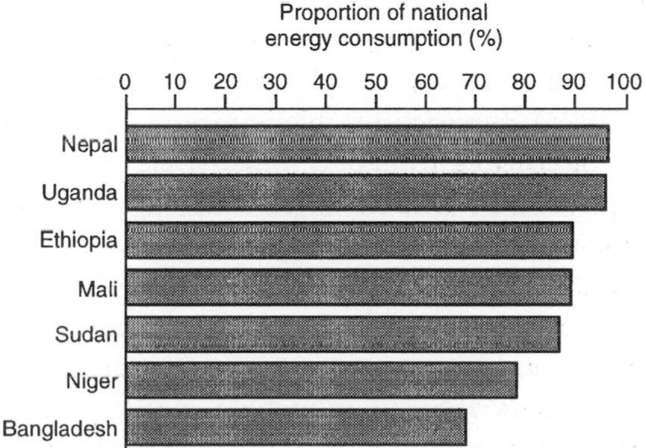

Figure 18.6 Wood fuel as a percentage of national energy consumption in selected developing countries (after UNEP, 1990; World Bank, 1996).

TABLE 18.2 *Biomass fuels as a percentage of national energy consumption in selected developed countries*

Country	Contribution of biomass to total energy consumption (%)
Austria	4.0
Belgium	0.2
Canada	3.0
Finland	14.0
New Zealand	0.4
Sweden	13.0
Switzerland	1.6
UK	0.3
USA	4.0

Source: after Sipilä (1993)

burnt in incinerators, and landfills are tapped for emissions of gaseous energy (see Chapter 17).

Sustainably managed, the use of biomass fuels produces no net emissions of carbon dioxide since the amount released into the atmosphere by burning will be taken up by growing plants. Although more effort can be made to use residues from other activities, and from the harvesting of forests, dedicated energy plantations are probably the largest potential source of biomass. A strong candidate for land to be used for such energy plantations is deforested or otherwise degraded land, which could be replanted with trees (Hall *et al.*, 1993). Monocultural plantations come with their own environmental impacts, of course (see Chapter 13), but using degraded lands for plantations would be a considerable improvement on their current ecologically impoverished state.

Although research and development is still needed to identify more accurately suitable plants to grow and where to grow them, biomass has the potential to make a very significant contribution to world energy needs. Some believe that with appropriate encouragement and effort, biomass has the potential to provide as much as 80 per cent of the 1985 world commercial energy use by the year 2050 (Hall *et al.*, 1993).

Tidal power

Energy in the oceans is found in the form of tides, waves, ocean currents, temperature differences, salt gradients and marine biomass, and although the total amounts available are large, only a small fraction is likely to be utilised in the foreseeable future, both because ocean energy is spread diffusely over a wide area and because much of the available energy is distant from centres of consumption. To date, only tidal power has been seriously investigated and technically proven. It is one of the oldest forms of energy exploited by human society – tide mills were operated on the coasts of Britain, France and Spain before AD 1100. Tides ebb and flow because of the gravitational interaction between the sun, the moon and Earth, and the centrifugal force of the Earth–Moon system. Their movements are therefore highly predictable, which gives tidal power a distinct advantage over many other renewable energy forms. Tidal energy is harnessed in much the same way as that of hydropower, by building a barrage across a suitable estuary. But whereas a hydroelectric dam depends on water flowing in one direction, a tidal barrage must allow the bay behind it to alternately fill and empty with the tides. As water flows out of an estuary from high tide to low tide, it passes through turbines, which generate electricity.

The amount of energy available from a site depends upon the range of the tides and the area of the enclosed bay. The number of sites suitable for modern commercial-scale energy production are limited, however, and only a few of these have been developed (Table 18.3). The largest is the world's first operational tidal power plant, associated with the barrage across the La Rance estuary on the Brittany coastline of northern France, which was built in the early 1960s. La Rance has an average tidal range of 8 m and the barrage's 24 turbines produce enough electricity for a city of 300 000 people.

TABLE 18.3 *Existing tidal energy plants*

Location	Mean tidal range (m)	Basin area (km2)	Installed capacity (MW)	Date of first use
La Rance (France)	8.0	17	240.0	1966
Kislaya (Russia)	2.4	2	0.4	1968
Jiangxia (China)	7.1	2	3.2	1980
Annapolis (Canada)	6.4	6	17.8	1984

Source: after Cavanagh *et al.* (1993)

Other locations where tidal energy could viably be harnessed include the Bay of Fundy in Canada, with an average tidal range of 10.8 m, the largest in the world, and the Severn Estuary in Britain where the average range is 8.8 m. Indeed, most of the potential for harnessing tidal energy in Europe lies on the British coast, where the exploitation of all practical sites could yield up to 20 per cent of the electricity needs of England and Wales. The Severn and the Mersey estuaries are among the UK's best sites, but the environmental effects of such schemes have featured highly in the debate over whether or not to go ahead with barrages across these estuaries. Despite the environmental advantages of tidally generated electricity, including the lack of pollutants generated and the protection barrages offer to coastlines against storm-surge tides, such schemes inevitably modify the hydrodynamics of their estuaries. Alterations to the tidal range, currents and the intertidal area within the barrage would affect several other environmental parameters, including sediment movement and water quality, which would in turn have an impact on the food chain, ultimately affecting bird and fish populations.

Geothermal energy

Geothermal energy is tapped from the heat that flows outwards from the Earth's interior energy, both from the core as it cools and from the decay of long-lived radioactive materials in rocks. On the global scale, it is most accessible along the boundaries between the Earth's crustal plates, but many geothermal possibilities also exist away from these areas of concentration.

Hot waters from the Earth have been used therapeutically and for their mineral salts since Roman times, when geothermal spring water was also used to heat bath houses. This energy has been exploited commercially, to provide heat, mechanical power and electricity, since the early 1900s. In 1930, for example, large-scale geothermal heating systems for buildings were built in Iceland, and similar operations subsequently constructed in France, Italy, New Zealand and the USA. In 1990, about 85 per cent of all residential buildings in Iceland were heated geothermally. These and most of the other prevailing geothermal systems are hydrothermal, tapping aquifers of hot water a few kilometres below the surface, while other approaches are largely in the research stage. One such alternative uses energy from hot, dry rocks, which is exploited by fracturing them at depth and pumping water down from the surface to be heated and then returned to the surface.

The environmental impacts associated with the use of geothermal energy can be divided into temporary ones due to drilling and exploration, and permanent ones resulting from well maintenance and power plant operation (Palmerini, 1993). Power plants take up land and may attract objections on aesthetic grounds. Air pollution is the greatest concern for most geothermal plants, since gases released during power production are not easily re-injected back into the thermal reservoir, although geothermal plants generally produce fewer gaseous emissions than their fossil fuel equivalents. Other environmental problems include:

- noise
- solid waste and residual water disposal
- micro-earthquakes
- subsidence.

Although geothermal energy is unlikely to contribute significantly to global energy needs in the foreseeable future, it could none the less play a key role for some countries where potential is high and energy demands are currently low (e.g. Djibouti, El Salvador, Kenya and the Philippines). The potential is well illustrated in Nicaragua where the plant at Momotombo came online in 1989, producing 40 per cent of the country's electricity needs. The plant cost US$80 million to build but displaced US$10m of oil imports in its first year of operation (Shillitoe, 1991).

The future for renewable energy

It is likely in the foreseeable future that energy demand will continue to rise as economic growth proceeds, despite the feeling in many quarters that truly sustainable development should aim to uncouple economic growth from rising energy use (see page 194). Increasing the efficiency with which energy is used can help to offset this rise, but is unlikely to absorb all the additional needs. Furthermore, the environmental impacts of conventional energy sources, notwithstanding their eventual depletion as non-renewable resources, will continue to play a role in deciding to what degree renewable energy sources contribute. Some believe that, given adequate support, renewables can meet much of the growing demand and gradually replace fossil fuels, at prices lower than those usually forecast for conventional energy sources, to be able to account for three-fifths of the world's electricity market by the middle of the twenty-first century (Johansson *et al.*, 1993). Renewable fuels for motor vehicles are also becoming more feasible technically, with hydrogen being the cleanest option (see page 277). Others are less optimistic about the future for renewables; the World Energy Council (WEC) highlights the degree and extent of change necessary to realise this scenario: 'It is difficult to believe that politicians and policies, energy consumers and behavioural patterns, technology and the capacity to manufacture it and put it into operation on the required scale, will change sufficiently within the required timescale' (WEC, 1993: 93).

These two views highlight the key aspect of renewable energy's future role. Essentially, the future pace of development will be set by political decisions – if we want greater use of renewables then governments need to encourage it. Table 18.4 summarises the main types of barrier to the deployment of renewable energy that have been identified in the EU, most of which can be traced back to a lack of sufficient political interest. The need for government support, particularly in the research and development phase, but also later in the energy marketplace was well illustrated in 1992 when the world's leading solar thermal electric corporation, Luz International, went bankrupt largely as a result of sudden changes in tax laws and regulations that virtually eliminated financial incentives for solar thermal electric technology in the USA.

However, even where the appropriate political decisions are made to move towards renewable energy sources, other difficulties may still arise because of the environmental effects of renewables. An interesting case in this respect is Sweden, where the long reliance on nuclear power is being reduced due to environmental and safety concerns. The strength of public feeling over the nuclear issue grew during the 1970s, to a level in 1976 when the ruling Social Democrats, who had governed the country for 44 years and were responsible for Sweden's nuclear energy programme, lost their majority in parliament. In a

TABLE 18.4 *Barriers and obstacles to renewable energy deployment in the European Union*

Barrier	Obstacle
Political	Lack of political motivation to support the market initiatives needed for the development of renewables
Legislative	Lack of an appropriate legal framework and legislation at EU and national levels that support the development of renewables Difficulties with linking electricity or heat from renewables into the existing electricity and heat networks
Financial	Lack of appropriate financing for long-term financial benefits
Fiscal	Renewable energy technologies suffer from distorted competition from conventional energy sources (e.g. coal, nuclear) in terms of final end-user prices
Administrative	Lack of practical support at the regional and local level to stimulate development of renewable energy projects
Technological	Technological obstacles related to research, development and demonstration
Information, education and training	Lack of awareness of the potential and possibilities for renewables

Source: EEA (2001)

pioneering move based on a 1980 referendum decision, Sweden shut its first nuclear reactor in 1999 as part of a planned complete withdrawal of nuclear power, but public feelings on what should replace nuclear energy are proving to be almost as strong (Löfstedt, 1998). Biomass is the most viable alternative, because solar power is not sufficient due to Sweden's geographical position at a high latitude; wind power's potential has been limited by public opposition and slow technical progress; and further development of hydroelectricity is constrained by the 1987 Natural Resources Law, which protects Sweden's remaining free-flowing rivers. But an increase in the use of biomass fuels has also attracted opposition from ornithologists and one environmental organisation, who argue that it would adversely affect biodiversity. The debate in Sweden emphasises the fact that all energy sources entail some form of environmental impact. A sustainable energy policy can only aim to minimise these impacts because they will not disappear.

NUCLEAR POWER

When energy generated by nuclear fission was first developed for civil uses in the 1950s, having grown out of a small number of national nuclear weapons programmes, it was heralded as cheap, clean and safe. The subsequent image of nuclear power has changed considerably since those times, and today it is one of the most controversial forms of energy from both the economic and environmental perspectives. By the end of 2000, nuclear reactors were in operation or under construction in more than 30 countries, and nuclear power stations have been supplying about 17 per cent of world electricity production since 1990

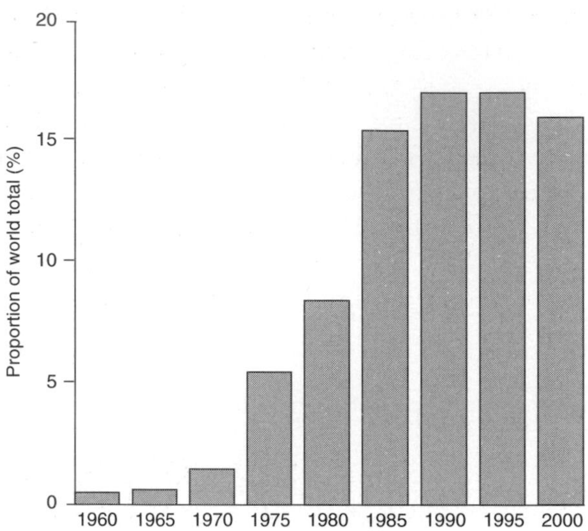

Figure 18.7 Proportion of world electricity generation produced by nuclear power in selected years, 1960–2000 (from data in UNEP, 1993; IAEA, 1991).

(Fig. 18.7). For some countries the proportion is much higher: four European states derived more than 50 per cent of their electricity in this way in 2001 (France 76 per cent, Lithuania 74 per cent, Belgium 57 per cent, Slovakia 53 per cent). Ultimately, nuclear power is a non-renewable form of energy production since uranium, the fuel used, is a mineral product. Advances in technology have meant that increasing amounts of uranium can be reprocessed and used again, perhaps doubling the lifetime of usable uranium from around 40 to 60 years of supply (WEC, 1993), but in the shorter term the future of nuclear power will be dependent largely upon the associated environmental concerns.

These concerns are based on the fact that nuclear material is radioactive and hence can be highly dangerous to living things, and that for some radioactive elements this radioactivity is very long-lasting. The potential for these materials to be released into the environment can be traced through the nuclear fuel cycle – from mining, conversion, preparation of reactor fuel, and the management of wastes – and can arise again with the decommissioning of old nuclear power plants. Considerable research has been focused on the dangers of increased mortality rates from cancers around operational nuclear plants. While studies of adults suggest that no such effect is associated with nuclear installations (Forman *et al.*, 1987), the situation is less clear-cut for children. The true causes of unusual clusters of childhood leukaemias around the UK reprocessing plants at Sellafield and Dounreay, for example, may well be linked to nuclear emissions, but there may also be an infective cause due to the sudden influx of newcomers to areas where local communities were previously well protected (Kinlen, 1988). Research into clusters of leukaemia and other cancers around nuclear installations will undoubtedly continue, but greater concerns abound over the disposal of nuclear wastes and the risk of major accidents.

Nuclear waste

Nuclear waste is commonly categorised into three classes (Table 18.5). In most of the countries that produce nuclear waste, low-level material makes up the large majority of waste by volume, but accounts for a minor proportion of the radioactivity in waste. High-level waste, by contrast, is volumetrically small but accounts for the largest share of radioactivity. Liquid low-level wastes have long been disposed of by adopting a 'dilute and disperse' philosophy. The British nuclear power complex at Sellafield on the Cumbrian coast, for example, has been discharging such low-level liquid waste into the Irish Sea since the early 1950s. The disposal of solid wastes, by contrast, has aimed to contain the material and then dispose of it. Until 1982, some low-level and selected intermediate-level wastes were contained in drums, which were dumped at sea, but this practice has been suspended and burial is now the preferred option. For low-level and short-lived, intermediate-level wastes, shallow burial is practised in such countries as the UK and the USA, while in France, surface storage is the preferred disposal option. In Sweden, however, a submarine repository has been constructed for this short-lived material below the bed of the Baltic Sea.

TABLE 18.5 *Categories of nuclear waste*

LOW-LEVEL WASTE Liquid and solids lightly contaminated by short-lived radionuclides (e.g. discarded protective clothing, contaminated building materials, uranium mine tailings)
INTERMEDIATE-LEVEL WASTE Materials contaminated by long-lived radionuclides such as plutonium and transuranic elements, most of which arise from the processes of energy production and reprocessing (e.g. fuel cladding, control rods, liquids used to store spent fuel before reprocessing)
HIGH-LEVEL WASTE Materials contaminated by highly active radionuclides with long half-lives, which may cause significant increases in the temperature of wastes (e.g. non-reprocessed spent fuel, liquid wastes produced during reprocessing of spent fuel)

Overall, the volumes of radioactive waste are small when compared to other waste products generated by human activity. Although estimates are difficult to come by, in part because of the political sensitivity that surrounds both the nuclear power and weapons industries, the total volume of high-level waste produced in the UK, for example, will probably have reached about 1200 m^3 by 2030, enough to fill two small houses. Substantially larger volumes can be expected from decommissioned reactors, but despite the amounts involved, disposal of high-level waste, which will remain dangerous for many thousands of years, is one of the key issues facing the nuclear industry. High-level waste is heat-generating and needs to be cooled and then managed safely for a period that is far longer than we believe human civilisation has existed to date. Although the nuclear industry is convinced that burial of high-level and long-lasting, intermediate-level waste material in suitable geological formations, at least several hundred metres deep, is the best method, opponents of the nuclear industry argue that this is not an acceptable solution. They advocate surface storage until a better management method can be devised. Public sensitivity over the issue of nuclear waste disposal is uniquely great, and although the technical aspects of high-level waste disposal have been solved to the satisfaction of most in the nuclear industry, the political aspects remain to be resolved.

Chernobyl

Concern at the potentially huge damaging effects of an accident at a nuclear power facility is the other major environmental issue surrounding nuclear power. Despite the stringent checks and safety measures employed at such facilities, accidents at nuclear reactors do happen, as evidenced by those at Chalk River, Ontario, Canada, in 1952, Windscale (now Sellafield), UK, in 1957, and Three Mile Island in Pennsylvania, USA, in 1979. These events were overshadowed on 26 April 1986 by the accident at the Chernobyl nuclear power station in the Ukraine (then part of the USSR), the most serious accident to have occurred at a nuclear power station to date, and an event that has haunted the world's nuclear industry since.

Radionuclides from the explosion at Chernobyl were dispersed throughout the northern hemisphere in trace amounts, with particular 'hotspots' in areas where rainfall washed radioactive material from clouds: in parts of the European Soviet republics, Austria, Bulgaria, Finland, Germany, Norway, Romania, Sweden, Switzerland, the UK and Yugoslavia. Most concern has focused on the biomedical dangers presented to humans by the radionuclide deposition. Initial fears over iodine-131 led to the destruction of fruit and vegetables grown in the open air and milk from cows grazing on contaminated grassland, but since iodine-131 has a half-life of only eight days, attention soon shifted to caesium-134 and -137, the latter with a half-life of 30 years. Caesium accumulates up the food chain from the soil through vegetation to contaminate meat, necessitating special measures to restrict the movement and sale for consumption of livestock as far from the accident as Scandinavia and Britain. Other dangerous radionuclides involved include strontium-90 (half-life, 29 years) and plutonium-239 (half-life, 24 000 years).

Restrictions on foodstuffs are still in place in some of the affected areas up to 3000 km from Chernobyl, because radioactive caesium from the fallout of the accident is lingering in the environment much longer than scientists had previously anticipated. Smith *et al.* (2000) found unexpectedly high levels of radioactivity in western Europe, which will last for 50 more years - 100 times longer than expected. The high levels of radioactive caesium were found in fish in lakes in Cumbria in northern England and Norway. During the first five years after Chernobyl, the concentration of radioactive caesium measured in most foods and in water declined by a factor of 10, but in the last few years they have changed very little. The environment has not been cleaning itself at the rate scientists previously thought.

Although the health risk to consumers is thought to be small, restrictions on foodstuffs from parts of Europe and the former Soviet Union will need to be maintained for at least another 10 to 15 years. In Britain, 389 farms still have restrictions on the sale and slaughter of sheep, which will have to continue for a total of 30 years after the accident. In the more contaminated parts of Ukraine and Belarussia, bans will have to continue for longer. Restrictions on the consumption of forest berries, fungi and fish, which contribute significantly to people's radiation exposure, will need to continue for at least a further 50 years.

The main overall biological effects of contamination from the Chernobyl catastrophe are shown in Table 18.6. Different ecosystems absorb radionuclides in different ways, and forests were some of the most contaminated. Foliage absorbed radionuclides in rainfall along with carbon-14 and tritium from the air by photosynthesis. When foliage fell in the autumn, most of these contaminants were transferred to the ground, concentrating in leaf litter and the thin upper soil layers. Coniferous forests kept radionuclides in their canopy longer, and such trees were consequently felled and disposed of in some heavily affected areas.

In meadows and pastures, radionuclides absorbed in the vegetation were also transferred to the upper soil layers in autumn. In agricultural areas, contaminated upper soil layers were turned over during ploughing and transferred to a depth

TABLE 18.6 *Main biological effects of the Chernobyl catastrophe*

Level	Short-term response	Long-term response
Biosphere	Global and local changes in radionuclide accumulation and dispersion	Global and local disturbances in genetic and phenetic structure of the biosphere
Ecosystem	Changes in ecosystem diversity, stability and development patterns	Changes in co-evolution process and ecosystem succession
Human population	Changes in birth and mortality rates	Changes in mutation rate, natural selection intensity and adaptive reaction
Human individual	Disturbances in physiology and behaviour	Changes in probability of cancer, hereditary abnormalities and other diseases

Source: after Savchenko (1991)

of 20–40 cm. Natural migration of radionuclides in soil depends upon their type and the speed at which they penetrate the soil, which is of the order of centimetres per year. Rich soils absorb them more easily and keep them for a longer period than poor, sandy soils. The solubility of radionuclides is another important factor: strontium-90 has the highest solubility, explaining its immediate uptake by plants.

The impacts of radionuclide contamination of soils depend upon their chemical forms and concentrations, soil type, the presence of normal atoms of radioactive isotopes in soils, their solubility and availability for roots. Caesium-137 moves very slowly through soils, sinking only a few millimetres per year. Strontium-90, by contrast, forms a weaker bond with soil particles and moves in the soil more rapidly. The generally slow movement of radioactive elements through soils has a beneficial side, in that it slows the transfer of contamination to groundwater.

There is generally considered to be a direct relationship between the levels of soil and plant contamination, and a direct relationship between the level of radioactive contamination of agricultural lands and contamination levels in animals, notably in meat and milk products. Concern over these effects has led to 257 000 ha of agricultural land being taken out of production in Belarus, for example, along with 1.34 million hectares of forest (Marples, 1992). Officially designated contaminated territories account for no less than 23 per cent of the surface area of Belarus, 5 per cent of Ukraine and 1.5 per cent of the Russian Federation. The population of these territories is around 6 million people (UNDP/UNICEF, 2002). Since the Chernobyl accident, about 350 000 people have been relocated away from the most severely contaminated areas.

The human health effects of the Chernobyl disaster are the subject of investigation and monitoring at both national and international levels. The International Atomic Energy Agency (IAEA) has studied radiation doses from the accident in highly contaminated areas where local food restrictions were in force, and found that the external radiation dose exceeded the background values in only 10 per cent of inhabitants (IAEA, 1991). Several assessments of the possible health risks have projected the excess cancer risk for the entire northern hemisphere to range from zero to 0.02 per cent. Although most of the long-term health effects of Chernobyl will be statistically undetectable, because they will be spread through a population of hundreds of millions over several decades, some evidence has emerged. The increase of thyroid cancer in children has increased markedly in areas close to Chernobyl, in northern Ukraine and in Belarus, mainly due to the accumulation in the thyroid gland of iodine-131 ingested in food and inhaled from the initial radioactive cloud. The disease is normally rare in children but by the end of 1998, 1800 cases of thyroid cancer in children residing in the affected area had been attributed to Chernobyl, although some claim that this is an underestimate (UNDP/UNICEF, 2002). Previous experience of the irradiation of the thyroid gland indicates that cases related to the exposure will continue to occur for at least 50 years after exposure. A conservative estimate of the number of cases of thyroid cancer occurring over the lifetimes of those exposed in childhood in the affected areas is 6000 to 8000 in the three countries.

Expected increases in the incidence of leukaemia have been less marked in the most heavily contaminated areas, although a study of children with leukaemia in Greece found that an increase in the incidence of the disease was associated with increased exposure to ionising radiation from the Chernobyl accident while in the womb (Petridou *et al.*, 1996).

An early estimate of the direct and indirect economic costs of the Chernobyl accident was calculated as at least US$15 billion, 90 per cent of which would be in countries of the former USSR (Tolba, 1992). Disruption of agricultural activities, relocation of people and consequent psychological stress were the main immediate consequences of the accident. More recent assessments of the costs are much greater. The government of Belarus estimated that losses over the 30

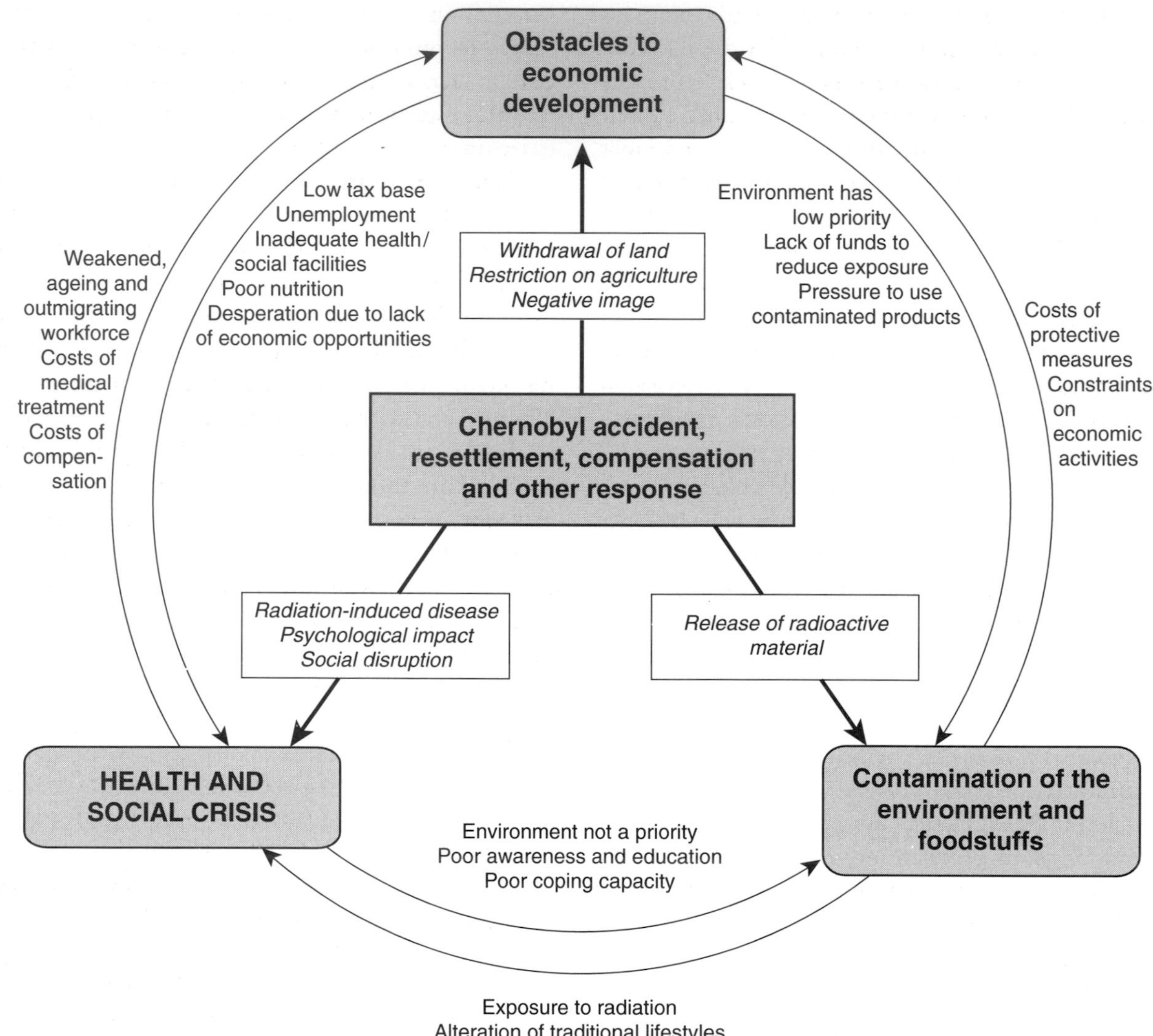

Figure 18.8 The downward spiral in communities affected by Chernobyl (after UNDP/UNICEF, 2002).

years following the accident will amount to US$235 billion while the Ukrainian Government estimated the loss as US$148 billion over the period from 1986 to 2000 (UNDP/UNICEF, 2002). These two countries have had to introduce a special Chernobyl emergency tax to help pay for the world's most severe nuclear disaster. The effects of the Chernobyl accident and the additional economic problems faced in the post-Soviet era have combined to produce a downward spiral of health and well-being in the communities affected (Fig. 18.8).

The future for nuclear power

Concern over the environmental effects associated with nuclear power means that the nuclear industry is conducted under some of the strictest controls and regulations of any major industry, and certainly more stringent than for other forms of power generation. While the risks of environmental damage from unsafe waste disposal or accidents are statistically small, these issues continue to rate highly in the public consciousness, and the lack of public confidence stemming from these concerns is perhaps the major problem facing the nuclear power industry today. The issues have consequently become subject to a high level of political, media and public relations activity. For some countries, perception of the risks associated with nuclear power have resulted in national programmes being stopped, scaled down or, in the case of Sweden, subject to a concerted effort to phase it out altogether (see above). Complete abandonment elsewhere is unlikely in the foreseeable future, however, given the very large economic investments nuclear power plants represent, and the widespread concern over carbon dioxide and other gaseous emissions from fossil fuel plants – environmental problems to which nuclear power plants do not contribute. As with renewable energy, the future use of nuclear power will be a political decision in which environmental aspects will undoubtedly play a significant role.

FURTHER READING

Johansson, T.B., Kelly, H., Reddy, A.K.N. and Williams, R.H. (eds) 1993 *Renewable energy: sources for fuels and electricity*. New York, Island Press. An encyclopaedic overview of the various forms of renewable energy, including technical, economic and political aspects.

Kaygusuz, K. 2002 Sustainable development of hydropower and biomass energy in Turkey. *Energy Conversion and Management* 43: 1099–120. An assessment of present and future use of two renewable energy sources.

Löfstedt, R. 1998 Sweden's biomass controversy: a case study of communicating policy issues. *Environment* 40(4): 16–20, 42–5. The debate over biomass in a country committed to phasing out nuclear power.

Nielsen, L. 1993 How to get the birds in the bush into your hand. *Energy Policy* November: 1133–44. A Danish research project demonstrating the potential electricity savings available in the home through behavioural and technical changes.

Roberts, L.E.J., Liss, P.S. and Saunders, P.A.H. 1990 *Power generation and the environment*. Oxford, Oxford University Press. This book contains a chapter on the history of power generation and its environmental impact, two detailed chapters on fossil fuels and nuclear fission, and a look at future prospects from renewables.

Schipper, L. and Meyers, S. 1992 *Energy efficiency and human activity*. Cambridge, Cambridge University Press. A detailed analysis of trends in energy use since 1970 in manufacturing, transport, the domestic and service sectors, with a look to the future, discussing how energy use can be restrained in order to meet goals of environmental and economic development.

Scientific American 1990 *Energy for planet Earth*. Special issue 263(3). A collection of articles that look at future energy needs for particular sectors, such as buildings and homes, industry, motor vehicles and particular world regions, as well as the future of fossil fuels, nuclear power and solar energy.

Websites

http://www.worldenergy.org/wec-geis/ the World Energy Council site contains information, data and reports on all aspects of energy.

http://www.caddet-ee.org/ specialises in the collection and dissemination of information on new energy-saving technologies.

http://www.eren.doe.gov/ the US Department of Energy's Energy Efficiency and Renewable Energy Network has much information and many links to related sites.

http://solstice.crest.org/ the Center for Renewable Energy and Sustainable Technology contains documents, data and discussion groups on many energy issues.

http://www.renewableenergy.com/ a comprehensive site with news and links covering all forms of renewables.

http://www.iaea.org/ the International Atomic Energy Agency site covers many aspects of atomic energy, including up-to-date data and reports.

POINTS FOR DISCUSSION

- How would an increase in our reliance on renewable energy sources, as against fossil fuels, contribute to solving other environmental issues?
- Human society's unsustainable use of fossil fuels is the root cause of virtually all other environmental issues. How far do you agree?
- Should governments do more to encourage the development of renewable energy sources or should they leave it to the power companies?
- Review the arguments for and against the banning of nuclear power.

19

MINING

TOPICS COVERED

Global economic aspects of mineral production
Environmental impacts of mining
Rehabilitation and reduction of mining damage
The true costs of mining

Key words: subsidence, crownhole, acid mine drainage, bioleach, full-cost pricing

We only need to look at the names used to describe key periods in the early development of societies – the Stone Age, the Bronze Age and the Iron Age – to realise the long importance of mining as a human activity. People have been using minerals from the Earth's crust since *Homo habilis* first began to fashion stone tools 2.5 million years ago. Today, we are more dependent than ever before upon the extraction of minerals from the Earth. Virtually every material thing in modern society is either a direct mineral product or the result of processing with the aid of mineral derivatives such as steel, energy or fertilisers.

The naturally occurring elements and compounds mined are usually classified into four groups:

1 metals (e.g. aluminium, copper, iron)
2 industrial minerals (e.g. lime, soda ash)
3 construction materials (e.g. sand, gravel)
4 energy minerals (e.g. coal, uranium, oil, natural gas).

In terms of volume extracted, construction materials are by far the largest product of the world's mining industry. They are found and extracted in every country. An estimated 11 billion tonnes of stone, and 9 billion tonnes of sand and gravel were taken from the ground in 1991. About 125 000 t of crushed rock, for example, is used to build 1 km of motorway. Metallic minerals are mined in smaller quantities. World production of iron ore in 1992 was 929 million tonnes, phosphate 144 million tonnes, and bauxite 107 million tonnes (Crowson, 1994).

GLOBAL ECONOMIC ASPECTS OF MINERAL PRODUCTION

The location of mineral concentrations in the Earth's crust is a fact of nature, but there are a host of human and natural factors that determine if and when a mineral is recognised as a resource, classified as a reserve, and eventually exploited. Such factors include global and local economic and political influences, as well as the local availability of basic requirements for mining such as energy and water. While the use of minerals dates from before the time of *Homo sapiens*, it was the Industrial Revolution that sparked their large-scale exploitation. Between 1750 and 1900, global mineral use increased ten-fold as the population doubled and, since 1900, use has increased more than thirteen-fold (Bosson and Varon, 1977).

The rapid increase in society's use of minerals has periodically sparked concern over their imminent depletion, a sentiment expressed strongly in the 1970s by the Club of Rome (Meadows *et al.*, 1972). Although the amount of minerals in the

TABLE 19.1 *Estimates of reserves at Palabora copper mine, South Africa*

	Defined at start-up in 1966	Full potential as seen in 1991
Reserves before mining:		
Tonnage (million tonnes)	259	854
Grade (% copper)	0.69	0.55
Contained copper (million tonnes)	1.79	4.67
Cut-off grade (% copper)	0.3	0.15
Mine life from start-up (years)	24	35

Source: Crowson (1992)

lithosphere is finite, improvements in technology and changing price/cost relationships, as well as shifting perceptions of the political risks of operating in particular countries, mean that estimates of reserves are constantly being re-evaluated. The dynamics of changing resources are illustrated at the level of one mine by the Palabora copper mine in South Africa, which at its start-up as an open-pit operation in 1966 was estimated to contain an ore deposit with 1.8 million tonnes of copper, giving the mine an operational life of 24 years. By 1991, however, the estimate had changed to 4.7 million tonnes, with a life of 35 years (Table 19.1). Improved technology and economic changes help to explain the increase in reserves, lowering the cut-off grade of exploitable ore. The estimated lifetime of the mine has since been extended still further, to 55 years from start-up, by the decision to develop an underground mine at Palabora. A further 2.3 million tonnes of contained copper were estimated to be economically accessible from the mine at the beginning of 1998 (Schickler, personal communication, 1998).

Another important factor is the simple fact that large areas of the world's land surface have not been mapped in detail for their minerals, let alone explored, particularly where underlying as opposed to surface geology is concerned. New ores are regularly discovered and new mines opened, such as the large copper mine at Neves Corvo in southern Portugal, which began operations in 1989 and produced more than 100 000 t of contained copper in 1997. These factors combine to balance the fears of those who suggest that the end is nigh for global mineral exploitation. In the words of one mining industry official: 'it is highly improbable that society will run out of minerals over the long term' (Crowson, 1992: iv). Indeed, the recycling of metals already taken from the lithosphere reduces some of the need for exploiting virgin resources, and Crowson suggests that these materials should be regarded as a renewable resource since most metals can be recycled indefinitely.

On the national scale, most countries have adequate reserves of minerals used in the construction industry and these materials (with the exception of cement) are seldom traded internationally. The use of other minerals, particularly metals, is very heavily concentrated in the rich countries, however, and their movement is an important component of international trade. Many developing countries rely heavily upon mineral exports as sources of foreign exchange, with at least 16 countries earning 40 per cent or more of their export income from non-fuel mineral exports in the early 1990s (Table 19.2). The sources of minerals for major industrial nations vary. Some industrialised countries have considerable domestic reserves of certain minerals: Australia mined 37 per cent of global bauxite production in 1990, for example, while for other minerals less developed countries are important

TABLE 19.2 *Non-fuel minerals as a percentage of the total value of exports from selected countries in recent years*

Country	Mineral(s)	Share (%)
Botswana	Diamonds, copper, nickel	87
Zambia	Copper	86
Guinea	Bauxite/alumina	82
Sierra Leone	Bauxite/alumina	80
Namibia	Diamonds, uranium, copper	76
Niger	Uranium	75
Zaire	Copper, diamonds	71
Surinam	Bauxite/alumina, aluminium	69
Papua New Guinea	Copper, gold	62
Liberia	Iron ore, diamonds	60
Jamaica	Bauxite/alumina	58
Togo	Phosphates	50
Central African Republic	Diamonds	46
Mauritania	Iron ore	41
Chile	Copper	41
Mongolia	Copper, molybdenum	40

Sources: Young (1992), World Bank (1992b)

sources. Generally, industrialised countries import their minerals from proximate areas of the developing world: from Latin America to the USA, Africa to western Europe, and Asia and Oceania to Japan. Most such ores are shipped from their country of origin with minimal processing, since trade barriers often act to prevent developing-country producers from adding value to their raw materials.

In 1990, the European Community, Japan and the USA accounted for 60 per cent of the world consumption of aluminium and refined copper, 58 per cent of refined lead, 55 per cent of tin and 50 per cent of zinc. However, since the Second World War, mineral use in the wealthier nations has not been growing as fast as in developing countries. Between 1950 and 1990, for example, the percentage of global aluminium used by developing countries rose from 2 per cent to 19 per cent, for zinc the percentage rose from 3 to 25 per cent over the same period, and for steel the rise was from 5 to 25 per cent (Wellner and Kürsten, 1992). This increase is largely accounted for by emerging industrial nations such as Brazil, China, India, Mexico and the 'Asian Tigers' of the Pacific rim.

The slow-down in demand in richer countries is explained by several factors. Industrial economies grew more slowly after the oil crisis of the mid-1970s, and structural change in these economies, with less emphasis on heavy industry, and moves towards services and high technology, has meant a slow-down in demand for physical materials. The growth of demand for virgin metals has also declined with more emphasis on recycling and the introduction of new materials such as plastics and ceramics.

ENVIRONMENTAL IMPACTS OF MINING

Concern at the environmental impacts of mining is by no means a recent phenomenon. Georgius Agricola (1556), author of the world's first

TABLE 19.3 *Environmental problems associated with mining*

Problem	Type of mining operation			
	Open pit and quarrying	Opencast (as in coal)	Underground	Dredging (as in tin or gold)
Habitat destruction	×	×	—	×
Dump failure/erosion	×	×	×	—
Subsidence	—	—	×	—
Water pollution	×	×	×	×
Air pollution*	×	×	×	—
Noise	×	×	—	—
Air/blast/ground vibration	×	×	—	—
Visual intrusion	×	×	×	×
Dereliction	×	—	×	×

×: Problem present
—: Problem unlikely
*: Can be associated with smelting, which may not be at the site of ore/mineral extraction
Source: modified after Whitlow (1990)

mining textbook, could have been writing today as he outlined the strongest arguments of mining's detractors in sixteenth-century Germany:

> *[T]he woods and groves are cut down, for there is need of an endless amount of wood for timbers, machines and the smelting of metals. And when the woods and groves are felled, then are exterminated the beasts and birds, very many of which furnish a pleasant and agreeable food for man. Further, when the ores are washed, the water which has been used poisons the brooks and streams, and either destroys the fish or drives them away.*
>
> *(Agricola, 1556: 8)*

Little appears to have changed in the more than 400 years since Agricola. In recent times, environmentalists have often quoted this passage as evidence of the long-standing environmental impact of mining. Interestingly, the quote is often used out of context, since Agricola in fact dismissed these anti-mining sentiments on the grounds that the environmental cost was far outweighed by the benefits to society:

> *If we remove metals from the service of man, all methods of protecting and sustaining health and more carefully preserving the course of life are done away with. If there were no metals, men would pass a horrible and wretched existence in the midst of wild beasts; they would return to the acorns and fruits and berries of the forest.*
>
> *(Agricola, 1556: 14)*

There can be no doubting that mining involves an environmental impact, and the destruction of habitats suggested in Agricola's textbook is perhaps the most obvious from among a diverse catalogue of ecological changes that the various types of mining operation can cause (Table 19.3). The question as to whether such impacts should be regarded as worth the benefits derived from the minerals extracted is a complex one, which will be returned to after consideration of the physical problems involved.

Habitat destruction

Vegetation is stripped and soil and rock moved, both for the extraction process itself and the

TABLE 19.4 *National areas of land disturbed by mining in the countries of the former Soviet Union*

Country	Area of disturbed land, beginning 1989 (thousand hectares)	Years required to reclaim
Russia	1180	11
Ukraine	198	8
Belarus	120	13
Uzbekistan	52	32
Kazakhstan	167	16
Georgia	3	7
Azerbaijan	21	19
Lithuania	32	27
Moldova	3	6
Latvia	44	44
Kyrgyzstan	7	16
Tajikistan	6	15
Armenia	8	82
Turkmenistan	5	7
Estonia	47	78

Source: after Bond and Piepenburg (1990)

concomitant building of plant, administration and housing facilities, and the history of mining activity has left considerable areas of disrupted habitat. The scale of damage is greatest at so-called opencast or strip mining sites. Several small coral islands in the Pacific Ocean exploited for their phosphate, the fossilised remains of centuries of birds' droppings, are among the worst affected. Opencast phosphate mining on the Pacific island state of Nauru has left a severely degraded landscape of highly irregular solution-pitted limestone over more than half of the island's 2100-ha area (Manner *et al.*, 1984).

Disposal of waste rock and/or 'tailings' – the impurities left after a mineral has been extracted from its ore – usually destroys still larger areas of natural ecosystems. This is an increasing problem, both because of the growth in demand for minerals and because as rich ores are mined out, so lower-grade deposits are worked, producing more waste per unit of mineral produced. While four centuries ago the average grade of copper ore mined was about 8 per cent (Bosson and Varon, 1977), the average grade in the 1990s was 0.91 per cent, which means about 990 million tonnes of waste generated for the 9 million tonnes of copper produced.

Similar product-to-waste ratios are found in the china clay mines of south-western England, where the production of 1 t of kaolin produces 1 t of mica, 2 t of undecomposed rock ('stent') and 6 t of quartz sand. In Cornwall and Devon, where china clay has been mined since the 1770s, these waste materials are dumped on surrounding land to produce a devastated landscape of heaps, pits and lagoons, with disused land in between, covering an area of 9100 ha (Bradshaw and Chadwick, 1980). On the national scale, such mining wastelands can cover a significant proportion of the national land area. Punning (1993) reports that excavation of Estonia's 4-billion-tonne oil shale reserves, the world's largest commercially exploited deposit,

Figure 19.1 Strip mining at the Cordero Rojo coal mine in the Powder River basin, north-western USA (Photo: Rio Tinto plc).

plus sand, gravel and peat extraction, has devastated at least 45 000 ha, or 1 per cent, of the national land area. Table 19.4 shows the areas disturbed in the countries of the former Soviet Union.

To date, however, the largest areas of strip mining in the world are for coal (Fig. 19.1). In the USA, which produces about 60 per cent of its coal by this method, it has affected vast areas of the Appalachian Mountains and large areas of the states of Ohio, West Virginia and Kentucky. On average, surface mining for coal consumes 10–30 ha of land for every million tonnes of coal produced in the USA, but in other parts of the world, local conditions and available technology can result in much higher land consumption-to-production ratios. In southern Russia's Kuznetsk Basin, for example, the figure is about 166 ha per million tonnes (Bond and Piepenburg, 1990).

In areas where smelting takes place near the extraction site, further damage can be caused by cutting trees to fuel smelters. In Copper Basin, Tennessee, USA, an area of 14 500 ha was deforested in the nineteenth century to smelt copper ore in vast open-air pits, and the charcoal requirements for iron ore smelting in the Grande Carajás Project in Pará state, Brazil, will spur the cutting of an estimated 50 000 ha of tropical forest each year during the mine's expected 250-year life (Fearnside, 1989). Although fuel for smelting was originally proposed to be grown on plantations, in practice, much of the charcoal has been made from natural forest.

The environmental impacts of mining are not just limited to terrestrial ecosystems. The mining of coral and sand for road and building materials is a widespread threat to coastlines in many parts of the world (Fig. 19.2) and particularly to island reefs in the Pacific Ocean. Apart from the physical destruction of reefs to provide limestone, and the introduction of toxic substances to the marine environment during the mining process, the dredging of sand causes beach erosion and alters water circulation, leading to sedimentation on reefs (IUCN, 1988; see also page 104).

Marine ecosystems can also be degraded by any form of mining in the coastal zone. A range of effects has been observed on the coral reefs fringing Misima Island in Papua New Guinea following the start of opencast mining for gold in 1989 (Fallon *et al.*, 2002). Increased soil erosion from the mine caused higher sedimentation rates on the reef, causing effects ranging from coral mortality due to smothering to contamination from metals contained in the sediment.

When exploitation of polymetallic (manganese) nodules from the deep-sea bed becomes a commercial proposition, although this is likely to be a long time in the future, it too will bring disturbances to the world's oceans. Impacts will occur at different levels in the water column: at the sediment surface due to collector impact, immediately above the sea bed due to the plumes of sediment raised, and at other levels as abraded nodule materials and sediments are discharged

Figure 19.2 Sand mining from beaches in Oman, in areas where building sand is scarce, is threatening breeding sites of endangered green and hawksbill turtles. Coastal erosion studies indicate that many of these beaches are ancient and slow to generate, hence removal of sand threatens their existence. Some beaches are earmarked for tourist development, another reason for their protection.

(Thiel and Schriever, 1990). The most commercially significant concentrations of nodules are found in the Pacific and Indian oceans.

Geomorphological impacts

Mining involves the movement of quantities of soil and rock, and thus by definition is a human-induced geomorphological process. The scale of the process has reached the level whereby estimates suggest that as much as a hundred times more material is stripped from the Earth by mining than by all the natural erosion carried out by rivers (Goudie, 1993b). The impacts of mining, however, are much more localised than the denudation caused by rivers. The opencast copper mine at Bingham Canyon in Utah, USA, reputed to be the largest human excavation in the world, has involved the removal of more than 3400 million tonnes of material over an area of 7.2 km^2, to a depth of 770 m. Nevertheless, such features can attain significant proportions on the national scale: surficial scars on the landscape of Kuwait, created by the oil-generated construction boom, cover large parts of the national land area, with some individual quarries measuring up to 40 km in length (Fig. 19.3). Reconstruction of the country's urban fabric after the Gulf War of the early 1990s has continued to enlarge these surface features.

New landforms are also created by the deposition of wastes, and these can create new geomorphological hazards, such as landslides and subsidence. The death of 150 people at Aberfan, South Wales, in 1966, when waste piles from the Merthyr Vale coal mine suddenly collapsed, graphically illustrated the dangers of unregulated waste tipping.

Surface subsidence is caused by underground workings, and is most commonly associated with coal mining where whole seams of material, frequently several metres thick, are removed, resulting in overlying strata collapsing downwards. Two major types of mining subsidence can be identified (Scott and Statham, 1998):

1 *crownholes* – localised crater-like holes that appear at the surface following collapse of strata into a mine

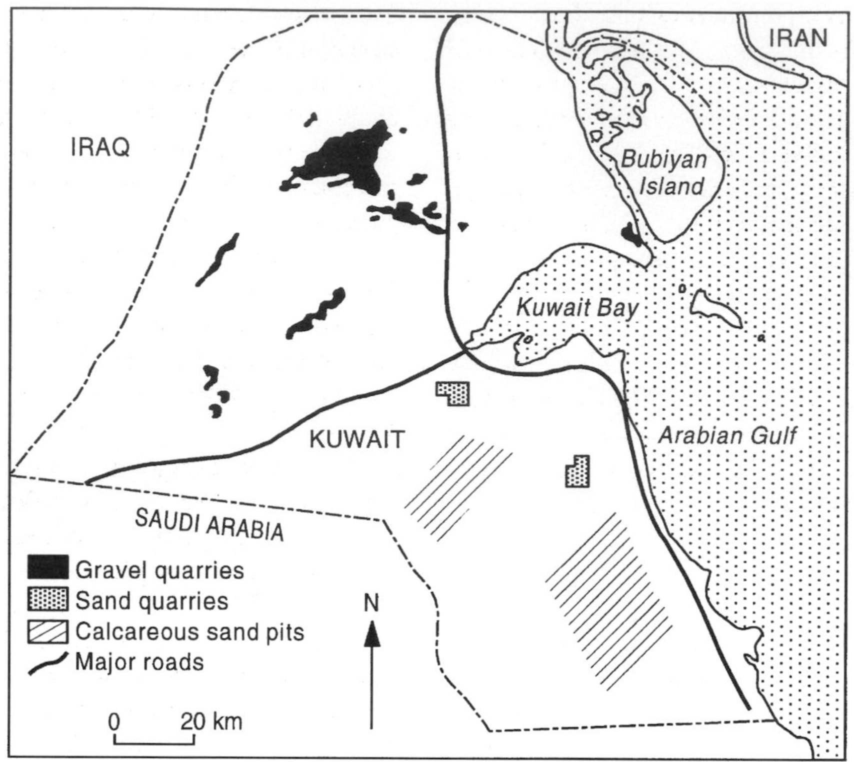

Figure 19.3 Distribution of sand and gravel quarries in Kuwait (after Khalaf, 1989).

2 *general subsidence* – settlement of the ground surface over a wide area resulting from the collapse of part of the mine.

Damage to mining equipment, buildings, communications and agriculture through the disruption of drainage systems are the most common results. In the USA, for example, subsidence due to underground coal mining affects 800 000 ha in 30 states (Gray and Bruhn, 1984). The problem is also common in Polish Silesia and in the Ruhr Valley cities of Germany. Subsidence into old gold mine workings meant that an entire street of houses had to be moved in the town of Malartic in Quebec, Canada (McCall, 1998).

Collapse also occurs when water is pumped

TABLE 19.5 *Examples of ground subsidence due to oil and gas abstraction*

Location	Area affected (km^2)	Maximum subsidence (m)	Period	Rate (mm/year)
Inglewood, USA	–	2.9	1917–63	63
Maracaibo, Venezuela	450	5.03	1929–90	82
Wilmington, USA	78	9.3	1928–71	216

Source: after Cooke and Doornkamp (1990); Goudie (1993b)

out of mine workings, lowering local water tables, which results in the drying out and shrinkage of materials such as clays, and through the reduction in underground fluid pressure that results from oil and natural-gas extraction. Blunden (1985) notes that marine inundation is a danger caused by subsidence from oil and gas extraction in areas near sea level and quotes examples from oil fields on the shore of Venezuela's Lake Maracaibo, coastal locations in Japan and the Po delta in Italy. A particularly dramatic example occurred in the Wilmington oilfield in southern California where abstraction between 1928 and 1971 caused 9.3 m of subsidence at Long Beach, Los Angeles (Table 19.5).

Pollution

Pollution from mineral extraction, transportation and processing can also present serious environmental problems, affecting soil, water and air quality. Examples of pollution emanating from the transport of oil are given elsewhere (see Chapters 6 and 7), while here the emphasis is on mine workings and smelting.

Down and Stocks (1977) note four major water pollution problems associated with mining, all of which can cause serious damage to aquatic life and, on occasion, to human health:

1 heavy metal pollution
2 acid mine drainage
3 eutrophication
4 deoxygenation.

Many tailings contain metal ores and other contaminants, formerly locked up in solid rock, which can be leached into soils and waterways, and blown into the surrounding atmosphere. Metals are also released into the environment during smelting. Contamination by heavy metals released by these activities has been very extensive from the earliest times. The concentration of copper found in Greenland ice layers began to exceed the natural background concentration about 2500 years ago due to smelting emissions during Roman and medieval times, particularly in Europe and China. Similarly, the lead content of Greenland ice was about four times the background concentration during the period 500 BC to AD 300 (Nriagu, 1996). Deposits of such metals in other parts of the environment may continue to act as significant secondary sources of contamination for hundreds of years after mining has ceased. Historic lead and zinc mining in the English Pennines has left a legacy of metals in alluvial deposits that are released under flood conditions, significantly affecting water quality in this part of northern England (Macklin *et al.*, 1997).

Some of the highest heavy metal concentrations are derived from uranium tailings and pumped water from iron mines, but one of the most notorious examples of deleterious effects on human health came from a lead–zinc mine in Toyama prefecture, Japan. Cadmium-polluted water from the mine, which was used in rice paddy fields, led to an outbreak of 'itai-itai' disease, a cadmium-induced bone disorder. About 100 deaths were reported from the disease among long-term residents of the area between 1947 and 1967 (Nriagu, 1981).

Acid mine drainage is responsible for water pollution problems in major coal-mining and metal-mining areas around the world. It was originally thought to be associated only with coal mining, but any deposit containing sulphide, and particularly pyrite, can be a source, and it is produced from the working of many other minerals. Reactions with air and water produce sulphuric acid, and pH values of polluted waterways may be as low as 2 or 3. Acid mine drainage is not just a problem of operational workings, but may continue as a pollution source for some years after mine closure (Robb, 1994). Surface tailings can also produce acid effluents; Hägerstrand and Lohm (1990) report that acid produced from sulphides in tailings around the Falun copper mine in central Sweden, which has been worked for 1000 years, have seriously damaged several lakes in the region and are still producing sulphuric acid. A large survey of freshwater bodies in acid-sensitive areas of the USA concluded that 3 per cent of lakes sampled and 26 per cent of streams had been acidified by acid mine drainage (Baker *et al.*, 1991).

When tailings dry out, dust blow can also cause localised pollution. Fine material blown from the quartzite dumps that surround Johannesburg, a city that grew up on the gold-mining activity in South Africa's Witwatersrand, used to stop work at food-processing factories and cause traffic chaos due to decreased visibility on windy days before the dumps were stabilised with vegetation (Bradshaw and Chadwick, 1980).

The processing of minerals can also release dangerous compounds into surrounding environments. In Brazil, where about 500 000 mostly small-scale prospectors were involved in gold mining in the late 1990s, the use of mercury as an amalgamate to separate fine gold particles from other minerals in river-bed sediments contaminates waterways in 'gold rush' zones. Miners released an estimated 100 t of mercury – an extremely toxic metal that accumulates in the food chain, and can cause birth defects and neurological problems – into the basin of the Madeira River, a tributary of the Amazon, each year. Fish eaten by the local population were found to contain mercury at concentrations up to five times the Brazilian safety limit (Malm *et al.*, 1990). Similar accumulations of mercury after use by artisanal gold miners have been reported from many other parts of South America, including Bolivia, Colombia, Peru and Venezuela. Its use in the 1950s and 1960s at the Discovery mine in the Canadian Northwest Territories means that nearby Giauque Lake is still designated a contaminated site by Environment Canada (Meech *et al.*, 1998). Globally, gold mining is thought to be responsible for about 10 per cent of all anthropogenic emissions of mercury (Lacerda, 1997).

Sulphur dioxide (SO_2) released from the open-air smelting of nickel ore at Sudbury, Ontario, Canada, in the early years of the twentieth century, destroyed vegetation over a 10 000 ha area. Construction of a closed plant with a 387.5 m stack in 1972 turned a local problem into a regional one, affecting lakes up to 60 km downwind, some with a pH of 5.5 or less. Nickel emissions from the stack have also been high. Nickel concentrations are naturally low in lake waters, with a global mean of 10 g/l. Some lakes in the Sudbury area have been found with nickel concentrations of 300 g/l. The combination of acid rain and heavy metal pollution has rendered many of the lakes in the area biologically dead (Blunden, 1985).

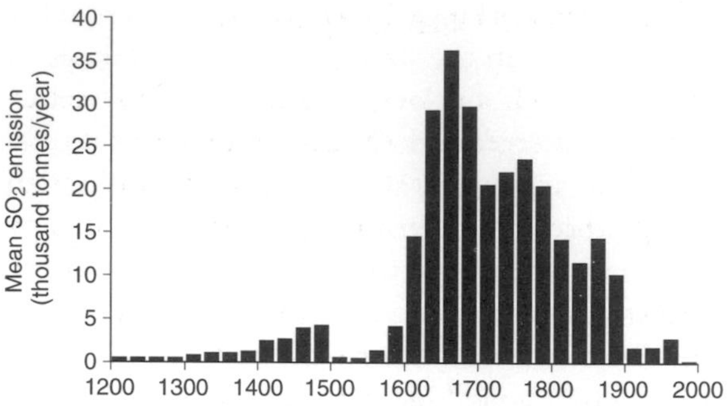

Figure 19.4 Estimated annual emissions of sulphur dioxide as 25-year means from the Falun mine, Sweden, 1200–2000 (Ek *et al.*, 2001).

Very long-lasting ecological effects of SO_2 emissions have also been reported from the Falun copper mine in Sweden where the first step in copper extraction was to roast the ore in the open air to reduce its high sulphur content. This process oxidised the sulphide, emitting sulphur to the air as SO_2. These emissions, which date from at least the thirteenth century, reached a maximum during the 1600s (Fig. 19.4) when Falun produced two-thirds of the world's copper supply. Deposition of SO_2 in the area around Falun exceeded the critical load in this region for centuries and acidification of local soils is still apparent, although it is limited to the most heavily polluted area 12 km to the north-west and to the south-east of the mine. Lakes in the Falun area also became more acidic dating from the seventeenth century, although the pH decrease has been moderate. However, despite 300 years of lowered emissions since the 1600s (the mine was closed in 1993), none of the acidified lakes in Falun show pH recovery due to a large store of sulphate in the soil, although land use change in the last 100 years is also likely to have had some effect (Ek *et al.*, 2001).

Miscellaneous impacts

The attraction of mining operations as sources of employment may have deleterious effects on other labour-intensive areas of the economy such as agriculture. Lewis and Berry (1988) cite Zambia's Copper Belt as an example, and Barrow (1991) suggests that Brazil's Pelada gold mine has had a similar effect. Itinerant miners venturing into little-explored terrain also carry with them new diseases and problems associated with alcohol, bringing often disastrous consequences for local populations. Gold prospectors in Amazonia, so-called *garimpeiros*, have spread virulent strains of malaria into the reserves occupied by the Yanomani indians in northern Amazonas. Local tribes with little resistance to the new strains of parasites have suffered severe outbreaks of malaria (Smith *et al.*, 1991), and met with previously unknown diseases such as measles.

The relationship between gold mining and malaria in Brazil is a complex one (Confalonieri, 1998) and the disease is rife among the miners themselves as well as local indigenous populations. The temporary shelters in mining camps provide little or no protection against mosquitoes, whose numbers are typically increase when miners destroy river banks, widening river beds to become swamplike habitats ideal for mosquito breeding. Costly treatment for malaria is beyond the means of many *garimpeiros* and those who can afford to buy anti-malarial drugs often stop taking the course when the fever subsides but before they are entirely cured. One result is the emergence of drug-resistant strains of the disease. Exposure among miners to mercury (see above) may also be playing a part in the high incidence of malaria, since mercury exposure is suspected to damage the immune system, perhaps increasing susceptibility to the disease. Malaria incidence in Brazil has risen steadily in recent decades, in line with the Amazonian gold rushes, from about 50 000 cases in 1970 to more than 0.5 million in the mid-1990s. The large majority of these cases are in the Amazon Basin, and indigenous populations have been severely hit. In the state of Roraima, where *garimpeiros* invaded the homeland of 10 000 Yanomani indians in 1987, malaria was responsible for 25 per cent of Yanomani deaths in the period 1991–95.

There are also cases where mining operations have had serious political ramifications. A classic example in this respect can be cited from the Panguna copper mine on the island of Bougainville in Papua New Guinea, where discontent over the distribution of mining revenues was a factor in precipitating a revolt out of a long-simmering dispute between the islanders and the PNG Government. Discontent over the unequal distribution of the income derived from mining was not helped by the fact that 130 000 t of metal-contaminated tailings were dumped into the Kawerong and Jaba rivers every day, severely depleting biological activity in the waters. The resulting conflict led to guerrillas taking over the island and the subsequent closure of the mine in 1989.

REHABILITATION AND REDUCTION OF MINING DAMAGE

Many techniques are available for the amelioration, curtailment or restoration of damage caused by mining and its associated activities (e.g. Down and Stocks, 1977; Bradshaw and Chadwick, 1980;

Figure 19.5 Rehabilitated grassland at the Cordero Rojo coal mine, north-western USA. Compare with Figure 19.1 (Photo: Rio Tinto plc).

Figure 19.6 An area of rehabilitated waste rock dumps at Bingham Canyon copper mine, Utah, USA (Photo: Rio Tinto plc).

Blunden, 1985), and there is little doubt that for a variety of reasons the mining industry in general has recently adopted a more benevolent approach to its environmental impact than in times past. Site surveys prior to mining, retention and replacement of topsoil after excavation, and careful reseeding with original species can be employed to return environments to something close to their original states (Fig. 19.5).

Waste dumps and tailings can be stabilised with vegetation both to ameliorate their visual impact and to reduce the risks of slope failure (Fig. 19.6), or flattened and properly drained as in the case of the waste tips at Aberfan. Acidic minespoil often requires the application of alkaline material before a vegetation cover can be successfully established and one method of so doing currently being developed uses waste by-products from dry flue gas desulphurisation scrubbers fitted to industrial chimneys to remove sulphur gases (Stehouwer *et al.*, 1995). The scrubbers use lime, limestone or dolomite sorbents to react with SO_2, forming anhydrite. The resulting by-products – a mixture of coal ash, anhydrite and residual sorbent – often have pH levels of 12 or greater due to the residual sorbent. The use of this material to help neutralise acidic mining wastes before rehabilitation is a good illustration of how the reappraisal of waste products can create new resources (see page 290). Disposal of scrubber waste has previously been in landfill sites, but the efforts being made in many countries to reduce acid rain emissions means that the volume of these by-products will rise substantially in coming years. Using them on acidic minespoil is therefore a much more sustainable method that helps to resolve two different environmental issues.

Mining wastes themselves are used for a variety of purposes in many countries, particularly in road construction and for landfill (Table 19.6), but transportation costs to areas of high demand often limit their use. Of the 27 million tonnes of china clay waste produced each year in Devon and Cornwall, about 1.5 million tonnes are used by local industries, but only a small proportion of the 12 million tonnes of china clay sands that have potential for use as aggregates are used, because local demand is low (DoE, 1992).

Many of the efforts made to reduce pollution originating in mining wastes have also stemmed from the realisation that the pollutants are themselves economically valuable. A 40 per cent reduction in SO_2 emissions from the Sudbury nickel smelter, for example, was achieved in the late 1960s when the gas was converted into sulphuric acid. Similarly, the use of new biotechnologies for recovering wasted resources, such as microbes to 'bioleach' metals from tailings, have also been developed in response to the lower grades of accessible ores and lower operating costs. Microbial mining of copper sulphide ores has been practised on an industrial scale since the late 1950s and, since then, bioleaching has also become common at uranium and gold mines. Another development in mining biotechnology is the use of bacterial cells to detoxify waste cyanide solution from gold-mining operations (Agate, 1996). The recycling of minerals, particularly metals, also helps to reduce environmental impacts, of course (see Chapter 17).

In recent decades, a growing public concern in

TABLE 19.6 *Production and use of mine and quarry wastes (waste rock and tailings) in selected countries*

Country/ mine type	Production/year (thousand tonnes)	Use
AUSTRALIA		
Coal	60	Small amounts for road construction and fill
BELGIUM		
Coal	4	Lightweight aggregate manufacture
FINLAND		
Various	8	Road construction; aggregate for concrete
INDIA		
Gypsum	2	Building plaster
Mica	>1	Mica insulation bricks
Coal	10	Mine backfilling (large percentage); road construction and fill
SOUTH AFRICA		
Gold ore	45	Road construction and aggregate for concrete (50 per cent); silicate bricks
UK		
China clay	22	Aggregate for concrete, silicate bricks, road construction and fill (<5 per cent)
Tin ore	0.5	Aggregate for concrete (minor percentage)
Coal	50	Road construction, bricks, cement, lightweight aggregate manufacture (10 per cent)
USA		
Copper	860	Road construction; bitumen filler
Iron ore	55	Road construction
Miscellaneous quarrying	68	Amenity banks

Source: after Blunden (1985)

the industrialised nations at the environmental cost of mining has led to increasing legislative controls on the activities of mining companies, including the requirement to conduct Environmental Impact Assessments (Ellis, 1989). The generally higher public profile of environmental issues and the pressures brought to bear by environmentalists, particularly in developed countries, has encouraged multinational mining companies to give environmental concerns a higher profile. The formation in 1991 of the International Council on Metals and the Environment, a group of more than 25 major metal-mining companies, is a clear indication of this trend. The Council aims to promote sound environmental and related health policies and practices to ensure the safe production, use, recycling and disposal of metals.

Increasingly, responsible mine operations have adopted the precautionary principle, with

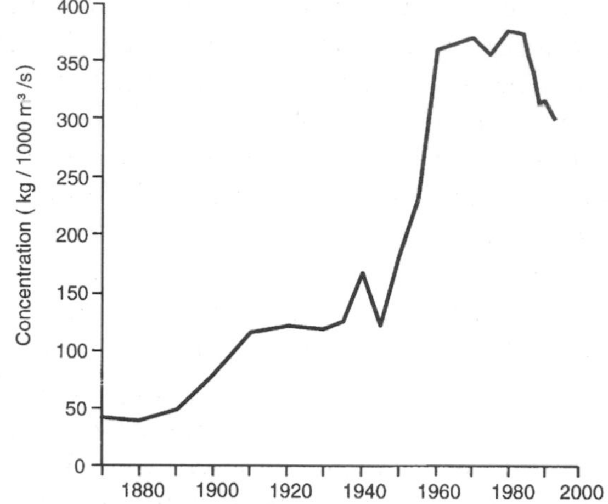

Figure 19.7 Chloride concentrations in the River Rhine at Kleve-Bimmen, 1875–1992 (after CIPRCP, 1984; Chapman, 1996).

considerable attention to the possible pollution effects, involving regular monitoring of groundwater, air quality and noise, to maintain acceptable standards. In some cases, such as at the Neves Corvo copper mine in Portugal, tailings are delivered to special ponds to reduce dust blow and to prevent oxidation of the contained sulphur, which could give rise to acidic water. The dam is operated on a no-discharge basis, with any overflow during storms caught and pumped back. Environmentally sound management and clean-ups cost money, however, and legal obligations are often necessary to force appropriate practices. There is little doubt that decades of environmentally unregulated coal mining in the USA resulted in widespread pollution of water bodies, subsidence effects, scarred landscapes and extensive waste dumps. Rehabilitation of this environmental degradation only became standard practice with the passing of the Surface Mining Control and Reclamation Act in 1977 (McGinley, 1992). The Act mandates that US coal-mining companies reclaim land in a fashion that renders the land as productive after mining as it was when mining started. While many similar legal obligations are now well established in most developed countries, they are often inadequate in developing countries where the importance of mining to the national economy (Table 19.2) makes the imposition of costly rehabilitation measures unattractive. The use of legally enforced pollution control equipment on copper smelters in the USA means that for each tonne of copper produced, 12 times less SO_2 is emitted than from smelters in Chile (UNECLA, 1991).

Legal instruments also cover international pollution problems. Massive salinisation of the River Rhine by chloride pollution (Fig. 19.7), 30 per cent of which is discharged by potash mines in the Alsace region of France, has become the subject of an international convention (Kiss, 1985). The Convention on the Protection of the Rhine against Chlorides was signed in 1976 by Switzerland, the former West Germany, France and the Netherlands but was not ratified by France until 1985 because of protests over proposals for the alternative disposal of the Alsatian salt. Under the convention, the French Government was required to install a subsoil injection system to dispose of the salt, paid for by the affected states in proportion to the degree to which they were polluted, in contradiction to the 'polluter pays' principle, but local farmers resisted the salt injection plan. The French now propose to store the salts above ground.

Care must be taken, when establishing clean-up targets, to set appropriate standards. Runnels *et al.* (1992) point out that levels of minerals and pH in rivers can be naturally high in areas of undisturbed mineral deposits due to weathering and leaching of metalliferous ores. Ironically, over the course of time, some areas previously damaged by mineral extraction have come to be regarded as sites of beauty and environmental importance with minimal active rehabilitation. Some former quarries and gravel pits in the UK are now designated nature reserves, such as the Felmersham Gravel Pits in Bedfordshire, a Site of Special Scientific Interest noted for its large number and wide variety of dragonflies, while the Norfolk Broads are entirely the work of medieval peat cutters (Lambert *et al.*, 1970).

In other situations, mining may even be advocated as an option for revitalising ecosystems damaged by other forms of development. Björk and Digerfeldt (1991) describe how peat extraction in two former wetland areas of Jamaica has allowed their re-establishment. The Black River and Negril wetlands were drained and canalised for the cultivation of rice and cannabis, respectively, but successful revitalisation of the wetland habitat after peat cutting on the Negril has led to similar proposals for the Black River.

THE TRUE COSTS OF MINING

The value to society of the products of mining are undeniable, providing as they do the basic building blocks of everyday life for virtually the entire human population. Mining inevitably entails disruption to the natural environment, although given minerals' importance to society and the relatively localised nature of the impacts, the disruption is on a far smaller scale than that of agriculture, for example.

Many positive steps are being taken to limit the environmental impacts of mining. The recycling of metals, for example, means less new damage and often makes economic sense even when the economics only take extraction, processing and distribution into account. Other new technologies for recovering wasted resources, such as the use of microbes to bioleach metals from tailings, for example, have also been developed in response to lower-grade ores. Operational mines, too, are increasingly incorporating environmental rehabilitation into their plans. Although such action often takes a higher priority in countries where legislation demands it, multinationals operating in countries with less stringent requirements also yield to pressures for more environmentally sustainable practices called for by shareholders and environmental pressure groups (Banks, 1993). For multinational companies whose largest stock in trade is their reputation, public feeling over the importance of the environment has undoubtedly influenced practice both directly and indirectly through legislation.

Ultimately, it is for society to decide whether the environmental cost is worth the societal benefits of mining output. Bingham Canyon, which has operated since 1904 and has another 30 years of life, has long provided about 15 per cent of US copper and made a significant contribution to the economy of the state of Utah. It is for American society to decide whether the environmental cost, in terms of a large hole in the ground and associated disruption to the local environment, is worth this contribution to society.

In this respect, as for numerous other environmentally damaging societal practices, there have been calls for the true costs, including those of clean-up and rehabilitation, to be incorporated into the price of the minerals themselves (e.g. von Below, 1993). In practice, the concept of 'full-cost pricing' is difficult to quantify, particularly for the mining industry where costs vary widely among producers, reflecting widely differing mineral endowments, and where prices set in international auction markets bear only a tenuous relationship to costs. Irrespective of the difficulties in putting economic values on certain aspects of the environment (see page 41), the cost–benefit equation that balances environmental and use value of minerals will vary from country to country and society to society. In practice, it is impossible to set objective 'correct' prices (Humphreys, 1993), although environmentalists are right to point out that some countries have a freer hand to assess this cost benefit than others, given the heavy reliance of some developing countries on income from their mineral exports.

Problems may occur, however, when those who benefit are not fully coincident with those who pay the costs. Banks (1993) points out that, generally, the greatest costs in environmental and social terms fall on local populations around mine sites, while the benefits, usually measured in economic and political terms, tend to be concentrated at the national and international scales (Fig. 19.8). When the cost–benefit equation is unbalanced, difficulties may arise, as the experience of Bougainville illustrates. Redressing the imbalances can also act retrospectively: a global precedent was set in 1993 when Australia paid off Nauru to settle the island's claim against the British Phosphate Commission for environmental degradation of two-thirds of its territory.

Figure 19.8 Collahuasi mine in Chile, reputed to be the world's fourth-largest copper mine, which exports most of its production. Situated in a remote part of the Andes, the mine has had little negative impact on a very sparse local population, but elsewhere local people bear the brunt of the costs of mining for benefits that tend to be concentrated at the national and international scales.

FURTHER READING

Banks, G. 1996 Compensation for mining: benefit or time-bomb? The Porgera gold mine. In Howitt, R., Connell, J. and Hirsch, P. (eds) *Resources, nations and indigenous peoples*. Melbourne, Oxford University Press: 223–35. An examination of the social changes brought about by a mine in Papua New Guinea.

Blunden, J. 1985 *Mineral resources and their management*. London, Longman. A comprehensive review of mineral resource management, including techniques for rehabilitation and reduction of mining's environmental impact.

Bosson, R. and Varon, B. 1977 *The mining industry and the developing countries*. New York, Oxford University Press. This book focuses on the role of mineral extraction in developing-world economies.

Down, C.G. and Stocks, J. 1977 *Environmental impact of mining*. London, Applied Science. A comprehensive review of the full range of environmental problems associated with mining and the techniques developed to combat them.

Ek, A.S., Löfgren, S., Bergholm, J. and Qvarfort, U. 2001 Environmental effects of one thousand years of copper production at Falun, central Sweden. *Ambio* 30: 96–103. A historical example of mining pollution with implications for more contemporary studies of biological recovery from acid rain.

Meech, J.A., Veiga, M.M. and Tromans, D. 1998 Reactivity of mercury from gold-mining activities in darkwater ecosystems. *Ambio* 27:

92–8. An overview of mercury pollution from artisanal gold mines, with particular emphasis on Brazil.

Young, J.E. 1992 *Mining the Earth*. Worldwatch Paper 109. Washington, DC, Worldwatch Institute. This pamphlet documents the environmental impacts of mining, and argues for more efficiency in mineral use, less government subsidy for mining and the need for developing countries to find less mineral-dependent modes of development.

Websites

http://moles.org/ news, information and campaigns from a non-governmental organisation that supports communities adversely affected by the mining and oil industries worldwide.

http://www.mpi.org.au/ the Mineral Policy Institute, an Australian body with a focus on social and environmental impacts of mining in Australia and the Asia Pacific region.

http://www.bullion.org.za/ the South African Chamber of Mines has mining data, educational resources and information on environmental policy.

http://www.riotinto.com/ one of the largest mining groups, Rio Tinto, presents educational resources, environmental reports and links to other mining companies.

http://www.globalmining.com/ the Global Mining Initiative aims to incorporate sustainable development objectives into the mining industry.

http://www.mineralresourcesforum.org/ a UNEP site with reports and news on many aspects of mining and the environment.

POINTS FOR DISCUSSION

- Is society eventually going to run out of minerals?
- Outline the main environmental impacts associated with mining.
- There are many techniques available to rehabilitate and reduce the environmental impact of mining, so why are they not always employed?
- Transnational mining companies should not be the target of environmental activists because the resources the companies exploit are for the benefit of everyone. How far do you agree?

20

WAR

TOPICS COVERED

Cost of being prepared
Direct wartime impacts
Indirect wartime impacts
Limiting the effects
Environmental causes of conflict

Key words: scorched earth policy, ecocide, nuclear winter

Aggression appears to be a fundamental characteristic of human nature, and violence has been used to resolve disputes since prehistoric times. Warfare is no less characteristic of the world today than in the past: since the end of the Second World War, more than 130 wars and violent internal conflicts have raged in more than 80 countries, most of these in the developing world. With the course of history, however, technological developments have greatly enhanced humankind's capacity for destruction and while the devastation to human life and civilisation is well appreciated, the ecological impacts of war are less well documented. They are, none the less, wide-ranging and not always obvious. In the broad perspective, war and the preparations for war are the antithesis to development, squandering, as they do, scarce resources and eroding the international confidence necessary to promote development. While competition for the control of resources has long been a reason for conflict, in recent years, resource degradation is increasingly being seen as a cause of war, and one that looks set to become more important with an increasing world population.

COST OF BEING PREPARED

Keeping armies operational and their weaponry updated represents a huge drain on economic, human and natural resources. Over the two decades of the 1970s and 1980s, the world spent about US$17 trillion on military activity at 1988 prices, averaging US$850 billion per year (SIPRI, 1990) and although military spending has fallen slightly since the mid-1980s in industrialised countries, it has continued to increase in most developing countries. The military employs between 60 and 80 million people worldwide, and arms imports by developing countries, half of which are financed by export credits, account for 30 per cent of the debt burden of those countries (United Nations, 1989b). The scale of military spending, some of its contradictions and trade-offs with social and environmental priorities are quoted by Tolba (1992):

- the UN Environment Programme spent US$450 million in the decade 1980–90, less than five hours' worth of global military spending
- one Apache helicopter costs US$12 million, a sum that could pay for the installation of 80 000 hand pumps to give Third World villages access to safe water
- one day spent on the 1991 war over Kuwait – US$1.5 billion – could have funded a five-year global child-immunisation programme against six deadly diseases, thereby preventing the death of 1 million children per year.

Although data on the military's consumption of resources are, like much information on the military, difficult to obtain, an indication of the scale of impact is given in the estimates collated by Renner (1991). The US military consumes 2–3 per cent of total US energy demand, while worldwide nearly 25 per cent of all jet fuel is used for military purposes. An estimate of the military consumption of non-fuel mineral resources is shown in Table 20.1. These estimates suggest that the quantities of aluminium, copper, nickel and platinum used by the military worldwide exceeds the entire developing world's demand for these minerals. Between 0.5 and 1 per cent of the world's land surface is thought to be used by the military during peacetime, and in many countries the military owns and/or manages very large areas of land for training and weapons testing, land that is for the most part closed to civilian access. In the Netherlands, 21 per cent of the largest area of continuous Dutch forest (the Veluwe) is in the possession of the military, as is a third of all Dutch heathland (Vertegaal, 1989). In some cases, such land is considered to have high conservation value. The ministry of defence in the UK is the custodian of more than 250 Sites of Special Scientific Interest (SSSIs), three of which form part of the Salisbury Plain training area, which contains the largest known expanse of unimproved chalk grassland in north-west Europe. This use of Salisbury Plain, some parts of which have been managed by the military since the late nineteenth century, has protected the grassland from the changes that have been wrought on most other lowland grasslands in Britain – urban development, agricultural improvement and industrial expansion.

However, military training on Salisbury Plain has its impacts too, of course. In a 50-year survey Hirst *et al.* (2000) found that in the 1990s habitat disturbance was occurring at a greater rate than natural regeneration following increased use of Salisbury Plain after the end of the Cold War, when many training facilities in Europe were closed. The higher levels of disturbance from heavy vehicles represent a significant threat to the chalk grassland through habitat loss and fragmentation.

TABLE 20.1 *Estimated military consumption of selected non-fuel minerals as a proportion of total global consumption in the early 1980s*

Mineral	Share (%)
Copper	11.1
Lead	8.1
Aluminium	6.3
Nickel	6.3
Silver	6.0
Zinc	6.0
Fluorspar	6.0
Platinum group	5.7
Tin	5.1
Iron ore	5.1

Source: after Kim (1984)

Military activities also generate large quantities of waste material, much of it hazardous. In the USA, the military is thought to be the country's largest generator of hazardous wastes, and more than 1500 military bases with consequent environmental problems have been identified. The largest of such contaminated sites, the Rocky Mountain arsenal in Colorado, has been used to dump 125 different dangerous chemicals over a 25-year period. Disposal of such material is no easy matter since its precise effects on the environment are not always well known. The USA is now disposing of many chemical weapons by incineration, but historically some of the methods used for their disposal have been less circumspect. Glasby (1997) notes that large quantities of chemical warfare agents were dumped into the Baltic Sea after the Second World War. These highly toxic chemicals were contained in wooden crates, which have probably disintegrated, allowing the widespread dispersal of chemicals. The long-term effects on the marine environment are not known.

Such problems are not confined to the national territories of the polluting armies. Soviet and

US forces, in particular, maintained numerous overseas bases during the Cold War years, many of which were vacated in the 1990s. Some 23 years of Soviet troop presence in the former Czechoslovakia, for example, has left a legacy of widespread groundwater contamination by waste oil, and similarly at sites in Moravia contaminated by poisons used in chemical warfare (Carter, 1993a). The question as to who should pay for clean-ups on these bases will remain a sensitive political issue for some time to come.

Some of the worst environmental degradation has been associated with the development of nuclear weapons. Testing of nuclear weapons in the atmosphere has caused the dispersion of radioactive material across the globe, adding a small increment to the background exposure of the world's population. A partial ban on atmospheric testing was signed in 1963, but it was not until 1980 that the last atmospheric test was made in north-west China. Underground testing of nuclear weapons, which often causes small to moderate earthquakes, still continues. The former Soviet nuclear test ground near Semipalatinsk in eastern Kazakhstan, which was closed in 2000 after nearly 40 years of testing, covers about 2000 km^2 of contaminated land. The location of the first nuclear explosion, carried out in July 1945 at Trinity Site in New Mexico, USA, is now considered safe enough to be open to visitors. However, visiting parties are strictly controlled, and children and pregnant women are actively discouraged. Part of the site will remain moderately radioactive for the foreseeable future (Szasz, 1995).

Serious contamination from weapons-producing plants has also come to light in recent times. By far the most serious accident involving radioactive wastes in the former Soviet Union, prior to Chernobyl, was an incident that occurred in 1957 at a secret nuclear weapons processing plant in the southern Urals known as Production Association Mayak, located between the towns of Kasli and Kyshtym, 80 km north-west of Chelyabinsk. The fact that the accident was not officially admitted to until 1989, when the then Soviet Government stated that 2 million curies of radioactive elements had been released from an exploded waste tank and deposited over an area of 90 000 ha, gives some indication of the secrecy surrounding such military operations. More than 10 000 people were evacuated from the area and parts of the region were still restricted more than 30 years later (Pryde, 1991).

It has also been revealed that Production Association Mayak routinely dumped medium-level and high-level radioactive wastes into the River Techa over the period 1949–56, totalling much greater doses of radioactivity for local populations, particularly those using the river for drinking water. Another serious incident occurred in 1967 when radionuclides from the dry banks of Lake Karachay, an open depot of radioactive waste from the plant, were blown into the surrounding region, severely contaminating 2700 km^2. The medical and ecological consequences of these incidents are the subject of continuing investigations (Akleyev and Lyubchansky, 1994).

Long-lasting environmental effects are also attributable to areas where biological weapons have been tested. A classic case is the island of Gruinard off the east coast of Scotland, site of experiments with highly contagious anthrax spores during the Second World War. The island remained uninhabited by government decree until 1988, but even now complete decontamination is difficult to guarantee (Szasz, 1995).

Anthrax was also one of numerous biological agents tested in the open air on Vozrozhdeniye Island, situated towards the western coast of the shrinking Aral Sea, site of a major testing ground for biological weapons during the Soviet era. Others include plague, Q fever, Venezuelan equine encephalitis and smallpox, many being special strains developed by the military to be resistant to conventional forms of treatment (Bozheyeva *et al.*, 1999).

Like Gruinard, Vozrozhdeniye was partly chosen for biological weapons testing because of its geographical isolation. Its insular location prevented the transmission of pathogenic

TABLE 20.2 *Military techniques for modifying the environment*

Technique	Military application	Feasibility
ATMOSPHERE		
Fog/cloud dispersion	Make target areas more visible	Relatively easy for supersaturated fog, difficult for warm fog
Fog/cloud generation	Protect target areas from attack and nuclear flash	Depends on availability of equipment and materials
Hailstone generation	Damage thin-skinned equipment, power/communication wires and antennas	Restricted to hail cloud conditions
Material to alter electrical properties	Interrupt communications and remote sensing	Unknown
Introduction of electric fields	Interrupt communications and remote sensing	Very high energies needed
Generating and directing destructive storms	Damage battlefields, ports, airfields	High energies needed but some success in hurricane dispersal
Rain and snow making	Inhibit mobility, block routes, hinder communications	Very dependent on cloud systems, highly localised, short duration
Lightning control	Start fires, destroy communications antennas	Conceptually may be possible
Climate modification	Strategic effect on food production and ecology	Very large energy input
Change of high atmosphere or ionosphere	Strategic effect on food production and possibly on human survival	Uncertain
OCEANS		
Change of physical, chemical and electrical parameters	Affect acoustic paths, possible effect on potential food supplies	Uncertain, but large energies may be necessary
Addition of radioactive material	Long-term effect on humans who may have ingested contaminated food	Uncertain and difficult
Tsunami generation	Destroy low-lying areas, surface fleets	Uncertain and difficult
LAND		
Earthquake/tsunami generation	Damage battlefields, major military bases and strategic facilities	Specific to certain locations and mechanism not fully understood

TABLE 20.2 *(continued)*

Technique	Military application	Feasibility
Burning of vegetation	Generally destructive	Easily started but dependent on flammability of vegetation, and weather conditions
Generation of avalanches and landslides	Disrupt communications	Only in mountainous areas or in regions of unstable soils
Modification of permafrost areas	Destroy roads, rail foundations, alteration of stream sources	Relatively easy by removal of insulating cover but effect limited to melting season
River diversions	Flooding, river navigation hazards	Large engineering effort necessary
Stimulation of volcanoes	Ash and gases may act as precipitation nuclei affecting certain communications and remote sensing	Uncertain, but results difficult to predict

Source: after Goldblat (1975)

micro-organisms to neighbouring mainland areas by animals or insects. It was also easy to protect: fast patrol boats guarded Vozrozhdeniye against intruders over nearly 40 years of testing. But the island's Soviet base was abandoned in 1992 after the break-up of the USSR and Vozrozhdeniye became connected to the mainland in 2001 as the Aral Sea continued to shrink. Connection to the mainland eliminated the remaining natural security benefits of the island, hence potentially infected animals living on Vozrozhdeniye now have easy access to the mainland, and vice versa, with unpredictable consequences for the local biosphere. The desiccation of the Aral Sea is well known as the world's worst human-induced environmental disaster (see page 130), but the eco-tragedy now has a new dimension.

In contrast to most of the above examples, positive aspects of incipient military conflict can also arise, as in the case of demilitarised areas maintained as buffer zones between opposing forces. The demilitarised zone separating North and South Korea, for example, has provided sanctuary to endangered and threatened animals and plants for more than 45 years. The 4 km-wide, 250 km-long corridor extends across the Korean peninsula, traversing a major river delta and old farmlands in the west and rugged mountains in the east. It has remained virtually untouched by human activity since its establishment in 1953, and thus represents a unique, rigidly protected nature reserve (Kim, 1997). In a similar vein, legal wranglings over the huge clean-up costs at the Rocky Mountain arsenal in Colorado, USA (see above), abated when bald eagles began to nest on the contaminated site, and the area has now been designated a nature reserve (Wiley and Rhodes, 1998).

DIRECT WARTIME IMPACTS

The relationship between people and their environment can be changed significantly during wartime. Priorities are altered, and certain resources are used more rapidly than during peacetime in order to fuel the war effort. In the time of Henry VIII, for example, many of England's oak trees were cut down to build warships. In addition to the destruction of agricultural land and woodland during the prolonged trench warfare on the coastal plains of France and Belgium during the First World War, wide areas of forest were felled beyond the battle zones to supply the war effort. In total, Belgium lost most of its remaining forests and France lost about 10 per cent of its forested area at this time (Graves, 1918).

Deliberate destruction and manipulation of the environment has also been used by armies to gain military advantage for centuries. The so-called 'scorched earth' policy, in which vegetation and crops are deliberately destroyed to prevent their use by the enemy, is an age-old military tactic: in Biblical times, for example, Samson tied burning torches to the tails of 300 foxes and drove them into the olive groves and grain fields of the Philistines. Similar tactics were used during the war sparked by the Soviet occupation of Afghanistan from 1979 to 1991. Vineyards, orchards, ornamental trees and shrubs were routinely felled for security reasons by the Soviet army and troops loyal to the former communist government of Afghanistan, and vegetation was widely cleared along highways in the provinces of Kabul and Parwan. The destruction of forests, both deliberate and inadvertent (as a result of wildfires set off by bombing) is thought to have reduced the forest cover from 3.4 per cent of the country to 2.6 per cent in just ten years (Formoli, 1995). The wide scope for environmental modification for military purposes in the modern era is indicated in Table 20.2, a list submitted during the negotiations for the 1977 Environmental Modification Convention (see below), although it should be noted that this list is by no means exhaustive in that it does not include the effects of the numerous biological and chemical weapons that have been developed.

A number of these modifications have been employed in conflicts. During the Viet Nam War, for example, the US army diverted water from the Mekong delta to drain the Plain of Reeds where

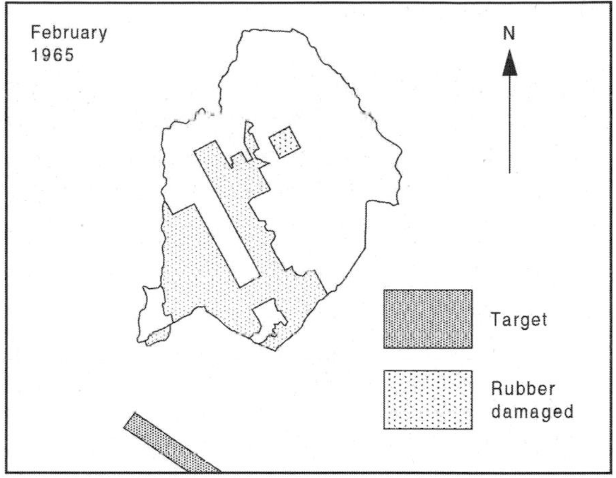

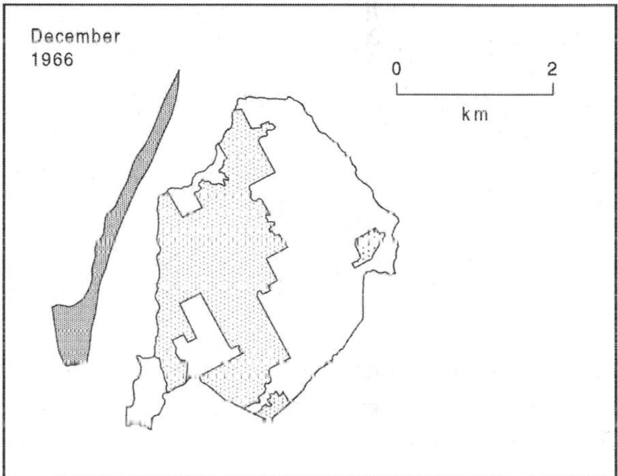

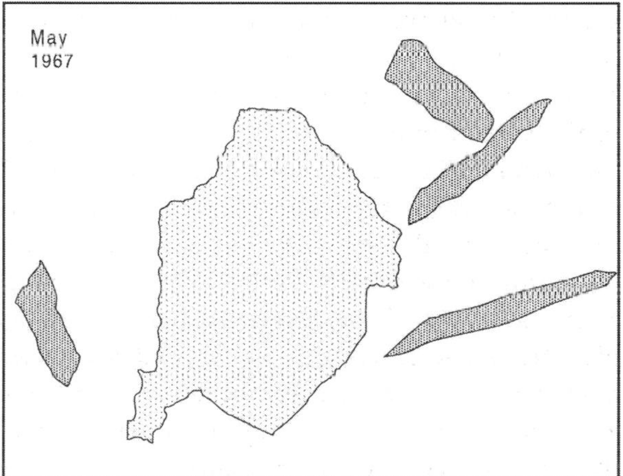

Figure 20.1 Damage to rubber trees on the Plantation de Dautieng by wind drift of aerially sprayed defoliants during the Viet Nam War (reprinted with permission from Orians and Pfeiffer, 1970. Copyright, 1970, American Association for the Advancement of Science).

an enemy base was located, and a classified rainmaking programme was conducted from 1967–72, seeding clouds from the air using silver and lead iodide to increase normal monsoon rainfall over the Ho Chi Minh trail, a vital supply line for North Vietnamese forces (Jasani, 1975). The Viet Nam War was also the scene of the most extensive systematic destruction of vegetation ever undertaken in warfare. About 72 million litres of herbicides were sprayed over thousands of square kilometres of the country between 1961 and 1971 in an effort to deprive North Vietnamese forces of the cover provided by dense jungle. Most spraying was targeted on forest but cropland was also sprayed, and although missions were only conducted during light winds to avoid non-target contamination, drifting sprays actually caused damage over much wider areas (Orians and Pfeiffer, 1970). An example of repeated wind-drift spray damage to rubber trees on the 9 km^2 Plantation de Dautieng is shown in Fig. 20.1.

About half of Viet Nam's mangrove forests were completely destroyed by aerial spraying, an area that has been slow to recover. Revegetation of the sprayed mangroves was essentially complete by the mid-1980s, but not always with the same commercially preferred species. The effects on wildlife are not well documented but a few studies and much anecdotal evidence suggest decreased abundance of many animal species, and herbicide-related illness in domestic livestock (Westing, 1984). In total, an estimated 22 000 km^2 of farmland and forest was destroyed, mainly in the south, by tactical spraying, intensive bombing and mechanical clearance of forest. A further 1170 km^2 of forest was destroyed by cratering from 13 million tonnes of bombs, and another 40 000 km^2 by bombardment (Collins *et al.*, 1991). The conflict left an estimated 26 million bomb and shell craters across 170 000 ha of Indochina, most in South Viet Nam, representing displacement of 2.6 billion m^3 of earth (Westing and Pfeiffer, 1972).

The military experience of defoliant chemical use in Viet Nam was subsequently repeated on a smaller scale in El Salvador during the 1980s, where government forces, with US backing, laid waste areas held by or sympathetic to anti-government guerrillas. The term 'ecocide' has been coined for such widespread destruction (Weinberg, 1991).

Another military method involving large-scale devastation to the natural environment, sometimes with that very aim in mind, is the destruction of various forms of infrastructure. Agricultural infrastructure is an obvious target. Afghanistan's irrigation systems were the subject of systematic destruction during the Afghan–Soviet war, leaving up to one-third of all irrigation systems damaged (Formoli, 1995). Dams have also been common targets in this respect. Allied bombing of the Möhne, Eder and Sorpe dams in the Ruhr valley of Germany during the Second World War, for example, aimed to cripple the German economic heartland. Water released from the damaged Möhne and Eder dams killed 1300 people, left 120 000 homeless and ruined 3000 ha of arable land as well as numerous factories and power plants. But the most devastating example of deliberate dam breaching occurred during the Second Sino-Japanese War of 1937–45 when the Chinese dynamited the Huayuankow dike on the Huang Ho River near Chengchow to stop the advance of Japanese forces. The mission was successful in this respect and several thousand Japanese troops were drowned, but the flow of water also ravaged major areas downstream: several million hectares of farmland were inundated, as well as 11 cities and more than 4000 villages. At least several hundred thousand Chinese drowned and several million were left homeless (Bergström, 1990).

Oilfields have also been the target of deliberate damage. Retreating Austrian forces systematically destroyed facilities on Romanian oilfields in 1916–17, as did Iraqi forces before being driven out of Kuwait in 1991. In the latter example, an estimated 7–8 million barrels of oil were discharged, more than twice the size of the world's previously largest spillage caused by a blow-out in the Ixtoc-1 well in the Bay of Campeche off the Mexican coast in 1979. The resulting oil slick contaminated large areas of Kuwait and a 460-km stretch of the northern coast of Saudi Arabia. An estimated 30 000 seabirds died in the immediate aftermath of the spills (Readman *et al.*, 1992), but the long-term effects on the Gulf's ecology are remarkably few. Just three and a half years later, there were no visible signs of immediate or delayed effects on coral reefs in Saudi waters (Vogt, 1995) and rapid recovery has been recorded in most other parts of the marine ecosystem. This finding is consistent with observations made during the Al-Nowruz oil well spill during the Iran–Iraq war, which suggests that the Gulf's ecology has high resilience to oil pollution (Saenger, 1994).

Fires started in 613 Kuwaiti oil wells caused the burning of 4–8 million barrels a day, and resulted in massive clouds of smoke and gaseous emissions, although fears of effects on the climate were unfounded. Other disturbances to the desert surfaces of Kuwait were caused by the large-scale movements of military vehicles and the building of an earth embankment along the border with Saudi Arabia (Fig. 20.2). Experience of similar surface destabilisation during the desert campaigns of North Africa in 1941 and 1942 saw wind erosion dramatically increase during the height of the fighting, but deflation subside back to pre-war levels within a few years of the end of the campaign (Oliver, 1945). Large military vehicles also have more long-lasting effects in desert areas, compacting soil, crushing and shearing vegetation, and altering the vertical and horizontal structure of plant communities. Compositional changes to soil and vegetation from a single tank pass during military manoeuvres in the Mojave Desert were found still to be evident 40 years after training had ceased (Prose, 1985).

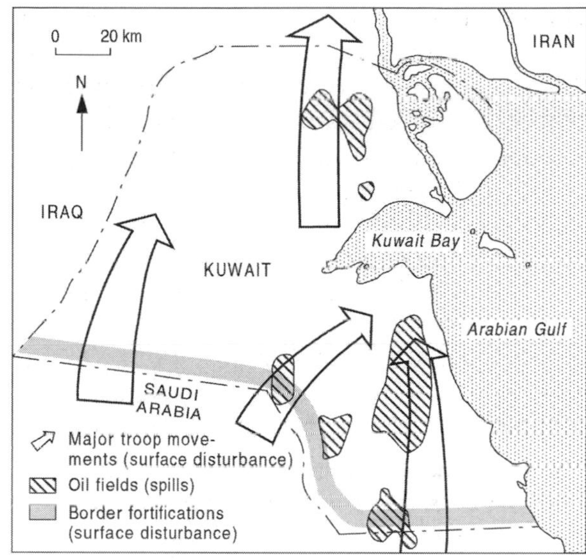

Figure 20.2 Military action in Kuwait 1990–91 and its impacts on surface terrain (after Middleton, 1991).

There are many other environmentally damaging consequences of warfare. In Croatia, for example, Richardson (1993) documented the numerous dangers caused by damage to industrial and municipal plants in the war that followed the break-up of Yugoslavia in the early 1990s. One of the most dangerous and possibly long-lasting hazards was caused by the release of PCBs from damaged transformers in many Croatian towns, including 50 in Dubrovnik alone. Contamination of soils, rivers and groundwater was expected. PCBs bioaccumulate up the food chain, they are highly toxic to fish, and there is some evidence linking them to the occurrence of certain cancers in humans.

Numerous impacts on wildlife have also been documented. Some mammal species have been brought close to and beyond extinction due to the direct and indirect effects of war. The last semi-wild Père David's deer was killed by foreign troops during the Boxer Rebellion in China in 1898–1900, although the species still survives in captivity. Similarly, the European bison was pushed close to extinction during the First World War by hunting to supply troops with food; it recovered in numbers during the interwar years but its numbers were again decimated in the Second World War. Many Pacific island endemic bird taxa were also lost or pushed towards extinction by warfare and its associated disturbances during the Second World War. On the twin islands of Midway, for example, extinctions of the Laysan finch and the Laysan rail occurred (Fisher and Baldwin, 1946). In Central Africa, war has several times in recent years threatened the

Figure 20.3 An area near Port Howard in the Falkland Islands still closed off to human use due to landmines, which remain uncleared since they were laid during the 1982 conflict in the South Atlantic (Photo: Mark Carwardine).

survival of the 600 mountain gorillas in the Virunga volcanoes, which straddle the borders of Rwanda, Zaire and Uganda.

Conversely, however, some species of wild animals can benefit during wartime because of decreased exploitation. Westing (1980) notes the cases of various North Atlantic fisheries during the Second World War and a number of fur-bearing mammals in northern and north-eastern Europe, such as the polar bear, red fox and wolf. During the Viet Nam War, tigers increased in abundance in some areas thanks to the ready availability of food provided by corpses.

Disruption and destruction does not necessarily cease with the end of hostilities. On average, 10 per cent of all munitions used in any war fail to explode, and remain to threaten humans, livestock and wildlife; they also impede post-war reconstruction and rehabilitation efforts in agriculture, forestry, fishing, mining and other related activities (Fig. 20.3). Landmines are a particular problem; they are present on a huge scale in some countries (Table 20.3) and can take many years to clear. In Poland, more than 84 million items of explosive ordnance from the Second World War (including more than 14 million mines) had been disposed of by the early 1980s, with 350 000 being located and destroyed every year (Westing, 1984).

TABLE 20.3 *Estimated number of landmines in the worst affected countries and territories, early 1990s*

Country	Number of land mines (million)
Afghanistan	9–10
Angola	9–15
Iraq	5–10
Kuwait	5
Cambodia	4–7
Western Sahara	1–2
Mozambique	1–2
Somalia	1
Bosnia and Herzegovina	1
Croatia	1

Sources: Ruel (1993) and UN Unit for the Coordination of Humanitarian Assistance

Nuclear war

There is little direct information on the possible environmental effects of a nuclear war, but some insight can be gained from the detonation of the only two nuclear devices used in warfare to date, in Japan at the end of the Second World War (Table 20.4), although these were relatively small devices by today's standards. The most important cause of death and physical destruction was the combined effects of blast and thermal energy. The fireball created by the blasts was intense enough to vaporise humans at the epicentre and burn human skin up to 4 km away. Ionising radiation comprised about 15 per cent of the explosive yield of the bombs, and among its effects was radiation sickness among many survivors of the blast.

There is a limited amount of information on the ecological effects of nuclear explosions, which has been gleaned from observations made after above-ground tests. Severe damage to vegetation caused by detonations decreases more or less geometrically with distance from the blast epicentre. After explosions, plants re-invade damaged areas at a rate of succession that would be expected after a severe disturbance, as Shields and Wells

TABLE 20.4 *Destruction caused by the first atom bombs dropped in wartime*

	Hiroshima	Nagasaki
Date of detonation	6 August 1945	9 August 1945
Type	Uranium 235	Plutonium
Height of explosion (m)	580	503
Yield (kiloton TNT)	12.5	22
Total area demolished (km^2)	13	6.7
Proportion of buildings completely destroyed (%)	67.9	25.3
Proportion of buildings partially destroyed (%)	24.0	10.8
Number of people killed (by 31 December 1945)	90 000–120 000	70 000

Source: Ohkita (1984)

(1962) documented for a Nevada Desert test site after numerous above-ground detonations.

Considerable research has been undertaken into the possible global effects of a major nuclear exchange, much of it focusing on the potential climatic effects. Although results vary with different scenarios, simulations using general circulation models (GCMs) have raised the spectre of a 'nuclear winter' in which the aftermath of a nuclear war would be characterised by darkened skies over large parts of the Earth due to smoke and dust injected into the atmosphere. Such effects could continue for weeks or months, blocking sunlight, reducing temperatures and affecting the global distribution of rainfall. Such climatic effects could have wide-ranging impacts on agriculture and major biomes, with serious implications for food production and distribution far beyond the areas of immediate devastation (SCOPE, 1989).

INDIRECT WARTIME IMPACTS

The dislocation of economies and societies caused by war can result in many perturbations to the way people interact with their environment. Environmental management, for instance, is often neglected as priorities change. During the 15-year civil war in Lebanon, the country's agricultural resource base was severely damaged because much of the extensive network of stone-wall terraces in upland areas was not properly maintained. Accelerated erosion and landslides have resulted in many such areas, and the cost of repairing the terraces probably far exceeds the budget of the ministry of agriculture (Zurayk, 1994). Such interruptions to conventional agricultural practices, along with forced mass migrations, disruption to transport systems and many other aspects of the economy can also leave people less able to cope with hazards of the natural environment, increasing their vulnerability to such disasters as famine. On occasion, deliberate destruction of the environment and infrastructure may lead to famine. A recent example is the mass starvation that swept southern Somalia in 1992–93 following a six-month period in which grain stores were plundered, and water pumps and other agricultural infrastructure and machinery were destroyed in the Bay region by forces loyal to former president Siad Barre (Drysdale, 2002).

Numerous indirect environmental impacts also occur due to the mass movements of refugees

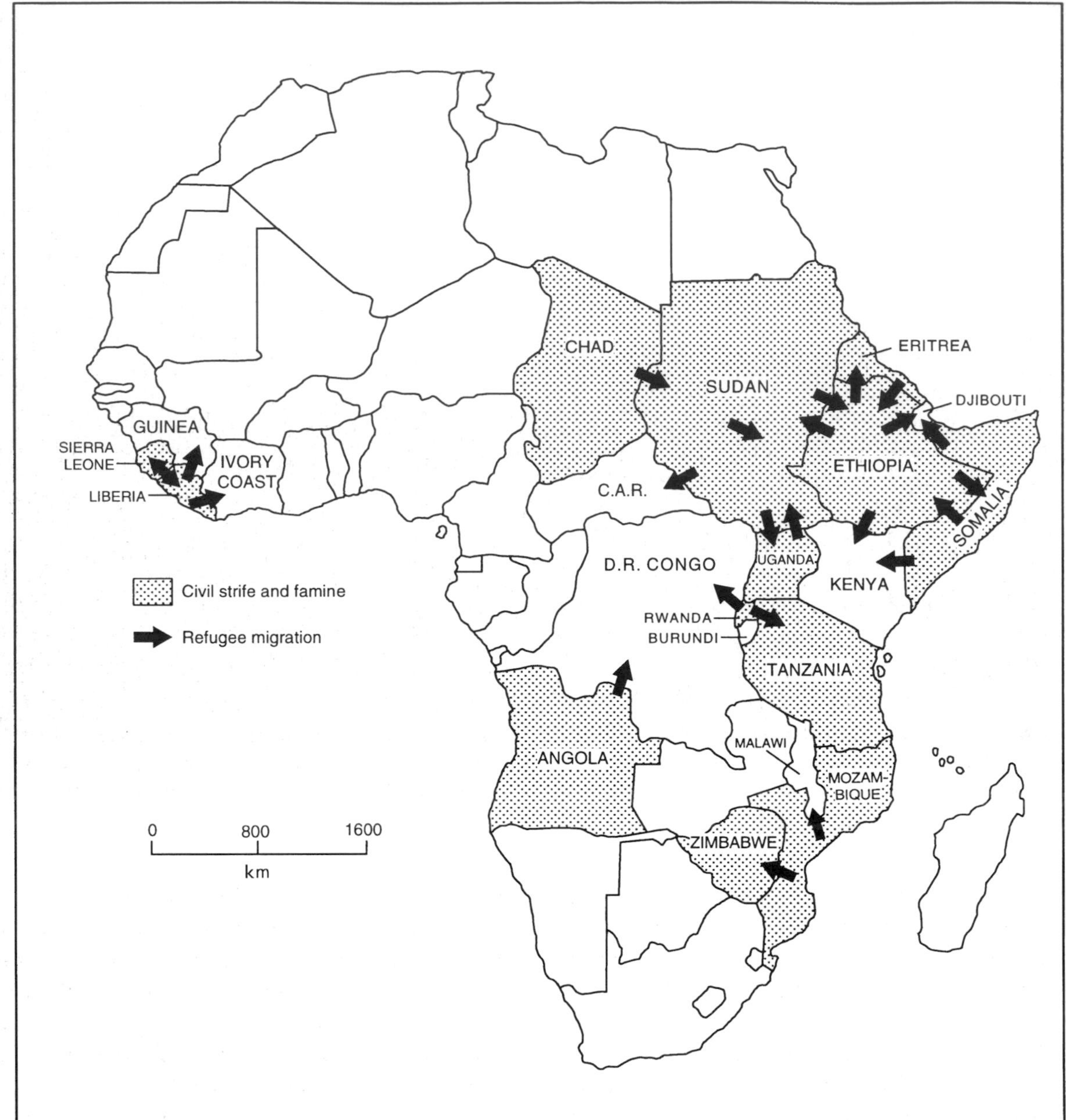

Figure 20.4 Refugee movements in sub-Saharan Africa due to civil strife and famine, 1980s and 1990s.

away from areas of conflict. During medieval Europe's Hundred Years' War, for example, whole regions were depopulated, leaving abandoned farmland to be colonised by brush and woodland. Refugees tend to concentrate in safer zones, in cities, towns or refugee camps in rural areas, exerting intense pressures on local resources. Deforestation for fuel and to build shelters, resultant soil erosion and local pollution, particularly of water sources, are impacts commonly associated with large concentrations of refugees. In Central America during the 1980s, for example, about 20 000 refugees fleeing the conflict in Nicaragua cleared extensive areas of rain forest

Figure 20.5 Collecting fuelwood near the village of Maúa in northern Mozambique. The civil war, which lasted from the mid-1970s to 1992, meant a great concentration of dislocated people around relatively safe settlements, dramatically increasing pressure on local resources.

for camps on the Miskito coast of Honduras, some of them decimating part of the Rio Platano forest reserve protected under the United Nations' Man and the Biosphere Programme (Weinberg, 1991).

Many such examples can also be quoted from sub-Saharan Africa, scene of numerous conflicts in recent decades. Some of the major resulting refugee movements in Africa are indicated in Fig. 20.4. Although some proportion of the refugees indicated on this map were fleeing areas affected by famine and drought, famine itself is often intimately linked to civil unrest as the example from Somalia cited above illustrates.

In lowland Ethiopia, clearance for shelter for 100 000 refugees is estimated to have deforested about 330 ha, plus the destruction caused by an annual fuelwood need of 85 000 t. Water demands for the 400 000 long-term Somali refugees in the arid Hararghe region of the country have necessitated the building of numerous earth dams and a truck delivery of 800 000 litres of water a day to the 210 000 refugees in the two camps at Hartisheikh, with unknown implications for local ecology (Tsehai, 1991).

In north-western Tanzania, the camp of 300 000 refugees from Burundi at Benaco absorbed 410 000 Rwandans fleeing civil war in a month in mid-1994. Suggestions from aid workers to ease the pressure on local forests included the trucking of refugees outside the 4- to 5-km radius round the camp, which was rapidly denuded in the quest for fuelwood, the marking of trees that would take a long time to regenerate, and the distribution of types of food aid that required less cooking, such as maize powder instead of maize grain (UNHCR, 1994). Although refugee assistance agencies tend to be aware of the potential environmental degradation around camps, such effective action to combat the problem is rarely forthcoming in practice (R. Black, 1994).

The scale of the refugee problem imposed on relatively safe neighbouring countries can reach high levels. Malawi in southern Africa, already one of the region's most densely populated countries, accommodated a million Mozambican refugees during the late 1980s and early 1990s, in addition to its own population of 10 million. Within Mozambique, where conflict was more or less continuous between 1964 and 1992, no less than 4.5 million people were estimated to have been displaced within the country by the time a peace accord was signed between the government and the opposition forces of Renamo. Natural resources around safer villages, towns and cities suffered for years from the increased population pressures (Fig. 20.5). Displaced people living in the capital city of Maputo had to travel up to 70 km into the dangerous surrounding countryside in the search for fuelwood, and little forest was left in the early 1990s in the Beira corridor area, which was protected by Zimbabwean troops during the war. On the coast, mangrove forests were severely exploited, presenting a serious threat to prawn and shrimp fisheries, one of Mozambique's major exports, which reproduce in the mangroves.

Cultivated land was not allowed adequate fallow periods, which traditionally last for up to seven years, due to high population densities. Additionally, the plots allocated to displaced people were often too small for a family, leading to further pressure and overutilisation (Dejene and Olivares, 1991). The war's impact on wildlife in Mozambique is also thought to have been severe. At independence in 1975, there were more than 65 000 elephants in the country; initial post-war estimates suggest that 15 000 remain. All the evidence points to Renamo having been heavily involved in the ivory trade, selling tusks to South Africa in return for arms.

Similar extreme degradation has been noted in restricted coastal zones and larger urban centres in Angola, another former Portuguese colony wracked by civil war in its post-independence years. Conversely, however, much of the countryside depopulated during the conflict has suffered little damage, with some exceptions such as the giant sable, only found in a small area of central Angola and now highly endangered since its habitat was a central area of conflict. Indeed, in the view of one international conservation agency, the likelihood of widespread environmental degradation will be much greater with ensuing peace than during the past two decades of war (IUCN, 1992). The economic paralysis caused by the war has functioned as a brake on the wholesale exploitation of certain natural resources, and greater danger lies in post-war rapid and unchecked economic exploitation. Nevertheless, there is no great competition between wildlife and people for land use in Angola - the key problem throughout much of the rest of Africa - and the clear valuation put on wildlife that is hunted by rural people for meat is a strong positive conservation aspect offering great potential.

LIMITING THE EFFECTS

Numerous treaties, conventions and agreements have been adopted to prevent utter human and environmental devastation in times of war, although they do not specifically cover nuclear weapons, and the effectiveness of such agreements as a deterrent and during the course of war are difficult to evaluate and enforce. Some of the most important relevant conventions are shown in Table 20.5 (Goldblat, 1990). Other parts of multilateral agreements also have a bearing on the environmental impacts of war, such as the two additional 1977 protocols to the Geneva Conventions of 1949, which bind belligerents to respect and protect the natural environment 'against widespread, long-term and severe damage', and prohibit the destruction of works and installations that harbour dangerous forces, including dams, dikes and nuclear generating stations. Such treaties are only binding on those states that are a party to them, however, and the

TABLE 20.5 *Major multilateral arms control agreements with a bearing on the environmental effects of warfare, 1970–90*

Agreement	Signed	Entry into force
Treaty prohibiting emplacement of nuclear weapons and other weapons of mass destruction on the sea bed	1971	1972
Bacterial and Toxin Weapons Convention prohibiting development, production and stockpiling of biological and toxin weapons, and/or their destruction	1972	1975
Environmental Modification Convention prohibiting military or other hostile use of environmental modification techniques	1977	1978
Inhumane Weapons Convention prohibiting or restricting use of certain conventional weapons deemed to be excessively injurious or to have indiscriminate effects	1981	1983
Treaty of Rarotonga (South Pacific nuclear-free treaty)	1985	1986

Source: after Goldblat (1990)

protocol that bans widespread, long-term and severe environmental damage has not been ratified by major powers such as the USA, France or the UK since these nations are concerned that it could be interpreted as outlawing nuclear weapons. Some treaties meant to exclude any military activities from certain areas have also been adopted. These include the Spitzbergen Treaty (1920), the Åland Convention (1921) and the Antarctic Treaty (1959).

The widespread ecological effects of the Gulf War prompted renewed calls from conservation agencies for a new Geneva Convention to limit the environmental effects of military conflicts, and the idea of a Green Cross – a new, impartial, international organisation, similar to the Red Cross and the Red Crescent, to assess environmental damage resulting from war – has also been suggested.

Although idealists might suggest that effective protection for the environment can only be achieved by global demilitarisation, this prospect seems unlikely. Two sovereign nations (Costa Rica and Japan) are non-militarised in a formal sense by virtue of their national constitutions, although in fact both support more or less potent armed forces. Indeed, the role played by environmental resources in causing conflict can be interpreted as a reason for increasing military might in a world of rising population.

ENVIRONMENTAL CAUSES OF CONFLICT

The environment has often been a cause of political tension and military conflict. Countries have fought for control over raw materials, energy supplies, land, river basins, sea passages and other environmental resources (Fig. 20.6). In the Second World War, for example, Japan sought control over oil, minerals and other resources in China and South-East Asia; rich phosphate deposits have been a root cause of the extended conflict in the western Sahara; Israel's long occupation of a large portion of the west bank of the River Jordan has been both for strategic military purposes and to secure access to large groundwater reserves, and if the Gulf region had not

Figure 20.6 This defunct tank near the border between Ethiopia and Eritrea is a relic of Ethiopia's civil war, which ended in 1991. Eritrea's independence in 1993, an outcome of this conflict, meant that Ethiopia became landlocked. This loss of access to the sea was a significant factor in the war between Ethiopia and Eritrea in 1998–2000.

been such a key supplier of oil supplies to the West it seems unlikely that western troops would have intervened to such an extent over the Iraqi invasion of Kuwait in 1990–91. Indeed, of the 50 or so armed conflicts fought in mid-1994, about 20 were considered (by Westing, 1994) to be environmentally induced in some way.

Some observers suggest that environmental change, particularly in the form of degradation and dwindling renewable resources, is set to become an increasingly potent cause of conflict as world population is likely to exceed 9 billion within the next 50 years and global economic output may increase five-fold (Homer-Dixon *et al.*, 1993). The degradation and loss of productive land, fisheries, forests and species diversity, the depletion and scarcity of water resources, and pressures brought about by stratospheric ozone loss and potentially significant climatic change may precipitate civil or international strife.

Such root causes of conflict have been identified from past eras. Pennell (1994), for example, has suggested that environmental degradation and economic isolation, and consequent impoverishment, made piracy an attractive option for inhabitants of the Guelaya Peninsula in north-western Morocco in the mid-nineteenth century. A key aspect of such scenarios is the role played by certain groups that are faced with dwindling finite resources. Such groups may be within a certain country or they may be identifiable on the global scale as countries with unequal access to resources. A proposed chain of factors in this equation, which leads to violence, is suggested in Fig. 20.7.

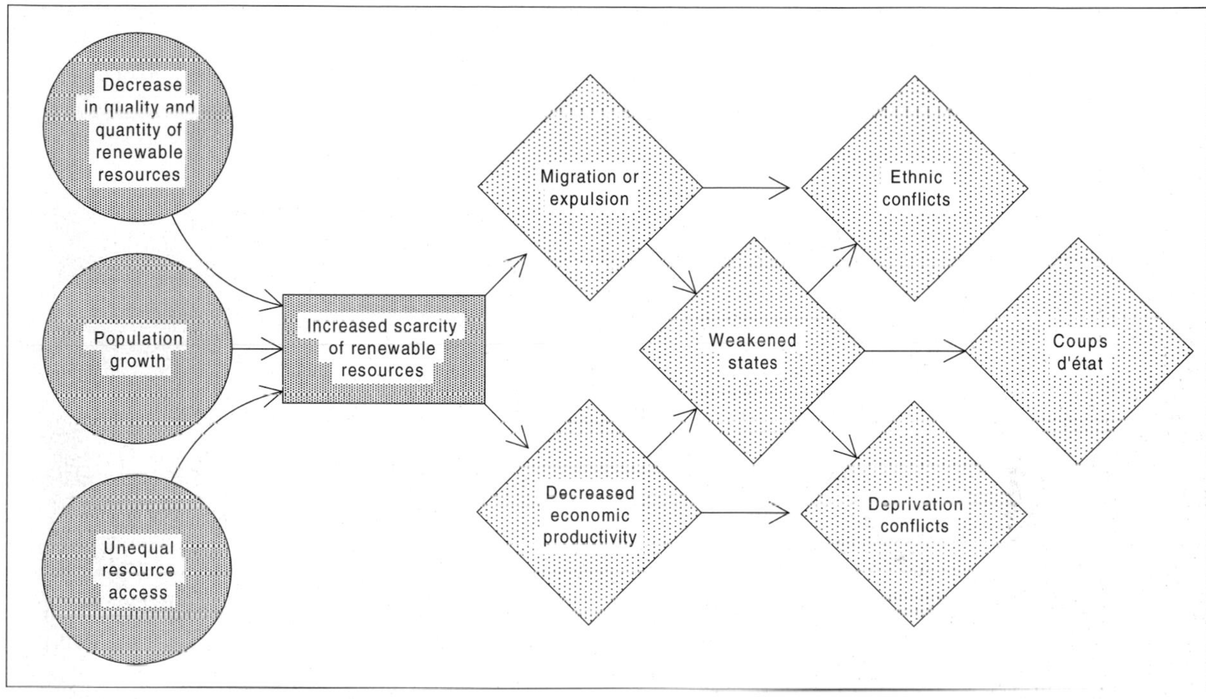

Figure 20.7 Some sources and consequences of renewable resource scarcity (Homer-Dixon *et al.*, 1993).

Some researchers have highlighted examples of such events in more recent history. Bennett (1991) presents a number of studies of African conflicts in this vein, while in Central America, Durham (1979) considers that the root causes of the 1969 'Soccer War' between El Salvador and Honduras ran much deeper than the national rivalry over the outcome of a football match that ignited the conflict. He traces the origins of a war in which several thousand people were killed in a few days to changes in agriculture and land distribution in El Salvador that began in the mid-nineteenth century, forcing poor farmers to concentrate in the uplands. Despite their efforts to conserve the land's resources, growing human pressure in a country with annual population growth rates of 3.5 per cent reduced land availability, and resulted in deforestation and soil erosion on the steep hillsides. Many farmers consequently moved to Honduras and it was their eventual expulsion that precipitated the war. In the years following the Soccer War, this competition for land was not addressed and *campesino* support for leftist guerrillas provided a powerful contribution to the country's subsequent ten-year civil war.

To end this section on a more positive note, the conservation value of the Korean demilitarised zone mentioned earlier in this chapter has been put forward as a possible mechanism to help build trust and understanding between rival powers. Kim (1997) suggests that joint maintenance of the ready-made nature reserve, already completely protected and clearly delimited thanks to its origin as a buffer zone, can foster mutual respect between North and South Korea, while helping to preserve the peninsula's biodiversity.

FURTHER READING

Gleick, P.H. 1994 Water, war and peace in the Middle East. *Environment* 36(3): 6–15, 35–42. A good overview of the region's hydropolitical issues as a potential cause of conflict.

Myers, N. 1993 *Ultimate security: the environmental basis of political stability*. New York, Norton.

An analysis of the importance of environmental issues in world politics.

Renner, M. 1991 Assessing the military's war on the environment. In *State of the world 1991*. New York, W.W. Norton. This article presents a catalogue of the military's environmental effects, mainly during peacetime.

Rubenson, S. 1991 Environmental stress and conflict in Ethiopian history: looking for correlations. *Ambio* 20: 179–82. This paper presents a case study of environmental stress and degradation in Ethiopia, and their links to social and political conflict.

SCOPE (Scientific Committee on Problems of the Environment) 1989 *Environmental consequences of nuclear war. Volume II: ecological and agricultural effects*, 2nd edn. SCOPE Report 28, Chichester, Wiley. A detailed compilation of the possible effects of a global nuclear exchange on the natural environment and our ability to produce food.

Warner, F. (ed.) 2000 *Nuclear test explosions: environmental and human impacts*. SCOPE Report 59, Chichester, Wiley. A series of papers on the medical and environmental effects of nuclear testing.

Wiley, K.B. and Rhodes, S.L. 1998 The transformation of the Rocky Mountain arsenal. *Environment* 40(5): 4–11, 28–35. Case study of how a severely contaminated military waste dump has become a nature reserve.

Websites

http://www.gci.ch/ one of Green Cross International's major programmes is concerned with the environmental impact of warfare.

http://www.library.utoronto.ca/pcs/state.htm site for the Environmental Scarcities, State Capacity and Civil Violence project, with case studies on water in China, cropland in India and Indonesian forests.

http://www.sipri.se/ the Stockholm International Peace Research Institute site contains information on military capabilities and expenditure.

http://www.opcw.org/ homepage of the Chemical Weapons Convention.

http://www.icbl.org/ the International Campaign to Ban Landmines site.

http://www.pmrma-www.army.mil US army site covering the contamination and remediation of Rocky Mountain arsenal.

http://www.mil.se/ovrigt/miljosek/miljo_e.html the Swedish Army's environmental division.

POINTS FOR DISCUSSION

- Does the human cost of warfare mean that the environmental cost is unimportant?
- Do you think that conflicts over natural resources are set to become more frequent in coming decades?
- Should atomic weapons be banned and is it realistic to do so?
- Can national armies be managed sustainably and should different rules apply during conflicts?

21

NATURAL HAZARDS

TOPICS COVERED

Hazard classifications
Response to hazards
Hazardous areas
Examples of natural disasters

Key words: risk, return period, disaster, tsunami, lahar, pyroclastic flow, levée

Natural hazards such as earthquakes, floods, tropical cyclones and disease epidemics are normal functions of the natural environment. They do, therefore, affect all living organisms but they are usually only referred to as disasters or catastrophes when they impact human society to cause social disruption, material damage and loss of life. As such, natural hazards should be defined and studied both in terms of the physical processes involved and the human factors affecting the vulnerability of certain groups of people to disasters.

Although some places are more hazardous than others, all locations are at risk – there is always the chance of a disaster – from some natural hazard or other. All places also have some natural advantages, however, and the presence or absence of human activities in any location is the result of weighing up the risks relative to the advantages. In many locations, the physical phenomenon responsible for a hazard also offers some of the advantages: a river, for example, will flood, which may be hazardous, but it is also a source of water and its floodplain is a location with flat land and fertile soils. Hence, the hazard should be seen as an occasionally disadvantageous aspect of a phenomenon that is beneficial to human activity over a different timescale. This is illustrated in Fig. 21.1 where the shaded zone represents an

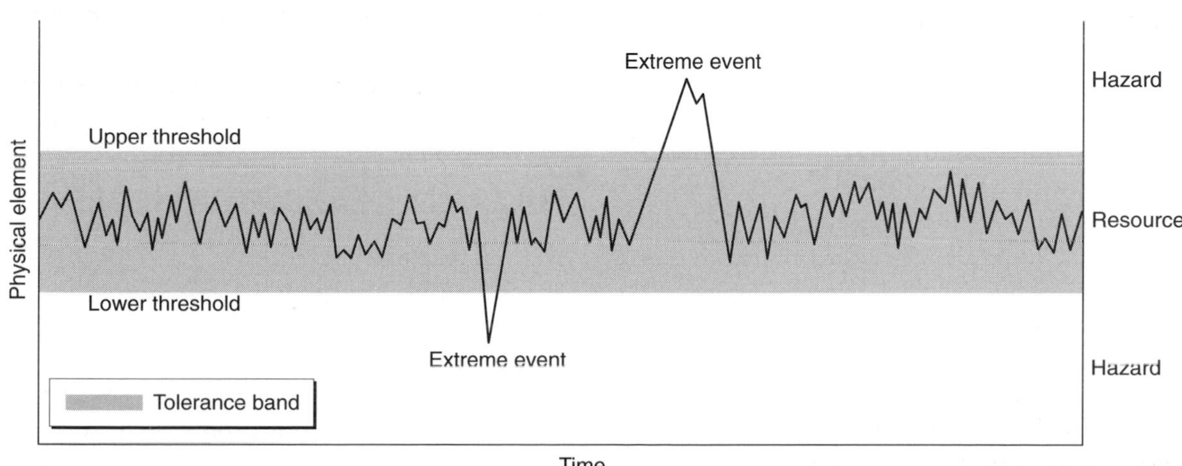

Figure 21.1 A physical element is perceived as a resource by society when its variations in magnitude are within certain limits, but is perceived as a hazard when they exceed certain thresholds (after Hewitt and Burton, 1971).

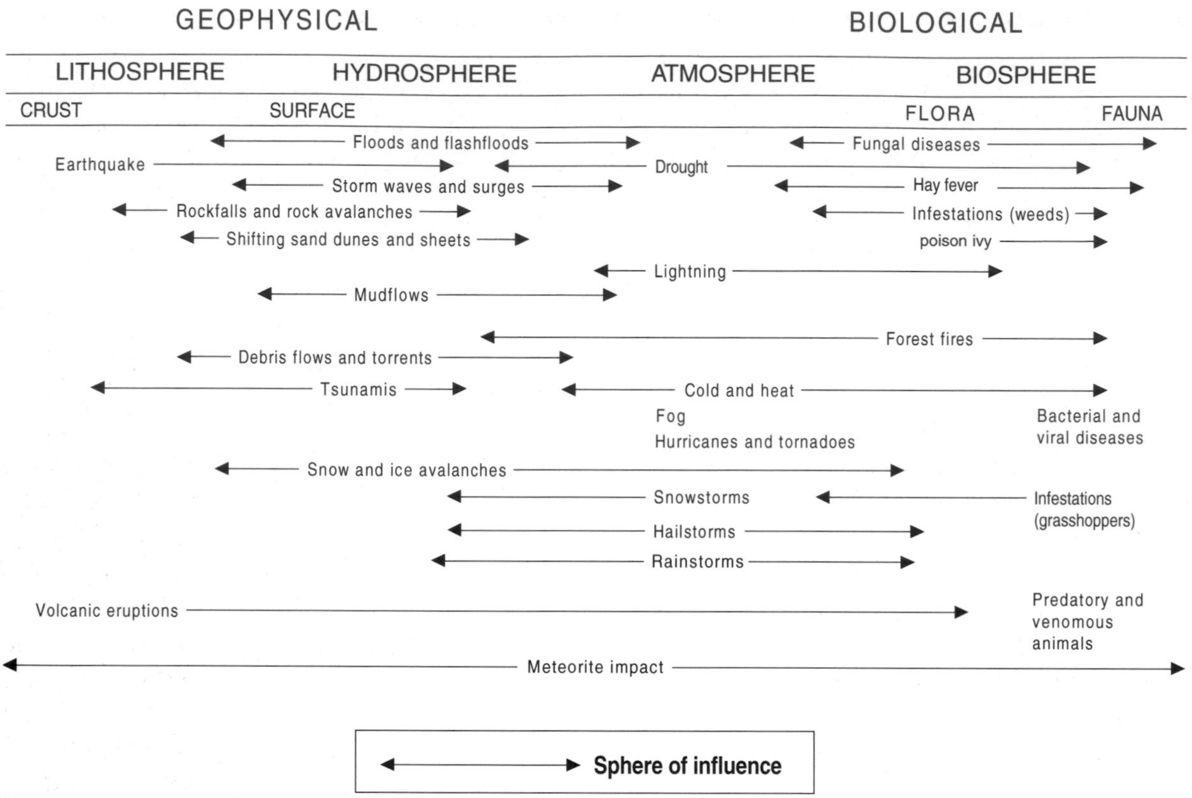

Figure 21.2 Types of natural hazard, by sphere of occurrence, medium and materials (after Gardner, 1993).

acceptable variability of a physical element: variations in its magnitude that society has adapted to and considers a resource. When this variability exceeds a certain threshold, however, the same physical element becomes a hazard for a period of time. In the case of the river, this could mean either a flood or a water shortage. Ironically, beyond the timescale of the hazardous event occurring, the very same hazard may contribute to the resource element of the location: when a flood happens, it presents a hazard to the occupants of the floodplain, but the floodplain is only there because the river floods occasionally.

To complicate matters, however, not all people who choose to live in a particular area enjoy the same degree of choice – a flexibility that may be dictated by the economic means at their disposal, among other things. To complicate matters further, there is a wide range of ways in which people respond to hazards and, again, the rich often have more options in this respect than poorer members of society. The study of natural hazards is very much a geographical subject, therefore, since it includes analysis of both the physical and human environment. It is also a fairly topical issue; the 1990s were declared the United Nations International Decade for Natural Disaster Reduction.

HAZARD CLASSIFICATIONS

Natural hazards can be classified in many different ways. Physical geographical approaches may divide them according to their geological, hydrological, atmospheric and biological origins, as shown in Fig. 21.2. They can also be classified spatially, as certain hazards only occur in certain regions: avalanches occur in snowy, mountainous areas, for example, and most volcanic eruptions and earthquakes occur at tectonic plate margins.

TABLE 21.1 *Natural disasters classified according to frequency of occurrence, duration of impact and length of forewarning*

Disaster	Frequency or type of occurrence	Duration of impact	Length of forewarning (if any)
Lightning	Random	Instant	Seconds–hours
Avalanche	Seasonal/diurnal; random	Seconds–minutes	Seconds–hours
Earthquake	Log-normal	Seconds–minutes	Minutes–years
Tornado	Seasonal	Seconds–hours	Minutes
Landslide	Seasonal/irregular	Seconds–decades	Seconds–years
Intense rainstorm	Seasonal/diurnal	Minutes	Seconds–hours
Hail	Seasonal/diurnal	Minutes	Minutes–hours
Tsunami	Log-normal	Minutes–hours	Minutes–hours
Flood	Seasonal; log-normal	Minutes–days	Minutes–days
Subsidence	Sudden or progressive	Minutes–decades	Seconds–years
Windstorm	Seasonal/exponential	Hours	Hours
Frost or ice storm	Seasonal/diurnal	Hours	Hours
Hurricane	Seasonal/irregular	Hours	Hours
Snowstorm	Seasonal	Hours	Hours
Environmental fire	Seasonal; random	Hours–days	Seconds–days
Insect infestation	Seasonal; random	Hours–days	Seconds–days
Fog	Seasonal/diurnal	Hours–days	Minutes–hours
Volcanic eruption	Irregular	Hours–years	Minutes–weeks
Coastal erosion	Seasonal/irregular; exponential	Hours–years	Hours–decades
Soil erosion	Progressive (threshold may be crossed)	Hours–millennia	Hours–decades
Drought	Seasonal/irregular	Days–years	Days–weeks
Crop blight	Seasonal/irregular	Weeks–months	Days–months
Expansive soil	Seasonal/irregular	Months–years	Months–years

Source: after Alexander (1993)

Another approach divides hazards into rapid-onset, intensive events (short, sharp shocks such as tornadoes) and slow-onset pervasive events, which often affect larger areas over longer periods of time (such as droughts). Table 21.1 is an attempt to classify hazards in this way, and gives some indication of the frequency and predictability of events in time.

Although these classifications provide a useful summary of hazard types, many disasters involve composite hazards. Hence, an earthquake may cause a tsunami wave at sea, landslides or avalanches on slopes, building damage and fires in urban areas, and flooding due to the failure of dams, as well as ground shaking and displacement along faults. Similarly, many natural disasters cause disruption to public hygiene and consequently result in heightened risks of disease transmission.

Other approaches to defining and classifying disasters focus on such factors as the areal extent or the scale of damage: Burton *et al.* (1978), for example, postulate that a 'disaster' must cause more than US$1 million in damage, or the death or injury of more than a hundred people. Such an absolute approach has disadvantages, however, since a certain level of damage will have a different impact on society depending upon the strength of that society to cope with the disaster, a recognition used to assess disaster-proneness at the national level (see below). There is a case for defining a disaster at any scale, from the personal to the regional, national and global, but in practice it is generally fair to say that very poor societies commonly suffer the highest numbers of casualties while very rich societies suffer the highest property damage. In a survey of natural disasters from 1977 to 1997, Alexander (1997) found that about 90 per cent of disaster-related deaths occurred in developing countries while some 82 per cent of economic losses were suffered by developed countries.

Another difficulty arises in distinguishing between purely 'natural' events and human-induced events. This chapter is not concerned with such obviously human-induced hazards and catastrophes as industrial accidents or pesticide poisonings. However, in one sense, all 'natural' disasters can be thought of as human-induced since it is the presence of people that defines whether or not an event creates a disaster. Many of the natural physical processes that cause disasters can also be triggered or made worse by human action: while a volcanic eruption is a purely natural process, for example, the failure of a slope, producing a landslide, can be induced by many human activities, such as road-cutting, construction or deforestation. The composite nature of many disasters also blurs the distinction: the major cause of death due to an earthquake is usually crushing beneath buildings, so is the disaster natural or human-induced? Some researchers believe that the difficulty of making this distinction has made the division pointless, and prefer to talk of 'environmental' hazards that refer to a spectrum with purely natural events at one end and distinctly human-induced events at the other (e.g. Smith, 2001). The phrase 'na-tech' (natural-technological) is also used.

RESPONSE TO HAZARDS

Responses can be made by individuals, particular groups or the whole society, and they are related to the information available on past events and the probability of recurrence, the perception of that information, and the awareness of particular opportunities. One form of response might be to change the land use in a particular area or to move an activity to another location. Other responses can be categorised into those that aim to reduce losses and those that accept losses. The range of possible adjustments in these categories that can be made in response to the hazard from volcanic lava flows is indicated in Table 21.2. In practice, the most effective responses often involve a combination of measures.

Some approaches take action to reduce the risk of impact by preventing or modifying events. Since there is no known way of altering the eruptive mechanism, modifying the event itself is not an option for lava flows, but other hazard events can be prevented or modified. Floods, for example, can be prevented over a long period by impounding water behind a dam, and frosts can be prevented by installing a heating system in an orchard. Other approaches aim to modify the loss potential through prediction, either in time (so issuing hazard warnings to enable evacuation) or in space (by producing hazard zone maps that can be used to regulate land use in hazard-prone areas). These examples illustrate the ways in which responses can operate over a variety of timescales, from the long-term view of building a dam or regulating land use, to the immediate efforts involved in evacuating a threatened area. Most of these responses are more generally characteristic of the more developed countries, which have sufficient economic resources, infrastructure and trained personnel to organise and carry out such actions.

Other responses effectively accept the losses imposed by a hazard and cope with these either by planning for them in the form of economic insurance, or by relying on help from a wider community when they happen, whether from fellow

TABLE 21.2 *Examples of possible adjustments to the hazard of lava flows from volcanoes*

Modify the event	Modify the risk	Modify the loss potential	Spread the losses	Plan for losses	Bear the losses
Not possible	Protect high-value installations	Introduce warning systems (requires monitoring)	Public relief from national and local government (available in most countries)	Individual family or company insurance (possible in more developed countries but still limited)	Individual family, company or community loss-sharing (traditional and still widely practised)
	Alter lava flow direction (e.g. by bombing or barriers)	Prepare for disaster through civil defence measures (emergency evacuation plans)	Government-sponsored and -supported insurance schemes (available in more developed countries)		
	Arrest forward motion (e.g. by watering flow margin)	Introduce land-use planning (based on hazard-zone maps)	International relief (greatest benefit in developing countries)		

Source: after Chester (1993)

members of a family or from local, national or foreign governments and aid agencies. Economic insurance is again an option usually available only in the more prosperous economies, and in some cases it can actually encourage individuals and societies to persist with activities in hazardous areas.

The adoption of a particular response is determined by assessment of the risk. Assessment may be made scientifically, by calculating the probability of a particular hazard causing a disaster, based on information of past event frequency and area of impact plus the current vulnerability of people and property. An important measure of frequency is the 'return period' for a particular magnitude of event. Engineers calculate the statistical probability of a certain size of event occurring, and hence speak of the 50-year flood, for example, and design structures to withstand the magnitude of hazard likely to affect the structure within its lifetime. It is not a fail-safe measure, of course, because land use changes in a catchment can alter its flood response characteristics, and because it is still possible that a bridge built to withstand a flood that happens on average only every 50 years is, in fact, hit by a 100-year flood. Conversely, assessments are often made on the basis of how a particular hazard is perceived. For many reasons, not least the lack of long-term records for many areas, perception is involved in most hazard risk assessments.

HAZARDOUS AREAS

While people have always been at risk from natural hazards, the perception that the world is becoming a more hazardous place is widespread. Proving this hypothesis is not possible since there is still no comprehensive worldwide database of hazards or disasters. The rising worldwide trend in the numbers of natural catastrophes compiled by the insurance industry (Fig. 21.3a), is at least partly due to improved reporting of events (Swiss Re, 1998), although there are well-founded fears that certain types of hazard, such as tropical cyclones, may be increasing in intensity with global warming (see Chapter 11). With an increase in world population, however, it is also logical to suppose that the number of people living in high-risk areas is also rising, although the effects of efforts to reduce the risks must also be considered.

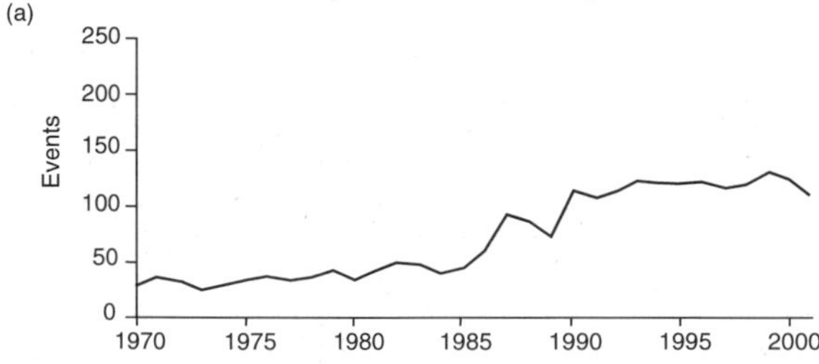

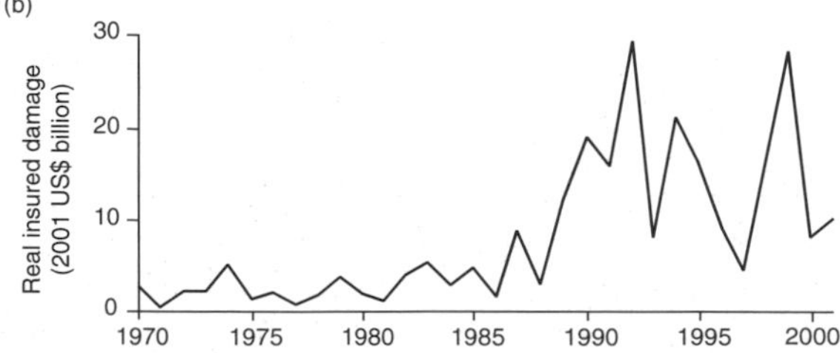

Figure 21.3 Global natural catastrophes, 1970–2001: (a) frequency; (b) insured losses at 2001 prices (after Swiss Re, 2002).

The resulting cost of damage caused by natural catastrophes has markedly increased on average since the late 1980s (Fig. 21.3b), a period characterised by particularly high losses in certain years. A small number of severe events pushed up these losses: the 1990 severe gales over Europe, Hurricane Andrew in 1992, the Northridge earthquake in southern California in 1994, and the winter storms across western Europe in 1999.

Some studies of particular locations have concluded that hazard potential is increasing. In a comprehensive account of the geophysical hazards facing Los Angeles County in California, where mid-winter storms are a main disaster catalyst, Cooke (1984) concluded that increasing risk was being driven mainly by rapid population growth and the spread of settlement into hills and mountains. In many individual cases, this spread is driven by the aesthetic desire for a less crowded environment with better views. Conversely, several examples of poorer sectors of society living in more hazardous urban zones, because they have no other option, are documented in Chapter 10.

On the national scale, some countries are perceived to be more prone to disasters than others. The impact of a particular event depends on where the event occurs and how prepared the area is to cope with the disaster. In the case of an earthquake in an urban area, for example, the impact will vary according to such factors as the quality of the buildings, the ability of the society to respond to any destruction, and the timing of the event. A significant decline in the number of flood fatalities in Australia since the 1850s has been put down to increased awareness of the flood hazard, better warning systems and the use of structural flood mitigation works (Coates, 1999). The success of preventing flood fatalities in Australia is reflected in the fact that death rates have fallen both in absolute and relative terms (i.e. despite the increase in population). The overall decadal death rate from flooding has fallen from nearly 24 per 100 000 population in the 1800s to 0.04 per 100 000 in the 1990s. An opposite trend has been identified in the USA, however, where

Smith and Ward (1998) documented a steady increase in the number of deaths from flooding over the period 1925–94, probably associated with the continuing problem of flashfloods.

In many ways, the overall impact of the disaster depends upon the economic strength of the country in which it occurs, because an event that affects a certain number of people and causes a certain amount of damage will have a greater overall impact in a poorer country than in a richer one. This logic has been followed in an attempt to quantify the risk associated with the effects of disasters on the national scale by the UN Disaster Relief Organisation (UNDRO). Using data on disasters over a 20-year period (1970–89), the study assessed the economic impact of disasters on the economies of 195 countries by expressing the damage costs of each disaster as a proportion of the annual gross national product (GNP) of the country concerned in a Disaster Proneness Index:

$$\text{Index} = \frac{\text{Damage (US\$)}}{\text{Annual GNP (US\$)}} \times 100$$

Only 'significant' disasters were included in the study: that is those causing damage in financial terms amounting to 1 per cent or more of a country's total annual GNP. The Total Index was then calculated as the sum of the individual indices for each significant disaster, providing a measure of the total effect of disasters over 20 years expressed as a percentage of annual GNP. An Average Index was also derived by calculating the mean of the individual indices for each significant disaster during the 20-year period, providing a measure of the impact of the average significant disaster that struck the country during the period, again as a percentage of the country's GNP (Kerpelman, 1990).

The most disaster-prone countries ranked according to the Total Index are shown in Table 21.3. Several small-island developing countries (the Cook Islands, Montserrat and Vanuatu) appear high in the ranking because they are particularly prone to tropical cyclones, and in two of these cases their high indices are based on a single disaster during the 20-year study period. Countries that are prone to frequent disastrous events, such as Nicaragua and Bolivia, are highly disaster prone according to their Total Index ranking, but the average impact of a significant disaster is not so great. Other countries prone to

TABLE 21.3 *The most disaster-prone countries according to the UNDRO Disaster Proneness Index*

Country	Number of disasters (1970–89)	Average Index (% 1980 GNP)	Total Index (% 1980 GNP)
Montserrat	1	1200	1200
Vanuatu	4	58	228
Nicaragua	8	26	207
Burkina Faso	4	48	191
Dominica	3	47	141
Cook Islands	1	119	119
Chad	5	18	92
Bolivia	10	8	84
St Lucia	2	41	81
Yemen	1	67	67

Source: Kerpelman (1990)

frequent disasters, such as India and the Philippines, do not appear in the top 70 countries according to their Total Index, indicating that their stronger economies are better able to absorb the effects of the disasters. Richer countries also have more to lose in economic terms, of course, but are generally better able to absorb these losses. Japan is a case in point here, experiencing the greatest absolute economic loss of any country during the study period (US$82 billion in two significant disasters) but this cost represented just 8 per cent of its 1980 GNP.

EXAMPLES OF NATURAL DISASTERS

The following sections take a closer look at a number of major natural causes of disaster. The selection is based on an analysis of the causes of deaths from all types of disasters over the period 1900–90 compiled by the US Office of Foreign Disaster Assistance (Blaikie *et al.*, 1994). Civil strife and famine were by far the largest causes of death, accounting for nearly 88 per cent of all casualties. Although natural events such as drought and flooding are often implicated as triggers for the onset of famine (see Chapter 5), the most significant direct natural causes of death were earthquakes, volcanoes, cyclones, epidemics and floods.

Earthquakes

Each year, around 1 million earth tremors are recorded over the Earth's surface, but the vast majority of these are so small that they are not felt

Figure 21.4 Part of the large-scale destruction of San Francisco in the USA that resulted from an earthquake along the San Andreas fault on 18 April 1906.

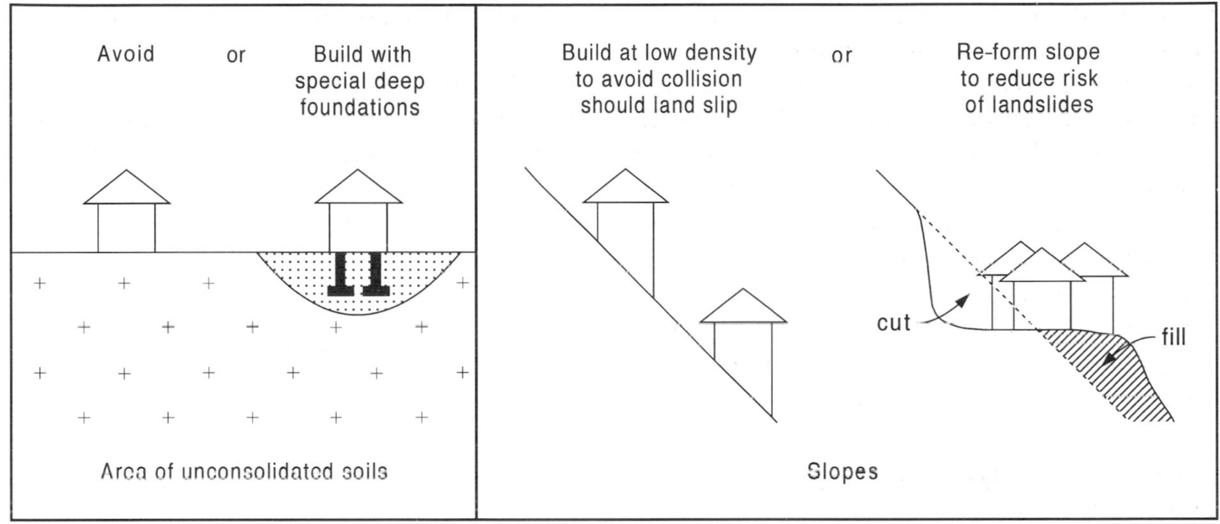

Figure 21.5 Some approaches to building on sites at high risk of disturbance during earthquakes (after Smith, 2001).

by people. The handful of major earthquakes that occur each year can cause widespread damage, however, causing some of the world's most devastating natural disasters. China has suffered some of the worst fatalities in major earthquakes: the 1556 event in Shensi caused more than 800 000 deaths, and the Tangshan earthquake of 1976 resulted in up to 250 000 fatalities.

Most earthquakes are caused by abrupt tectonic stress release around the margins of the Earth's lithospheric plates, although they also occur at weak spots near the centre of plates. Whilst the locations of plate boundaries and most associated faults are well known, the precise timing and location of individual earthquakes are virtually impossible to predict in any reliable way. A combination of seismic monitoring and observations of natural phenomena, including animal behaviour, were used in the only successful prediction of a large-scale earthquake, which devastated the Chinese city of Haicheng in Liaoning Province in February 1976. An evacuation of Haicheng and two other cities was ordered 48 hours before the main earthquake, almost certainly saving many thousands of lives (Adams, 1975).

The main environmental hazard created by seismic earth movements is ground shaking, which is related to the magnitude of the shock and normally assessed on the Richter scale, a complex logarithmic scale that measures the vibrational energy of a shock. In April 1906 an earthquake along the San Andreas fault in California, USA, caused slippage by up to 6 m along 300 km of the fault, and the transient waves unleashed by the slip devastated large parts of San Francisco (Fig. 21.4). Local site conditions have an important effect on ground motion, with greater structural damage usually found in areas underlain by unconsolidated material as opposed to rock. Particularly severe damage was caused during the earthquake of 1985 in the central parts of Mexico City, which is built on a dried lake bed. Similarly, a relatively modest-magnitude earthquake, registering 5.4 on the Richter scale, in San Salvador in 1986, caused unusually large destruction since most of the city is built on volcanic ash up to 25 m thick.

Another serious earthquake hazard associated with soft, water-saturated sediments is soil liquefaction, in which intense shaking causes sediments to temporarily lose strength and behave as a fluid. Loss of bearing strength can cause buildings to subside and soils can flow on slopes of greater than 3°. Mudflows, landslides, and rock and snow avalanches triggered by ground vibrations often play a major role in earthquake disasters, particularly in mountainous areas. In a study of major earthquakes in Japan, landslides were found to cause more than half of all earthquake-related deaths (Kobayashi, 1981). Tectonic displacement of the sea bed is the main cause of large sea waves, or tsunamis, which can travel thousands of kilometres at velocities greater than 900 km/h and cause great damage when they hit coastlines. Tsunamis are most common in the Pacific where one of the largest and most damaging, measuring 24 m in height, drowned 26 000 people in Sanriku, Japan, in 1896.

Some attempts have been made to regulate earthquake intensity by injecting pressurised water along faults to stimulate many small tremors and thus reduce the probability of major movements, but the technique still runs the risk of itself inducing a major earthquake. Hence, responses to the hazards associated with earthquakes focus on reducing risk and coping with the losses. Since the failure of structures is a major cause of injury during seismic shaking, much effort has gone into building location and design. Earthquake-hazard zoning maps have been produced for many countries and some of the approaches used for building in particularly hazardous sites are shown in Fig. 21.5. Building codes are only effective if enforced, however, and several recent examples of major damage and loss of life in urban areas have occurred in cities where such codes have not been followed (see page 172).

Even in places where strict building codes are adequately enforced, major damage and loss of life can still occur because more vulnerable older buildings have not been upgraded. The earthquake that hit south-central Japan in January 1995 resulted in the total destruction of 56 000 buildings and severe damage to another 110 000 in the small city of Kobe, causing more than 6000 deaths. Most of these buildings were constructed before the introduction of new Japanese earthquake code standards in 1981, while damage to post-1981 structures was minimal. Although the cost of upgrading all older buildings in a large Japanese city to these new standards would be enormous (perhaps US$50–100 billion), the lesson of Kobe suggests that such a scheme would still make financial sense. The total direct economic loss from the Kobe disaster has been estimated conservatively at US$130 billion, making it the most costly natural disaster in human history and the most devastating to hit a developed country. Additional direct and indirect losses from the disruption to business and loss of productivity is likely to raise the overall cost to more than US$200 billion (Chandler, 1997). The technology to upgrade older buildings is available, but to date the political will to implement such a programme has been lacking.

Volcanic eruptions

Volcanic eruptions have killed around 250 000 people in the last 400 years due to a range of associated hazards, which include falls of rock and ash, the force of lateral blasts, emission of poisonous gases, debris avalanches, lava flows, mudflows known as 'lahars', and 'pyroclastic flows' made up of suspended rock froth and gases (Chester, 1993), and their environmental effects can extend over very great distances (Table 21.4). Active volcanoes may also present more persistent hazards to human health through long-term exposure to carbon dioxide, radon and other

TABLE 21.4 *Hazards associated with volcanic eruptions*

Distance from volcano	Hazard
Up to 10 km	Lava flow Pyroclastic flows and surges Gases
Up to 100 km	Atmospheric blast Lahars Debris avalanches
Up to 1000 km	Tephra fall
Up to 10 000 km	Tsunami
Planet-wide	Climatic effects

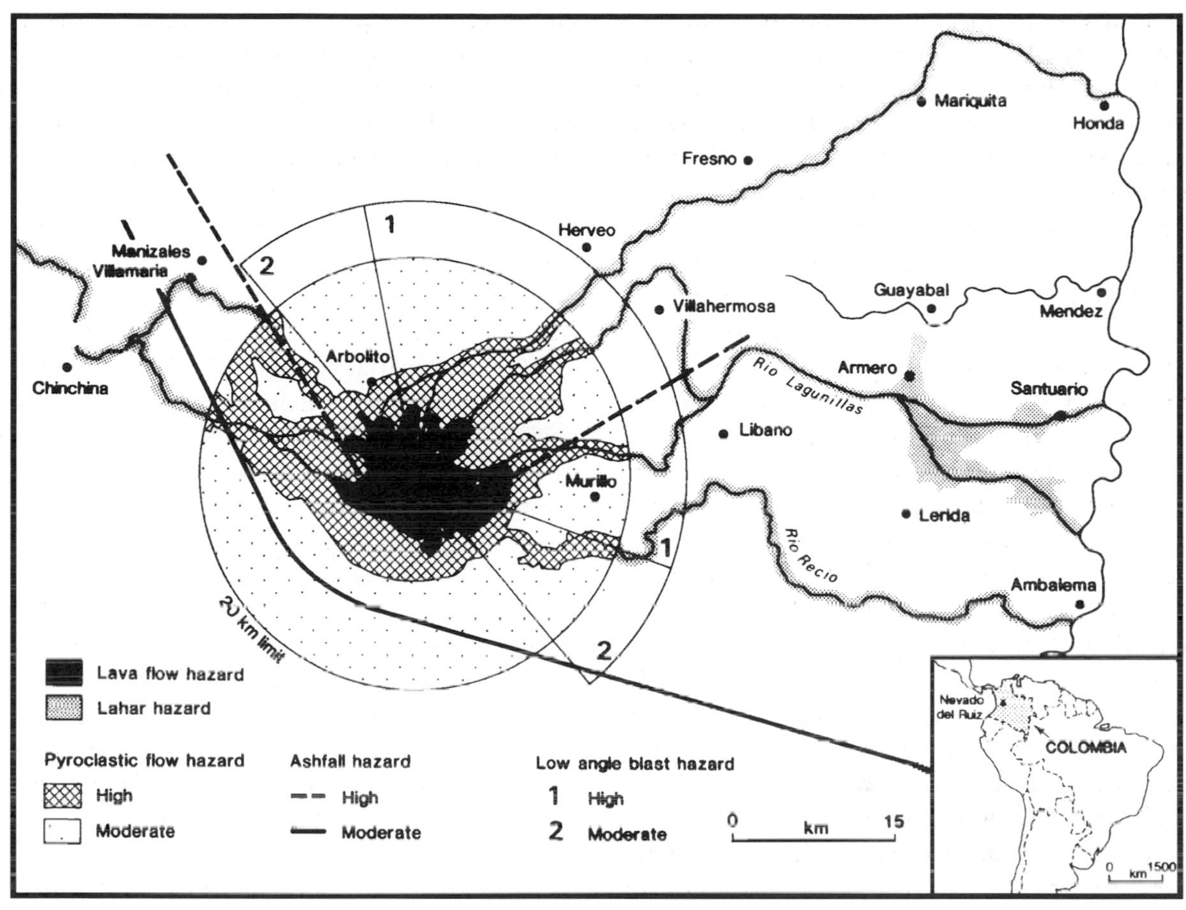

Figure 21.6 Volcanic hazards from Nevado del Rúiz, Colombia (Chester, 1993).

pollutants (Baxter, 2000). Only about 500 volcanoes are thought to have actually erupted in historical time, although the traditional categories of 'active', 'dormant' and 'extinct' are not infrequently found wanting when supposedly inactive volcanoes erupt, sometimes causing widespread destruction (e.g. Mount Pinatubo, the Philippines, 1991). Like earthquakes, the distribution and activity of volcanoes is controlled by global tectonics.

Lava flows, one of the most characteristic volcanic hazards, are rarely fast enough to cause death but can completely destroy anything in their path. The distinction has been made above between responses to the lava flow hazards typically employed in more developed countries and those more common in less developed countries (Table 21.2). Developing countries tend to rely upon hazard warning and evacuation, whilst some wealthier nations have attempted to modify the hazard by bombing to divert flows, erecting barriers to protect inhabited areas, and controlling the flow using water (e.g. in Hawaii, Iceland, Italy and Japan), although these countries also employ warning and evacuation measures. Such a distinction can also be made with regard to the whole suite of volcanic hazards. Chester suggests that 'many deaths are now avoidable often through the application of fairly simple methods of prediction' (1993: 186) since the activity of most volcanoes builds up to a major eruption and the build-up can be detected, given adequate monitoring using seismometers, tiltmeters and gas sample analysers. None the less, the fact remains that a great deal of infrastructure worldwide lies in the path of potential volcanic disasters, not least in major cities located near volcanoes, including Mexico City (Popocatepetl), Bandung (Tangkuban Parahu), Naples (Vesuvius) and Managua (Masaya Nindiri and Apoyeque), all of which lie within 50 km of their respective volcanoes. Urbanisation, particularly in developing countries, has led to increasing global exposure to the risks posed to large cities by volcanic eruptions (Chester *et al.*, 2001).

However, a stark contrast can be drawn between the successful prediction and evacuation of the area around Mount Pinatubo on Luzon Island in the Philippines in 1991, and the failure to evacuate in time for the 1985 eruption of Nevado del Rúiz in the Colombian Andes. Seismic and other anomalies in activity were reported from Nevado del Rúiz for almost a year before the eruption, and numerous geologists and seismologists were called in to assess the situation, but poor communication and bureaucratic inefficiency prevented action being taken in time. A hazard map was prepared (Fig. 21.6), the final version of which was due to be presented to the authorities the day after the eruption started on 13 November. The eruption melted part of the summit's snow and ice cap, and a torrent of meltwater and pyroclastic debris flowed from the summit, coalescing in lower channels to entrain vegetation, water and mud to form rapidly flowing lahars. Nearly 70 per cent of the population of Armero (about 20 000 people) was killed in the mudflow and another 1800 died on the volcano's western flanks. A total of 50 schools, two hospitals, 58 industrial plants, roads, bridges and power lines were destroyed or damaged, and 60 per cent of the region's livestock perished. The total cost to the economy of Colombia was estimated at US$7.7 billion, or 20 per cent of the country's GNP in that year. As Voight (1990: 151) concludes: 'the catastrophe was not caused by technological ineffectiveness or defectiveness, not by an overwhelming eruption, or by an improbable run of bad luck, but rather by cumulative human error – by misjudgement, indecision and bureaucratic short-sightedness'. At the time, the Colombian Government had other problems on its mind. On 6 November, the president ordered troops to storm Bogota's Palace of Justice, which had been occupied by guerrillas; 100 people were killed, including ten senior judges.

Tropical cyclones

Tropical cyclones – which are also known as hurricanes in the Atlantic and typhoons in the western Pacific – are among the most destructive of all natural hazards. They are generated over warm tropical oceans and rely on a complex combination of processes to grow, mature and eventually die. Warm ocean water is needed to fuel the tropical cyclone's heat engine, so they only form when the temperature of the top 50 m or so of the ocean surface is at least 26.5 °C. The air above has to be marked by a fairly rapid drop in temperature with height, allowing thunderstorms to form initially. It is thunderstorm activity that allows the heat from the ocean waters to be liberated, helping the tropical cyclone to develop. These conditions also need to occur at least 500 km from the Equator. This is because the force exerted by the spinning Earth (the Coriolis force) has to be at a certain strength for tropical cyclone formation, and this strength drops off towards the Equator.

The destruction caused by tropical cyclones occurs in a combination of very strong winds (the threshold wind speed is 33 m/s but winds can exceed 80 m/s), torrential rainfall (which can cause flooding, landslides and mudflows), high waves, and particularly by storm surge, which results from the shearing effect of wind on water to cause local rises in sea level and inundation of coasts where cyclones make landfall. Around 15 per cent of the world's population is considered to be at risk, Smith, K. (1997) placing tropical cyclones second only to floods according to the number of hazard-related deaths, most from drowning in the storm surge. An indication of the economic cost of tropical cyclone damage for some of the worst recent events is given in Table 21.5, which is compiled from a list of the insur-

TABLE 21.5 *Some of the most damaging tropical cyclones, 1970–2001*

Date	Name	Country	Victims (dead and missing)	Insured damage (2001 US$ billion)
23 Aug 1992	Hurricane Andrew	USA, Bahamas	38	20.2
27 Sept 1991	Typhoon Mireille	Japan	51	7.3
15 Sept 1989	Hurricane Hugo	North America	61	6.0
22 Sept 1999	Typhoon Bart	Japan	26	4.3
20 Sept 1998	Hurricane Georges	USA, Caribbean	600	3.8
10 Sept 1999	Hurricane Floyd	USA, Bahamas	70	2.5
1 Oct 1995	Hurricane Opal	USA	59	2.4
11 Sept 1992	Hurricane Iniki	USA, N. Pacific	4	2.0
12 Sept 1979	Hurricane Frederic	USA	–	1.8
5 Sept 1996	Hurricane Fran	USA	39	1.8
18 Sept 1974	Tropical cyclone Fifi	Honduras	2000	1.8
3 Sept 1995	Hurricane Luis	Caribbean	116	1.7
10 Sept 1988	Hurricane Gilbert	Jamaica	350	1.7
4 Aug 1970	Hurricane Celia	USA, Cuba	31	1.4
12 Sept 1998	Typhoon Vicki	Japan, Philippines	12	1.4
19 Oct 1991	Hurricane Grace	USA	2	1.3

Source: after Swiss Re (2002)

ance industry's most costly losses from both natural and human-induced events between 1970 and 2001: 16 tropical cyclones were among the top 40 most damaging events, and the insured loss from Hurricane Andrew was the most costly of all, exceeding that of the terrorist attacks in New York and Washington on 11 September 2001.

Since attempts to control, alter or destroy tropical cyclones have not been successful, most attention is focused on reducing society's vulnerability through being prepared. Meteorological satellites provide some of the most important instruments for detecting and monitoring tropical cyclones, and early warning systems are the most widely used response. Improvements in forecasting, warnings and evacuation plans have reduced the human death toll in many countries in recent decades, although property damage has risen over the period. In the USA, 70 million people are estimated to be at risk from hurricanes, nearly 10 per cent of them resident in Florida. Several observers have noted the marked difference in impact between developed and developing countries, the former tending to be better organised and able to cope with communicating warnings and facilitating evacuation.

The devastating impact of tropical cyclones on the economies of some small-island developing countries has been shown above (Table 21.3), but one of the countries most frequently and severely affected is Bangladesh, which was hit by 61 severe cyclones, 32 of which were accompanied by storm surges, between 1797 and 1997 (Khalil, 1992; Swiss Re, 1998). Some of these events have been characterised by huge fatalities. The cyclone of October 1876, which produced a storm surge 12 m high, resulted in the loss of 400 000 lives; 300 000 people were killed in a 10 m storm surge in November 1970, and 140 000 people lost their lives in tropical cyclone Gorky in April 1991. Bangladesh's high population density has meant that survivors quickly re-occupy the coastal lowlands after cyclone disasters strike, and the Cyclone Protection Project, part of Bangladesh's Flood Control Action Plan (see below), aims to rehabilitate coastal embankments built in the 1960s and 1970s to provide some protection for these zones. However, eradication of the threat from cyclones and their associated storm surges is impossible, and much more effort has gone into communicating warnings to the most rural populations so that the additional 3400 multipurpose community shelters that are to be constructed will be used. More effective protection and replanting of threatened areas of mangrove forest in the Sundarbans (see page 117) and elsewhere along the coast has also been called for, to provide a quasi-natural defence against storm surges (Khalil, 1992). The long-term effectiveness of these measures is questioned, however, by some of the predicted changes in the region due to climate and sea level change (see page 191).

Epidemics

Infectious diseases, many of which are preventable and curable, are the greatest cause of morbidity and mortality on the global scale. They account for nearly half of all deaths in developing countries, where children under five are the most susceptible; whereas, in developed countries, diseases of old age and abundance, such as heart disease and cancers, are the most common killers. This clear difference is linked to the fact that far greater numbers of people in developing countries suffer from malnutrition, inadequate water supply and sanitation, poor hygiene practices and overcrowded living conditions. Indeed, the global trend towards urbanisation, with many of the fastest urban growth rates being recorded in the poorer countries of the world, has been cited as a major factor in the increased frequency of many infectious diseases (Olshansky *et al.*, 1997).

An infectious disease is caused by a host being invaded by a parasite or other pathogen, which might be a bacterium, virus or worm. The pathogen carries out part of its lifecycle inside the host and the disease is often a by-product of these activities. Many pathogens also need a vector, an 'accomplice', to help them spread from host to host, a role often played by an insect. Many of the major diseases of the developing world are associated with water, an association in which three

TABLE 21.6 *Classification of diseases associated with water*

Role of water	Comments and examples
Waterborne	Diseases arise from presence in water of human or animal faeces or urine infected with disease pathogens, which are transmitted when water is used for drinking or food preparation (e.g. diarrhoea, dysentery, cholera, typhoid, guinea worm)
Water hygiene	Diseases result from inadequate use of water to maintain personal cleanliness (e.g. waterborne diseases above as well as infestation with lice or mites)
Water habitat	Diseases for which water provides the habitat for vectors. The pathogen may enter the human body from the water directly (e.g. bilharzia is transmitted when a fluke leaves a snail vector and penetrates human skin as people wash or swim), or indirectly via a fly (e.g. trypanosomiasis – sleeping sickness) or mosquito (e.g. malaria)

Source: after GEMS (1988)

typical situations can be distinguished (Table 21.6), although some diseases may fall under more than one category. Although any disease represents a natural hazard to individuals and groups of people, the emphasis here will be on epidemics, relatively sudden outbreaks of a disease in a certain area. Two diseases associated with water will be examined: cholera and malaria.

Cholera is caused by a bacterium known as Vibrio, which can survive outside the human gut in such environments as water, moist foodstuffs, human faeces and soiled hands, all of which can act as media for its transmission. The disease originated in South Asia, notably in the Ganges–Brahmaputra delta and appears to spread and recede geographically in cycles. It first reached Europe in the nineteenth century, causing six major pandemics, and realisation of the key transmission role played by dirty water was a central force behind the public health movement in many European countries in the nineteenth century. These measures have helped to all but eradicate cholera from most developed countries but a seventh pandemic spread from Indonesia in the early 1960s, reaching Africa and southern Europe within ten years. Its rapid spread through Africa caused many deaths after a period of about 75 years when the disease was a relatively minor ailment (Stock, 1976). The pandemic reached South America at Lima, Peru, in January 1991 where it is thought to have been brought from Asia in a ship's ballast water. The warming of coastal and inshore waters by an El Niño/Southern Oscillation event at the time may have stimulated the growth of a plankton harbouring the cholera bacterium (Reid, 1995). The bacterium contaminated fish and shellfish, which were consumed by the local human population, and the spread of the disease was facilitated when it entered Lima's drinking water supply via inadequately treated wastewater. From its landfall in Peru, where over 1700 cases per day were reported during the first four months, the disease spread throughout much of South America within a year (WHO, 1992). The outbreak of a more recent epidemic in Bangladesh at the end of 1992, involving a new, hardier strain of the cholera Vibrio first identified in a coastal algal bloom, has raised fears of an eighth pandemic (Levins *et al.*, 1994). Although it may not be possible to prevent totally the emergence of new cholera strains and the cyclic spread of the disease, there is no doubt that the introduction of basic sanitary measures, the provision of safe drinking water and the promotion of less crowded living conditions in developing countries would do a great deal towards limiting the spread of this often fatal disease. It is the lack of these basic provisions that makes cholera a common killer in refugee camps in many tropical areas.

Unlike cholera, the transmission of malaria requires a vector. The disease is caused by a single-celled pathogen that is transmitted through a number of species of blood-sucking mosquito. Until just a few decades ago malaria was a virtually global disease, common in parts of Europe, for example, such as the Rhine delta and low-lying parts of Mediterranean countries until after the Second World War. In England, the disease was known as 'marsh ague' or 'marsh fever', and it was a common cause of death among many inhabitants of wetlands in the nineteenth century (Dobson, 1980). Since 1945, however, there have been significant successes in malaria eradication. In 1950, 140 countries reported cases of the disease, but by 1990 that number had fallen to 90 (Diesfeld, 1997). The widespread drainage of wetlands, an ideal habitat for mosquitoes, along with concerted campaigns using DDT in the 1950s, has helped to eradicate malaria from most temperate and many subtropical areas, and the disease has been all but wiped out in Europe and North America (although in more recent times, increasing international travel to the tropics has resulted in cases among some temperate-zone residents).

Malaria, however, is still endemic in most parts of the developing world and despite numerous attempts at global eradication since the 1950s, has reached epidemic proportions in a number of countries in association with high population densities, poor sanitation, environmental degradation and civil unrest. Accurate assessment of the malaria burden is difficult because most deaths from the disease occur at home, the clinical features of malaria are very similar to those of many other infectious diseases and good-quality microscopy is available in just a few centres. However, worldwide, malaria is thought to account for at least a million deaths annually, most in young children, among about 500 million clinical cases. Its main impact is in sub-Saharan Africa where at least 90 per cent of deaths from malaria occur (Greenwood and Mutabingwa, 2002). Resistance of the pathogen to drugs and resistance of mosquitoes to insecticides, as well as the rudimentary level of health services in many areas, have combined to increase the malaria problem in sub-Saharan Africa over the past three decades. Much research into antimalarial drugs was conducted in the USA during the Viet Nam War, but since then, less effort has been focused on a disease that is just one of many that continue to plague the inhabitants of the poorer parts of the world.

Another influence on the geographical pattern of malaria incidence is global warming, since even subtle changes in temperature can have significant effects on the distribution of mosquitoes. Models of global climate change suggest that the current level of cases per year could increase significantly by 2100 (Stone, 1995). The greatest changes in transmission potential occur in areas of seasonal malaria and around the margins of existing transmission, such as highland areas. Such an expansion of the malarial zone has been evident in Rwanda in recent years where substantial increases in temperature since the early 1960s peaked in the late 1980s, and malaria became established in areas where it had previously been rare or absent. The incidence of the disease in high-altitude zones rose by 500 per cent, and most of this increase has been a function of higher temperatures and higher rainfall (Loevinsohn, 1994). Elsewhere in the highlands of East Africa, however, recent increases in the incidence of malaria have too hastily been attributed to global warming (Hay *et al.*, 2002). The rise in antimalarial drug resistance, population migrations, and the breakdown of health services and vector control operations have all contributed to the resurgence of malaria in East African areas where no significant climate change has been detected.

Floods

Floods are one of the most common natural hazards and are experienced in every country. They occur when land not normally covered by water becomes inundated, and they can be classified into coastal floods, most of which are associated with storm surges (see above and Chapter 7), and river channel floods. Atmospheric phenomena are usually the primary cause of river flooding. Heavy rainfall is most commonly responsible, but melting snow and the temporary damming of rivers by floating ice can be important seasonally. The likelihood of these factors leading to a flood is affected by many different characteristics of the drainage basin, such as topography, drainage density and vegetation cover, and human influence often plays a role – especially by modifying land use (including the effects of urbanisation) and through conscious attempts to modify the flood hazard.

Techniques for predicting floods are well developed. Most floods have a seasonal element in their occurrence and they are often forecasted using meteorological observations, with the lag time to peak flow of a particular river in response to a rainfall event being calculated using a flood hydrograph. Return periods for particular flood magnitudes can be calculated for engineering purposes, although long periods of historical data are required, which are by no means available for all river basins, and developments can alter a river's flood characteristics. Flood hazard maps are commonly utilised for land use zoning.

Many of the human responses to the flood hazard shown in Table 21.7 (see also Table 8.3) have been put into practice on the Mississippi River in the USA, one of the world's largest and most intensively managed rivers. Channel straightening, for example, was initiated in the

TABLE 21.7 *Flood hazards and their abatement*

Critical flood characteristics	Adjustments to floods	Flood hazard abatement	
		Structural	Non-structural
Depth	Accept the loss	Levées, dikes, embankments, flood walls	State laws
Duration ordinances	Public relief	Channel capacity alteration (width, depth, slope and roughness)	Zoning
Area inundated	Abatement and control	Removing channel obstructions	Subdivision regulations
Flow velocity	Land evaluation and other structural change	Small headwater dams	Building codes
Frequency and recurrence	Emergency action and rescheduling	Large mainstream dam	Urban renewal
Lag time	Land-use regulation	Gully control, bank stabilisation and terracing	Permanent evacuation
Seasonality	Flood insurance		Government acquisition of land and property (creation of floodway and public open space)
Peak flow	Controlled flooding of low-value land		Fiscal methods: building financing and tax assessment
Shape of rising and recession limbs			Warning signs and notices
Sediment load			Flood insurance

Source: after Alexander (1993)

1930s to increase the gradient and velocity of stretches of the Mississippi so that river water would erode and deepen the channel, thus increasing its flood capacity. Engineering structures have also been widely employed; more than 4500 km of the Mississippi are lined with embankments or levées designed to confine floodwaters.

Despite the numerous attempts at hazard mitigation, the Mississippi still overflows its banks with devastating consequences. The 1993 flood on the river's upper reaches is the most recent severe example. High soil-moisture content from heavy rains the previous year enhanced runoff during the period between April and August 1993, when rainfall totals were well above average in the north-central states. As levées failed in more than 1000 places, record flooding occurred along many stretches of the Mississippi, Missouri and several tributaries (Fig. 21.7). Floodwaters washed over an area of about 4 million hectares, destroying or seriously damaging more than 40 000 buildings (Williams, 1994), with the period of inundation prolonged in many places by the levées, which prevented the return of water to the channel once the peak had passed. The cost

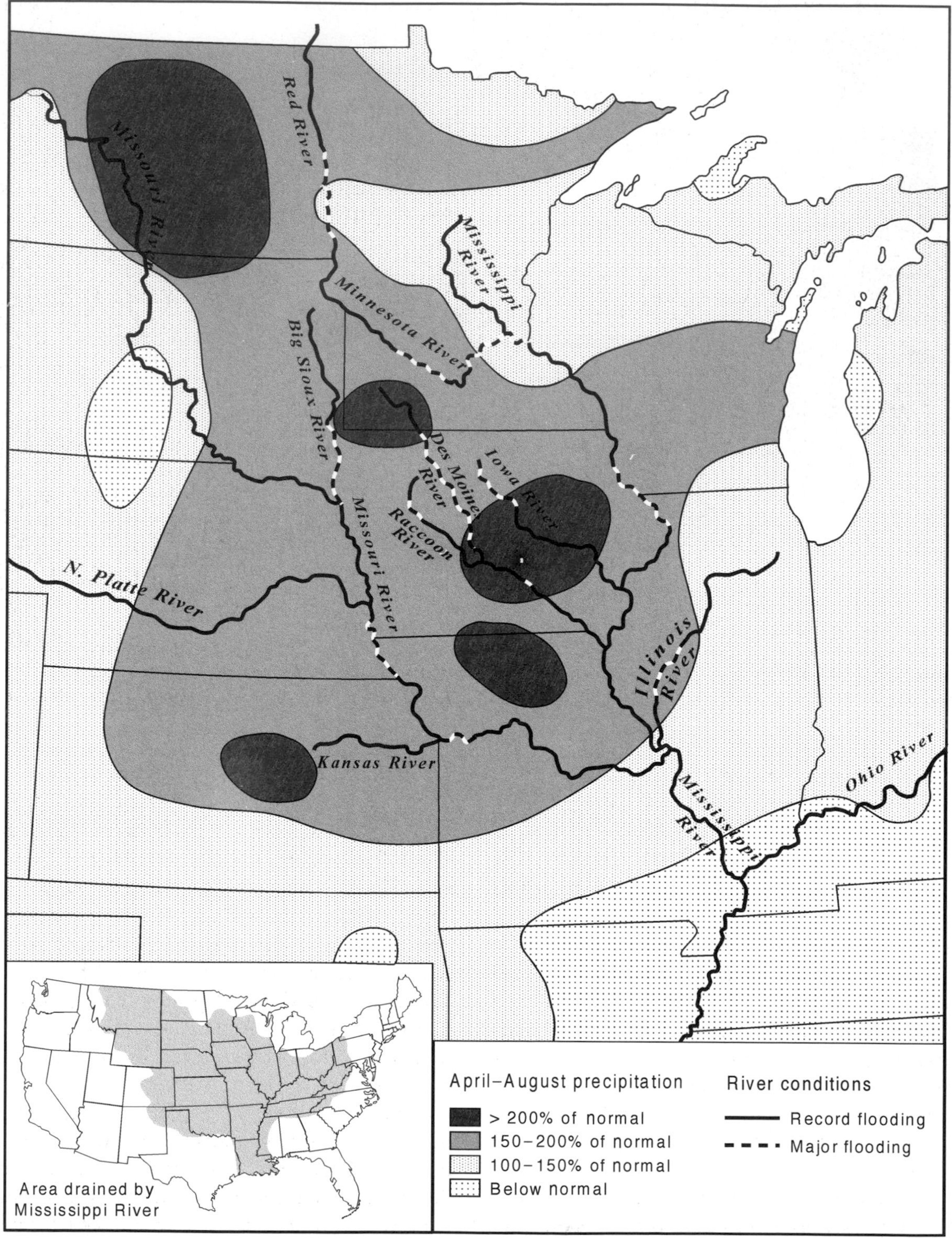

Figure 21.7 Precipitation anomalies and flooding on the upper Mississippi and tributaries in 1993 (after Williams, 1994).

Figure 21.8 Not all floods are considered as hazards by rural Bangladeshis, since some enhance crop production. Yields from both the paddy rice being planted here and the taller jute plants behind benefit from a period of inundation.

of the damage to insurance companies totalled US$755 million, while the total damage is estimated to amount to US$12 billion. A total of 45 people were killed (Swiss Re, 1994). Worries that changes in the frequency and return periods of flooding on the Mississippi may be an effect of global warming have been highlighted by Knox (1993), who found a strong relationship between these variables by studying a 7000-year record of flood sediments on tributaries of the upper Mississippi. Flood magnitude and frequency varied greatly during periods when annual temperature changed by only 1–2°C and mean annual precipitation varied by just 10–20 per cent.

The inability of structures to prevent river flooding has been highlighted in discussions over a controversial proposal to build large embankments along stretches of the major rivers of the Bangladesh delta. About 80 per cent of the national territory is floodplain, prone both to cyclonic flooding and floods from the major rivers (Ganges, Brahmaputra and Meghna), which flow through the country. Recent major river floods occurred in 1987 and 1988, inundating 40 per cent and 57 per cent of the country respectively (Rasid and Pramanik, 1990). These events had return periods of 100 years or more, and their effects inspired the proposal as part of a comprehensive flood control action plan coordinated by the World Bank.

Not least among the arguments put forward against the embankments – an expensive high-tech, top-down solution – is the fact that they are designed to prevent all floods. Yet Bangladeshi villagers distinguish between beneficial rainy season floods, known as *barsha* in Bengali, and harmful floods of abnormal depth and timing,

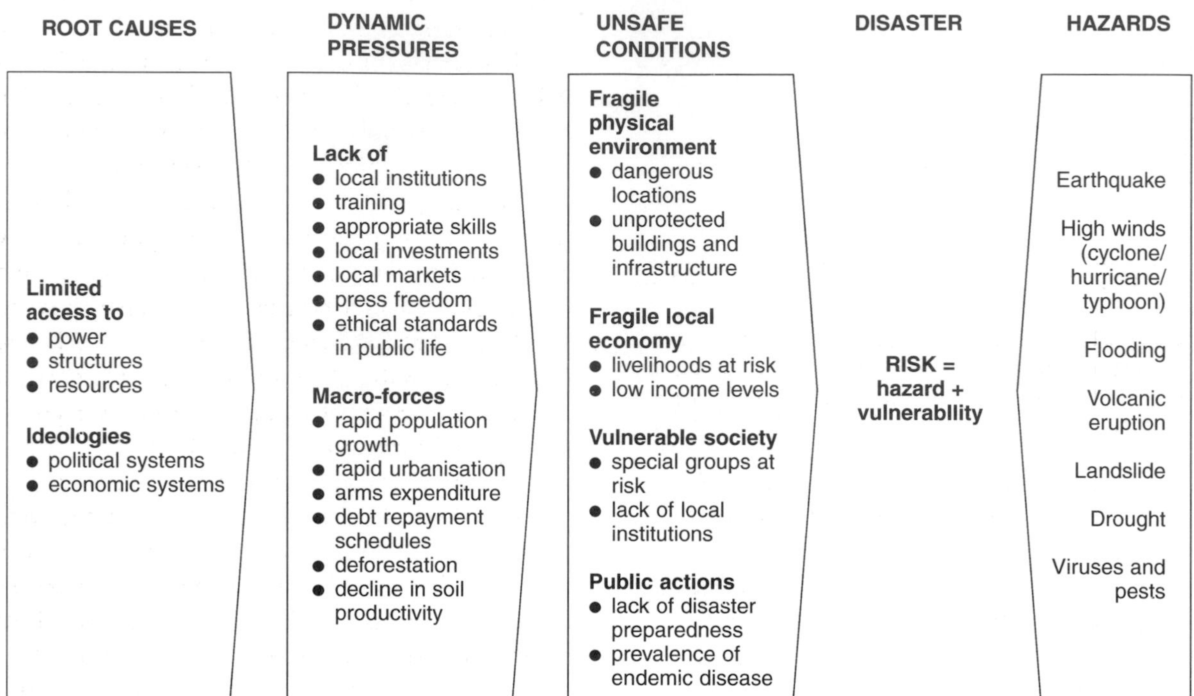

Figure 21.9 Model of pressures that create vulnerability and result in disasters (after Blaikie *et al.*, 1994).

which are termed *bonna*. *Barsha* floods water the paddy fields and enhance soil fertility both through sediment inputs and nitrogen-fixing algae, which thrive in the floodwater (Fig. 21.8). Fish caught in flooded fields are also the main source of animal protein for rural inhabitants (Boyce, 1990).

While there is widespread agreement that Bangladesh needs to improve resource management, opponents of the embankment proposal suggest that less technological, more ecocentric approaches, which aim to improve flood management and actively involve local inhabitants, would be more appropriate. New ponds could be constructed for fishing and water storage, and low embankments would check flooding only during the early growth stages of the rice crop. Better preparation for unusually high and rapid flooding would complement these approaches. In addition to these measures, others advocate a more deep-rooted, less technocentric, strategy that seeks to alleviate the factors that cause people to be vulnerable to flooding (Fig. 21.9). Better access to land, replacements for land lost to erosion, and compensation for animals and other assets lost to natural disasters would all protect livelihoods and reduce vulnerability, as would better health care and facilities (Blaikie *et al.*, 1994).

Despite the controversies in Bangladesh surrounding the desirability of physical structures to protect societies from flooding, such projects remain popular methods for flood control. Many of the objections to a permanent structure can be addressed by a movable flood barrier like the Thames Barrier in London, designed to protect the UK's capital from unusually high tides caused by storm surges. These occur when very strong winds and waves associated with an area of low pressure in the North Sea force water towards Britain's east coast. When a storm surge coincides with a high tide the result is often flooding on a

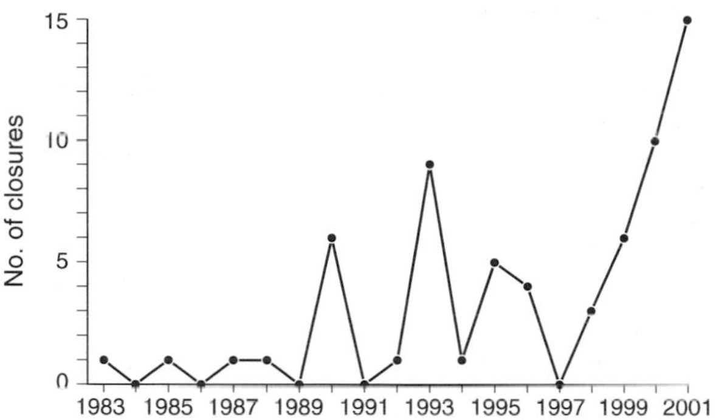

Figure 21.10 The number of times the Thames Barrier has been closed to protect London from flooding, 1983–2001 (data from the UK Environment Agency).

large scale, as in 1953 when parts of the English coast were inundated from Yorkshire to Dover, drowning more than 300 people.

For most of the time the Thames Barrier is open, allowing shipping to operate normally, and minimising interference with the natural flow of the river. An early warning system operates to predict exceptional water levels from surge, tide and river flows several hours in advance, allowing time for the Barrier to be closed. Following completion in 1982, the trend over its first 20 years or so has been towards more frequent closures (Fig. 21.10), most to protect against tidal flooding but on occasion to assist in preventing fluvial flooding from upriver by stopping sea water from enhancing already unusually high river levels. The designers of the Thames Barrier envisaged that the number of closures would increase to around ten per year on average in the first two decades of the twenty-first century because of predicted sea level rise due both to ground subsidence in south-east England and the effects of global warming.

FURTHER READING

Alexander, D. 1993 *Natural disasters*. London, University College London Press. After documenting the range of geophysical hazards, the author looks at the impacts on human society and the options for disaster management.

Blaikie, P., Cannon, T., Davis, I. and Wisner, B. 1994 *At risk: natural hazards, people's vulnerability and disasters*. London, Routledge. This book emphasises the importance of the human factors behind hazards that become disasters, looking at the social, political and economic causes of people's vulnerability to natural events.

Chandler, A.M. 1997 Engineering design lessons from Kobe. *Nature* 387: 227–9. An appraisal of the Kobe earthquake, arguing the need to update old buildings to make them more earthquake-resistant.

Chester, D.K., Degg, M., Duncan, A.M. and Guest, J.E. 2001 The increasing exposure of cities to the effects of volcanic eruptions: a global survey. *Environmental Hazards* 2: 89–103. A risk assessment of volcanic hazards in major cities.

McCall, G.J.H., Laming, D.J.C. and Scott, S.C. (eds) 1992 *Geohazards: natural and man-made hazards*. London, Chapman & Hall. A collection of papers focusing on geophysical hazards, including those associated with earthquakes, volcanoes and soils.

McGuire, B., Mason, I. and Kilburn, C. 2002 *Natural hazards and environmental change*. London, Arnold. With a focus on rapid-onset geophysical events this book highlights the links between natural hazards and environmental change.

Smith, K. 2001 *Environmental hazards: assessing risk and reducing disaster*, 3rd edn. London, Routledge. All major rapid-onset events of both the natural and human environments – including seismic, mass movement, atmospheric, hydrological and technological hazards – are covered in this book, which emphasises the physical aspects of hazards and their management.

Thompson, P.M. and Sultana, P. 1996 Distributional and social impacts of flood control in Bangladesh. *The Geographical Journal* 162: 1–13. A study of the effects of flood protection structures on the rural poor.

Websites

http://geohazards.cr.usgs.gov/ site run by the US Geological Survey, covers landslides, earthquakes and volcanoes with an emphasis on the USA.

http://www.eri.u-tokyo.ac.jp/ the Earthquake Research Institute at the University of Tokyo.

http://www.solar.ifa.hawaii.edu/Tropical/ tropical storms worldwide, current and past.

http://www.ffwc.net/ the Bangladesh Flood Forecasting and Warning Centre has data satellite imagery and management information.

http://www.swissre.com includes annual reports on hazards and catastrophes compiled by the insurance industry.

http://www.cidi.org/ the Centre for International Disaster Information has situation reports on numerous natural disasters.

http://www.unisdr.org/ the International Strategy for Disaster Reduction aims to help all societies to become resilient to the effects of natural and technological hazards and disasters.

http://www.disasterlinks.net/ contains links to a wide variety of disaster-related websites.

POINTS FOR DISCUSSION

- Can natural hazards ever be construed as good for human society?
- Since the precise timing and location of earthquakes are virtually impossible to predict, why do people live in zones that are renowned for their earthquake activity?
- Are some countries naturally more hazardous than others?
- If floods are the most common natural hazard, why aren't we better at coping with them?

22

CONCLUSIONS

TOPICS COVERED
Regional perspectives
National efforts
Global dimensions
The sustainable future

Key words: systemic and cumulative change, multi-pollutant/multi-effect approach, command and control approach, debt-for-nature swap

Environmental issues are not new phenomena. People have always interacted with the natural world and these interactions have always thrown up challenges to human societies. Mistakes have been made and solutions often found to problems encountered or created. However, as human society has become more complex, as our numbers have increased, so the range and scale of environmental issues have multiplied. Although there are still many sizeable portions of the Earth that show little obvious evidence of human impact (e.g. many hyper-arid deserts, the deep oceans, parts of the polar regions and some of the tropical rain forests), there is nowhere that is not affected to some extent by changes in the chemical make-up of the atmosphere and associated changes in climate and pollution levels (Fig. 22.1). Likewise, the nature of relatively unaffected parts of the planet also has an impact on those regions that are directly used by human populations, through their effects on global climate and biogeochemical cycles.

It is the global nature of so many environmental issues that has thrust them to prominence in our minds. Environmental scientists have recognised two types of global environmental change: 'systemic' change through a direct impact on globally functioning systems (e.g. emissions of greenhouse gases that affect global climate) and 'cumulative' changes that attain global significance through their worldwide distribution (e.g. species loss) and/or because of their effects on a large proportion of a global resource (e.g. soil degradation on prime agricultural land). The impacts of both are recognised as being of global proportions, but the appropriate responses to systemic and cumulative change vary. Addressing a systemic global environmental issue like global warming requires a truly global strategy: all countries must take action together to combat the potential dangers. Addressing a cumulative global environmental issue can also benefit from a worldwide approach, through the exchange of data and comparison of strategies, for example. However, many cumulative global environmental issues need to be tackled at the local level because the precise causes of problems are often different in different places.

Several common themes emerge from the preceding chapters on individual issues. Numerous issues have arisen as a result of deliberate attempts to manipulate the natural environment (e.g. widespread soil erosion on agricultural land, the Aral Sea tragedy, the impacts of big dams). Other issues have arisen as unexpected repercussions of activities conducted to improve human societies (e.g. contamination by fertiliser residues, global warming, acid rain). The complexities of

Figure 22.1 This landscape in northern Greenland shows little obvious sign of human impact, although the Inuit population has hunted wildlife here for about 4500 years, albeit at sustainable levels. However, cores taken from Greenland's ice sheet show contamination transported through the atmosphere from Europe began to exceed natural background levels as early as Roman times.

many environmental issues are also clear. Cause and effect are not always obvious (e.g. the many factors affecting forest decline), and human activities and natural forces can sometimes combine to give a synergistic effect (see Table 11.5). Effects on the natural environment are also not always immediate or expected, due to the operation of thresholds, feedbacks and time lags. Many environmental issues overlap and interact. Deforestation in the tropics is a major cause of biodiversity loss, and global warming has been seen to have all sorts of potential knock-on effects for many other issues. A full understanding of many issues is hampered by our imperfect knowledge of the physical environment, and as knowledge and understanding improves, so our perspective on certain issues can change. Reappraisal of the Sahelian fuelwood crisis, widely feared in the 1970s but now thought to be less of a concern, is a case in point here. We can only base our approach to managing environmental issues on existing knowledge, while continuing to conduct research to improve our understanding. But in the meantime we must be aware of uncertainties, hence the emergence of important tenets such as the precautionary principle and 'no regrets' approaches.

Some elements of human society emerge as common to several issues. Ownership, or a lack of ownership, is a pertinent issue in many atmospheric and marine environmental problems. Economic factors are also important. Poverty can drive people to degrade their resources, or force individuals to live in marginal and hazardous areas. Societies and governments may not tackle environmental problems because they lack the

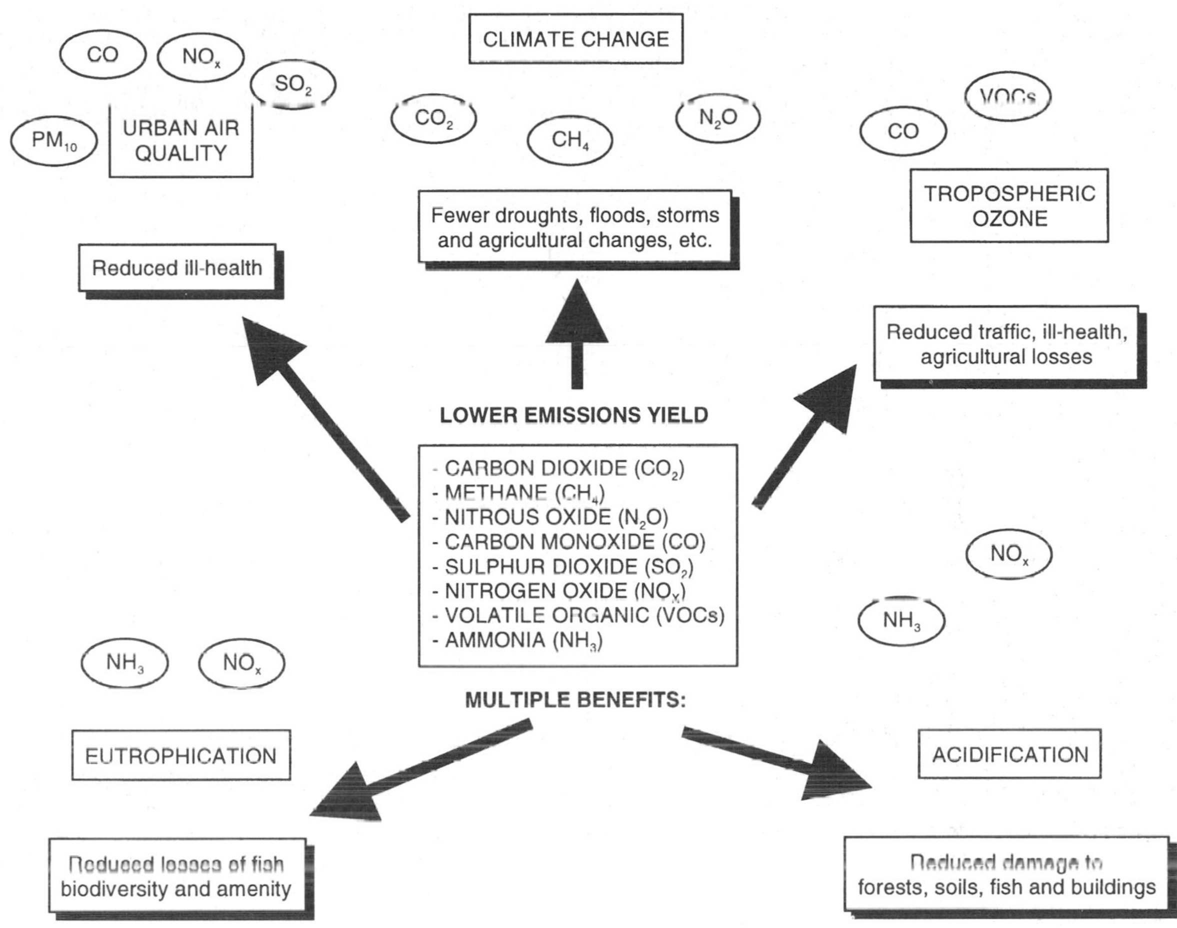

Figure 22.2 The multipollutant/multi-effect approach (after EEA, 1997).

necessary economic resources even though workable solutions to their problems may be well known. This is not to say, however, that poor people necessarily cause environmental problems, nor that richer sectors of society do not cause them. Political factors also play a role. Many wars and other conflicts arise due to competition over resources, and inadequate enforcement of conservation measures may result from political priorities that lie elsewhere or through the weakening of civil authority, as several examples from the former Soviet Union indicate.

Supposedly advanced societies can also learn useful lessons about environmental management from indigenous cultures, which often remain 'closer to nature' than urban, industrialised groups. However, again, this is not necessarily to say that societies operating with rudimentary levels of technology are always better at avoiding environmental problems. Examples to the contrary have been highlighted in this book: accelerated soil erosion rates in central Mexico were as great before the arrival of the plough brought Spanish invaders, as after, and evidence suggests that at least half of Hawaiian bird species became extinct in the pre-European period.

This book has also highlighted many areas where environmental issues have been tackled successfully. A general trend towards improving urban air quality has been documented with higher development levels, and there are many examples of waste disposal problems being

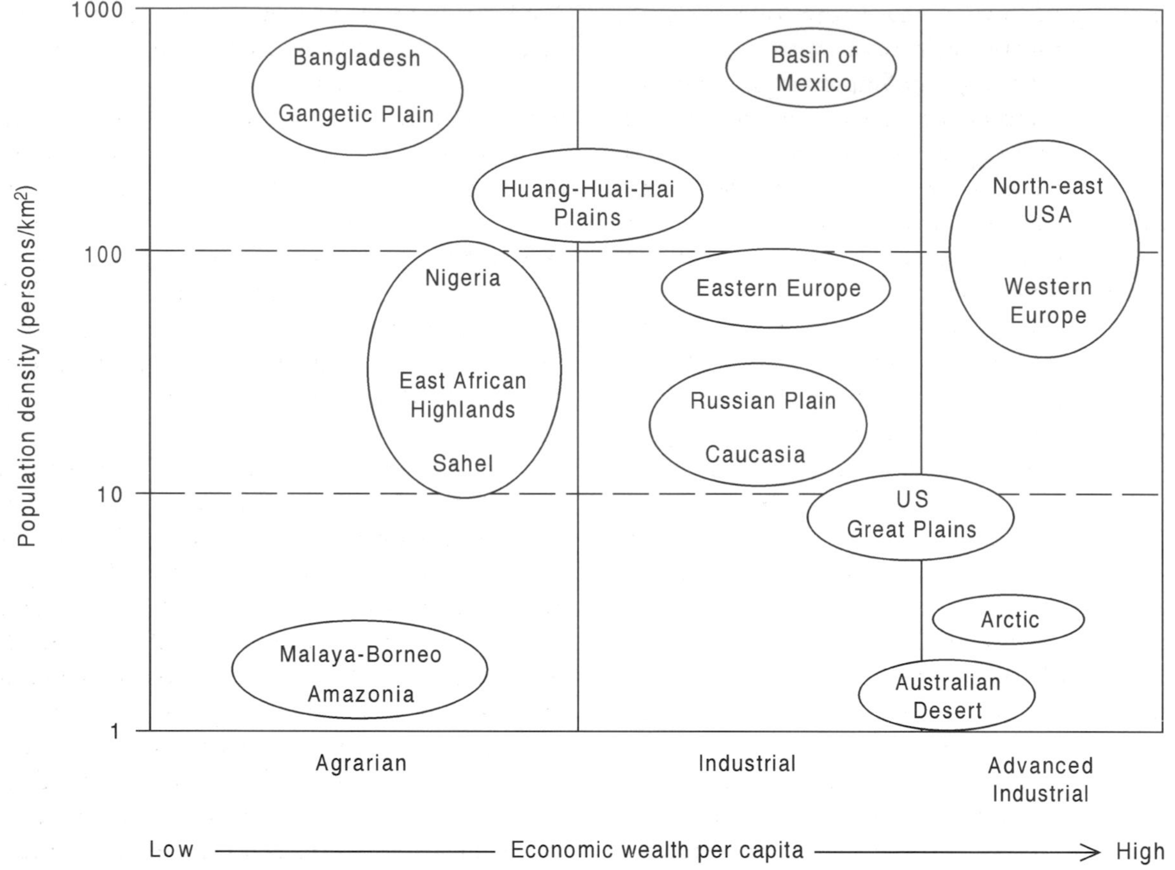

Figure 22.3 Characteristics of regional environmental transformation (after Clark, 1989; Kates *et al.*, 1990).

ameliorated by viewing society's waste products as resources waiting to be used. Methods for rehabilitating and protecting the environment are constantly being developed, and the fact that many issues are interrelated means that efforts to tackle one issue often have beneficial effects for others. As Fig. 22.2 shows, efforts to reduce air pollutants brings numerous benefits to many aspects of the physical and human environments.

Of course the very rise in interest in sustainable development can be seen as an appropriate response by society to the scale and critical nature of many environmental issues. This book has been organised on an issue-by-issue basis, but the remainder of this chapter is devoted to a more regional approach, looking at the ways in which particular societies are reacting to the many environmental issues that affect them.

REGIONAL PERSPECTIVES

A distinction can be made between three global regions that should have different priorities in their contributions to global environmental sustainability (Goodland *et al.*, 1993a). The North must concentrate on reducing its long history of environmental damage due to overconsumption and affluence; the main priority in the South should be to stabilise population growth; and in the countries of the East, the former communist bloc, modernisation of wasteful and polluting technology should be the central priority. To promote these strategies, it is in the self-interest of the North to accelerate the transfer of technology to the East and South.

Measures of population density and relative affluence (which also has a major bearing on technological capacity) have been used to classify some major world regions, as shown in Fig. 22.3. More detailed design and implementation of sustainable development strategies for these regions will vary according to their conditions and characteristics (Clark, 1989). Low-income, low-density areas such as Amazonia and Malaya–Borneo include many of the world's remaining settlement frontiers where large-scale clearance for agriculture, grazing and timber has begun only recently, after a longer history of resource use in shifting cultivation, small-scale plantation and mining sites. The widespread poverty of farmers engaged in land clearance, and the poorly developed state of local institutions, which might guide sustainable development, mean that appropriate management in such areas will be particularly difficult to attain.

Regions of low income and high population density, such as on the Ganges–Brahmaputra floodplains of the Indian subcontinent and the Huang–Huai–Hai plains of China, by contrast, have long histories of human agricultural endeavour. This agricultural character has been augmented in recent decades by rapid industrial development and urbanisation, bringing the types of pollution problems that have been faced by industrial European countries within the last 100 years. The introduction of employment opportunities that can relieve pressure on agricultural land, but that do not simultaneously exacerbate urbanisation and industrial pollution problems, is the critical management challenge.

Other developing-world areas, such as the Basin of Mexico, with an agricultural history over hundreds of years, have experienced very rapid population growth, speedy industrialisation and a burgeoning consumer culture in recent decades. The grave problems of pollution and congestion are typical, if extreme, for primate cities of the developing world. Development in such cases can only be made sustainable, and problems eased, by a concerted national effort at decentralisation.

Areas that bear the more direct mark of environmental impact from the relatively wealthy countries include such frontier regions as the Arctic and the Australian desert – environments perceived as 'harsh' and 'sensitive' by the residents of more temperate climes. Exploitation, primarily for mineral resources, in these areas of low population density involves sophisticated technology and large economic investments. Although our knowledge of such environments remains poor, and hence the consequences of inappropriate actions can be great, the involvement of relatively few powerful corporate and governmental institutions means that the potential for introducing sustainable development practices is relatively good.

The greatest potential and the largest responsibilities for appropriate change lie in the densely populated wealthy industrialised regions of the world. Such regions have perpetrated a disproportionately large environmental impact, both locally and at the continental and global levels, through such pathways as emissions of acid rain precursors and greenhouse gases, and through the export of environmentally damaging practices to other parts of the world. In recent times these areas have also achieved some successes in improving environmental quality locally. Some of the leads these countries have taken are considered in more detail in the following section.

NATIONAL EFFORTS

While some of the technologies and efforts to curb environmental problems have been introduced in the preceding chapters, numerous instances of environmental improvements being hampered by economic and political forces have also been noted. Hence, it is appropriate to look at some of the ways in which governments and other institutions have worked to promote sustainable practices. The rise in interest in environmental issues and sustainable development has inspired numerous efforts to gather information on national environmental situations and to target development efforts more appropriately. These include country status quo reports on environment and development issues, submitted to the UN Conference on Environment and

TABLE 22.1 *Evolution of environmental policies in the Netherlands since 1970*

Period	Policy focus	Style of decision processes	Instruments
Delimiting the ecological arena (1970–83): improving national conditions for human health	Cleaning up national stocks of air and water; starting soil clean-up operations	Top-down national legislation and quality standards; no stakeholder involvement	National and European laws; regulatory licensing enforced by local authorities; add-on technology
Encouraging pollution prevention (1984–89): protecting national health conditions for people and ecosystems	Pollution prevention to preserve national stocks of air, water, soil and biodiversity; initial attention to acidification and ozone depletion	Start of stakeholder involvement in design and implementation of programmes; allowing room for technology choices and flexible timing	Emission reduction targets; environmental care; EIA; financial incentives; legal liability; stricter enforcement
Enhancing eco-efficiency (1990–99): building on previous actions to include international responsibility for global air quality	More attention to transnational issues of acidification, global warming and ozone depletion	Greater autonomy granted to local authorities and private enterprise to set objectives; more flexible design and implementation	Targets; target groups; negotiated agreements; economic, technological, fiscal and social instruments
Super-optimisation for sustainable development (2000–30): opportunities for multisectoral benefits	More attention to limiting national use and improving management of global stocks of biodiversity, energy and minerals	Provoking integration of ecological, economic and social interests nationally and internationally; new processes to develop joint objectives	Target global resources; incentives for producers and consumers; breakthrough technologies; get prices right; quality consumption

Source: after Keijzers (2000)

Development (UNCED) from both developed and developing nations, and several schemes aimed primarily at developing countries, such as National Environmental Action Plans prepared with World Bank assistance, National Conservation Strategies prepared with assistance from IUCN, an international NGO, and a number of similar initiatives sponsored by bilateral aid agencies, such as the US Agency for International Development's Country Environmental Profiles.

Data collection has provided the basis for a set of environmental indicators, which are being developed on the national level (e.g. Environment Canada, 1991), and by member countries of organisations such as the Organization for Economic Co-operation and Development

(OECD, 1991b). Such indicators are being used to monitor the sustainability of resource use. In the forestry sector, for example, an index of intensity of forest use is calculated as the ratio of the total annual harvest to the annual growth of national forest stock. Hence, a ratio of one or less reflects a sustainable use of forest resources.

These and other types of procedure have been used to produce the first national strategy for achieving sustainable development in the Netherlands. The Dutch National Environmental Policy Plan, first introduced in 1989 and subsequently updated, provides the framework for a systematic revision of national policies to drive the country towards sustainability. Since 1970 Dutch environmental policy has evolved through three distinct phases (Table 22.1) and has expanded from a focus on protecting human health and ecosystem integrity to ensuring a sustainable supply of natural resources at national and international levels. This expansion has been achieved by changing the style of policies, while management of the physical environment has evolved from a focus on eliminating pollution after it occurred to avoiding it altogether. At the beginning of the new century, Dutch environmental policy is shifting from pollution prevention to the sustainable management of resource stocks (Keijzers, 2000).

The first period identified by Keijzers, from 1970 to 1983, focused on cleaning up national stocks of air and water in the Netherlands primarily by a 'command and control' approach of setting national quality standards enforced by environmental laws. This legislation and the policies it contained were based on a number of general principles, which have continued to underpin subsequent development of policy. They are:

- *the polluter pays principle* (polluters are liable for the costs of clean-up and the prevention of pollution)
- *stand still principle* (polluted areas should not be polluted further and clean areas must remain clean)
- *principle of isolation and control* (pollution should be controlled at its source, not exported elsewhere through the air or water)
- *principle of priority for pollution abatement at source* (as opposed to end-of-pipe solutions)
- *the use of 'best technical means' technology* (to eliminate emissions when serious risks to public health are at stake regardless of economic costs, and the use of 'best practical means' technology when health effects are limited, so enabling industry to consider economic costs)
- *the principle of avoiding unnecessary pollution.*

This approach proved to be effective at dealing with the most obvious pollution problems. Air and water quality improved thanks to significant reductions in emissions of pollutants including heavy metals, SO_2, and phosphates from detergents. However, the excessive use of fertilisers and pesticides in agriculture, and emissions from the transport sector continued to cause serious pollution.

The second period, 1984 to 1989, saw a shift in emphasis from reacting to pollution problems to actively trying to prevent pollution before it becomes a problem. This shift was accompanied by a change in policy, away from the command and control approach of environmental laws, to a structure of economic incentives, social institutions and self-regulation to achieve sustainable development. Policies aimed to encourage pollution prevention by starting to involve all stakeholders, including business, government and local communities, in the design and implementation of programmes. This focus on preventing pollution at source spawned new instruments such as environmental care programmes for businesses and environmental impact assessments. This process enabled a shift in perception from seeing environmental standards as being in conflict with economic interests (because achieving the standards cost money) towards viewing a clean environment as a necessary precondition for a sound economy. The results from this period were also encouraging as chemical industries developed substitutes for phosphate detergents, agreements were reached to reduce packaging materials and the production of mercury

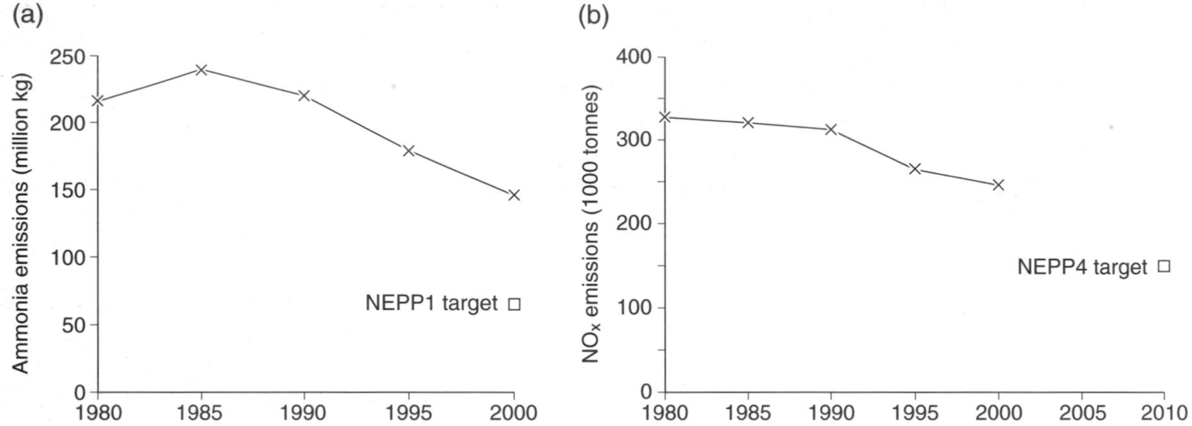

Figure 22.4 Recent performance and targets under NEPPs in the Netherlands (from data at http://www.rivm.nl/environmentaldata/, accessed April 2002) for (a) ammonia emissions by agriculture (from manure and fertilisers); (b) NO_x emissions from motor traffic.

batteries was stopped. Industry and energy suppliers made further drastic reductions in emissions of heavy metals and SO_2, and pollution from agriculture began to decline.

The third period in the evolution of Dutch environmental policy, from 1990 to 1999, saw the introduction of the National Environmental Policy Plans (NEPPs): NEPP1 in 1989, followed by NEPP2 in 1993 and NEPP3 in 1998. These NEPPs have continued the policy of emissions reductions, which aim to safeguard the quality of air, water and soil by setting targets for pollution prevention. The focus has been on eight themes: dispersal of toxic substances, waste disposal, disturbance by noise and external safety from hazards, eutrophication, acidification, climate change (global warming and ozone depletion), groundwater depletion, and squandering of resources. By this time a clear premise of the Dutch environmental management philosophy had emerged: that a high-quality environment cannot be achieved through conventional pollution-control measures alone. A mixture of new, cleaner technologies and structural changes in patterns of production and consumption are also needed. The NEPPs have emphasised the need to improve eco-efficiency (the level of emissions and resources used per unit of production), and further encouraged the integration of ecological and economic concerns.

The style of policies to implement the NEPPs has also continued to develop away from environmental legislation and regulations towards more open decision-making with more flexibility and autonomy for local authorities and enterprises to define their own ways of achieving the targets set. When launching NEPP3, the Dutch Government announced further significant achievements since NEPP1: energy efficiency in industry had increased by more than 10 per cent, the sale of CFCs ceased in 1995, the proportion of waste reused increased from 61 per cent to 72 per cent in the period 1990–96, and acidification had decreased by 50 per cent since 1985.

Contrasting recent performances for two sources of pollution are shown in Fig. 22.4. The agricultural emissions of ammonia (largely from manure), a source of acidification in the Netherlands, rose through the early 1980s as the total agricultural manure production sharply increased, but has declined as low emission spreading techniques (injection into the subsoil) have increasingly been applied since 1990. In 2000, emissions were 68 per cent of the 1980 level, whereas the NEPP1 set a target of 30 per cent of 1980 levels. The progress towards the most recent target for nitrogen oxides (NO_x) emissions from motor traffic, set for 2010, is shown in Fig. 22.4b. The emission of NO_x has been declining since

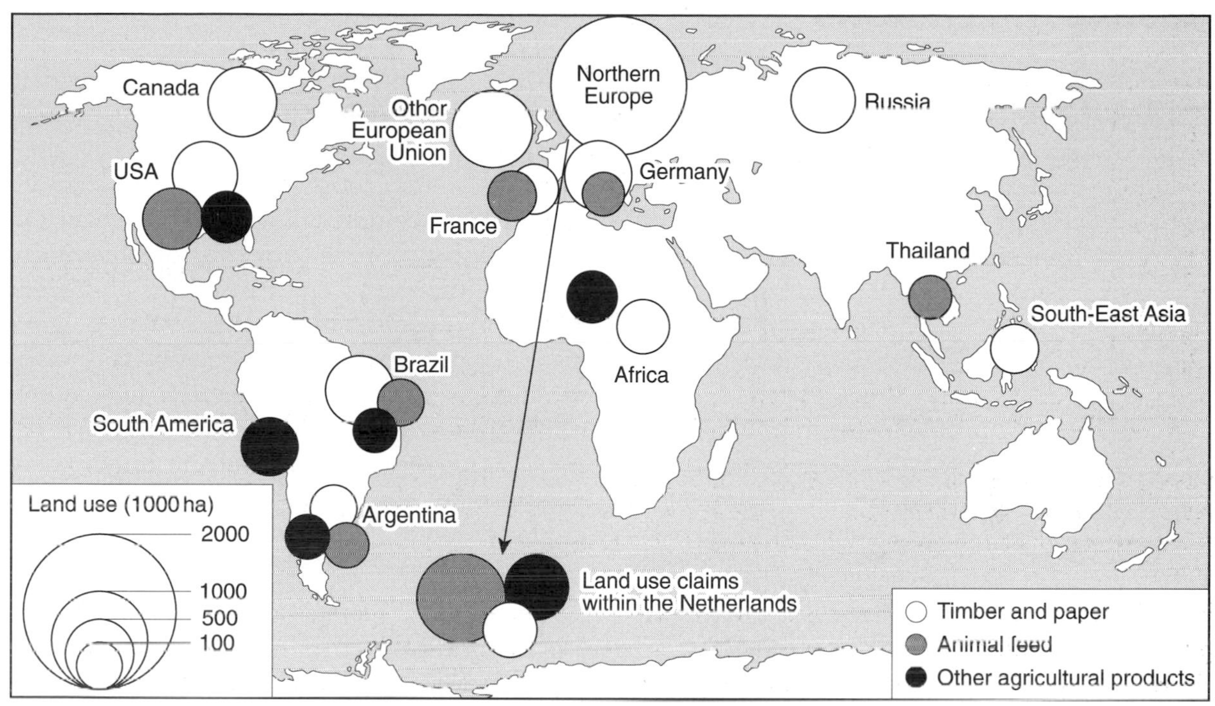

Figure 22.5 Land use claims relating to consumption patterns for food and timber production imported into the Netherlands in 1995 (after RIVM, 1999).

1990, thanks to the use of three-way catalytic converters in passenger cars and delivery vans, and to the application of increasingly clean diesel engines.

Since the introduction of NEPP1 in 1989, the Netherlands has succeeded in achieving economic growth with a reduction in environmental pressure in many areas. Keijzers (2000) indicates that continued improvements in environmental performance have been seen in several sectors since NEPP1, including the construction and energy industries, waste disposal, sewage and water cleaning sectors. However, policies to curb energy use, to reduce emissions of CO_2 and NO_x, and to minimise resource use were still well away from sustainable levels as described in NEPP1. Key areas for attention in NEPP4, launched in 2002, include managing CO_2 emissions, controlling future infrastructure development, minimising resource use and reducing the burden on biodiversity.

None the less, the Netherlands, like all countries, does not operate in a vacuum. The country is affected by pollutants that arrive from outside its borders, such as water pollution in the River Rhine and acid rain precursors from many parts of Europe. The low-lying country is also particularly vulnerable to any rise in sea level that may be consequent upon human-induced global warming. Dutch exports produced under a sustainable regime may be less competitively priced than equivalent products derived from unsustainable methods, and Dutch imports of resources from other countries may effectively contribute to environmentally damaging practices elsewhere. The scale of impact overseas is indicated in Fig. 22.5, which shows the land use claims relating to consumption in the Netherlands of imported food and timber. While the NEPP is undoubtedly a very significant step in the right direction at the national level, its measures can only work effectively if paralleled by similar initiatives elsewhere. Hence, it is appropriate to look again at international efforts to promote sustainable development.

GLOBAL DIMENSIONS

International attempts to promote conservation and sustainable development have been mentioned already in this and other chapters. These include the establishment of the UN Environment Programme, the recommendations of organisations such as the World Commission on Environment and Development, and many global international agreements, such as the conventions on biological diversity and climate change negotiated at UNCED (the so-called Earth Summit held in Rio de Janeiro in 1992). There are also many other regional agreements and conventions.

Table 22.2 shows the major global conventions that address environmental issues. Such conventions typically arise as concern at the environmental effects of certain activities reaches a threshold at which global action is considered necessary. They reflect the fact that a sufficient number of countries recognise a common interest in trying to combat certain issues. The international community can act quickly when issues are perceived to be particularly acute. The Chernobyl accident in April 1986 precipitated two conventions on nuclear accident notification and assistance in the same year.

The years indicated in the table are when the conventions were first agreed. A country becomes a signatory to a treaty when someone given authority by a national government signs it. Usually a signatory has no duty to perform the obligations under the treaty until the treaty comes into force in that country, which means the national government has to ratify the treaty, or otherwise adopt its provisions in national law, and the treaty itself is ratified by a prescribed number of countries. This can take a long time. The UN Convention on the Law of the Sea, for example, was first signed in 1982, but did not come into force until 1994. Countries can still join conventions after the first signing session. The Convention on Wetlands of International Importance Especially as Waterfowl Habitats (the Ramsar Convention) was first signed by 18 countries in 1971, but by early 2003 the original 18 had been joined by 118 other contracting parties.

These conventions represent a significant step forward at the international level, but international agreements and conventions still have their weaknesses. Enforcement is a particular difficulty. While non-compliance with an agreement can lay a country open to criticism, compliance can rarely be enforced by law. Part of this problem is related to the difficulties of translating agreements in principal into national laws. Another weakness is structural. Any agreement requires compromise between the parties, and international agreements, often involving hundreds of nation-states, can result in provisions that are imprecise and not very constraining.

Agenda 21 is not shown in Table 22.2 because it is not a binding treaty. It was, none the less, an important outcome of the Earth Summit, a sort of global action plan for achieving sustainable development during the twenty-first century. Agenda 21 is designed to reconcile conservation and development, and addresses general development issues, issues relating to specific resources, the roles of various groups, and means of implementation. Such implementation is the concern of individual countries and although it has no legal force, Agenda 21 has considerable political authority.

These and other initiatives have all made a positive contribution, but they have done little to address one of the major underlying difficulties in achieving global sustainability: poverty and the unequal distribution of resources. Inequalities exist at all levels (see Chapter 2), but on the global scale an obvious imbalance is evident between North and South. Some minor, if significant, attempts have been made to address both debt problems and environmental problems in developing countries by converting part of the external debt of a country into a domestic obligation to support a specific programme. Some examples of existing 'debt-for-nature swaps' are detailed in Table 22.3. An international conservation group, or in some cases a national government, purchases part of a debtor country's foreign debt on the

TABLE 22.2 *Major global environmental conventions*

Topic	Convention	Agreed	Main aims
Antarctica	Antarctic Treaty and Convention	Washington, DC 1959	Ensure Antarctica is used for peaceful purposes, such as scientific research
Wetlands	Convention on Wetlands of International Importance Especially as Waterfowl Habitats	Ramsar 1971	Stem progressive encroachment on and loss of wetlands
World heritage	Convention Concerning the Protection of the World Cultural and Natural Heritage	Paris 1972	Establish a system of collective protection of cultural and natural heritage sites of outstanding universal value
Ocean dumping	Convention on the Prevention of Marine Pollution by Dumping of Wastes and Other Matter	London, Mexico City, Moscow and Washington, DC 1972	Control marine pollution by prohibiting the dumping of certain materials and regulating ocean disposal of others, encouragement of regional agreements and establishment of mechanisms to assess liability and settle disputes
Biological and toxin weapons	Convention on the Prohibition of the Development, Production, and Stockpiling of Bacteriological (Biological) and Toxin Weapons, and on their Destruction	London, Moscow and Washington DC 1972	Prohibit acquisition and retention of biological agents and toxins not justified for peaceful purposes, and the means of delivering them for hostile purposes
Endangered species	Convention on International Trade in Endangered Species of Wild Fauna and Flora	Washington, DC 1973	Protect endangered species from overexploitation by controlling trade in live or dead individuals and in their parts through a system of permits
Ship pollution	Protocol of 1978 Relating to the International Convention for the Prevention of Pollution from Ships, 1973	London 1978	Modify the 1973 convention to eliminate international pollution by oil and other harmful substances and to minimise accidental discharge of such substances
Migratory species	Convention on the Conservation of Migratory Species of Wild Animals	Bonn 1979	Promote international agreements to protect wild animals that migrate across international borders
Southern Ocean	Convention on the Conservation of Antarctic Marine Living Resources	Canberra 1980	Safeguard the environment and protect marine ecosystems surrounding Antarctica

TABLE 22.2 *continued*

Topic	Convention	Agreed	Main aims
Law of the sea	UN Convention on the Law of the Sea	Montego Bay 1982	Establish a comprehensive legal regime for the seas and oceans, including environmental standards and rules to control marine pollution
Ozone layer	Vienna Convention for the Protection of the Ozone Layer	Vienna 1985	Protect human health and the environment by promoting research, monitoring of the ozone layer and control measures against activities that produce harmful affects
Nuclear accident notification	Convention on Early Notification of a Nuclear Accident	Vienna 1986	Provide relevant information about nuclear accidents to minimise transboundary radiological consequences
Nuclear accident assistance	Convention on Assistance in the Case of a Nuclear Accident or Radiological Emergency	Vienna 1986	Facilitate prompt provision of assistance following a nuclear accident or radiological emergency
CFC control	Protocol on Substances that Deplete the Ozone Layer	Montreal 1987	Require developed states to cut consumption of CFCs and halons with allowances for increases in consumption by developing countries
Hazardous waste movement	Basel Convention on the Control of Transboundary Movements of Hazardous Wastes and their Disposal	Basel 1989	Establish obligations to reduce transboundary movements of wastes, minimise generation of hazardous wastes and ensure their environmentally sound management
Biodiversity	UN Convention on Biological Diversity	Nairobi 1992	Protect biological resources, and regulate biotechnology firms' access to and ownership of genetic material, and compensation to developing countries for extraction of their genetic materials
Climate change	Convention on Climate Change	New York 1993	Stabilise atmospheric concentrations of greenhouse gases
Desertification	UN Convention to Combat Desertification	Paris 1994	Promote sustainable development of drylands
Persistent Organic Pollutants (POPs)	Convention on Persistent Organic Pollutants (POPs)	Stockholm 2001	International control of 12 chemicals

TABLE 22.3 *Examples of established debt-for-nature agreements*

Debtor	Date	Debt face value (US$ million)	Cost (US$ million)	Investor	Terms and aims
Bolivia	7/87	0.65	0.1	Conservation International	Full legal protection of Beni Biosphere Reserve, set-up of buffer zones and local management fund
Costa Rica	4/89	24.5	3.5	Government of Sweden	Expansion and protection of Guanacaste National Park
Madagascar	1/91	0.12	0.06	Conservation International	Interest from endowment fund for ecosystem management programmes, species inventories and environmental education
Poland	1/90	0.05	0.01	WWF-International and WWF-Sweden	Clean-up of River Vistula and development of Biebrza National Park
Sudan	12/88	0.8	Bank donation		Midland Bank donated US$0.8 million in loans to UNICEF to provide ten village water wells with hand pumps

Source: modified after WCMC (1992: Table 32.11)

secondary market, at a fraction of the theoretical face value of the debt. The low cost of the original debt on the secondary market is a reflection of the fact that creditors have a low expectation of the debt itself being repaid. Once the debt has been acquired, the 'investing' conservation organisation agrees a rate of exchange with the debtor country so that the debt is converted into local currency, which is then used to finance the intended aims.

Supporters of debt-for-nature swaps point out that they are agreements in which all parties involved stand to gain. Conservation organisations increase the spending power of their money, through the difference between the secondary market cost and the face value of the debt purchased, and are able to exert influence on conservation policies in developing countries. The debtor gains by reducing its foreign debt, reducing the government's need to raise foreign currency to service its debts, and also wins finance to support conservation programmes, which the government has some control over. For some debtor countries, however, such agreements may be seen as an infringement upon national sovereignty, passing some control and influence over national resources to foreign interests.

Despite debt-for-nature swaps, part of the packages of development aid that flow from North to South, the problems of indebtedness and unequal trading relations remain essentially unchanged. Redistribution from rich to poor

Figure 22.6 Mowing the lawn in front of the Indian parliament building in New Delhi: an image that symbolises the fact that the world's developing countries have much to offer richer nations in terms of appropriate use of nature and technology.

countries on any significant scale still appears to be politically impossible, yet it is the resolution or otherwise of this issue, perhaps more than any other, that will dictate the pace of global change towards sustainability.

THE SUSTAINABLE FUTURE

It is important to realise that sustainable development does not mean no human impact on the environment. Such a situation is impossible to achieve so long as there are people on the planet. The ideal scenario to strive for is one in which all environmental impacts can be undertaken consciously, in the full knowledge of the costs and consequences, even though this situation is a long way off, not least because we still have much to learn about the operation of nature. And even when and if such a status quo is attained, accidents will still happen. Some 17 years before the supertanker *Exxon Valdez* released 36 000 t of oil into Prince William Sound in Alaska in March 1989, causing widespread environmental damage, the author of a report on human activities in the region declared that: 'the volume of tanker traffic in this area [Prince William Sound at Valdez] makes occasional massive oil spills a distinct likelihood' (Price, 1972: 77). Despite stringent safety procedures, which can reduce the chances of catastrophe, the best-laid human plans can malfunction for unexpected reasons, among them human error. Hence, there will always be a need for continual vigilance and the existence of contingency plans for when the unexpected does occur.

However, one of the central themes of this book has been the need to change the ways in which socio-economic systems work in order to reduce environmental impacts that feed back on the operation of society. The emphasis on economic growth, which relies on increasing the amount of resources channelled through society, must be replaced by an emphasis on sustainable development – the qualitative improvement in human welfare. There is no doubt that great modifications to society are necessary to achieve a globally sustainable future. Some suggest that the scale of alteration is comparable to only two other changes in the history of humankind: the agricultural revolution of the late Neolithic period and the Industrial Revolution of the eighteenth and nineteenth centuries (Ruckelshaus, 1989). However, while these previous revolutions were gradual and largely unconscious, the sustainable revolution will have to be a conscious one. It needs to be constantly flexible, guided by scientific and economic information and interpretation of events, and steered by governmental and private policy.

A basic prerequisite for this revolution is a fundamental revision of the way in which we view human society and the operation of nature. The need to evaluate natural systems properly, in economic and other terms, has been mentioned, as has the need to realise that maintaining a sustainable environment is a global challenge that must be based on a long-term perspective, and is required for the entire world's population, not just a select and privileged few. If a large portion of the members of our species is poor, we cannot hope to live in a peaceful and sustainable world, and if the transfer of technology and wealth from rich to poor is not forthcoming on a much greater scale than at present, then efforts by less developed nations to improve their conditions along the same lines as the rich nations have followed will result in increasing worldwide ecological damage. At the same time, however, many lessons can be learnt by the richer countries – which are often over-reliant upon polluting and unnecessary technology – from more appropriate innovations in developing countries (Fig. 22.6).

Another fundamental change that many have

Figure 22.7 Natural materials on sale for medicinal, magical and religious uses in Agadez, Niger. The contrast between this scene and its equivalent in urban industrialised societies illustrates the real and perceived gap that has grown between many people and nature. Sustainable development requires a return to the realisation that we are a part of, and not apart from, nature.

to make is the need to return to the realisation that humans are part of nature and not separate from it. Industrial and urban societies have benefited greatly from their histories of economic development, but one of the casualties has been this basic truism. Traditional societies have maintained their environments through spiritual connections with plants and animals and other aspects of the natural world, which although not lost by all (Fig. 22.7), has been widely replaced by a notion of human domination of nature. This notion must change. In simple and not overly dramatic terms, the continuance of the human race depends upon our ability to abstain from destroying the natural systems that regenerate our planet.

These changes in values will also have to be encouraged, cajoled and in some cases enforced. Existing powerful institutional forces thus need to change their direction too, and be supplemented with new ones. Money, trade and national defence are the concerns of today's most influential institutions, but the priorities of governments change and the existence of organisations such as the World Bank, transnational corporations and NATO, has a relatively short history. As realisation of the importance of environmental concerns continues to grow, historical experience suggests that such power bases can adjust accordingly. The alternative, which also has historical precedents, is institutional collapse.

Perhaps, above all, it is important to realise that a sustainable future lies in our hands. The need to alter values, beliefs and behaviour should by now be clear. Anyone who has read this book as far as this point should be well aware of that.

FURTHER READING

Kasperson, J.X., Kasperson, R.E. and Turner II, B.L. (eds) 1995 *Regions at risk: comparisons of threatened environments*. Tokyo, United Nations University Press. Nine in-depth regional case studies of critical environmental regions from international teams, which analyse the physical problems, social causes and hopes for remediation.

Keijzers, G. 2000 The evolution of Dutch environmental policy: the changing ecological arena from 1970–2000 and beyond. *Journal of Cleaner Production* 8: 179–200. An outline of how policy has developed in the Netherlands.

Meyer, W.B. 1996 *Human impact on the Earth*. Cambridge, Cambridge University Press. A good, thorough overview of society's effects on the physical environment.

Turner II, B.L., Clark, W.C., Kates, R.W., Richards, J.F., Mathews, J.T. and Meyer, W.B. (eds) 1990 *The Earth as transformed by human action*. Cambridge, Cambridge University Press. A landmark international compendium that considers changes in society and nature from thematic and regional perspectives.

Vitousek, P.M., Mooney, H.A., Lubchenco, J. and Melillo, J.M. 1997 Human domination of Earth's ecosystems. *Science* 277: 494–9. A succinct precis of the scale of human influence on the structure and functioning of the planet.

World Bank 1992 *World development report 1992: development and the environment*. New York, Oxford University Press. This book takes a realistic look at the environmental problems associated with economic development, assesses the current situation, and suggests appropriate policies to combat them.

WRI 2003 *World resources 2002–2004*. Washington, World Resources Institute. The World Resources Institute produces a compact and wide-ranging update on environmental issues every two or three years.

Websites

For ease of reference, sites are shown in tables. Tables 22.4 and 22.5 give selections of national and international sites that contain information on environmental programmes, data and state-of-the-environment reports. Table 22.6 is a selection of sites with large directories and news on all types of environmental issues.

TABLE 22.4 *National environment websites*

Country	Agency	Site address
Australia	Environment Australia	http://www.erin.gov.au/
Austria	Federal Environment Agency	http://udk.bmu.gv.at/
Bangladesh	Centre for Advanced Studies	http://www.bcas.net/
Brazil	Environment Ministry (in Portuguese)	http://www.ibama.gov.br/
Chile	National Environment Commission	http://www.conama.cl/
China	China Environmental Protection	http://www.zhb.gov.cn/
Canada	Environment Canada's Green Lane	http://www.ec.gc.ca/
Denmark	Environment Ministry	http://www.mem.dk/
Eire	Department of the Environment	http://www.environ.ie/
Finland	Environmental Administration	http://www.vyh.fi/
France	Environment Institute	http://www.ifen.fr/
Germany	Federal Ministry of Environment	http://www.bmu.de/
Iceland	Environment Ministry	http://www.environment.is
India	Ministry for Environment and Forests	http://envfor.nic.in/
Indonesia	Environment Control Agency	http://www.bapedal.go.id/
Israel	Environment Ministry	http://www.environment.gov.il/
Italy	Environment Ministry (in Italian)	http://www.minambiente.it/
Jamaica	National Environment and Planning Agency	http://www.nrca.org/
Japan	Environment Ministry	http://www.env.go.jp/
Lebanon	Environment Information Centre	http://www.spnl.org/
Mauritius	Environment Ministry	http://ncb.intnet.mu/environ.htm
Mexico	Environment Ministry (in Spanish)	http://www.semarnat.gob.mx/
Netherlands	Ministry of Housing, Spatial Planning and the Environment	http://www.vrom.nl/
New Zealand	Environment Ministry	http://www.mfe.govt.nz/
Norway	Pollution Control Authority	http://www.sft.no/
Panama	National Association for Conservation of Nature (in Spanish)	http://www.ancon.org/
Poland	Environment Ministry (in Polish)	http://www.mos.gov.pl/
Singapore	Environment Ministry	http://www.env.gov.sg/
South Africa	Environmental Affairs	http://www.environment.gov.za/
South Korea	Environment Ministry	http://www.me.go.kr/
Spain	Environment Ministry (in Spanish)	http://www.mma.es/
Sweden	Environmental Protection Agency	http://www.environ.se/
Switzerland	Environment Agency	http://www.umwelt-schweiz.ch/
Thailand	Thailand Environment Institute	http://www.tei.or.th/
Uganda	National Environment Management Authority	http://www.nemaug.org/
UK	Environment Agency	http://www.environment-agency.gov.uk/
USA	Environmental Protection Agency	http://www.epa.gov/
Venezuela	Environment Ministry (in Spanish)	http://www.marnr.gov.ve/

TABLE 22.5 *International environment websites*

Region	Agency	Site address
Baltic	BALLERINA: the Baltic Sea region online Environmental Information Resources for Internet Access	**http://www.baltic-region.net/**
Central Europe	Central European Environmental Data Request Facility	**http://www.cedar.at/**
Europe	European Environment Agency	**http://www.eea.eu.int/**
Global	UNEP's Information Unit for Conventions	**http://www.unep.ch/conventions/**
Global	UNEP's Global Resource Information Database	**http://www.grid.unep.ch/**
Global	World Resources Institute	**http://www.wri.org/**
Global	Earthtrends	**http://www.earthtrends.wri.org**
Global	Food and Agriculture Organization	**http://www.fao.org/**
South Pacific	South Pacific Regional Environment Programme	**http://www.sprep.org.ws/**

TABLE 22.6 *Directories and news sites on all types of environmental issues*

Topics	Site address
All topics directory	**http://www.lib.kth.se/~lg/envsite.htm**
All topics directory	**http://www.envirolink.org/**
All topics directory and virtual library	**http://earthsystems.org/**
Environment news	**http://www.planetark.org/**
Environment news	**http://www.worldenvironment.com/**
Environment news and links	**http://www.Eco-Portal.com/**
Environment news	**http://ens.lycos.com/**
Pollution news and links	**http://pollution.com/**

POINTS FOR DISCUSSION

- Will any environmental issues become more critical in coming decades? Conversely, will any become less so?
- Using specific examples, assess the differences between 'systemic' and 'cumulative' global environmental changes.
- Do you think that the best approach to resolving environmental issues is technocentric or ecocentric, or should the approach vary depending on the issue?
- What do you consider to be the most urgent environmental issue facing your country? Is the government doing enough to resolve it?

GLOSSARY

abiotic: non-biological (see also **biotic**)

afforestation: the planting of a forest in an area where no forest originally existed

albedo: the reflectivity of a surface to sunlight. High albedo means that the majority of the incoming radiation is reflected (e.g. snow); low albedo means that the majority of the incoming radiation is absorbed (e.g. water)

algae: photosynthetic, mostly aquatic plant group, including single-cell **phytoplankton** and multicellular forms such as kelp

anadromous: fish that migrate from the sea into fresh water for breeding (e.g. salmon)

anaerobic: the term applied to organisms that live, or processes that occur, in the absence of gaseous or dissolved oxygen. Anaerobic bacteria, for example, obtain oxygen by breaking down such things as vegetable matter

anthropogenic: produced as the result of human action

atmosphere: the gaseous layer surrounding a planet. The Earth's atmosphere is divided into four parts distinguished according to the rate of change of temperature with height: the **troposphere**, **stratosphere**, mesosphere and thermosphere

bacteria: single-celled microscopic organisms that lack chlorophyll and an obvious nucleus. Bacteria play a significant role in the decomposition of organic matter

benthic: applied to organisms living close to the bottom of a lake, river or sea

bioaccumulation: the build-up in concentration of a toxic substance in the body of an individual organism over time that occurs when an organism is unable to excrete the substance or break it down within the body. (Confusingly, some authors also use the term for the increasing concentration of a substance up a food chain, a process better referred to as biomagnification)

biochemical oxygen demand (BOD): the amount of oxygen used in biological/chemical processes that decompose organic material in waste effluent. BOD is often used as an indicator of the polluting capacity of wastes because the amount of dissolved oxygen in water has a profound effect on the plants and animals living in it

biodegradation: the decomposition or breakdown of organic matter by micro-organisms, particularly by oxygen-using or 'aerobic' bacteria

biodiversity: a term that refers to the number, variety and variability of living organisms. It is commonly defined in terms of genes, species and ecosystems, corresponding to three fundamental levels of biological organisation

biogeochemical cycle: the chemical interactions that take place amongst the atmosphere, biosphere, hydrosphere, and lithosphere

biological classification: the arrangement of organisms into a hierarchy of groups that reflects evolutionary relationships. Usually the smallest group, or **taxon** (plural: taxa), is the **species**, although species may be subdivided into subspecies and varieties. Similar species are grouped into **genera**, genera into **families**, families into **orders**, orders into classes, classes into phyla (singular: phylum) for animals – the equivalent for plants is divisions – and these into kingdoms. The formal Latin name given to an organism has two parts: the European crested newt, for example, is known as *Triturus cristatus*, indicating that it is the species *cristatus* in the genus *Triturus*

biomass: the total mass of living organisms, usually expressed in dry weight per unit area

biome: ecosystem of a large geographical area with characteristic plants and typical climate (e.g. **tundra**, **savanna**)

biosphere: the part of the Earth and its atmosphere in which organisms live

biotic: living or biological in origin, as opposed to abiotic

boreal forest: coniferous forest occurring in the low Arctic

business as usual: a situation in which individuals, industries and societies continue to act as they have in the past, with no changes made to deal with some existing or anticipated problem, such as global warming

carbon cycle: an example of a **biogeochemical cycle**: the complex series of reactions by which carbon passes through the Earth's atmosphere, biosphere, hydrosphere, and lithosphere. For example, plants remove carbon in the form of carbon dioxide from the atmosphere and use it to produce carbohydrates in living organisms (**photosynthesis**). When those organisms die, the carbon is returned to the Earth as carbon dioxide, as fossil fuels (during decay), or as inorganic compounds such as calcium carbonate (limestone)

carcinogen: a substance with the potential to cause cancer

carrying capacity: the maximum number of a species that can be sustained in an area. If the carrying capacity is exceeded, **degradation** of the habitat will occur

cations: atoms that have lost one or more electrons and are thus left with a net positive charge

cetacean: a whale, dolphin or porpoise. The cetaceans are an order of aquatic mammals

CFCs (chlorofluorocarbons): organic compounds that include atoms of carbon, chlorine and fluorine. CFCs have been created for use in industrial processes and manufacturing, and have become a cause for environmental concern due to their inadvertent release into the atmosphere. Breakdown of CFCs in the stratosphere releases chlorine, which is thought to contribute to **ozone** depletion

cogeneration: the process by which two different and useful forms of energy are produced at the same time (e.g. when boiling water to generate electricity the leftover steam can be sold for heating or industrial processes)

coliform organisms: a group of bacteria abundant in the intestines of humans and other warm-blooded animals. Their concentration in water is used as an indicator of water quality

commodity: any useful material or product that might be traded

contamination: the occurrence of a relatively large amount of a toxic substance, relative to the normal ambient condition. Contamination does not necessarily imply ecological damage, but further contamination may result in **pollution**

critical load: a damage threshold for pollutant deposition

cryosphere: the portion of Earth's surface that is frozen throughout the year

curie: the former unit of measurement for radioactivity, now replaced by the becquerel (1 curie = 3.7 × 1010 becquerels)

defoliation: the removal of plant foliage by mechanical means, chemical spray or excessive eating by herbivores

deforestation: the removal of forests from an area of land. People commonly cut down trees to use as fuel or for building materials, or to clear land for agricultural, developmental or other purposes

degradation: a reduction in quality or decline in usefulness of an environmental resource. Human causes of degradation occur when a resource is used in an unsustainable manner

denitrification: the breakdown of **nitrates** by soil bacteria resulting in the release of nitrogen

desert: a biome of a predominantly dry climate with relatively little plant life. The term is often also used in a looser sense to indicate a relative lack of fresh water and/or organic life

dioxins: a family of chlorinated organic compounds that are highly toxic. They are by-products of the manufacture of certain classes of herbicides, disinfectants and bleaches, and can also be formed when chlorine compounds are burned at low temperatures in waste incinerators

DNA (deoxyribonucleic acid): the principal material of inheritance, found in chromosomes

ecocide: an intentional, anti-environmental action carried out over a large area (e.g. as a military strategy)

ecosystem: a community of organisms and the **abiotic** environment they inhabit and interact with. Ecosystems are recognised at many different geographical scales from a pond, for example, up to the scale of a **biome**

endemism: an endemic species is one found only in a particular geographical region, due to such factors as isolation, particular soils or climate

Environmental Impact Assessment (EIA): an interdisciplinary process by which the possible environmental consequences of proposed actions are predicted and considered. An EIA may include assessments of ecological, anthropological, sociological, geomorphological and other environmental effects

environmental monitoring: repeated measurements of environmental parameters, including inorganic, ecological, social and economic factors, with a view to documenting and often predicting important changes

epidemic: an unusual outbreak of a disease affecting a number of people in a relatively short time

epiphyte: a plant that grows on the outside of another plant, using it for support and not as a source of nutrients (e.g. lichen on trees)

erodibility: the susceptibility of soil to erosion

erosivity: the power of an agent of erosion (water, wind, ice, gravity) to move soil material

eutrophication: the addition of mineral **nutrients** to an ecosystem, so raising primary productivity. The process occurs naturally, particularly in saltwater and freshwater bodies over long periods, but can also be human-induced, in which case it is sometimes referred to as 'cultural eutrophication'. Cultural eutrophication occurs by the, often inadvertent, addition of nutrients in sewage and detergents, and from runoff contaminated with fertilisers

evapotranspiration: combined term for water lost as a vapour from soil or open water (evaporation) and water lost from the surface of a plant, mainly through stomata (transpiration)

extinction: an event resulting in the complete loss of all surviving individuals of a species or other taxon

extirpation: the localised loss of a species or other taxon from an area, with the taxon still surviving elsewhere

family: a group of similar **genera** (see **biological classification**)

fossil fuel: a mined fuel comprising the remains of plant or animal life from a previous geological period primarily composed of hydrocarbons (e.g. coal, oil, natural gas)

genera (singular: genus): groups of similar species (see **biological classification**)

general circulation model (GCM): a computer program that uses fundamental laws of physics and chemistry to analyse the interaction of temperature, pressure, solar radiation and other climatic factors in order to predict climatic conditions for the past, present or future

glacial maximum: the maximum extent of glaciation at the height of an **Ice Age**

GNP (gross national product): a measure of the market value of an economy's production over a given period (usually a year). GNP is adjusted to give GDP (gross domestic product) by removing the value of profits from overseas investments and those profits from the economy that go to foreign investors

greenhouse effect: the process by which atmospheric gases trap long-wave radiation emitted from the surface of a planet to warm its atmosphere

heavy metals: a broad group of metals of atomic weight higher than that of sodium and having a specific gravity in excess of 5.0. Those recognised as being of environmental concern, when present at certain concentrations that are harmful to living organisms, include copper, cadmium, mercury, tin, lead, antimony, vanadium, chromium, molybdenum, manganese, cobalt and nickel

humus: decomposed organic matter in soils

hydrosphere: the water present at or near the Earth's surface as a liquid, solid or gas, including oceans, lakes, rivers, streams, groundwater and water vapour in the atmosphere

Ice Age (glacial period, glacial epoch): recurring periods in Earth's history when the climate was colder and glaciers expanded to cover larger areas of the Earth's surface

interglaciation (interglacial period): the time between glaciations when Earth's climate is warmer and the ice sheets have retreated from large areas of the continents. The climate and extent of ice today is considered an interglacial period

***K*-strategy:** a species lifestyle in which the reproduction rate is relatively low and individuals relatively long-living, which allows a species to persist in a particular place for a long period (cf. ***r*-strategy**)

keystone species: a species whose removal from the **ecosystem** of which it forms a part leads to a series of extinctions in that ecosystem

leachate: solution formed when water percolates through a permeable medium

leaching: removal of soil materials in solution

lichen: a symbiotic association of **algae** and fungus: the fungus provides protection and moisture while the algae provide food for the fungus

lithosphere: the Earth's crust and the upper part of the mantle, consisting of rocks, soil and **sediments**

macrobenthic: large **benthic** organisms

mammal: a member of a class of animals that are warm-blooded, have vertebrae, secrete milk to their young through mammary glands, and have hair on their bodies (e.g. humans, whales, tigers, bats)

mass extinction: the extinction of an unusually large number of taxa in a relatively short period of time

Mesolithic (Middle Stone Age): a transitional period in the development of human societies between the **Palaeolithic** and **Neolithic** periods. The Mesolithic occupies different timespans in different places

methylation: the addition of a methyl group (CH_3) to an element or compound, usually through the activity of **anaerobic** bacteria. The process of methylation has important implications when it involves **heavy metals** since it releases them from materials, so enabling them to enter the food chain

monoculture: the cultivation of a single-species crop

morbidity rate: the number of people in a population with physical injury and disease in a given time interval

mortality rate: the number of deaths in a population in a given time interval

mycorrhizae: structures formed by an association of certain fungi and the roots of some plants

Neolithic (New or Late Stone Age): a level of human development characterised by the use of polished stone tools and the beginnings of settled agriculture. The Neolithic occupies different timespans in different places

net primary production: the amount of organic material produced by living organisms from inorganic sources minus that used in respiration

NIMBY syndrome: an individual or collective reaction to new developments that involve environmental change; stands for 'Not In My Back Yard'

nitrate: a chemical compound containing nitrogen, an essential nutrient for all living things. Nitrates are formed naturally in the soil by micro-organisms and are also produced industrially for use as fertilisers

non-point source: a term applied to a source of pollution meaning a dispersed source, such as a field, as opposed to a 'point source' such as an industrial waste pipe

nutrient: an element or compound needed by a living organism. A distinction is often made between 'macronutrients', which are needed in relatively large quantities (e.g. carbon, hydrogen, oxygen, nitrogen, potassium and phosphorus), and 'micronutrients' needed in smaller amounts (e.g. iron, manganese, zinc, copper)

onchocerciasis: a tropical disease caused by the worm *Onchocerca volvulus* and transmitted by a biting fly that breeds in fast-flowing streams. The worm's larvae can enter the eye, causing blindness, hence the disease's common name 'river blindness'

order: see **biological classification**

organic matter: living and dead **biomass**

organochlorines: organic compounds containing chlorine, which tend to **bioaccumulate** and are often long-lasting in the environment. They include several types of pesticides (e.g. aldrin, dieldrin, DDT) and **PCBs**

overexploitation: the unsustainable use of a renewable natural resource

oxisol: a relatively infertile loamy and clayey soil found only in the tropics

ozone: a chemical compound made up of three oxygen atoms. It is a natural constituent of the atmosphere with highest concentrations in the stratosphere (the 'ozone layer'), where it plays an important role in screening the Earth's surface from ultraviolet solar radiation. Elevated concentrations of ozone also sometimes occur in the troposphere above urban areas due to photochemical reactions between oxides of nitrogen and hydrocarbons emitted by motor vehicles in bright sunlight. In this situation ozone is considered a pollutant, a common constituent of some smogs

Palaeolithic (Early Stone Age): the stage in the development of a human society when people obtain food by hunting, fishing and gathering wild plants, as opposed to engaging in settled agriculture. The Palaeolithic occupies different timespans in different places

palaeontology: the study of fossils

pandemic: a disease prevalent over a whole country of the world. Commonly used to describe global **epidemics**

PCBs (polychlorinated biphenyls): a group of chlorinated hydrocarbons (see **organochlorines**) used mainly in high-voltage transformers. There is evidence that these compounds are highly persistent and toxic when released into the environment

pelagic: organisms that inhabit the open water at sea or in a lake

permafrost: permanently frozen ground where temperatures below 0 °C have persisted for at least two consecutive winters and the intervening summer. Permafrost covers about 26 per cent of the Earth's land surface

pest: an organism considered to be undesirable to human populations or their activities

pesticide: a chemical or a microbial pathogen that is toxic to pests. The term is used to group insecticides, herbicides and fungicides

pheromones: chemicals used by an animal as a social cue

photosynthesis: a set of coordinated biochemical reactions that occur in green plants, blue-green **algae** and **phytoplankton**. They are powered by sunlight absorbed by chlorophyll and other pigments, in which organic compounds are synthesised from carbon dioxide and water, with the release of oxygen as a waste product

phytoplankton: see **plankton**

phytotoxic: poisonous to green plants

plankton: animals and plants, many of them microscopically small, that float or swim in fresh or salt water. The animals (zooplankton) are mainly protozoa, small crustacea and larval stages of molluscs. The plants (phytoplankton) are almost all algae

podsolic soils (podsols): a soil type formed at an advanced stage of leaching, which has left an acid humus layer

pollution: the occurrence of toxic substances or energy in a quantity that exceeds the tolerance of an ecological community or species, resulting in damage (cf. **contamination**)

POPs (persistent organic pollutants): a wide range of environmentally hazardous chemicals that tend to be long-lived and accumulate up the food chain. POPs include PCBs, polychlorinated dioxins and pesticides such as aldrin, chlordane, DDT and dieldrin

primary/secondary forest: primary forest is that which has not been disturbed by human action, while secondary forest is a forest ecosystem that has regrown following human disturbance. In practice, it can be difficult to distinguish in the field between the two types of forest, particularly where secondary forest is more than around 60 years old

***r*-strategy:** a species lifestyle in which the reproduction rate is relatively high and individuals relatively short-living, which means that a species is a good coloniser but tends to persist in a particular place for a short period (cf. ***K*-strategy**)

radionuclide: an unstable nucleus of an atom that undergoes spontaneous radioactive decay, thereby emitting radiation and eventually changing from one element into another

rain forest: a mature forest with large biomass in parts of the temperate or tropical zone where precipitation totals are high

rangeland: any extensive area of native herbaceous or shrubby vegetation that is grazed by wild or domesticated animals (e.g. steppe, **savanna**, **tundra**)

raptor: a bird of prey that hunts by day (e.g. eagle, goshawk, kite, osprey, peregrine falcon, vulture)

resilience: the ability of a system to recover from a disturbance

resistance: the ability of a system to tolerate a disturbance without experiencing a significant change of state

savanna: a tropical or subtropical grassland with scattered trees or shrubs

schistosomiasis: any one of a group of diseases caused by infestation by blood flukes of humans and some other mammals. The diseases are common in low latitudes and are also known as bilharzia

sediment: unconsolidated particles, usually produced by the breakdown of rocks, ranging from clay-size to boulders that may be carried by natural agents (wind, water, ice) and eventually deposited to form sedimentary deposits. Organisms and chemical precipitation can also produce sediment. In arid regions, evaporation can result in the deposition of salts, another type of sediment

smog: an atmospheric condition of poor visibility and high concentration of air pollutants. The word is derived from 'smoke' and 'fog'

species: groups of living individuals that look alike and can interbreed, but cannot interbreed with other species (see **biological classification**)

Stone Age: the period in the development of human societies during which tools and implements were made of stone, bone or wood, but no metals were used. The dates of the period vary widely from place to place, and the term is sometimes used to describe present-day pre-agricultural peoples, although they may use imported metal implements. See also **Palaeolithic**, **Mesolithic** and **Neolithic**

stratosphere: the layer of the **atmosphere** above the **troposphere**, which extends on average between 10 km and 50 km above the Earth's surface

succession: the natural development of vegetation over time involving the progressive replacement of earlier plant communities with others towards a so-called 'climax' in which vegetation is supposedly in equilibrium with the existing environmental conditions. A distinction is made between 'primary' succession, which takes place on land with no previous vegetation, and 'secondary' succession, which occurs on land where vegetation has been partially or wholly destroyed

symbiotic: situation in which two species live together and both of them benefit from the relationship

synergy: the combined effect of two or more environmental influences that exceeds the sum of their individual effects

taxon: see **biological classification**

technology: a term used to describe the total system of means by which people interact with their environment and each other. It includes tools, information and knowledge, and the organisation of resources for productive activity

tectonic: relating to the forces and movements of the Earth and its crust. Earthquakes, volcanoes, and mountain building are related to tectonic activity

temperate: mild or moderate. The middle latitudes, between polar and tropical climate regimes, are generally referred to as having a temperate climate

terrestrial: land above sea level. Terrestrial flora and fauna live within land-based environments, not within the aquatic realm

topography: surface relief of the land

toxic: poisonous

transnational corporations (TNCs): large business organisations whose activities span international boundaries

troposphere: the lowest layer of the Earth's **atmosphere**, which varies in depth between 10 and 12 km over the poles to 17 km over the Equator. It is the layer in which most weather features occur

trypanosomiasis: a group of debilitating, long-lasting diseases caused by infestation with microscopic single-celled organisms. In Africa, sleeping sickness among humans and ngana in cattle, both transmitted by tsetse flies, are common forms of trypanosomiasis. A common form in the Americas is the incurable Chagas' disease

tsunami: a sea wave caused by a submarine earthquake, underwater volcanic explosion or massive slide of sea bed sediments

tundra: treeless vegetation that occurs at high latitude or high altitude

ultisol: a relatively infertile loamy and clayey tropical soil

vector: an organism that conveys a parasite from one host to another (e.g. the *Anopheles* mosquito, which transmits malaria)

volatile organic compounds (VOCs): the term used for a class of atmospheric pollutants, also known as hydrocarbons, which can produce **ozone** in combination with NO_x and sunlight. There are many hundreds of VOCs (e.g. methane, propane, toluene, ethanol) emitted into the atmosphere from both natural and human sources

volatilisation: the collective term for evaporation and sublimation

wetland: any habitat that is regularly saturated by water at or near the surface (e.g. bogs, fens, marshes, salt marshes, swamps)

zooplankton: see **plankton**

BIBLIOGRAPHY

Abrol, I.P., Yadav, J.S.P. and Massoud, F.I. 1988 *Salt-affected soils and their management*. FAO Soils Bulletin 39.

Abu-Zeid, M. 1989 Environmental impacts of the Aswan High Dam. *Water Resources Development* 5: 147–57.

Acheson, J.M. 1975 The lobster fiefs: economic and ecological effects of territoriality in the Maine lobster industry. *Human Ecology* 3: 183–207.

Adams, J. 1993 The emperor's old clothes: the curious comeback of cost-benefit analysis. *Environmental Values* 2: 247–60.

Adams, J., Maslin, M. and Thomas, E. 1999 Sudden climate transitions during the Quaternary. *Progress in Physical Geography* 23: 1–36.

Adams, R. 1975 The Haicheng earthquake of 4 February 1975: the first successfully predicted major earthquake. *Earthquake Engineering and Structural Dynamics* 4: 423–37.

Agate, A.D. 1996 Recent advances in microbial mining. *World Journal of Microbiology and Biotechnology* 12: 487–95.

Agnew, C. and Anderson, E. 1992 *Water resources in the arid realm*. London, Routledge.

Ågren, C. 1993 SO_2 emissions: the historical trend. *Acid News* 5: 1–4.

Agricola, G. 1556 *De re metallica*. Translated by Hoover, H.C. and Hoover, L.H. 1950. New York, Dover Publications.

Ahmad, M. and Kutcher, G.P. 1992 *Irrigation planning with environmental considerations: a case study of Pakistan's Indus basin*. World Bank Technical Paper 166. Washington, World Bank.

Akiner, S., Cooke, R.U. and French, R.A. 1992 Salt damage to Islamic monuments in Uzbekistan. *Geographical Journal* 158: 257–72.

Akleyev, A.V. and Lyubchansky, E.R. 1994 Environmental and medical effects of nuclear weapon production in the southern Urals. *Science of the Total Environment* 142: 1–8.

Alexander, D. 1993 *Natural disasters*. London, University College London Press.

Alexander, D. 1997 The study of natural disasters, 1977–1997: some reflections on a changing field of knowledge. *Disasters* 21: 284–304.

Al-Kharabsheh, A., Al-Weshah, R. and Shatanawi, M. 1997 Artificial groundwater recharge in the Azraq Basin (Jordan). *Dirasat Agricultural Sciences* 24: 357–70.

Al-Saleh, M.A. 1992 Declining groundwater level of the Minjur Aquifer, Tebrak area, Saudi Arabia. *Geographical Journal* 158: 215–22.

Alström, K. and Åkerman, A.B. 1992 Contemporary soil erosion rates on arable land in southern Sweden. *Geografiska Annaler* 74(A): 101–8.

Amin, M.A. 1977 Problems and effects of schistosomiasis in irrigation schemes in Sudan. In Worthington, E.B. (ed.), *Arid land irrigation in developing countries*. Oxford, Pergamon: 407–11.

Andersen, M.S. 1998 Assessing the effectiveness of Denmark's waste tax. *Environment* 40(4): 10–15, 38–41.

Anderson, A.B. and Ioris, E.M. 1992 Valuing the rain forest: economic strategies by small-scale forest extractivists in the Amazon estuary. *Human Ecology* 20: 337–69.

Anderson, D.M. 1994 Red tides. *Scientific American* 271(2): 52–8.

Angold, P.G. 1997 The impact of a road upon adjacent heathland vegetation: effects on plant species composition. *Journal of Applied Ecology* 34: 409–17.

Archer, L.J. 1993 *Aircraft emissions and the environment: CO_x, SO_x, HO_x, and NO_x*. Oxford, Oxford Institute for Energy Studies.

Arrossi, S. 1996 Inequality and health in metropolitan Buenos Aires. *Environment and Urbanization* 10: 167–86.

Asher, J., Warren, M., Fox, R., Harding, P., Jeffcoate, G. and Jeffcoate, S. 2001 *The millennium atlas of butterflies in Britain and Ireland*. Oxford, Oxford University Press.

Aubreville, A. 1949 *Climats, forêts et désertification de l'Afrique tropicale*. Paris, Société d'éditions géographiques maritimes et coloniales.

Baer, K.E. and Pringle, C.M. 2000 Special problems of urban river conservation: the encroaching megalopolis. In Boon, P.J., Davies, B.R. and Petts, G.E. (eds) 2000 *Global *perspectives on river conservation: science, policy and practice.* Chichester, Wiley: 385–402.

Bahn, P. and Flenly, V. 1992 *Easter Island.* London, Thames & Hudson.

Bailey, R.S. and Steele, J.H. 1992 North Sea herring fluctuations. In Glantz, M.H. (ed.), *Climatic variability, climate change and fisheries.* Cambridge, Cambridge University Press: 213–30.

Baker, L.A., Herlihy, A.T., Kaufmann, P.R. and Eilers, J.M. 1991 Acidic lakes and streams in the United States: the role of acidic deposition. *Science* 252: 1151–4.

Bakir, F., Damlaji, S., Amin-Zaki, L., Murtadha, M., Khalidi, A., Al-Rawi, N., Tikriti, S., Dhakir, H., Clarkson, T., Smith, J. and Doherty, R. 1973 Methylmercury poisoning in Iraq. *Science* 181: 230–41.

Bakir, H.A. 2001 Sustainable wastewater management for small communities in the Middle East and North Africa. *Journal of Environmental Management* 61: 319–28.

Baltz, D.M. 1991 Introduced fishes in marine systems and inland seas. *Biological Conservation* 56: 151–77.

Banks, G. 1993 Mining multinationals and developing countries: theory and practice in Papua New Guinea. *Applied Geography* 13: 313–27.

Barker, J.R., Thurow, T.L. and Herlocker, D.J. 1990 Vegetation of pastoralist campsites within the coastal grassland of Somalia. *African Journal of Ecology* 28: 291–7.

Barlaz, M.A., Eleazer, W.E. and Whittle, D.J. 1993 Potential to use waste tires as a supplemental fuel in pulp and paper mill boilers, cement kilns and in road pavement. *Waste Management and Research* 11: 463–80.

Barrow, C.J. 1981 Health and resettlement consequences and opportunities created as a result of river impoundment in developing countries. *Water Supply and Management* 5: 135–50.

Barrow, C.J. 1991 *Land degradation.* Cambridge, Cambridge University Press.

Bartone, C. 1990 Economic and policy issues in resource recovery from municipal solid wastes. *Resources, Conservation and Recycling* 4: 7–23.

Bates, T.S., Lamb, B.K., Guenther, A., Dignon, J. and Stoiber, R.E. 1992 Sulphur emissions to the atmosphere from natural sources. *Journal of Atmospheric Chemistry* 14: 315–37.

Battarbee, R.W. 1994 Surface water acidification. In Roberts, N. (ed.), *The changing global environment.* Oxford, Blackwell: 213–41.

Baxter, P.J. 2000 Impacts of eruptions on human health. In Sigurdsson, H., Houghton, B.F., McNutt, S.R., Rymer, H. and Stix, J. (eds) *Encyclopedia of volcanoes.* San Diego, Academic Press: 1035–43.

Bayfield, N.G., Barker, D.H. and Yah, K.C. 1992 Erosion of road cuttings and the use of bioengineering to improve slope stability in Peninsular Malaysia. *Singapore Journal of Tropical Geography* 13: 75–89.

BCAS (Bangladesh Centre for Advanced Studies) 1993 Bangladesh vulnerability profile to climate change and sea level rise. *Bangladesh Environmental Newsletter* 4(4): 3.

Beaumont, P., Blake, G.H. and Wagstaff, J.M. 1988 *The Middle East: a geographical study*, 2nd edn. London, David Fulton.

Been, V. 1994 Locally undesirable land uses in minority neighbourhoods: disproportionate siting or market dynamics? *The Yale Law Journal* 103: 1383–422.

Beg, M.A.A. 1990 *Report on status of air pollution in Karachi, past, present and future.* Karachi, Pakistan Council of Scientific and Industrial Research.

Behnke, R.H. and Scoones, I. 1993 Rethinking range ecology: implications for range management in Africa. In Behnke, R.H., Scoones, I. and Kerven, C. (eds), *Range ecology at disequilibrium.* London, Overseas Development Institute: 1–30.

Behnke, R.H., Scoones, I. and Kerven, C. (eds) 1993 *Range ecology at disequilibrium.* London, Overseas Development Institute.

Bennett, E.L. and Reynolds, C.J. 1993 The value of a mangrove area in Sarawak. *Biodiversity and Conservation* 2: 359–75.

Bennett, O. (ed.) 1991 *Greenwar: environment and conflict* London, Panos.

Benstead, J.P., Stiassny, M.L.J., Loiselle, P.V., Riseng, K.J. and Raminosoa, N. 2000 River conservation in Madagascar. In Boon, P.J., Davies, B.R. and Petts, G.E. (eds) *Global perspectives on river conservation: science, policy and practice.* Chichester, Wiley: 205–31.

Bentley, N. 1998 An overview of the exploitation, trade and management of corals in Indonesia. *TRAFFIC Bulletin* 17: 67–78.

Bergström, M. 1990 The release in war of dangerous forces from hydrological facilities. In Westing, A.H. (ed.), *Environmental hazards of war.* London, Sage: 38–47.

Berkes, F. (ed.) 1989 *Common property resources: ecology and community-based sustainable development.* London, Belhaven.

Bertrand, F. 1993 *Contribution à l'étude de l'environnement et de la dynamique des mangroves de Guinée.* Paris, ORSTOM.

Berz, G.A. 1991 Global warming and the insurance industry. *Nature and Resources* 27(1): 19–28.

Betts, R.A., Cox, P.M., Lee, S.E. and Woodward, F.I. 1997 Contrasting physiological and structural vegetation feedbacks in climate change simulations. *Nature* 387: 796–9.

Bhatti, N., Streets, D.G. and Foell, W.K. 1992 Acid rain in Asia. *Environmental Management* 16: 541–62.

Bird, E.F.C. 1985 *Coastline changes: a global review*. Chichester, Wiley.

Biswas, A.K. 1990 Watershed management. In Thanh, N.C. and Biswas, A.K. (eds), *Environmentally-sound water management*. Delhi, Oxford University Press: 155–75.

Björk, S. and Digerfeldt, G. 1991 Development and degradation, redevelopment and preservation of Jamaican wetlands. *Ambio* 20: 276–84.

Black, M. 1994 *Mega-slums: the coming sanitary crisis*. London, Wateraid.

Black, R. 1994 Environmental change in refugee-affected areas of the Third World: the role of policy and research. *Disasters* 18: 107–16.

Blaikie, P. 1985 *The political economy of soil erosion*. London, Longman.

Blaikie, P., Cannon, T., Davis, I. and Wisner, B. 1994 *At risk: natural hazards, people's vulnerability and disasters*. London, Routledge.

Blum, E. 1993 Making biodiversity conservation profitable: a case study of the Merck/INBio agreement. *Environment* 35(4): 16–45.

Blunden, J. 1985 *Mineral resources and their management*. London, Longman.

Boardman, J. 1990a Soil erosion on the South Downs: a review. In Boardman, J., Foster, I.D.L. and Dearing, J.A. (eds), *Soil erosion on agricultural land*. Chichester, Wiley: 87–105.

Boardman, J. 1990b *Soil erosion in Britain: costs, attitudes and policies*. Social Audit Paper 1. University of Sussex, Brighton, Education Network for Environment and Development.

Bond, A.R. and Piepenburg, K. 1990 Land reclamation after surface mining in the USSR: economic, political, and legal issues. *Soviet Geography* 31: 332–65.

Bormann, F.H. and Likens, G.E. 1979 *Pattern and process in a forested ecosystem*. New York, Springer Verlag.

Boserüp, E. 1965 *The conditions of agricultural growth: the economics of agrarian change under population pressure*. London, Allen & Unwin.

Bosson, R. and Varon, B. 1977 *The mining industry and the developing countries*. New York, Oxford University Press.

Boutron, C.F., Görlach, U., Candelone, J.-P., Bolshov, M.A. and Delmas, R.J. 1991 Decrease in anthropogenic lead, cadmium and zinc in Greenland snows since the late 1960s. *Nature* 353: 153–6.

Bowonder, B., Prasad, S.S.R. and Unni, N.V.M. 1988 Dynamics of fuelwood prices in India. *World Development* 16: 1213–29.

Boyce, J.K. 1990 Birth of a megaproject: political economy of flood control in Bangladesh. *Environmental Management* 14: 419–28.

Bozheyeva, G., Kunakbayev, Y. and Yeleukenov, D. 1999 *Former Soviet biological weapons facilities in Kazakhstan: past, present, and future*. Monterey Institute of International Studies, Center for Nonproliferation Studies Occasional Papers 1.

Bradshaw, A.D. and Chadwick, M.J. 1980 *The restoration of land*. Oxford, Blackwell Scientific.

Brandon, C. and Ramankutty, R. 1993 *Toward an environmental strategy for Asia*. World Bank Discussion Paper 224.

Braun, S. and Fluckiger, W. 1984 Increased population of the aphis *Aphis pomi* at a motorway. Part 2 the effect of drought and deicing salt. *Environmental Pollution (series A)* 36: 261–70.

Briscoe, J. 1987 A role for water supply and sanitation in the child survival revolution. *Bulletin of the Pan American Health Organization* 21: 92–105.

Broad, R. 1994 The poor and the environment: friends or foe? *World Development* 22: 811–22.

Brookes, A. 1985 River channelization: traditional engineering methods, physical consequences, and alternative practices. *Progress in Physical Geography* 9: 44–73.

Brown, S. and Lugo, A.E. 1990 Tropical secondary forests. *Journal of Tropical Ecology* 6: 1–32.

Bryant, D., Burke, L., McManus, J. and Spalding, M. 1998 *A map-based indicator of threats to the world's coral reefs*. Washington, DC, World Resources Institute.

Bryson, R.A. and Barreis, D.A. 1967 Possibility of major climatic modifications and their implications: northwest India, a case for study. *Bulletin of the American Meteorological Society* 48: 136–42.

Bucher, E.H. 1992 The causes of extinction of the passenger pigeon. *Current Ornithology* 9: 1–36.

Burke, L., Kura, Y., Kassem, K., Revenga, C., Spalding, M. and McAllister, D. 2000 *Pilot analysis of global ecosystems (PAGE): Coastal ecosystems*. Washington, DC, World Resources Institute.

Burt, T. 1993 From Westminster to Windrush: public policy

in the drainage basin. *Geography* 78: 388–400.

Burt, T. 1994 Long-term study of the natural environment – perceptive science or mindless monitoring. *Progress in Physical Geography* 18: 475–96.

Burt, T., Heathwaite, A.L. and Trudgill, S.T. 1993 *Nitrate: processes, patterns and management*. Chichester, Wiley.

Burton, I., Kates, R.W. and White, G.F. 1978 *The environment as hazard*. New York, Oxford University Press.

Buschiazzo, D.E., Aimar, S.B. and Garcia Queijeiro, J.M. 1999 Long-term maize, sorghum and millet monoculture effects on an Argentina typic ustipsamment. *Arid Soil Research and Rehabilitation* 13: 1–15.

Caddy, J.F. and Gulland, J.A. 1983 Historical patterns of fish stocks. *Marine Pollution* 7: 267–78.

Capra, F. 1982 *The turning point: science, society and the rising culture*. New York, Simon & Schuster.

Carapico, S. 1985 Yemeni agriculture in transition. In Beaumont, P. and McLachlan, K. (eds), *Agricultural development in the Middle East*. Chichester, Wiley: 241–54.

Carreira, J.A. and Neill, F.X. 1995 Mobilization of nutrients by fire in a semiarid gorse-scrubland ecosystem of southern Spain. *Arid Soil Research and Rehabilitation* 9: 73–89.

Carson, R. 1962 *Silent spring*. Boston, Houghton Mifflin.

Carter, F.W. 1993a Czechoslovakia. In Carter, F.W. and Turnock, D. (eds), *Environmental problems in Eastern Europe*. London, Routledge: 63–88.

Carter, F.W. 1993b Poland. In Carter, F.W. and Turnock, D. (eds), *Environmental problems in Eastern Europe*. London, Routledge: 107–34.

Caughley, G. 1979 What is this thing called carrying capacity? In Boyce, M.S. and Hayden-Wing, L.D. (eds), *North American elk: ecology, behaviour and management*. Laramie, University of Wyoming Press: 2–8.

Cavanagh, J.E., Clarke, J.H. and Price, R. 1993 Ocean energy systems. In Johansson, T.B., Kelly, H., Reddy, A.K.N. and Williams, R.H. (eds), *Renewable energy: sources for fuels and electricity*. New York, Island Press: 513–47.

Caviedes, C.N. and Fik, T.J. 1992 The Peru–Chile eastern Pacific fisheries and climatic oscillation. In Glantz, M.H. (ed.), *Climatic variability, climate change and fisheries*. Cambridge, Cambridge University Press: 355–75.

Chakela, Q. and Stocking, M. 1988 An improved methodology for erosion hazard mapping. Part II, Application to Lesotho. *Geografiska Annaler* 70(A): 181–9.

Chandler, A.M. 1997 Engineering design lessons from Kobe. *Nature* 387: 227–9.

Chao, B.F. 1995 Anthropological impact on global geodynamics due to water impoundment in major reservoirs. *Geophysical Research Letters* 22: 3533–6.

Chapman, D. (ed.) 1996 *Water quality assessments*, 2nd edn. London, E. & F.N. Spon.

Charlson, R.J. and Wigley, T.M.L. 1994 Sulfate aerosol and climatic change. *Scientific American* 270(2): 28–35.

Charney, J., Stone, P.H. and Quirk, W.J. 1975 Drought in the Sahara: a bio-geophysical feedback mechanism. *Science* 187: 434–5.

Chen, R.Z., Peng, G.Y. and Hunag, F.X. 1990 Effect of simulated acid rain on the growth and yield of soybean and peanut. *Journal of Ecology (China)* 9: 58–60.

Chester, D. 1993 *Volcanoes and society*. London, Edward Arnold.

Chester, D.K., Degg, M., Duncan, A.M. and Guest, J.E. 2001 The increasing exposure of cities to the effects of volcanic eruptions: a global survey. *Environmental Hazards* 2: 89–103.

Chien, N. 1985 Changes in river regime after the construction of upstream reservoirs. *Earth Surface Processes and Landforms* 10: 143–59.

Christensen, B. 1983 Mangroves – what are they worth? *Unasylva* 35: 2–15.

CIDA (Canadian International Development Agency) 2000 *Canada Climate Change Development Fund: management framework and business plan*. Ottawa, CIDA, 13 July.

Cincotta, R.P., Wisnewski, J. and Engelman, R. 2000 Human population in the biodiversity hotspots. *Nature* 404: 990–2.

CIPRCP (Commission Internationale pour la Protection du Rhin Contre la Pollution) 1984 *Rapport Annual 1984*. Koblenz, CIPRCP.

CIRIA (Construction Industry Research and Information Association) 1989 *The engineering implications of rising ground water levels in the deep aquifer beneath London*. London, CIRIA.

Clark, R.B. 1992 *Marine pollution*, 3rd edn. Oxford, Clarendon Press.

Clark, W.C. 1989 Managing planet Earth. *Scientific American* 261(3): 19–26.

Clayton, K. 1991 Scaling environmental problems. *Geography* 76: 2–15.

Clements, F.E. 1916 *Plant succession: an analysis of the development of vegetation*. Publication 242. Carnegie Institute, Washington, DC.

Coates, L. 1999 Flood fatalities in Australia, 1788–1996. *Australian Geographer* 30: 391–408.

Cohen, M.N. 1977 *The food crisis in prehistory: overpopulation and the origins of agriculture*. New Haven, Yale University Press.

Cohn, J.P. 1990 Elephants: remarkable and endangered. *BioScience* 40: 10–14.

Colby, M.E. 1989 *The evolution of paradigms of environmental management in development*. World Bank Strategic Planning and Review Discussion Paper. Washington, DC, World Bank.

Colinvaux, P. 1993 *Ecology 2*. New York, Wiley.

Collar, N.J., Crosby, M.J. and Stattersfield, A.J. 1994 *Birds to watch 2. The world list of threatened birds*. Cambridge, BirdLife International.

Collins, C.O. and Scott, S.L. 1993 Air pollution in the Valley of Mexico. *Geographical Review* 83: 119–33.

Collins, N.M., Sayer, J.A. and Whitmore, T.C. (eds) 1991 *The conservation atlas of tropical forests: Asia and the Pacific*. London, Macmillan/IUCN.

Confalonieri, U. 1998 Malaria in the Brazilian Amazon. In WRI, *World resources 1998–99*. New York, Oxford University Press: 48–9.

Connell, J.H. 1978 Diversity in tropical rain forests and coral reefs. *Science* 199: 1302–10.

Conway, G.R. and Pretty, J.N. 1991 *Unwelcome harvest: agriculture and pollution*. London, Earthscan.

Cooke, R.U. 1984 *Geomorphological hazards in Los Angeles*. London, Allen & Unwin.

Cooke, R.U. 1989 Geomorphological contributions to acid rain research: studies of stone weathering. *Geographical Journal* 155: 361–6.

Cooke, R.U. and Doornkamp, J.C. 1990 *Geomorphology in environmental management*, 2nd edn. Oxford, Clarendon Press.

Corte-Real, J., Sorani, R. and Conte, M. 1998 Climate change. In Mairota, P., Thornes, J.B. and Geeson, N. (eds), *Atlas of Mediterranean environments in Europe: the desertification context*. Chichester, Wiley: 34–6.

Coull, J.R. 1993 Will a blue revolution follow the green revolution? The modern upsurge of aquaculture. *Area* 25: 350–7.

Crocker, R.L. and Major, J. 1955 Soil development in relation to vegetation and surface age at Glacier Bay, Alaska. *Journal of Ecology* 43: 427–48.

Crowder, B.M. 1987 Economic costs of reservoir sedimentation: a regional approach to estimating cropland erosion damages. *Journal of Soil and Water Conservation* 42: 194–7.

Crowson, P. 1992 *Mineral resources: the infinitely finite*. Ottawa, The International Council on Metals and the Environment.

Crowson, P. 1994 *Minerals handbook 1994–95*. New York, Stockton Press.

Csirke, J. 1988 Small shoaling pelagic fish stocks. In Gulland, J.A. (ed.), *Fish population dynamics*, 2nd edn. Chichester, Wiley: 277–84.

Curtis, D., Hubbard M. and Shepherd, A. (eds) 1988 *Preventing famine: policies and prospects for Africa*. London, Routledge.

D'Yakanov, K.N. and Reteyum, A.Y. 1965 The local climate of the Rybinsk reservoir. *Soviet Geography* 6: 40–53.

Daly, H.E. 1987 The economic growth debate: what some economists have learned but many others have not. *Journal of Environmental Economics and Management* 14: 323–36.

Daly, H.E. 1993 The perils of free trade. *Scientific American* 269(5): 24–9.

Davies, B.R., Boon, P.J. and Petts, G.E. 2000 River conservation: a global imperative. In Boon, P.J., Davies, B.R. and Petts, G.E. (eds) *Global perspectives on river conservation: science, policy and practice*. Chichester, Wiley: xi–xvi.

Davis, M.B. 1976 Erosion rates and land use history in southern Michigan. *Environmental Conservation* 3: 139–48.

Davis, T.J. (ed.) 1993 *Towards the wise use of wetlands*. Gland, Wise Use Project, Ramsar Convention Bureau.

de Jong, J. and Wiggens, A.J. 1983 Polders and their environment in the Netherlands. In *Polders of the world, an international symposium: final report*. Wageningen, International Institute for Land Reclamation and Improvement: 221–41.

De Noni, G., Trujillo, G. and Viennot, M. 1986 L'érosion et la conservation des sols en Equateur. *Cahiers ORSTOM Série Pédologie* 22: 235–45.

de Vries, H.J.M. 1989 *Sustainable development*. Groningen, Netherlands, ISRIC.

DECADE (Domestic Equipment and Carbon Dioxide Emissions) 1995 *Second year report*. Oxford, University of Oxford Environmental Change Unit, Energy and Environment Programme.

Décamps, H. and Fortuné, M. 1991 Long-term ecological research and fluvial landscapes. In Risser, P.G. (ed.), *Long-term ecological research*. SCOPE Report 47. Chichester, Wiley: 135–51.

Deegan, J. 1987 Looking back at Love Canal. *Environmental*

Science and Technology 21: 328–31.

Deichmann, U. and Eklundh, L. 1991 *Global digital datasets for land degradation studies: a GIS approach*. UNEP/GEMS GRID Case Study Series 4. Nairobi, UNEP.

Dejene, A. and Olivares, J. 1991 *Integrating environmental issues into a strategy for sustainable agricultural development: the case of Mozambique*. World Bank Technical Paper 146.

Diamond, J.M. 1984 Historic extinctions: a Rosetta Stone for understanding prehistoric extinctions. In Martin, P.S and Klein, R.G. (eds), *Quaternary extinctions: a prehistoric revolution*. Tucson, AZ, University of Arizona Press: 824–62.

Diarra, D.C. and Akuffo, F.O. 2002 Solar photovoltaic in Mali: potential and constraints. *Energy Conversion and Management* 43: 151–63.

Dickinson, G., Murphy, K. and Springuel, I. 1994 The implications of the altered water regime for the ecology and sustainable development of Wadi Allaqi, Egypt. In Millington, A.C. and Pye, K. (eds), *Environmental change in drylands: biogeographical and geomorphological perspectives*. Chichester, Wiley: 379–91.

Dierber, F.E. and Kiattisimkul, W. 1996 Issues, impacts, and implications of shrimp aquaculture in Thailand. *Environmental Management* 20: 649–66.

Diesfeld, H.J. 1997 Malaria auf dem Vormarsch? *Geographische Rundschau* 49: 232–9.

Dijkema, G.P.J., Reuter, M.A. and Verhoef, E.V. 2000 A new paradigm for waste management. *Waste Management* 20: 633–8.

Dittrich, V., Hassan, S.O. and Ernst, G.H. 1985 Sudanese cotton and the whitefly: a case study of the emergence of a new primary pest. *Crop Protection* 4: 161–76.

Dixon, J.A., Talbot, L.M. and Le Moigne, J.-M. 1989 *Dams and the environment*. World Bank Technical Paper 110.

DME (Danish Ministry of Energy) 1990 *Energy 2000: a plan of action for sustainable development*. Copenhagen, DME.

Dobson, M. 1980 'Marsh fever' – the geography of malaria in England. *Journal of Historical Geography* 6: 357–89.

DoE (Department of the Environment) 1992 *The UK environment*. London, HMSO.

Dolk, H., Vrijheid, M., Armstrong, B., Abramsky, L., Banchi, F., Garne, E., Nelen, V., Robert, E., Scott, J.E.S., Stone, D. and Tenconi, R. 1998 Risk of congenital abnormalities near hazardous waste landfill sites in Europe: the EUROHAZCON study. *Lancet* 352: 423–7.

Doolette, J.B. and Smyle, J.W. 1990 Soil and moisture conservation technologies: review of literature. In Doolette, J.B. and Magrath, W.B. (eds), *Watershed development in Asia: strategies and technologies*. World Bank Technical Paper 127.

Döös, B.R. 1994 Why is environmental protection so slow? *Global Environmental Change* 4: 179–84.

Dottridge, J. and Abu Jaber, N. 1999 Groundwater resources and quality in northeastern Jordan: safe yield and sustainability. *Applied Geography* 19: 313–23.

Doumenge, F. 1986 La révolution aquacôle. *Annales de Géographie* 530: 445–82, 527–86.

Dove, M.R. 1993 A revisionist view of tropical deforestation and development. *Environmental Conservation* 20: 17–24, 56.

Dovers, S.R. and Handmer, J.W. 1993 Contradictions in sustainability. *Environmental Conservation* 20: 217–22.

Dovers, S.R. and Handmer, J.W. 1995 Ignorance, the precautionary principle, and sustainability. *Ambio* 24: 92–7.

Down, C.G. and Stocks, J. 1977 *Environmental impact of mining*. London, Applied Science.

Drake, J.A., Mooney, H.A., di Castri, F., Groves, R.H., Kruger, F.J. and Williamson, M. (eds) 1989 *Biological invasions: a global perspective*. SCOPE Report 37. Chichester, Wiley.

Dresch, J. 1986 Degradation of natural ecosystems in the countries of the Maghreb as a result of human impact. In Glantz, M.H. (ed.), *Arid land development and the combat against desertification*. Moscow, UNEP: 65–7.

Driscoll, C.T., Lawrence, G.B., Bulger, A.J., Butler, T.J., Cronan, C.S., Eager, C., Lambert, K.F., Likens, G.E., Stoddard, J.L. and Weathers, K.C. 2001 Acidic deposition in the Northeastern United States: sources and inputs, ecosystem effects, and management strategies. *BioScience* 51: 180–98.

Drysdale, J. 2002 *Whatever happened to Somalia?*, 2nd edn. London, Haan.

Dudley, N. 1986 Acid rain and British pollution control policy. In Goldsmith, E. and Hildyard, N. (eds), *Green Britain or industrial wasteland?* Cambridge, Polity Press: 95–107.

Durham, W.H. 1979 *Scarcity and survival in Central America: ecological origins of the Soccer War*. Stanford, Stanford University Press.

Durning, A. 1991 Asking how much is enough. In Brown, L.R. (ed.), *State of the world 1991*. New York, W.W. Norton: 153–69.

Durning, A. and Brough, H.B. 1991 *Taking stock: animal farming and the environment*. Worldwatch Paper 103. Worldwatch Institute, Washington, DC.

Earthquest 1991 *Science capsule*, Vol. 5(1). Washington, DC, Office for Interdisciplinary Earth Studies.

EEA (European Environment Agency) 1997 *Air pollution in Europe*. EEA Environmental Monograph 4.

EEA 2001 *Renewable energies: success stories*. Environmental issue report No 27. Copenhagen, EEA.

Ehrenfeld, D.W. 1988 Why put a value on biodiversity? In Wilson, E.O. and Peter, F.M. (eds), *Biodiversity*. Washington, DC, National Academy Press: 212–16.

Ehrlich, P. 1968 *The population bomb: population control or race oblivion*. New York, Ballantine.

Ehrlich, P. and Ehrlich, A.H. 1981 *Extinction: the causes and consequences of the disappearance of species*. New York, Random House.

Ek, A.S., Löfgren, S., Bergholm, J. and Qvarfort, U. 2001 Environmental effects of one thousand years of copper production at Falun, central Sweden. *Ambio* 30: 96–103.

El-Hinnawi, E. and Hashmi, M.H. 1987 *The state of the environment*. London, Butterworth.

Elkington, J. and Burke, T. 1987 *The green capitalists*. London, Victor Gollancz.

Ellis, D. 1989 *Environments at risk: case histories of impact assessment*. Berlin, Springer-Verlag.

Elsom, D. 1996 *Smog alert: managing urban air quality*. London, Earthscan.

El-Swaify, S.A., Dangler, E.W. and Armstrong, C.L. 1982 *Soil erosion by water in the Tropics*. Research Extension Series 024. Honolulu, College of Tropical Agriculture and Human Resources, University of Hawaii.

Elvingson, P. 1993 Younger stands now affected. *Acid News* 5: 8–9.

Elvingson, P. 1997 Still many trees damaged. *Acid News* 4–5: 14–15.

Elvingson, P. 2001 Forest damage: largely unchanged. *Acid News* 4: 18.

EMEP (European Monitoring and Evaluation Programme) 2000 *Transboundary acidification and eutrophication in Europe: EMEP summary report 2000*. Oslo, EMEP.

Endler, J.A. 1982 Pleistocene forest refuges: fact or fantasy? In Prance, G.T. (ed.) *Biological diversification in the tropics*. New York, Columbia University Press: 641–57.

Environment Canada 1991 *A report on Canada's progress towards a national set of environmental indicators*. SOE Report 91-1. Ottawa, Environment Canada.

Eurostat 1997 *Environmental statistics 1996*. Luxembourg, European Commission.

Evans, R. 1990 Assessment of soil erosion risk in England and Wales. *Soil Use and Management* 1: 127–31.

Ezcurra, E. 1990 The Basin of Mexico. In Turner II, B.L., Clark, W.C., Kates, R.W., Richards, J.F., Mathews, J.T. and Meyer, W.B. (eds), *The Earth as transformed by human action*. Cambridge, Cambridge University Press: 577–88.

Fallon, S.J., White, J.C. and McCulloch, M.T. 2002 *Porites* corals as recorders of mining and environmental impacts: Misima Island, Papua New Guinea. *Geochimica et Cosmochimica Acta* 66: 45–62.

FAO (Food and Agriculture Organization) 1991 *Environment and sustainability in fisheries*. Rome, FAO Committee on Fisheries.

FAO 1993 *The state of food and agriculture*. Rome, FAO.

FAO 1994 *Land degradation in South Asia: its severity, causes and effects on people*. World Soil Resources Report 78.

FAO 1995 *Forest resources assessment 1990: global synthesis*. FAO Forestry Paper 124.

FAO 1997a *Review of the state of world fishery resources: marine fisheries*. FAO Fisheries Circular 920.

FAO 1997b *Review of the state of world aquaculture*. FAO Fisheries Circular 886.

FAO 2000 *The challenges of sustainable forestry development in Africa*. Rome, FAO.

FAO 2001 *Global forest resources assessment 2000*. FAO Forestry Paper 140.

Farman, J.C., Gardiner, B.G. and Shanklin, J.D. 1985 Large losses in total ozone in Antarctica reveal seasonal ClO_x/NO_x interaction. *Nature* 315: 207–10.

Farrington, J. 1992 Transport, environment and energy. In Hoyle, B.S. and Knowles, R.D. (eds), *Modern transport geography*. London, Belhaven: 51–66.

Fearnside, P.M. 1989 The charcoal of Carajás: a threat to the forests of Brazil's eastern Amazon region. *Ambio* 18: 141–3.

Fearnside, P.M. 1990 Environmental destruction in the Brazilian Amazon. In Goodman, D. and Hall, A. (eds), *The future of Amazonia: destruction or sustainable development?* Basingstoke, Macmillan: 179–225.

Fearnside, P.M. 1995 Hydroelectric dams in the Brazilian Amazon as sources of 'greenhouse' gases. *Environmental Conservation* 22: 7–19.

Fernando, C.H. 1991 Impact of fish introductions in tropical Asia and America. *Canadian Journal of Fisheries and Aquatic Sciences* 48: 24–32.

Feshbach, M. and Friendy, A. 1992 *Ecocide in the USSR: health and nature under siege*. New York, Basic Books.

Fey, M.V., Manson, A.D. and Schutte, R. 1990 Acidification of the pedosphere. *South African Journal of Science* 86: 403–6.

Fickett, A.P., Gellings, C.W. and Lovins, A.B. 1990 Efficient use of electricity. *Scientific American* 263(3): 29–36.

Fischer, G., Frohberg, K., Parry, M.L. and Rosenzweig, C. 1994 Climate change and world food supply, demand and trade: who benefits, who loses? *Global Environmental Change* 4: 7–23.

Fisher, H.I. and Baldwin, P.H. 1946 War and the birds of Midway Atoll. *Condor* 48: 3–15.

Fleischer, S., Andersson, G., Brodin, Y., Dickson, W., Herrmann, J. and Muniz, I. 1993 Acid water research in Sweden – knowledge for tomorrow? *Ambio* 22: 258–63.

Folke, C. and Jansson, A.M. 1992 The emergence of an ecological economics paradigm: examples from fisheries and aquaculture. In Svedin, U. and Aniansson, B. (eds), *Society and the environment: a Swedish perspective*. Dordrecht, Kluwer: 69–87.

Forman, D., Cook-Mozaffari, P., Darby, S., Davey, G., Stratton, I., Doll, R. and Pike, M. 1987 Cancer near nuclear installations. *Nature* 329: 499–505.

Formoli, T.A. 1995 Impacts of the Afghan–Soviet war on Afghanistan's environment. *Environmental Conservation* 22: 66–9.

Forslund, J. 1986 Groundwater quality today and tomorrow. *World Health Statistical Quarterly* 39: 81–92.

Franzen, L.G. 1994 Are wetlands the key to the ice age cycle enigma? *Ambio* 23: 300–8.

Freeman, D.B. 1992 Prickly pear menace in eastern Australia. *Geographical Review* 82: 411–29.

Freer-Smith, P.H. 1998 Do pollutant-related forest declines threaten the sustainability of forests? *Ambio* 27: 123–31.

Fricke, H. and Hissman, K. 1990 Natural habitat of the coelacanths. *Nature* 346: 323–4.

Fritsch, J.M. and Sarrailh, J.M. 1986 Les transports solides dans l'écosystème forestier tropical humide guyanais: effets du défrichement et de l'aménagement de pâturages. *Cahiers ORSTOM Série Pédologie* 22: 209–22.

Fryrear, D.W. 1981 Long-term effect of erosion and cropping on soil productivity. In Péwé, T.L. (ed.), *Desert dust: origins, characteristics and effects on man*. Geological Society of America Special Paper 186: 253–9.

Fujita, M.S. and Tuttle, M.D. 1991 Flying foxes (*Chiroptera: Pteropodidae*): threatened animals of key ecological and economic importance. *Conservation Biology* 5: 455–63.

Gabbay, S. 1998 *The environment in Israel*. Jerusalem, Ministry of the Environment.

Gammelsrød, T. 1992 Improving shrimp production by Zambezi River regulation. *Ambio* 21: 145–7.

Gandy, M. 1996 Crumbling land: the postmodernity debate and the analysis of environmental problems. *Progress in Human Geography* 20: 23–40.

Gardner, J.S. 1993 Mountain hazards. In French, H.M. and Slaymaker, O. (eds), *Canada's cold environments*. Montreal, McGill-Queen's University Press: 247–67.

Geertz, C. 1963 *Agricultural involution: the process of change in Indonesia*. Berkeley, CA, University of California Press.

GEMS (Global Environment Monitoring System) 1988 *Assessment of freshwater quality*. Nairobi, UNEP/WHO.

George, D.J. 1992 Rising groundwater: a problem of development in some urban areas of the Middle East. In McCall, G.J.H., Laming, D.J.C. and Scott, S.C. (eds), *Geohazards: natural and man-made hazards*. London, Chapman & Hall: 171–82.

Gerhard, J. and Haynie, F.H. 1974 *Air pollution effects on catastrophic failure of metals*. EPA-650/3-74-009, Research Triangle Park, North Carolina.

GESAMP (Joint Group of Experts on the Scientific Aspects of Marine Environmental Protection) 1990 *The state of the marine environment*. Oxford, Blackwell Scientific.

GESAMP 2001 *A sea of troubles*. GESAMP Report Study 70. UNEP Nairobi.

Ghassemi, F., Jakeman, A.J. and Nix, H.A. 1995 *Salinisation of land and water resources: human causes, extent, management and case studies*. Sydney, University of New South Wales Press.

Gibbon, D., Lake, A. and Stocking, M. 1995 Sustainable development: a challenge for agriculture. In Morse, S. and Stocking, M. (eds), *People and environment*. London, UCL Press: 31–68.

Gibbons, J.R.H. and Clunie, F.G.A.U. 1986 Sea level changes and Pacific prehistory. *Journal of Pacific History* 21: 58–82.

Gilland, B. 1993 Cereals, nitrogen and population: an assessment of the global trends. *Endeavour* 17: 84–7.

Gillis, A.M. 1992 Keeping aliens out of paradise. *Bioscience* 42: 482–5.

Glantz, M.H. and Orlovsky, N. 1983 Desertification: a review of the concept. *Desertification Control Bulletin* 9: 15–22.

Glantz, M.H., Rubinstein, A.Z. and Zonn, I. 1993 Tragedy in the Aral Sea basin: looking back to plan ahead. *Global Environmental Change* 3: 174–98.

Glasby, G.P. 1997 Disposal of chemical weapons in the Baltic Sea. *Science of the Total Environment* 206: 267–73.

Glazovsky, N.F. 1995 The Aral Sea basin. In Kasperson, J.X., Kasperson, R.E. and Turner II, B.L. (eds), *Regions at risk: comparisons of threatened environments*. Tokyo, United Nations University Press: 92–139.

Gleick, P.H. (ed.) 1993 *Water in crisis: a guide to the world's fresh water resources*. New York, Oxford University Press.

Godoy, R., Lubowski, R. and Markandya, A. 1993 A method for the economic valuation of non-timber tropical forest products. *Economic Botany* 47: 220–33.

Goldblat, J. 1975 The prohibition of environmental warfare. *Ambio* 4: 186–90.

Goldblat, J. 1990 The mitigation of environmental disruption by war: legal approaches. In Westing, A.H. (ed.), *Environmental hazards of war*. London, Sage: 48–60.

Goldemberg, J., Johansson, T.B., Reddy, A.K.N. and Williams, R.H. 1988 *Energy for a sustainable world*. New Delhi, Wiley.

Goldschmidt, T. 1996 *Darwin's dreampond: drama on Lake Victoria*. Boston, MIT Press.

Goldsmith, E. and Hildyard, N. 1984 *The social and environmental effects of large dams*, vol. 1. Wadebridge, Cornwall, Wadebridge Ecological Centre.

Gómez-Pompa, A., Flores, J.S. and Sosa, V. 1987 The 'pet kot': a man-made tropical forest of the Maya. *Interciencia* 12: 10–15.

Goodland, R. 1986 Hydro and the environment: evaluating the tradeoffs. *Water Power and Dam Construction* November: 25–9.

Goodland, R. 1990 The World Bank's new environmental policy for dams and reservoirs. *Water Resources Development* 6: 226–39.

Goodland, R.J.A. and Irwin, H.S. 1974 An ecological discussion of the environmental impact of the highway construction program in the Amazon Basin. *Landscape Planning* 1: 123–54.

Goodland, R.J.A., Daly, H.E. and Serafy, S. El 1993a The urgent need for rapid transformation to global environmental sustainability. *Environmental Conservation* 20: 297–309.

Goodland, R.J.A., Juras, A. and Pachauri, R. 1993b Can hydro-reservoirs in tropical moist forests be environmentally sustainable? *Environmental Conservation* 20: 122–30.

Goodrich, J.A., Lykins, B.W. and Clark, R.M. 1991 Drinking water from agriculturally contaminated groundwater. *Journal of Environmental Quality* 20: 707–17.

Gorham, E. 1958 The influence and importance of daily weather conditions in the supply of chloride, sulphate and other ions to freshwaters from atmospheric precipitation. *Philosophical Transactions of the Royal Society, London* 241B: 147–78.

Goriup, P.D. 1989 Acidic air pollution and birds in Europe. *Oryx* 23: 82–6.

Goudie, A.S. 1993a *The nature of the environment*, 3rd edn. Oxford, Blackwell.

Goudie, A.S. 1993b *The human impact on the natural environment*, 4th edn. Oxford, Blackwell.

Goudie, A.S. 1993c Human influence on geomorphology. *Geomorphology* 7: 37–59.

Goudie, A.S. and Middleton, N.J. 1992 The changing frequency of dust storms through time. *Climatic Change* 20: 197–225.

Goudie, A.S. and Viles, H.A. 1997 *Salt weathering hazards*. Chichester, Wiley.

Government of Iraq 1980 Desertification in the Greater Mussayeb Project, Iraq. In *UNESCO/UNEP/ UNDP case studies on desertification*. Natural Resources Research Series XVIII. Paris, UNESCO: 176–213.

Grainger, A. 1993a *Controlling tropical deforestation*. London, Earthscan.

Grainger, A. 1993b Rates of deforestation in the humid tropics: estimates and measurements. *Geographical Journal* 159: 33–44.

Granéli, E. and Haraldson, C. 1993 Can increased leaching of trace metals from acidified areas influence phytoplankton growth in coastal waters? *Ambio* 22: 308–11.

Grasl, H. 1990 Possible climatic effects of contrails and additional water vapour. In Schumann, U. (ed.), *Air traffic and the environment*. Berlin, Springer-Verlag: 124–37.

Graveland, J., van der Wal, R., van Balen, J.H. and van Noordwijk, A.J. 1994 Poor reproduction in forest passerines from decline of snail abundance on acidified soils. *Nature* 368: 446–8.

Graves, H.S. 1918 Effect of the war on forests of France. *American Forestry* 24: 707–17.

Gray, R.E. and Bruhn, R.W. 1984 Coal mine subsidence: eastern United States. *Geological Society of America, Reviews in Engineering Geology* 6: 123–49.

Great Barrier Reef Marine Park Authority 2001 *Water quality: a threat to the Great Barrier Reef*. Townsville, Queensland, Great Barrier Reef Marine Park Authority.

Greenwood, B. and Mutabingwa, T. 2002 Malaria in 2002. *Nature* 415: 670–2.

Grenon, M. and Batisse, M. (eds) 1989 *Futures for the Mediterranean basin: the Blue Plan*. New York, Oxford University Press.

Grigg, D.B. 1992 *The transformation of agriculture in the West.* Oxford, Blackwell.

Grigg, D.B. 1993 The role of livestock products in world food consumption. *Scottish Geographical Magazine* 109: 66–74.

Groombridge B. and Jenkins M.D. 2000 *Global Biodiversity: Earth's living resources in the 21st century.* Cambridge, WCMC.

Grossman, L.S. 1992 Pesticides, caution, and experimentation in Saint Vincent, Eastern Caribbean. *Human Ecology* 20: 315–36.

Grove, R. 1990 The origins of environmentalism. *Nature* 345: 11–14.

Grubb, M.J. and Meyer, N.I. 1993 Wind energy: resources, systems and regional strategies. In Johansson, T.B., Kelly, H., Reddy, A.K.N. and Williams, R.H. (eds), *Renewable energy: sources for fuels and electricity.* New York, Island Press: 157–212.

Gunn, J.M. and Keller, W. 1990 Biological recovery of an acid lake after reductions in industrial emissions of sulphur. *Nature* 345: 431–3.

Hägerstrand, T. and Lohm, U. 1990 Sweden. In Turner II, B.L., Clark, W.C., Kates, R.W., Richards, J.F., Mathews, J.T. and Meyer, W.B. (eds), *The Earth as transformed by human action.* Cambridge, Cambridge University Press: 605–22.

Hall, D.O. and Rosillo-Calle, F. 1991 *Biomass in developing countries.* Report to the Office of Technology Assessment, Washington, DC.

Hall, D.O., Rosillo-Calle, F., Williams, R.H. and Woods, J. 1993 Biomass for energy: supply prospects. In Johansson, T.B., Kelly, H., Reddy, A.K.N. and Williams, R.H. (eds), *Renewable energy: sources for fuels and electricity.* New York, Island Press: 595–651.

Hall, D.R. 1993 Albania. In Carter, F.W. and Turnock, D. (eds), *Environmental problems in Eastern Europe.* London, Routledge: 7–37.

Halls, A.J. (ed.) 1997 *Wetlands, biodiversity and the Ramsar Convention: the role of the Convention on Wetlands in the conservation and wise use of biodiversity.* Gland, Ramsar Convention Bureau.

Hanan, N.P., Prevost, Y., Diouf, A. and Diallo, O. 1991 Assessment of desertification around deep wells in the Sahel using satellite imagery. *Journal of Applied Ecology* 28: 173–86.

Hansen, J.R., Hansson, R. and Norris, S. 1996 *The State of the European Arctic environment.* EEA Environmental Monograph 3.

Hardin, G. 1968 The tragedy of the commons. *Science* 162: 1243–8.

Hardoy, J.E., Mitlin, D. and Satterthwaite, D. 1992 *Environmental problems in third world cities.* London, Earthscan.

Hardoy, J.E., Mitlin, D. and Satterthwaite, D. 2001 *Environmental problems in an urbanizing world.* London, Earthscan.

Harper, P.P. 1992 La Grande Rivière: a subarctic river and hydroelectric megaproject. In Calow, P. and Petts, G.E. (eds), *The rivers handbook,* vol. 1. Oxford, Blackwell Scientific: 411–25.

Hartig, J.H. and Vallentyne, J.R. 1989 Use of an ecosystem approach to restore degraded areas of the Great Lakes. *Ambio* 18: 423–8.

Hassan, F.A. 1980 Prehistoric settlement along the Main Nile. In Williams, M.A.J. and Faure, H. (eds), *The Sahara and the Nile.* Rotterdam, Balkema: 421–50.

Hawksworth, D.L. 1990 The long-term effects of air pollutants on lichen communities in Europe and North America. In Woodwell, G.M. (ed.), *The Earth in transition: patterns and processes of biotic impoverishment.* Cambridge, Cambridge University Press: 45–64.

Hay, S.I., Cox, J., Rogers, D.J., Randolph, S.E., Stern, D.I., Shanks, G.D., Myers, M.F. and Snow, R.W. 2002 Climate change and the resurgence of malaria in the East African highlands. *Nature* 415: 905–9.

Hays, J.D., Imbrie, J. and Shackleton, N.J. 1976 Variations in the earth's orbit: pacemaker of the ice ages. *Science* 235: 1156–67.

Heath, J., Pollard, E. and Thomas, J.A. 1984 *Atlas of butterflies in Britain and Ireland.* Harmondsworth, Viking.

Heliotis, F.D., Karandinos, M.G. and Whiton, J.C. 1988 Air pollution and the decline of the fir forest in Parnis National Park, near Athens, Greece. *Environmental Pollution* 54: 29–40.

Hellawell, J.M. 1988 River regulation and nature conservation. *Regulated Rivers: Research and Management* 2: 425–43.

Hellden, U. 1988 Desertification monitoring: is the desert encroaching? *Desertification Control Bulletin* 17: 8–12.

Hellden, U. 1991 Desertification – time for an assessment? *Ambio* 20: 372–83.

Henderson-Sellers, A. 1994 Numerical modelling of global climates. In Roberts, N. (ed.), *The changing global environment.* Oxford, Blackwell: 99–124.

Hengeveld, R. 1989 *Dynamics of biological invasions.* London, Chapman & Hall.

Hernández Tejeda, T. and de Bauer, L.I. 1986 Photochemical oxidant damage on *Pinus hartwegii* at the Desierto de los Leones, Mexico DF. *Phytopathology* 76: 377.

Hewitt, K. and Burton, I. 1971 *The hazardousness of a place: a regional ecology of damaging events*. Toronto, University of Toronto Geography Department.

Hilz, C. and Radka, M. 1991 Environmental negotiation and policy: the Basel Convention on transboundary movement of hazardous wastes and their disposal. *International Journal of Environment and Pollution* 1: 55–72.

Hinrichsen, D. and Láng, I. 1993 Hungary. In Carter, F.W. and Turnock, D. (eds), *Environmental problems in Eastern Europe*. London, Routledge: 89–106.

Hira, P.R. 1969 Transmission of schistosomiasis in Lake Kariba, Zambia. *Nature* 2124: 670–2.

Hirst, R.A., Pywell, R.F. and Putwain, P.D. 2000 Assessing habitat disturbance using an historical perspective: The case of Salisbury Plain military training area. *Journal of Environmental Management* 60: 181–93.

Hobbelink, H. 1991 *Biotechnology and the future of world agriculture*. London, Zed Books.

Hodgson, D.A. and Johnston, N.M. 1997 Inferring seal populations from lake sediments. *Nature* 387: 30–1.

Hofmann, D.J. 1996 Recovery of the Antarctic ozone hole. *Nature* 384: 222–3.

Holdgate, M.W. 1991 Conservation in a world context. In Spellerberg, I.F., Goldsmith, F.B. and Morris, M.G. (eds), *The scientific management of temperate communities for conservation*. Oxford, Blackwell Scientific: 1–26.

Holling, C.S. 1995 What barriers? What bridges? In Gunderson, L.H., Holling, C.S. and Light, S.S. (eds) *Barriers and bridges to the renewal of ecosystems and institutions*. New York, Columbia University Press: 10–20.

Holmberg, J. 1991 *Poverty, environment and development: proposals for action*. Stockholm, Swedish International Development Agency.

Homer-Dixon, T.F., Boutwell, J.H. and Rathjens, G.W. 1993 Environmental change and violent conflict. *Scientific American* 268(2): 16–23.

Houghton, J.T., Ding, Y., Griggs, D.J., Noguer, M., van der Linden, P.J. and Xiaosu, D. (eds) 2001 *Climate change 2001: the scientific basis*. Cambridge, Cambridge University Press.

Houghton, J.T., Jenkins, G.J. and Ephraums, J.J. (eds) 1990 *Climate change: the IPCC scientific assessment*. Cambridge, Cambridge University Press.

Howard, K.W.F. and Haynes, J. 1993 Groundwater contamination due to road de-icing chemicals. *Geoscience Canada* 20: 1–8.

Howells, G. 1990 *Acid rain and acid waters*. London, Ellis Horwood.

Hoyt, E. 2000 *Whale watching 2000: worldwide tourism numbers, expenditures, and expanding socioeconomic benefits*. Cape Cod, MA, International Fund for Animal Welfare.

Hudson, N.W. 1991 *A study of the reasons for success or failure of soil conservation projects*. FAO Soils Bulletin 64.

Hughes, T.P. 1994 Catastrophes, phase shifts, and large-scale degradation of a Caribbean coral reef. *Science* 265: 1547–51.

Hulme, M. 2001 Climatic perspectives on Sahelian desiccation: 1973–1998. *Global Environmental Change* 11:19–29.

Humborg, C., Ittekkot, V., Cociasu, A. and Bodungen, B.V. 1997 Effect of Danube River dam on Black Sea biogeochemistry and ecosystem structure. *Nature* 386: 385–8.

Humphreys, D. 1993 *The phantom of full cost pricing*. London, Rio Tinto Zinc.

Hurni, H. 1993 Land degradation, famine, and land resource scenarios in Ethiopia. In Pimental, D. (ed.), *World soil erosion and conservation*. Cambridge, Cambridge University Press: 27–61.

Hurst, P. 1992 Pesticide reduction programs in Denmark, the Netherlands, and Sweden. *International Environmental Affairs* 4: 234–53.

IAEA (International Atomic Energy Agency) 1991 *The International Chernobyl Project: summary brochure, assessment of radiological consequences and evaluation of protective measures*. Report by International Advisory Committee, IAEA, Vienna.

IGN 1992 Mali: transect methodology to assess ecosystem change. In *UNEP World Atlas of Desertification*. London, Edward Arnold: 62–5.

Innes, J.L. 1992 Forest decline. *Progress in Physical Geography* 16: 1–64.

Intergovernmental Panel on Climate Change (IPCC) 1992 *Climate change 1992: the supplementary report to the IPCC scientific assessment*. Report by working group I. Cambridge, Cambridge University Press.

International Energy Outlook 1998. Washington, US Department of Energy.

Irish Peatland Conservation Council (IPCC) 1998 *Towards a conservation strategy for the bogs of Ireland*. Dublin, IPCC.

IUCN (International Union for Conservation of Nature and Natural Resources) 1988 *Coral reefs of the world*. Volume 3:

Central and Western Pacific. Cambridge, IUCN.

IUCN 1992 *Angola: environmental status quo assessment*. Harare, IUCN Regional Office Southern Africa.

IUCN 1993 Fears for the tiger. *IUCN Bulletin* 2: 5.

IUCN 2000 *The 2000 Red List of threatened species*. Gland, IUCN.

IUCN/UNEP 1986a *Review of the protected areas system in the Indo-Malayan realm*. Gland, IUCN.

IUCN/UNEP 1986b *Review of the protected areas system in the Afrotropical realm*. Gland, IUCN.

IUCN/UNEP/WWF 1980 *World conservation strategy: living resource conservation for sustainable development*. Gland, IUCN.

IUCN/UNEP/WWF 1991 *Caring for the Earth: a strategy for sustainable living*. Gland, IUCN.

Ives, J.D. and Messerli, B. 1989 *The Himalayan dilemma: reconciling development and conservation*. London, Routledge.

Iwama, G.K. 1991 Interactions between aquaculture and the environment. *Critical Reviews in Environmental Control* 21: 177–216.

Jackson, J., Kirby, M., Berger, W., Bjorndal, K., Dotsford, L., Bourque, B., Brabury, R., Cooke, R., Erlandson, J., Estes, J., Hughes, T., Kidwell, S., Lange, C., Lenihan, H., Pandolfi, J., Peterson, C., Steneck, R., Tigner, M. and Warner, R. 2001 Historical overfishing and the recent collapse of coastal ecosystems. *Science* 293: 629–38.

Jagtap, T.G., Chavan, V.S. and Untaawale, A.G. 1993 Mangrove ecosystems of India: a need for protection. *Ambio* 22: 252–4.

Janzen, J. 1994 Somalia. In Glantz, M.H. (ed.), *Drought follows the plow*. Cambridge, Cambridge University Press: 45–57.

Jasani, B. 1975 Environmental modification – new weapons of war? *Ambio* 4: 191–8.

Jenkins, R. 1987 *Transnationals and uneven development*. London, Croom Helm.

Jepson, P., Harvie, J.K., Mackinnon, K. and Monk, K.A. 2001 The end for Indonesia's lowland forests? *Science* 292: 859–61.

Jeyaratnam, J. 1990 Acute pesticide poisoning: a major global health problem. *World Health Statistics Quarterly* 43: 139–44.

Jickells, T.D., Carpenter, R. and Liss, P.S. 1990 Marine environment. In Turner II, B.L., Clark, W.C., Kates, R.W., Richards, J.F., Mathews, J.T. and Meyer, W.B. (eds), *The Earth as transformed by human action*. Cambridge, Cambridge University Press: 313–34.

Jimenez, R.D. and Velasquez, A. 1989 Metropolitan Manila: a framework for its sustained development. *Environment and Urbanization* 1: 51–8.

Jodha, N.S. 1992 *Common property resources*. World Bank Discussion Paper, August 1992, Washington, DC.

Johansson, T.B., Kelly, H., Reddy, A.K.N. and Williams, R.H. 1993 Renewable fuels and electricity for a growing world economy: defining and achieving the potential. In Johansson, T.B., Kelly, H., Reddy, A.K.N. and Williams, R.H. (eds), *Renewable energy: sources for fuels and electricity*. New York, Island Press: 1–71.

Johns, A.D. 1992 Species conservation in managed tropical forests. In Whitmore, T.C. and Sayer, J.A. (eds), *Tropical deforestation and species extinction*. London, Chapman & Hall: 15–53.

Johns, A.G. and Johns, B.G. 1995 Tropical forests and primates: long-term co-existence? *Oryx* 29: 205–11.

Johnston, G.H. 1981 *Permafrost engineering, design and construction*. Toronto, Wiley.

Jones, D.K.C., Cooke, R.U. and Warren, A. 1986 Geomorphological investigation, for engineering purposes, of blowing sand and dust hazard. *Quarterly Journal of Engineering Geology* 19: 251–70.

Jones, P.D., Briffa, K.R., Barnett, T.P. and Tett, S.F.B. 1998 High-resolution palaeoclimatic records for the last millennium: interpretation, integration and comparison with General Circulation Model control-run temperatures. *The Holocene* 8: 455–71.

Jutila, E. 1992 Restoration of salmonid rivers in Finland. In Boon, P.J., Calow, P. and Petts, G.E. (eds), *River conservation and management*. Chichester, Wiley: 353–62.

Kaiser, J. 1996 Acid rain's dirty business: stealing minerals from soil. *Science* 272: 198.

Kajak, Z. 1992 The River Vistula and its floodplain valley (Poland): its ecology and importance for conservation. In Boon, P.J., Calow, P. and Petts, G.E. (eds), *River conservation and management*. Chichester, Wiley: 35–49.

Kallend, A.S., Marsh, A.R.W., Pickles, J.H. and Proctor, M.V. 1983 Acidity of rain in Europe. *Atmospheric Environment* 17: 127–37.

Kashulina, G., Reimann, C., Finne, T.E., Halleraker, J.H., Äyräs, M. and Chekushin, V.A. 1997 The state of the ecosystems in the central Barents Region: scale, factors and mechanism of disturbance. *Science of the Total Environment* 206: 203–25.

Kates, R.W., Turner II, B.L. and Clark, W.C. 1990 The great transformation. In Turner II, B.L., Clark, W.C., Kates,

R.W., Richards, J.F., Mathews, J.T. and Meyer, W.B. (eds), *The Earth as transformed by human action*. Cambridge, Cambridge University Press: 1–17.

Kawasaki, T. 1983 Why do some pelagic fishes have wide fluctuations in their numbers? Biological basis of fluctuation from the viewpoint of evolutionary ecology. In Sharp, G.D. and Csirke, J. (eds), Reports of the expert consultation to examine changes in abundance and species composition of neritic fish resources. *FAO Fisheries Report* 291: 1065–80.

Keijzers, G. 2000 The evolution of Dutch environmental policy: the changing ecological arena from 1970–2000 and beyond. *Journal of Cleaner Production* 8: 179–200.

Kerpelman, C. 1990 *Preliminary study on the identification of disaster-prone countries based on economic impact*. Geneva, Office of the United Nations Disaster Relief Organisation.

Khakimov, F.I. 1989 *Soil reclamation conditions of delta desertification: tendencies of transformation and spatial differentiation*. Pushchino, Nauchnyi Tsentr Biologicheskikh Issledovaniy AN SSR (in Russian).

Khalaf, F.I. 1989 Desertification and aeolian processes in Kuwait. *Journal of Arid Environments* 12: 125–45.

Khalil, G.M. 1992 Cyclones and storm surges in Bangladesh: some mitigative measures. *Natural Hazards* 6: 11–24.

Khan, A.U. 1994 History of decline and present status of natural tropical thorn forest in Punjab. *Biological Conservation* 67: 205–10.

Khordagui, H.K. 1992 A conceptual approach to selection of a control measure for residual chlorine discharge in Kuwait Bay. *Environmental Management* 16: 309–16.

Kiersch, G.A. 1965 The Vaiont reservoir disaster. *Mineral Information Service* 18: 129–38.

Kim, K.C. 1997 Preserving biodiversity in Korea's demilitarized zone. *Science* 278: 242.

Kim, S.S. 1984 *The quest for a just world*. Boulder, CO, Westview Press.

Kinlen, L.J. 1988 Evidence for an infective cause of childhood leukaemia: comparison of a Scottish new town with nuclear reprocessing sites in Britain. *Lancet* 10 December: 1323–7.

Kirkitsos, P. and Sikiotis, D. 1996 Deterioration of Pentelic marble, Portland limestone and Baumberger sandstone in laboratory exposures to NO_2: a comparison with exposures to gaseous HNO_3. *Atmospheric Environment* 30: 941–50.

Kiss, A. 1985 The protection of the Rhine against pollution. *Natural Resources Journal* 25: 613–32.

Kivinen, E. and Pakarinen, P. 1981 Geographical distribution of peat resources and major peatland complex types in the world. *Annals, Academy Sciencia Fennicae* A132: 1–28.

Klein, B.C. 1989 Effects of forest fragmentation on dung and carrion beetle communities in central Amazonia. *Ecology* 70: 1715–25.

Klein, D.R. 1971 Reaction of reindeer to obstructions and disturbances. *Science* 173: 393–8.

Klinger, L.F. and Erickson III, D.J. 1997 Geophysical coupling of marine and terrestrial ecosystems. *Journal of Geophysical Research* 102(D21): 25359–70.

Knox, J.C. 1993 Large increases in flood magnitude in response to modest changes in climate. *Nature* 361: 430–2.

Knox, P. and Agnew, J. 1994 *The geography of the world economy*, 2nd edn. London, Edward Arnold.

Knutson, T.R., Tuleya, R.E. and Kurihara, Y. 1998 Simulated increase of hurricane intensities in a CO_2-warmed climate. *Science* 279: 1018–20.

Kobayashi, Y. 1981 Causes of fatalities in recent earthquakes in Japan. *Journal of Disaster Science* 3: 15–22.

Koike, K. 1985 Japan. In Bird, E.C.F. and Schwartz, M.L. (eds), *The world's coastline*. Stroudsburg, PA, Van Nostrand Reinhold: 843–55.

Kreimer, A., Lobo, T., Menezes, B., Munasinghe, M. and Parker, R. 1993 *Towards a sustainable urban environment: the Rio de Janeiro study*. World Bank Discussion Paper 195.

Krimgold, F. 1992 Modern urban infrastructure: the Armenian case. In Kreimer, A. and Munasinghe, M. (eds), *Environmental management and urban vulnerability*. World Bank Discussion Paper 168: 263–5.

Kroonenberg, S.B., Badyukovab, E.N., Stormsa, J.E.A., Ignatovb, E.I. and Kasimov, N.S. 2000 A full sea-level cycle in 65 years: barrier dynamics along Caspian shores. *Sedimentary Geology* 134: 257–74.

Kummer, D.M. 1991 *Deforestation in the postwar Philippines*. Chicago, University of Chicago Press.

Lacerda, L.D. 1997 Global mercury emissions from gold and silver mining. *Water, Air and Soil Pollution* 97: 209–21.

Lahlou, A. 1996 Environmental and socio-economic impacts of erosion and sedimentation in North Africa. In Walling, D.E. and Webb, B.W. (eds), *Erosion and sediment yield: global and regional perspectives*. Wallingford, International Association of Hydrological Sciences Publication no. 236: 491–500.

Lal, R. 1993 Soil erosion and conservation in West Africa. In Pimental, D. (ed.), *World soil erosion and conservation*. Cambridge, Cambridge University Press: 7–25.

Lambert, J.H., Jennings, J.N., Smith, C.T., Green, C. and Hutchinson, J.N. 1970 *The making of the Broads: a reconsideration of their origin in the light of new evidence*. Royal Geographical Society Research Series 3. London, Royal Geographical Society.

Lamprey, H.F. 1975 *Report on the desert encroachment reconnaissance in northern Sudan, October 21–November 10, 1975*. Khartoum, National Council for Research, Ministry of Agriculture, Food and Resources.

Landsberg, H.E. 1981 *The urban climate*. New York, Academic Press.

Langdale, G.W., Mills, W.C. and Thomas, A.W. 1992 Conservation tillage development for soil erosion control in the southern Piedmont. In Hurni, H. and Kebede, T. (eds) *Erosion, conservation and small-scale farming*. Bern, Geographia Bernensia: 453–8.

Langford, T.E.L. 1990 *Ecological effects of thermal discharges*. London, Elsevier Applied Science.

Larson, W.E., Pierce, F.J. and Dowdy, R.H. 1983 The threat of soil erosion to long-term crop production. *Science* 219: 458–65.

Laurance, W.F., Delamônica, P., Laurance, S.G., Vasconcelos, H.L. and Lovejoy, T.E. 2000 Rainforest fragmentation kills big trees. *Nature* 404: 836.

Layrisse, M. 1992 The 'holocaust' of the Amerindians. *Interciencia* 17: 274.

Le Ble, S. and Cuignon, R. 1988 Mise en évidence de l'influence du canal du Dique sur l'archipel du Rosaire (Colombie): circulation des eaux et dispersion des rejets en suspension. *Bulletin Institut de Geologie du Bassin d'Aquitaine* 44: 5–13.

Leach, G. and Mearns, R. 1988 *Beyond the fuelwood crisis: people, land and trees in Africa*. London, Earthscan.

Lee, K.N. 1989 The Columbia River basin: experimenting with sustainability. *Environment* 31: 6–11, 30–3.

Lelek, A. 1989 The Rhine river and some of its tributaries under human impact in the last two centuries. *Canadian Special Publication of Fisheries and Aquatic Science* 106: 469–87.

Leprun, J.-C., da Silveira, C.O. and Sobral Filho, R.M. 1986 Efficacité des pratiques culturales antiérosives testées sous différents climats brésiliens. *Cahiers ORSTOM Série Pédologie* 22: 223–33.

Lerner, D.N. and Tellam, J.H. 1993 The protection of urban groundwater from pollution. In Currie, J.C. and Pepper, A.T. (eds), *Water and the environment*. Chichester, Ellis Horwood: 322–35.

Levins, R., Awerbuch, T., Brinkman, U., Eckardt, I., Epstein, P., Makhoul, N., Albuquerque de Possas, C., Puccia, C., Spielman, A. and Wilson, M.E. 1994 The emergence of new diseases. *American Scientist* 82: 52–60.

Lewis, L.A. and Berry, L. 1988 *African environments and resources*. London, Allen & Unwin.

Lisk, D.J. 1991 Environmental effects of landfills. *Science of the Total Environment* 100: 415–68.

Liu, J., Linderman, M., Ouyang, Z., An, L., Yang, J. and Zhang, H. 2001 Ecological degradation in protected areas: the case of Wolong Nature Reserve for giant pandas. *Science* 292: 98–101.

Lloyd, G.O. and Butlin, R.N. 1992 Corrosion. In Radojevic, M. and Harrison, R.M. (eds), *Atmospheric acidity: sources, consequences and abatement*. London, Elsevier Applied Science: 405–34.

Lockeretz, W. 1978 The lessons of the dust bowl. *American Scientist* 66: 560–9.

Lockeretz, W. 1988 Open questions in sustainable agriculture. *American Journal of Alternative Agriculture* 3. 174–81.

Loeb, V., Siegel, V., Holm-Hansen, O., Hewitt, R., Fraser, W., Trivelpiece, W. and Trivelpiece, S. 1997 Effects of sea-ice extent and krill or salp dominance on the Antarctic food web. *Nature* 387: 897–900.

Loevinsohn, M.E. 1994 Climatic warming and increased malaria incidence in Rwanda. *Lancet* 343: 714–17.

Löffler, E. and Kubinok, J. 1988 Soil salinization in north-east Thailand. *Erdkunde* 42: 89–100.

Löfstedt, R. 1998 Sweden's biomass controversy: a case study of communicating policy issues. *Environment* 40(4): 16–20, 42–5.

Lonergan, S.C. 1993 Impoverishment, population, and environmental degradation: the case for equity. *Environmental Conservation* 20: 328–34.

Lorius, C., Jouzel, J., Raynaud, D., Hansen, J. and Le Treut, H. 1990 The ice-core record: climate sensitivity and future greenhouse warming. *Nature* 347: 139–45.

Lottermoser, B.G. and Morteani, G. 1993 Sewage sludge: toxic substances, fertilizers, or secondary metal resources? *Episodes* 16: 329–33.

Lovelock, J.E. 1989 *The ages of Gaia*. Oxford, Oxford University Press.

Lowe, M.D. 1991 *Shaping cities: the environmental and human dimensions*. Worldwatch Paper 105. Worldwatch Institute, Washington, DC.

Lowe, M.D. 1994 Reinventing transport. In Brown, L.R. (ed.), *State of the world 1994*. New York, W.W. Norton: 81–98.

Lugo, A.E. 1988 Estimating reductions in the diversity of tropical forest species. In Wilson, E.O. and Peter, F.M. (eds), *Biodiversity*. Washington, DC, National Academy Press: 58–70.

Lundgren, L.W. 1999 *Environmental geology*, 2nd edn. Upper Saddle River, Prentice Hall.

Lupinacci, J. 2000 Creating corporate value and environmental benefit with improved energy performance. *Environmental Quality Management* 10(2): 11–17.

MacArthur, R.H. and Wilson, E.O. 1967 *The theory of island biogeography*. Princeton, Princeton University Press.

Macklin, M.G., Hudson-Edwards, K.A. and Dawson, E.J. 1997 The significance of pollution from historic metal mining in the Pennine orefields on river sediment contaminant fluxes to the North Sea. *Science of the Total Environment* 194–195: 391–7.

McCabe, J.T. 1990 Turkana pastoralism: a case against the tragedy of the commons. *Human Ecology* 18: 81–103.

McCall, G.J.H. 1998 Geohazards and the urban environment. In Maund, J.G. and Eddlestone, M. *Geohazards in engineering geology*. London, Geological Society: 309–18.

McCauley, J.F., Breed, C.S., Grolier, M.J. and Mackinnon, D.J. 1981 The US dust storm of February 1977. In Péwé, T.L. (ed.), *Desert dust: origins, characteristics and effects on man*. Geological Society of America Special Paper 186: 123–47.

McCully, P. 1996 *Silenced rivers: the ecology and politics of large dams*. London, Zed Books.

McFarland, M. 1989 Chlorofluorocarbons and ozone. *Environmental Science and Technology* 23: 1203–8.

McGinley, P.C. 1992 Regulation of the environmental impacts of coal mining in the USA: market economics, cost–benefit analysis and mistakes of the past. *Natural Resources Forum* 16: 261–70.

McGuffie, K., Henderson-Sellers, A. and Zhang, H. 1998 Modelling climatic impacts of future rainforest destruction. In Maloney, B.K. (ed.), *Human activities and the tropical rainforest*. Dordrecht, Kluwer: 169–93.

McNeely, J. 1988 *Economics and biological diversity*. Gland, IUCN.

McNeely, J. 1994 Lessons from the past: forests and biodiversity. *Biodiversity and Conservation* 3: 3–20.

Maddox, J. 1990 Clouds and global warming. *Nature* 247: 329.

Magrath, W. and Arens, P. 1989 *The costs of soil erosion in Java: a natural resource accounting approach*. World Bank Environment Department Working Paper 18.

Maitland, P.S. 1991 Conservation of fish species. In Spellerberg, I.F., Goldsmith, F.B. and Morris, M.G. (eds), *The scientific management of temperate communities for conservation*. Oxford, Blackwell Scientific: 129–48.

Maki, A.W. 1991 The Exxon oil spill: initial environmental impact assessment. *Environmental Science and Technology* 25: 24–29.

Malm, O., Pfeiffer, W.C., Souza, C.M.M. and Reuther, R. 1990 Mercury pollution due to gold mining in the Madeira River basin, Brazil. *Ambio* 19: 11–15.

Malthus, T.R. 1798 *An essay on the principle of population*. London, Johnson.

Mann, R.H.K. 1988 Fish and fisheries of regulated rivers in the UK. *Regulated Rivers: Research and Management* 2: 411–24.

Manner, H.I., Thaman, R.R. and Hassall, D.C. 1984 Phosphate mining induced changes on Nauru Island. *Ecology* 65: 1454–65.

Maragos, J.E. 1993 Impact of coastal construction on coral reefs in the US-affiliated Pacific Islands. *Coastal Management* 21: 235–69.

Margarita, M. and Loyarte, G. 1996 Climatic fluctuations as a source of desertification in a semi-arid region of Argentina. *Desertification Control Bulletin* 28: 8–14.

Marples, D.R. 1992 Post-Soviet Belarus and the impact of Chernobyl. *Post-Soviet Geography* 33: 419–31.

Marsh, G.P. 1874 *The Earth as modified by human actions*. New York, Sampson Low.

Marshall, B.E. and Junor, F.J.R. 1981 The decline of *Salvinia molesta* on Lake Kariba. *Hydrobiologia* 83: 477–84.

Martin, P.S. and Klein, R.G. 1984 *Pleistocene extinctions*. Tucson, AZ, University of Arizona Press.

Mather, A.S. 1990 *Global forest resources*. London, Belhaven.

May, R.M. 1978 Human reproduction reconsidered. *Nature* 272: 491–5.

Meade, R.B. 1991 Reservoirs and earthquakes. *Engineering Geology* 30: 245–62.

Meadows, D.H., Meadows, D.L., Randers, J. and Behrens III, W.W. 1972 *The limits to growth: a report to the Club of Rome's project on the predicament of mankind*. New York, Potomac Associates.

Meadows, P.S. and Campbell, J.I. 1988 *An introduction to marine science*. London, Blackie & Son.

Mee, L.D. 1992 The Black Sea crisis: a need for concerted international action. *Ambio* 21: 278–86.

Meech, J.A., Veiga, M.M. and Tromans, D. 1998 Reactivity of mercury from gold mining activities in darkwater ecosystems. *Ambio* 27: 92–8.

Meierding, T.C. 1993 Marble tombstone weathering and air pollution in North America. *Annals of the Association of American Geographers* 83: 568–88.

Mellquist, P. 1992 River management – objectives and applications. In Boon, P.J., Calow, P. and Petts, G.E. (eds), *River conservation and management*. Chichester, Wiley: 1–8.

Menzel, A. and Fabian, P. 1999 Growing season extended in Europe. *Nature* 397: 659.

Meybeck, M. 1982 Carbon, nitrogen and phosphorus transport by world rivers. *Science* 282: 401–50.

Meybeck, M., Chapman, D. and Helmer, R. 1989 *Global freshwater quality: a first assessment*. Oxford, Blackwell.

Meynell, P.-J. 1993 Developing collaboration to protect Saudi Arabia's wetlands. *IUCN Wetlands Programme Newsletter* 7: 11–12.

Middleton, N.J. 1985 Effect of drought on dust production in the Sahel. *Nature* 316: 431–4.

Middleton, N.J. 1991 *Desertification*. Oxford, Oxford University Press.

Middleton, N.J. and Thomas, D.S.G. 1997 *World atlas of desertification*, 2nd edn. London, Arnold.

Middleton, N.J. and van Lynden, G.W.J. 2000 Secondary salinization in South and Southeast Asia. *Progress in Environmental Science* 2: 1–19.

Milliman, J.D. 1990 Fluvial sedimentation in coastal seas: flux and fate. *Nature and Resources* 26(4): 12–22.

Milly, P.C.D., Wetherald, R.T., Dunne, K.A. & Delworth, T.L. 2002 Increasing risk of great floods in a changing climate. *Nature* 415: 514 –17.

Mistry, J.F. 1989 Salinity ingress in the coastal area of Saurashtra-Gujarat State (India). *ICID Bulletin* 32: 52–60.

Mittermeir, R., Konstant, B., Nicoll, M. and Langrand, O. 1992 *Lemurs of Madagascar: an action plan for their conservation 1993–1999*. Gland, IUCN.

Moffatt, I. 1999 Edinburgh: a sustainable city? *International Journal of Sustainable Development and World Ecology* 6: 135–48.

Moleele, N.M., Ringrose, S., Matheson, W. and Vanderpost, C. 2002 More woody plants? the status of bush encroachment in Botswana's grazing areas. *Journal of Environmental Management* 64: 3–11.

Molina, M.J. and Rowland, F.S. 1974 Stratospheric sink chlorofluoromethanes: chlorine atom catalyzed destruction of ozone. *Nature* 249: 810–14.

Monteiro, C.A. de F. 1989 Environmental quality in the great national metropolis and its industrial portuary appendix. In Gerasimov, I.P. (ed.), *Environmental problems in cities of developing countries*. Moscow, UNEP: 26–39.

Moore, D. and Driver, A. 1989 The conservation role of water supply reservoirs. *Regulated Rivers: Research and Management* 4: 203–12.

Moore, N.W., Hooper, M.D. and Davis, B.N.K. 1967 Hedges, I. Introduction and reconnaissance studies. *Journal of Applied Ecology* 4: 201–20.

Moran, E.F. 1981 *Developing the Amazon*. Bloomington, Indiana University Press.

Moreira, J.R. and Poole, A.D. 1993 Hydropower and its constraints. In Johansson, T.B., Kelly, H., Reddy, A.K.N. and Williams, R.H. (eds), *Renewable energy: sources for fuels and electricity*. New York, Island Press: 73–119.

Mortimore, M. 1987 Shifting sands and human sorrow: social response to drought and desertification. *Desertification Control Bulletin* 14: 1–14.

Motz, H. and Geiseler, J. 2001 Products of steel slags an opportunity to save natural resources. *Waste Management* 21: 285–93.

Mukherjee, M.D. 1996 Pisciculture and the environment: an economic evaluation of sewage-fed fisheries in east Calcutta. *Science, Technology and Development* 14: 73–99.

Murray, I. 1994 Time and tide rip into the frontier of old England. *Times* 23 March: 7.

Murray, J.E., Johnson, M.S. and Clarke, B. 1988 The extinction of Partula on Moorea. *Pacific Science* 42: 150–3.

Musannif, B. 1992 Integrated approach to energy efficient housing refurbishment. *Energy Management* July/August: 22–3.

Myers, N. 1979 *The sinking ark: a new look at the problem of disappearing species*. Oxford, Pergamon.

Myers, N., Mittermeier, R.A., Mittermeier, C.G., Da Fonseca, G.A.B. and Kent, J. 2000 Biodiversity hotspots for conservation priorities. *Nature* 403: 853–8.

Mylona, S. 1993 *Trends of sulphur dioxide emissions, air concentrations and depositions of sulphur in Europe since 1880*. EMEP/MSC-W Report 2/93. Oslo, EMEP.

Naylor, R. 1996 Invasions in agriculture: assessing the cost of the golden apple snail in Asia. *Ambio* 25: 443.

Naylor, R.L., Goldburg, R.J., Primavera, J.H., Kautsky, N.,

Beveridge, M.C.M., Clay, J., Folke, C., Lubchenco, J., Mooney, H. and Troell, M. 2000 Effect of aquaculture on world fish supplies. *Nature* 405: 1017–24.

Nelson, P.M. (ed.) 1987 *Transportation noise reference handbook*. London, Butterworth.

NEPA (National Environmental Protection Agency) 1997 *1996 report on the state of the environment*. Beijing, NEPA.

Neumann, A.C. and Macintyre, I. 1985 Reef response to sea level rise: keep-up, catch-up or give-up. *Proceedings of the 5th International Coral Reef Congress, Tahiti* 3: 105–10.

Newcombe, K. 1984 *An economic justification for rural afforestation: the case of Ethiopia*. World Bank Energy Department Paper 16.

Nichol, J.E. 1989 Ecology of fuelwood production in Kano region, northern Nigeria. *Journal of Arid Environments* 16: 347–60.

Nicol, S. and de la Mare, W. 1993 Ecosystem management and the Antarctic krill. *American Scientist* 81: 36–47.

Nicholls, N. 1997 Increased Australian wheat yield due to recent climate trends. *Nature* 387: 484–5.

Nilsson, K. 1991 Emission standards for waste incineration. *Waste Management and Research* 9: 224–7.

Nishida, K., Nagayoshi, Y., Ota, H. and Nagasawa, H. 2001 Melting and stone production using MSW incinerated ash. *Waste Management* 21: 443–9.

Nixon, S.W. 1993 Nutrients and coastal waters: too much of a good thing? *Oceanus* 36(2): 38–47.

Noble, A.G. 1980 Noise pollution in selected Chinese and American cities. *GeoJournal* 4: 573–5.

Nortcliff, S. and Gregory, P.J. 1992 Factors affecting losses of soil and agricultural land in tropical countries. In McCall, G.J.H., Laming, D.J.C. and Scott, S.C. (eds), *Geohazards: natural and man-made hazards*. London, Chapman & Hall: 183–90.

Norton, D.A. 1991 *Trilepidea adamsii*: an obituary for a species. *Conservation Biology* 5: 52–7.

Nriagu, J. 1981 *Cadmium in the environment: health effects*. New York, Wiley.

Nriagu, J. 1996 A history of global metal pollution. *Science* 272: 223–4.

Nriagu, J., Blankson, M.L. and Ocran, K. 1996 Childhood lead poisoning in Africa: a growing public health problem. *Science of the Total Environment* 181: 93–100.

Nunn, P.D. 1990 Recent environmental changes on Pacific islands. *Geographical Journal* 156: 125–40.

Nunn, P.D. 1994 *Oceanic islands*. Oxford, Blackwell.

Nurse, L.A., McLean, R.F. and Suarez, A.G. 1998 Small island states. In IPCC *The regional impacts of climate change: an assessment of vulnerability*. Cambridge, Cambridge University Press: 331–54.

O'Hara, S.L., Street-Perrot, F.A. and Burt, T.P. 1993 Accelerated soil erosion around a Mexican highland lake caused by prehispanic agriculture. *Nature* 362: 48–51.

O'Riordan, T. and Turner, R.K. 1983 *An annotated reader in environmental planning and management*. Oxford, Pergamon Press.

Obeng, L. 1978 Environmental impacts of four African impoundments. In Gunnerson, C.G. and Kalbermatten, J.M. (eds), *Environmental impacts of international civil engineering projects and practices*. New York, American Society of Civil Engineers.

Odén, S. 1968 *The acidification of air precipitation and its consequences in the natural environment*. Energy Committee Bulletin 1. Stockholm, Swedish Natural Sciences Research Council.

Odum, E.P. 1979 The value of wetlands: a hierarchical approach. In Greeson, P.E., Clark, J.R. and Clark, J.E. (eds), *Wetland functions and values: the state of our understanding*. Minneapolis, MN, American Water Resources Association Technical Publication: 16–25.

OECD (Organisation for Economic Co-operation and Development) 1991a *The state of the environment*. Paris, OECD.

OECD 1991b *Environmental indicators: a preliminary set*. Paris, OECD.

OECD 1993 *OECD environmental data compendium*. Paris, OECD.

Ohkita, T. 1984 Health effects on individuals and health services of the Hiroshima and Nagasaki bombs. In WHO, *Effects of nuclear war on health and health service*. Geneva, WHO.

Oldeman, L.R., Hakkeling, R.T.A. and Sombroek, W.G. 1990 *World map of the status of human-induced soil degradation. An explanatory note*. Wageningen, ISRIC/UNEP.

Oliver, F.W. 1945 Dust storms in Egypt and their relation to the war period, as noted in Maryut, 1939–45. *Geographical Journal* 106: 26–49.

Olshansky, S.J., Carnes, B., Rogers, R.G. and Smith, L. 1997 Infectious diseases: new and ancient threats to world health. *Population Bulletin* 52(2).

Olson, S.L. and James, H.F. 1984 The role of Polynesians in the extinction of the avifauna of the Hawaiian Islands. In

Martin, P.S and Klein, R.G. (eds), *Quaternary extinctions: a prehistoric revolution*. Tucson, AZ, University of Arizona Press: 768–80.

Olsson, L. 1993 On the causes of famine – drought, desertification and market failure in the Sudan. *Ambio* 22: 395–403.

Opie, J. 1993 *Ogallala: water for a dry land*. Lincoln, NE, University of Nebraska.

Orians, G.H. and Pfeiffer, E.W. 1970 Ecological effects of the war in Vietnam. *Science* 168: 544–54.

Ortiz, N., Pires, M.A.F. and Bressiani, J.C. 2001 Use of steel converter slag as nickel adsorber to wastewater treatment. *Waste Management* 21: 631–5.

OSPAR Commission 2000 *Quality Status Report 2000*. London, OSPAR Commission.

Ostro, B. 1994 *Estimating the health effects of air pollutants: a method with an application to Jakarta*. World Bank Policy Research Working Paper 1301.

Otzen, U. 1993 Reflections on the principles of sustainable agricultural development. *Environmental Conservation* 20: 310–16.

Palmerini, C.G. 1993 Geothermal energy. In Johansson, T.B., Kelly, H., Reddy, A.K.N. and Williams, R.H. (eds), *Renewable energy: sources for fuels and electricity*. New York, Island Press: 549–91.

Pape, R. 1993 Air pollution in the Taiga. *Acid News* 2: 13–14.

Paskett, C.J. and Philoctete, C.-E. 1990 Soil conservation in Haiti. *Journal of Soil and Water Conservation* 45: 457–9.

Patrick, S.T., Timberlid, J.A. and Stevenson, A.C. 1990 The significance of land-use and land management change in the acidification of lakes in Scotland and Norway: an assessment utilizing documentary sources and pollen analysis. *Philosophical Transactions of the Royal Society, London* 327(1240): 363–7.

Pauly, D., Christensen, V., Dalsgaard, J., Froese, R. and Torres, F. 1998 Fishing down marine food webs. *Science* 279: 860–3.

Pearce, D. 1993 *Economic values and the natural world*. London, Earthscan.

Pearce, D. and Mäler, K. 1991 Environmental economics and the developing world. *Ambio* 20: 52–4.

Pearce, D. and Turner, R.K. 1992 Packaging waste and the polluter pays principle: a taxation solution. *Journal of Environmental Planning and Management* 35: 5–15.

Pearce, D., Barde, J.-P. and Lambert, J. 1984 Estimating the cost of noise pollution in France. *Ambio* 13: 27–8.

Peck, A.J. and Williamson, D.R. 1987 Effects of forest clearing on groundwater. *Journal of Hydrology* 94: 47–65.

Pennell, C.R. 1994 The geography of piracy: northern Morocco in the mid-nineteenth century. *Journal of Historical Geography* 20: 272–82.

Penner, J.E., Lister, D.H., Griggs, D.J., Dokken, D.J. and McFarland, M. (eds) 1999 *Aviation and the global atmosphere*. Geneva, IPCC.

Perkins, J.S. and Thomas, D.S.G. 1993 Environmental responses and sensitivity to permanent cattle ranching, semi-arid western central Botswana. In Thomas, D.S.G. and Allison, R.J. (eds), *Landscape sensitivity*. Chichester, Wiley: 273–86.

Pernetta, J.C. 1989 Projected climate change and sea-level rise: a relative impact rating for the countries of the Pacific Basin. In Pernetta, J.C. and Hughes, P.J. (eds), *Implications of expected climate changes in the South Pacific: an overview*. UNEP Regional Seas Reports and Studies 128. Nairobi, UNEP: 14–24.

Pernetta, J.C. 1992 Impacts of climate change and sea-level rise on small island states: national and international responses. *Global Environmental Change* 2: 19–31.

Pernetta, J.C. and Sestini, G. 1989 *The Maldives and the impacts of expected climate changes*. UNEP Regional Seas Reports and Studies 104. Nairobi, UNEP.

Peters, C.M., Gentry, A.H. and Mendelsohn, R.O. 1989 Valuation of an Amazonian rain forest. *Nature* 339: 655–6.

Peters, R.L. and Lovejoy, T.E. 1990 Terrestrial fauna. In Turner II, B.L., Clark, W.C., Kates, R.W., Richards, J.F., Mathews, J.T. and Meyer, W.B. (eds), *The Earth as transformed by human action*. Cambridge, Cambridge University Press: 353–69.

Petridou, E., Trichopoulos, D., Dessypris, N., Flytzani, V., Haidas, S., Kalmanti, M., Koliouskas, D., Kosmidis, H., Piperopoulou, F. and Tzortzatou, F. 1996 Infant leukaemia after *in utero* exposure to radiation from Chernobyl. *Nature* 382: 352–3.

Petts, G.E. 1984 *Impounded rivers: perspectives for ecological management*. Chichester, Wiley.

Petts, G.E. 1988 Regulated rivers in the United Kingdom. *Regulated Rivers: Research and Management* 2: 201–20.

Pfaff, A., Broad, K. and Glantz, M. 1999 Who benefits from climate forecasts? *Nature* 397: 645–6.

Phantumvanit, D. and Liengcharernsit, W. 1989 Coming to terms with Bangkok's environmental problems. *Environment and Urbanization* 1: 31–9.

Pimental, D. 1984 Energy flows in food systems. In Pimental,

D. and Hall, C. (eds), *Food and energy resources*. New York, Academic Press: 1–23.

Pimental, D. 1991 Diversification of biological control strategies in agriculture. *Crop Protection* 10: 243–53.

Pimental, D. and Levitan, L. 1988 Pesticides: where do they go? *Journal of Pesticide Reform* 7(4): 2–5.

Pimm, S.L., Russell, G.R., Gittleman, J.L. and Brooks, T.M. 1995 The future of biodiversity. *Science* 269: 347–50.

Plowden, C. and Bowles, D. 1997 The illegal market in tiger parts in northern Sumatra, Indonesia. *Oryx* 31: 59–66.

Plucknett, D.L. and Smith, N.J.H. 1986 Sustaining agricultural yields. *BioScience* 36: 40–5.

Poore, D. and Sayer, J. 1987 *The management of tropical moist forest lands: ecological guidelines*. Gland, IUCN.

Postel, S.L., Daily, G.C. and Ehrlich, P.R. 1996 Human appropriation of renewable fresh water. *Science* 271: 785–8.

Potter, G.L., Ellsaesser, H.W., MacCracken, M.C. and Luther, F.M. 1975 Possible climatic impact of tropical deforestation. *Nature* 258: 697–8.

Prance, G.T. 1990 Flora. In Turner II, B.L., Clark, W.C., Kates, R.W., Richards, J.F., Mathews, J.T. and Meyer, W.B. (eds), *The Earth as transformed by human action*. Cambridge, Cambridge University Press: 387–91.

Price, L.W. 1972 *The periglacial environment, permafrost, and man*. Commission on Geographical Resources Paper 14. Washington, DC, Association of American Geographers.

Prose, D.V. 1985 Persisting effects of armoured military manoeuvres on some soils of the Mojave Desert. *Environmental Geology and Water Sciences* 7: 163–70.

Pryde, P.R. 1972 *Conservation in the Soviet Union*. Cambridge, Cambridge University Press.

Pryde, P.R. 1991 *Environmental management in the Soviet Union*. Cambridge, Cambridge University Press.

Punning, J.-M. 1993 Environmental problems in Estonia. In Punning, J.-M. and Hult, J. (eds), *Human impact on environment*. Tallinn, Estonian Academy of Sciences: 7–23.

Rabinovitch, J. 1992 Curitiba: towards sustainable urban development. *Environment and Urbanization* 4: 62–73.

Rabinowitz, A. 1993 Estimating the indochinese tiger *Panthera tigris corbetti* population in Thailand. *Biological Conservation* 65: 213–17.

Rackham, O. 1986 *The history of the countryside*. London, Dent.

Rapin, F., Blanc, P. and Corvi, C. 1989 Influence des apports sur le stock de phosphore dans le lac Léman et sur son eutrophisation. *Revue des Sciences de l'Eau* 2: 721–37.

Rasid, H. and Pramanik, M.A.H. 1990 Visual interpretation of satellite imagery for monitoring floods in Bangladesh. *Environmental Management* 14: 815–21.

Raskin, P.D. 1995 Methods for estimating the population contribution to environmental change. *Ecological Economics* 15: 225–33.

Readman, J.W., Fowler, S.W., Villeneuve, J.-P., Cattini, C., Oregioni, B. and Mee, L.D. 1992 Oil and combustion product contamination of the Gulf marine environment following the war. *Nature* 358: 662–5.

Reid, W.V. 1992 How many species will there be? In Whitmore, T.C. and Sayer, J.A. (eds), *Tropical deforestation and species extinction*. London, Chapman & Hall: 55–73.

Reid, W.V. 1995 Biodiversity and health: prescription for progress. *Environment* 37: 36.

Reijnen, R. and Foppen, R. 1994 The effects of car traffic on breeding bird populations in woodland. 1 Evidence of reduced habitat quality for willow warblers breeding close to a highway. *Journal of Applied Ecology* 31: 85–94.

Reinking, R.F., Mathews, L.A. and St-Amand, P. 1975 *Dust storms due to desiccation of Owens Lake*. International Conference on Environmental Sensing and Assessment. Las Vegas, IEEE.

Renberg, I., Korsman, T. and Birks, H.J.B. 1993 Prehistoric increases in the pH of acid-sensitive Swedish lakes caused by land-use change. *Nature* 362: 824–7.

Renner, M. 1991 Assessing the military's war on the environment. In *State of the world 1991*. New York, W.W. Norton.

Rhoades, J.D. 1990 Soil salinity – causes and controls. In Goudie, A.S. (ed.) *Techniques for desert reclamation*. Chichester, Wiley: 109–34.

Richards, J.F. 1990 Agricultural impacts in tropical wetlands: rice paddies for mangroves in south and southeast Asia. In Williams, M. (ed.), *Wetlands: a threatened landscape*. Institute of British Geographers Special Publication 25. Oxford, Blackwell: 217–33.

Richardson, M.L. 1993 The assessment of hazards and risks to the environment caused by war damage to industrial installations in Croatia. Paper presented at the International Conference on the Effects of War on the Environment, Zagreb, 15–17 April.

Riebsame, W.E. 1990 The United States Great Plains. In Turner II, B.L., Clark, W.C., Kates, R.W., Richards, J.F., Mathews, J.T. and Meyer, W.B. (eds), *The Earth as transformed by human action*. Cambridge, Cambridge University Press: 561–75.

Ringrose, S. and Matheson, W. 1992 The use of Landsat MSS imagery to determine the areal extent of woody vegetation cover change in the west-central Sahel. *Global Ecology and Biogeography Letters* 2: 16–25.

RIVM 1999 *Milieubalans 1999*. Alphen aan den Rijn, Samsom.

Robb, G.A. 1994 Environmental consequences of coal mine closure. *Geographical Journal* 160: 33–40.

Roberts, C.M., McClean, C.J., Veron, J.E.N., Hawkins, J.P., Allen, G.R., McAllister, D.E., Mittermeier, C.G., Schueler, F.W., Spalding, M., Wells, F., Vynne, C. and Werner, T.B. 2002 Marine biodiversity hotspots and conservation priorities for tropical reefs. *Science* 295: 1280–4.

Roberts, L. 1988 Conservationists in Panda-monium. *Science* 241: 529–31.

Robinson, R.A. and Sutherland, W.J. 2002 Post-war changes in arable farming and biodiversity in Great Britain. *Journal of Applied Ecology* 39: 157–76.

Rosenberry, P., Knutson, R. and Harmon, L. 1980 Predicting effects of soil depletion from erosion. *Journal of Soil and Water Conservation* 35: 123–34.

Ruckelshaus, W.D. 1989 Toward a sustainable world. *Scientific American* 261(3): 114–20C.

Ruel, S. 1993 The scourge of land mines. *UN Focus* October.

Rukang, F. 1989 Environment and cancer in Shanghai. *Journal of Environmental Science (China)* 1: 1–9.

Runnels, D.D., Shepherd, T.A. and Angino, E.E. 1992 Metals in water. *Environmental Science and Technology* 26: 2316–23.

Ryding, S.-O. and Rast, W. (eds) 1989 *The control of eutrophication of lakes and reservoirs*. UNESCO Man and the Biosphere Series, vol. 1. Paris, Parthenon.

Saenger, P. 1994 Cleaning up the Arabian Gulf: aftermath of an oil spill. *Search* 25: 19–22.

Saheed, S.M. 1995 Country report – Bangladesh. Paper presented at the Expert Consultation of the Asian Network on Problem Soils, 23–27 October, Manila, Philippines.

Salo, J., Kalliola, R., Häkkinen, I., Mäkinen, Y., Niemalä, P., Puhakka, M. and Coley, P.D. 1986 River dynamics and the diversity of Amazon lowland forest. *Nature* 322: 254–8.

Salvat, B. 1992 Coral reefs – a challenging ecosystem for human societies. *Global Environmental Change* 2: 12–18.

Sanford, R.L., Saldarriaga, J., Clark, K.E., Uhl, C. and Herrera, R. 1985 Amazon rain-forest fires. *Science* 227: 53–5.

Sandford, S. 1977 *Dealing with drought and livestock in Botswana*. London, Overseas Development Institute.

Sapozhnikova, S.A. 1973 *Map diagram of the number of days with dust storms in the hot zone of the USSR and adjacent territories*. Report HT-23-0027. Charlottesville, VA, US Army Foreign and Technology Center.

Sather, J.M. and Smith, R.D. 1984 *An overview of major wetland functions and values*. Washington, DC, Fish and Wildlife Service, FWS/OBS-84/18.

Savchenko, V.K. 1991 The Chernobyl catastrophe and the biosphere. *Nature and Resources* 27(1): 37–46.

Sayer, J.A., Harcourt, C.S. and Collins, N.M. (eds) 1992 *The conservation atlas of tropical forests: Africa*. London, Macmillan/IUCN.

Schindler, D.W., Curtis, P.J., Parker, B.R. and Stainton, M.P. 1996 Consequences of climate warming and lake acidification for UV-B penetration in North American boreal lakes. *Nature* 379: 705–8.

Schlesinger, W.H. 1991 *Biogeochemistry: an analysis of global change*. San Diego, Academic Press.

Schlesinger, W.H., Reynolds, J.F., Cunningham, G.L., Huenneke, L.F., Jarrell, W.M., Virginia, R.A. and Whitford, W.G. 1990 Biological feedbacks in global desertification. *Science* 247: 1043–8.

Schmidheiny, S., Chase, R. and DeSimone, L.D. 1997 *Signals of change*. Geneva, World Business Council for Sustainable Development.

Schneider, S.H. 1989a *Global warming: are we entering the greenhouse century?* San Francisco, Sierra Club Books.

Schneider, S.H. 1989b The changing climate. *Scientific American* 261(3): 38–47.

Schteingart, M. 1989 The environmental problems associated with urban development in Mexico City. *Environment and Urbanization* 1: 40–50.

Schwartz, J. 1994 Low level lead exposure and children's IQ: a meta-analysis and search for a threshold. *Environmental Research* 65: 42–55.

Schwarz, H.E., Emel, J., Dickens, W.J., Rogers, P. and Thompson, J. 1990 Water quality and flows. In Turner II, B.L., Clark, W.C., Kates, R.W., Richards, J.F., Mathews, J.T. and Meyer, W.B. (eds), *The Earth as transformed by human action*. Cambridge, Cambridge University Press: 253–70.

Scoones, I. and Graham, O. 1994 New directions for pastoral development in Africa. *Development in Practice* 4: 188–98.

SCOPE (Scientific Committee on Problems of the Environment) 1989 *Environmental consequences of nuclear war. Volume II: Ecological and agricultural effects*, 2nd edn. SCOPE Report 28. Chichester, Wiley.

Scott, M.J. and Statham, I. 1998 Development advice: mining subsidence. In Maund, J.G. and Eddlestone, M. *Geohazards in engineering geology*. London, Geological Society: 391–400.

Sen, A. 1981 *Famines and poverty*. Oxford, Oxford University Press.

Sestini, G. 1992 Implications of climatic change for the Po delta and Venice Lagoon. In Jeftic, L., Milliman, J.D. and Sestini, G. (eds), *Climatic change and the Mediterranean*, vol. 1. London, Edward Arnold: 428–94.

Sevaldrud, I.H., Muniz, I.P. and Kalvenes, S. 1980 Loss of fish populations in southern Norway: dynamics and magnitude of the problem. In Drablos, D. and Tollan, A. (eds), *Ecological impact of acid precipitation*. Norway, SNSF-Project: 350–1.

Shahgedanova, M. and Burt, T.P. 1994 New data on air pollution in the former Soviet Union. *Global Environmental Change* 4: 201–27.

Shaw, D.G. 1992 The Exxon Valdez oil spill: ecological and social consequences. *Environmental Conservation* 19: 253–8.

Sheail, J. 1988 River regulation in the United Kingdom: an historical perspective. *Regulated Rivers: Research and Management* 2: 221–32.

Sheeline, L. 1993 Pacific fruit bats in trade: are CITES controls working? *Traffic USA* 12(1): 1–4.

Shields, L.M. and Wells, P.V. 1962 Effects of nuclear testing on desert vegetation. *Science* 135: 38–40.

Shiklomanov, I.A. 1993 World fresh water resources. In Gleik, P.H. (ed.), *Water in crisis: a guide to the world's fresh water resources*. New York, Oxford University Press: 13–24.

Shillitoe, S. 1991 Alternative energy in Nicaragua. *The Chemical Engineer* 491: 17–19.

Sibley, C.G. and Monroe, B.L. 1990 *Distribution and taxonomy of birds of the world*. New Haven, Yale University Press.

Siebe, C. and Cifuentes, E. 1995 Environmental impact of wastewater irrigation in Central Mexico: an overview. *International Journal of Environmental Health Research* 5: 161–73.

Siegert, F., Ruecker, G., Hinrichs, A. and Hoffmann, A.A. 2001 Increased damage from fires in logged forests during droughts caused by El Niño. *Nature* 414, 437–40.

Simon, J.L. 1981 *The ultimate resource*. Oxford, Martin Robertson.

Sipilä, K. 1993 New power-production technologies: various options for biomass cogeneration. *Bioresource Technology* 46: 5–12.

SIPRI (Stockholm International Peace Research Institute) 1990 *World armaments and disarmament, SIPRI Yearbook 1990*. Oxford, Oxford University Press.

Skaf, R. 1988 A story of a disaster: why locust plagues are still possible. *Disasters* 12: 122–7.

Skjaerseth, J.B. 1993 The 'effectiveness' of the Mediterranean Action Plan. *International Environmental Affairs* 5: 313–34.

Slaymaker, O. and French, H.M. 1993 Cold environments and global change. In French, H.M. and Slaymaker, O. (eds), *Canada's cold environments*. Montreal, McGill-Queen's University Press: 313–34.

Slingerland, M. and Masdewel, M. 1996 Mulching on the central plateau of Burkina Faso. In Reij, C., Scoones, I. and Toulmin, C. *Sustaining the soil: indigenous soil and water conservation in Africa*. London, Earthscan: 85–9.

Smith, B. 1997 Water: a critical resource. In King, R., Proudfoot, L. and Smith, B. (eds) *The Mediterranean: environment and society*. London, Arnold: 227–51.

Smith, D.R. 1992 Salinization in Uzbekistan. *Post-Soviet Geography* 33: 21–33.

Smith, J.T., Comans, R.N.J., Beresford, N.A., Wright, S.M., Howard, B.J. and Camplin, W.C. 2000 Chernobyl's legacy in food and water. *Nature* 405: 141.

Smith, K. 1997 Climatic extremes as hazards to humans. In Thompson, R.D. and Perry, A. (eds) *Applied climatology: principles and practice*. London, Routledge: 304–16.

Smith, K. 2001 *Environmental hazards: assessing risk and reducing disaster*, 3rd edn. London, Routledge.

Smith, K. and Ward, R. 1998 *Floods: physical processes and human impacts*. Chichester, Wiley.

Smith, N.J.H., Alvim, P., Homma, A., Falesi, I. and Serrão, A. 1991 Environmental impacts of resource exploitation in Amazonia. *Global Environmental Change* 1: 313–20.

Smith, R.A. 1852 On the air and rain of Manchester. *Memoirs and Proceedings of the Manchester Literary and Philosophical Society* 2: 207–17.

Smith, S.J., Pitcher, H. and Wigley, T.M.L. 2001 Global and regional anthropogenic sulfur dioxide emissions. *Global and Planetary Change* 29: 99–119.

Smyth, C.G. and Royle, S.A. 2000 Urban landslide hazards: incidence and causative factors in Niterói, Rio de Janeiro State, Brazil. *Applied Geography* 20: 95–118.

Soboleva, O.V. and Mamadaliev, U.A. 1976 The influence of Nurek Reservoir on local earthquake activity. *Engineering Geology* 10: 293–305.

Sorensen, J.C. and McCreary, S.T. 1990 *Coasts: institutional arrangements for managing coastal resources and environments*. Washington, DC, US Department of the Interior and US Agency for International Development.

Soutar, A. 1967 The accumulation of fish debris in certain California coastal sediments. *Californian Cooperative Ocean Fisheries Investment Report* 11: 136–9.

Spencer, J.W. and Kirby, K.J. 1992 An inventory of ancient woodland for England and Wales. *Biological Conservation* 62: 77–93.

Srivastava, J.P.L. and Kaul, R.N. 1995 Desertification in the Aravalli Hills of Haryana: progress towards a viable solution. *Desertification Control Bulletin* 26: 37–43.

Stanley, D.J. 1996 Nile delta: extreme case of sediment entrapment on a delta plain and consequent coastal land loss. *Marine Geology* 129: 189–95.

Stanley, D.J. and Warne, A.G. 1993 Nile delta: recent geological evolution and human impact. *Science* 260: 628–34.

Stehouwer, R.C., Sutton, P. and Dick, W.A. 1995 Minespoil amendment with dry flue desulfurization by-products: plant growth. *Journal of Environmental Quality* 24: 861–9.

Stiassny, M.L.J and Raminosoa, N. 1994 The fishes of the inland waters of Madagascar. *Annales du Musée Royal de l'Afrique Centrale Zoologie* 275: 133–49.

Stock, R.F. 1976 *Cholera in Africa: diffusion of the disease 1970–75, with special reference to West Africa*. African Environment Special Report 3, London, International African Institute.

Stocking, M.A. 1987 Measuring land degradation. In Blaikie, P.M. and Brookfield, H.C. (eds), *Land degradation and society*. London, Routledge: 49–63.

Stolzenburg, W. 1992 The mussels' message. *Nature Conservancy* 42 (Nov./Dec.): 16–23.

Stone, R. 1995 If the mercury soars, so may health hazards. *Science* 267: 957–8.

Stonich, S.C. and DeWalt, B.R. 1996 The political ecology of deforestation in Honduras. In Sponsel, L.E., Headland, T.N. and Bailey, R.C. (eds), *Tropical deforestation: the human dimension*. New York, Columbia University Press: 187–215.

Sullivan, S. 1999 The impacts of people and livestock on topographically diverse open wood- and shrub-lands in arid north-west Namibia. *Global Ecology and Biogeography* 8: 257–77.

Svidén, O. 1993 Clean fuel and engine systems for twenty-first-century road vehicles. In Giannopoulos, G. and Gillespie, A. (eds), *Transport and communications innovation in Europe*. London, Belhaven: 122–48.

Swanson, T. (ed.) 1995 *Intellectual property rights and biodiversity conservation: an interdisciplinary analysis of the values of medicinal plants*. Cambridge, Cambridge University Press.

Swearingen, W. 1994 Northwest Africa. In Glantz, M.H. (ed.), *Drought follows the plow*. Cambridge, Cambridge University Press: 117–33.

Swinnerton, C.J. 1984 Protection of groundwater in relation to waste disposal in Wessex Water Authority. *Quarterly Journal of Engineering Geology* 17: 3–8.

Swiss Re 1994 *Natural catastrophes and major losses in 1993: insured damage drops significantly*. sigma 2/94. Zurich, Swiss Reinsurance Company.

Swiss Re 1998 *Natural catastrophes and major losses in 1997: exceptionally few high losses*. sigma 3/98. Zurich, Swiss Reinsurance Company.

Swiss Re 2002 *Natural catastrophes and man-made disasters in 2001: man-made losses take on a new dimension*. sigma 1/02. Zurich, Swiss Reinsurance Company.

Szasz, F.M. 1995 The impact of World War II on the land: Gruinard Island, Scotland and Trinity Site, New Mexico as case studies. *Environmental History Review* 19(4): 15–30.

Tacon, A.G.J. 1996 Feeding tomorrow's fish. *World Aquaculture* 27: 20–32.

Takeuchi, K., Katoh, K., Nan, Y. and Kou, Z. 1995. Vegetation cover change in desertified Kerqin sandy lands, Inner Mongolia. *Geographical Reports, Tokyo Metropolitan University* 30, 1–24.

Tamang, D. 1995 An overview of drought and desertification in Nepal. In European Environment Bureau, *Implementation of the Convention on Desertification*. Brussels, EEB: 104–11.

Temple, S.A. 1977 Plant-animal mutualism: coevolution with dodo leads to near extinction of plant. *Science* 197: 885–6.

Terborgh, J. 1992 Why American songbirds are vanishing. *Scientific American* 266(5): 56–62.

Teufel, D. 1989 Die Zukunft des Autoverkehrs. Heidelberg, Bericht 17 Umwelt- und Prognose Institut (unpublished).

Thanh, N.C. and Tam, D.M. 1990 Water systems and the environment. In Thanh, N.C. and Biswas, A.K. (eds), *Environmentally-sound water management*. Delhi, Oxford University Press: 1–29.

Thiel, H. and Schriever, G. 1990 Deep-sea mining, environmental impact and the DISCOL Project. *Ambio* 19: 145–50.

Thomas, D.S.G. and Middleton, N.J. 1994 *Desertification: exploding the myth*. Chichester, Wiley.

Thomas, J.A. 1991 Rare species conservation: case studies of European butterflies. In Spellerberg, I.F., Goldsmith, F.B. and Morris, M.G. (eds), *The scientific management of temper-*

ate communities for conservation. Oxford, Blackwell Scientific: 149–97.

Thompson, D.R., Becker, P.H. and Furness, R.W. 1993 Long-term changes in mercury concentrations in herring gulls *Larus argentatus* and common terns *Sterna hirundo* from the German North Sea coast. *Journal of Applied Ecology* 30: 316–20.

Thomson, J.A.M. and Foster, S.S.D. 1986 Effect of urbanisation on groundwater of limestone islands: an analysis of the Bermuda case. *Journal of Institute of Water Engineering Science* 40: 527–40.

Thrupp, L.A. 1991 Sterilization of workers from pesticide exposure: causes and consequences of DBCP-induced damage in Costa Rica and beyond. *International Journal of Health Services* 21: 731–9.

Tiffen, M., Mortimore, M. and Gichuki, F. 1994 *More people, less erosion: environmental recovery in Kenya*. Chichester, Wiley.

Tipping, E., Bass, J.A.B., Hardie, D., Haworth, E.Y., Hurley, M.A. and Wills, G. 2002 Biological responses to the reversal of acidification in surface waters of the English Lake District. *Environmental Pollution* 116: 137–46.

TMG (Tokyo Metropolitan Government) 1985 *Protecting Tokyo's environment*. Tokyo, TMG.

Tolba, M.K. 1990 Building an environmental institutional framework for the future. *Environmental Conservation* 17: 105–10.

Tolba, M.K. 1992 *Saving our planet*. London, Chapman & Hall.

Tolba, M.K. and El-Kholy, O.A. (eds) 1992 *The world environment 1972–1992*. London, Chapman & Hall.

Toledo, V.M., Batis, A.I., Becerra, R., Martinez, E. and Ramos, C.H. 1995 La selva util: etnobotánica cuantitativa de los grupos indígenas del trópico húmedo de México. *Interciencia* 20: 177–87.

Tolmazin, D. 1985 Changing coastal oceanography of the Black Sea. I: Northwestern shelf. *Progress in Oceanography* 15: 217–76.

Tolouie, E., West, J.R. and Billam, J. 1993 Sedimentation and desiltation in the Sefid-Rud reservoir, Iran. In McManus, J. and Duck, R.W. (eds), *Geomorphology and sedimentology of lakes and reservoirs*. Chichester, Wiley: 125–38.

Touloumi, G., Pocock, S.J., Katsouyanni, K. and Trichopoulos, D. 1994 Short-term effects of air pollution on daily mortality in Athens: a time-series analysis. *International Journal of Epidemiology* 23: 957–67.

Transnet 1990 *Energy, transport and the environment*. London Transnet.

Trudgill, S.T., Viles, H.A., Inkpen, R., Moses, C., Gosling, W., Yates, T., Collier, P., Smith, D.I. and Cooke, R.U. 2001 Twenty-year weathering remeasurements at St Paul's Cathedral, London. *Earth Surface Processes and Landforms* 26: 1129–42.

Tsehai, F.M. 1991 Refugees in Ethiopia: some reflections on the ecological impact. *IDOC Internazionale* 22(2): 2–6.

Tucker, C.J., Dregne, H.E. and Newcombe, W.W. 1991 Expansion and contraction of the Sahara Desert from 1980 to 1990. *Science* 253: 299–301.

Turner II, B.L., Kasperson, J.X., Kasperson, R.E., Dow, K. and Meyer, W.B. 1995 Comparisons and conclusions. In Kasperson, J.X., Kasperson, R.E. and Turner II, B.L. (eds) *Regions at risk: comparisons of threatened environments*. Tokyo, United Nations University Press: 519–86.

Turner II, B.L., Moss, R.H. and Skole, D.L. 1993 *Relating land use and global land-cover change: a proposal for an IGBP-HDP core project*. International Geosphere-Biosphere Programme Report 24.

Turnock, D. 1993 Romania. In Carter, F.W. and Turnock, D. (eds), *Environmental problems in Eastern Europe*. London, Routledge: 135–63.

UNDP (UN Development Programme) 1993 *Human development report*. New York, UNDP.

UNDP/UNICEF 2002 *The human consequences of the Chernobyl nuclear accident: a strategy for recovery*. Vienna, IAEA.

UNECLA (UN Economic Commission for Latin America and the Caribbean) 1990 *The water resources of Latin America and the Caribbean: planning, hazards, and pollution*. Santiago, UNECLA.

UNECLA 1991 *Sustainable development: changing production patterns, social equity and the environment*. Santiago, UNECLA.

UNEP (UN Environment Programme) 1984 *Socio-economic activities that may have an impact on the marine and coastal environment of the East African region*. UNEP Regional Seas Reports and Studies 41. Nairobi, UNEP.

UNEP 1987 *Environmental data report*. Oxford, Blackwell.

UNEP 1989a *State of the Mediterranean marine environment*. Athens, UNEP.

UNEP 1989b Sustainable water development. *Water Resources Development* 5: 223–51.

UNEP 1990 *Green energy: biomass fuels and the environment*. Nairobi, UNEP.

UNEP 1992a *World atlas of desertification*. Sevenoaks, Edward Arnold.

UNEP 1992b *The Aral Sea: diagnostic study for the development of* an action plan for the conservation of the Aral Sea. Nairobi, UNEP.

UNEP 1993 *Environmental data report 1993/94.* Oxford, Blackwell.

UNEP 1995 *Global biodiversity assessment.* Cambridge, Cambridge University Press.

UNEP 2001 *The Mesopotamian Marshlands: demise of an ecosystem.* Early Warning and Assessment Technical Report, UNEP/DEWA/TR.01-3 Rev. 1. Nairobi, United Nations Environment Programme.

UNEP IE/PAC 1993 *Cleaner production worldwide.* Paris, IE/PAC.

UNEP/UNESCO/UN-DIESA 1985 *Coastal erosion in west and central Africa.* UNEP Regional Seas Reports and Studies 67. Nairobi, UNEP.

UNEP/WHO 1988 *Assessment of freshwater quality.* Nairobi, UNEP.

UNEP/WHO 1992 *Urban air pollution in megacities of the world.* Oxford, Blackwell.

UNESCO 1978 *World water balance and water resources of the Earth.* Paris, UNESCO.

UNFPA (UN Population Fund) 2001 *The state of world population 2001.* New York, UNFPA.

UNHCR (UN High Commissioner for Refugees) 1994 Environmental issues in Benaco refugee camp. Press release, 21 June.

United Nations 1989a *Prospects of world urbanization.* Population Studies 112.

United Nations 1989b *Study on the economic and social consequences of the arms race and military expenditures.* Disarmament Study Series 19. New York, UN.

United Nations 2001 *Johannesburg summit 2002: world summit on sustainable development.* New York, UN.

USEPA (US Environmental Protection Agency) 1983 *Chesapeake Bay Program: findings and recommendations.* Philadelphia, EPA.

USEPA 1990 *National water quality inventory. 1988 Report to Congress, Office of Water.* EPA 440-4-90-003. Washington, DC.

Valencia, R., Balslev, H. and Paz y Miño, G. 1994 High tree alpha-diversity in Amazonian Ecuador. *Biodiversity and Conservation* 3: 21–8.

Van Lynden, G.W.J. and Oldeman, L.R. 1997 *The assessment of the status of human-induced soil degradation in South and Southeast Asia.* Nairobi, UNEP, and Wageningen, ISRIC.

Vertegaal, P.J.M. 1989 Environmental impact of Dutch military activities. *Environmental Conservation* 16: 54–64.

Viet Nam 1985 *Viet Nam: National conservation strategy.* Ho Chi Minh City, Committee for Rational Utilisation of National Resources and Environmental Protection Programme 52-02.

Viner, A. 1992 Biodiversity, fisheries, and the future of Lake Victoria. *IUCN Wetlands Programme Newsletter* 6: 3–6.

Viras, L.G., Paliatsos, A.G. and Fotopoulos, A.G. 1996 Nine-year trend of air pollution by CO in Athens, Greece. *Environmental Monitoring and Assessment* 40: 203–14.

Vitousek, P.M., Aber, J.D., Howarth, R.W., Likens, G.E., Matson, P.A., Schindler, D.W., Schlesinger, W.H. and Tilman, D.G. 1997a Human alteration of the global nitrogen cycle: sources and consequences. *Ecological Applications* 7: 737–50.

Vitousek, P.M., Mooney, H.A., Lubchenco, J. and Melillo, J.M. 1997b Human domination of Earth's ecosystems. *Science* 277: 494–9.

Vogt, H.P. 1995 Coral reefs in Saudi Arabia: 3.5 years after the Gulf War oil spill. *Coral Reefs* 14: 271–3.

Voight, B. 1990 The 1985 Nevado del Rúiz volcano catastrophe: anatomy and retrospection. *Journal of Volcanology and Geothermal Research* 42: 151–88.

von Below, M.A. 1993 Sustainable mining development hampered by low mineral prices. *Resources Policy* 19: 177–81.

Wali, A. 1988 *Kilowatts and crisis: a study of development and social change in Panama.* Boulder, CO, Westview.

Walker, H.J. 1990 The coastal zone. In Turner II, B.L., Clark, W.C., Kates, R.W., Richards, J.F., Mathews, J.T. and Meyer, W.B. (eds), *The Earth as transformed by human action.* Cambridge, Cambridge University Press: 271–94.

Walther, G.-R., Post, E., Convey, P., Menzel, A., Parmesan, C., Beebee, T.J.C., Fromentin, J. M., Hoegh-Guldberg, O. and Bairlein, F. 2002 Ecological responses to recent climate change. *Nature* 416: 389–95.

Wang, W.C., Yung, Y.L., Lacis, A.A., Mo, T. and Hansen, J.E. 1976 Greenhouse effect due to manmade perturbations of other gases. *Science* 194: 685–90.

Warren, A. and Agnew, C. 1988 *An assessment of desertification and land degradation in arid and semi-arid areas.* Drylands paper 2. London, International Institute for Environment and Development.

Warren, A. and Khogali, M. 1992 *Assessment of desertification and drought in the Sudano-Sahelian region 1985–1991.* New York, United Nations Sudano-Sahelian Office.

Watt, W.D., Scott, C.D. and White, W.J. 1983 Evidence of acidification of some Nova Scotia rivers and its impact on

Atlantic salmon, *Salmo salar*. *Canadian Journal of Fisheries and Aquatic Science* 40: 462–73.

WCED (World Commission on Environment and Development) 1987 *Our common future*. Oxford, Oxford University Press.

WCED 1992 *Our common future reconvened*. London, WCED.

WCMC (World Conservation Monitoring Centre) 1992 *Global biodiversity: status of the Earth's living resources*. London, Chapman & Hall.

Weber, P. 1994 Safeguarding oceans. In Brown, L.R. (ed.), *State of the world 1994*. New York, W.W. Norton: 41–60.

WEC (World Energy Council) 1992 *1992 survey of energy resources*. London, WEC.

WEC 1993 *Energy for tomorrow's world: the realities, the real opinions and the agenda for achievement*. London, Kogan Page.

Weinberg, B. 1991 *War on the land: ecology and politics in Central America*. London, Zed Books.

Weinberg, C.J. and Williams, R.H. 1990 Energy from the sun. *Scientific American* 263(3): 98–106.

Weiss, H., Courty, M.-A., Wetterstrom, W., Guichard, F., Senior, L., Meadow, R. and Curnow, A. 1993 The genesis and collapse of third millennium north Mesopotamian civilisation. *Science* 261: 995–1004.

Wellner, F.-W. and Kürsten, M. 1992 International perspective on mineral resources. *Episodes* 15: 182–94.

Westing, A.H. 1980 *Warfare in a fragile world: military impact on the human environment*. London, Taylor & Francis.

Westing, A.H. (ed.) 1984 *Herbicides in war: the long-term ecological and human consequences*. London, Taylor & Francis.

Westing, A.H. 1994 Environmental security for the Horn of Africa: an overview. In Polunin, N. and Burnett, J. (eds), *Surviving with the biosphere*. Edinburgh, Edinburgh University Press: 354–7.

Westing, A.H. and Pfeiffer, E.W. 1972 The cratering of Indochina. *Scientific American* 226(5): 21–29.

Westlake, K. 1997 Sustainable landfill – possibility or pipe-dream? *Waste Management and Research* 15: 453–61.

Whitelegg, J. 1992 Transport and the environment. *Geography* 77: 91–3.

Whitelegg, J. 1993 *Transport for a sustainable future: the case for Europe*. London, Belhaven.

Whitelegg, J. 1994 Transportation: for a sustainable policy. *Acid News* 1 (February): 8–9.

Whitlow, R.J. 1990 Mining and its environmental impacts in Zimbabwe. *Geographical Journal of Zimbabwe* 21: 50–80.

Whitmore, T.C. 1998 *An introduction to tropical rain forests*, 2nd edn. Oxford, Oxford University Press.

Whittaker, R.H. and Likens, G.E. 1973 Carbon in the biota. In Woodwell, G.M. and Pecan, E.V. (eds), *Carbon and the biosphere*. Washington, DC, US Department of Commerce: 281–302.

WHO (World Health Organization) 1992 Cholera in the Americas. *Weekly Epidemiological Record* 67: 33–40.

Wigley, T., Richels, R. and Edmonds, J. 1996 Economic and environmental choices in the stabilization of atmospheric CO_2 concentrations. *Nature* 379: 240–3.

Wijkman, A. and Timberlake, L. 1984 *Natural disasters: acts of God or acts of man?* London, Earthscan.

Wiley, K.B. and Rhodes, S.L. 1998 The transformation of the Rocky Mountain arsenal. *Environment* 40(5): 4–11, 28–35.

Williams, E.H. and Bunkley-Williams, L. 1990 The worldwide coral reef bleaching cycle and related sources of coral mortality. *Atoll Research Bulletin* 335: 1–71.

Williams, J. 1994 The great flood. *Weatherwise* 47: 18–22.

Williams, M. 1990a Understanding wetlands. In Williams, M. (ed.), *Wetlands: a threatened landscape*. Institute of British Geographers Special Publication 25. Oxford, Blackwell: 1–41.

Williams, M. 1990b Agricultural impacts in temperate wetlands. In Williams, M. (ed.), *Wetlands: a threatened landscape*. Institute of British Geographers Special Publication 25. Oxford, Blackwell: 181–216.

Williams, M.A.J., Dunkerley, D.L., De Deckker, P., Kershaw, A.P. and Stokes, T. 1993 *Quaternary environments*. London, Edward Arnold.

Williams, W.D. and Aladin, N.V. 1991 The Aral Sea: recent limnological changes and their conservation significance. *Aquatic Conservation* 1: 3–24.

Willson, B. 1902 *Lost England: the story of our submerged coasts*. London, George Newnes.

Wilson, E.O. 1989 Threats to biodiversity. *Scientific American* 261(3): 60–6.

Wilson, J.S. 1858 The general and gradual desiccation of the Earth and atmosphere. *Report of the Proceedings of the British Association for the Advancement of Science*: 155–6.

Wirawan, N. 1993 The hazard of fire. In Brookfield, H. and Byron, Y. (eds), *South-East Asia's environmental future*. Tokyo, UN University Press: 242–60.

Wischmeier, W.H. 1976 Use and misuse of the Universal Soil Loss Equation. *Journal of Soil and Water Conservation* 31: 5–9.

Wood, L.B. 1982 *The restoration of the tidal Thames*. London, Hilger.

Woodruff, N.P. and Siddoway, F.H. 1965 A wind erosion equation. *Proceedings of the Soil Science Society of America* 29: 602–8.

Woodwell, G.M., Wurster, C.F. and Isaacson, P.A. 1967 DDT residues in an east coast estuary: a case of biological concentration of a persistent insecticide. *Science* 156: 821–4.

World Bank 1985 *Desertification in the Sahelian and Sudanian zones of West Africa*. Washington, DC, World Bank.

World Bank 1989 Philippines: environmental and natural resource management study. Washington, DC, World Bank.

World Bank 1992a *World development report 1992: development and the environment*. New York, Oxford University Press.

World Bank 1992b *Strategy for African mining*. World Bank Technical Paper 181

World Bank 1996 *Review of policies in the traditional energy sector. Regional report: Burkina Faso, Mali, Niger, Senegal, The Gambia*. Washington, DC, World Bank.

World Bank 2000 *A review of the World Bank's 1991 forest strategy and its implementation*. Washington, World Bank.

World Commission on Dams 2000 *Dams and development: a new framework for decision-making*. London, Earthscan.

Worster, D. 1979 *Dust bowl*. New York, Oxford University Press.

Worthington, E.B. 1978 Some ecological problems concerning engineering and tropical diseases. *Progress in Water Technology* 11: 5–11.

WRI (World Resources Institute) 1986 *World resources 1986*. New York, Basic Books.

WRI 1987 *World resources 1987*. New York, Basic Books.

WRI 1990 *World resources 1990–91*. New York, Oxford University Press.

WRI 1992 *World resources 1992–93*. New York, Oxford University Press.

WRI 1994 *World resources 1994–95*. New York, Oxford University Press.

WRI 1996 *World resources 1996–97*. New York, Oxford University Press.

WRI 1998 *World resources 1998–99*. New York, Oxford University Press.

WRI 2000 *World resources 2000–2001*. Washington, DC, WRI.

Wright, D.H. 1990 Human impacts on energy flow through natural ecosystems, and implications for species endangerment. *Ambio* 19: 89–194.

Wright, R.F. and Hauhs, M. 1991 Reversibility of acidification: soils and surface waters. *Proceedings of the Royal Society of Edinburgh* 97B: 169–91.

WWF (World Wide Fund for Nature) 1998 *The WWF tiger status report: 1998 year of the tiger*. Gland, WWF.

Yakowitz, H. 1993 Waste management: what now? what next? an overview of policies and practices in the OECD area. *Resources, Conservation and Recycling* 8: 131–78.

Yan, N.D., Keller, W., Scully, N.M., Lean, D.R.S. and Dillon, P.J. 1996 Increased UV-B penetration in a lake owing to drought-induced acidification. *Nature* 381: 141–3.

Yhdego, M. 1991 Scavenging solid wastes in Dar es Salaam, Tanzania. *Waste Management and Research* 9: 259–65.

Young, A. and Mitchell, N. 1994 Microclimate and vegetation edge effects in a fragmented podocarp-broadleaf forest in New Zealand. *Biological Conservation* 67: 63–72.

Young, J.E. 1992 *Mining the Earth*. Worldwatch Paper 109. Washington, DC, Worldwatch Institute.

Zaika, B.E. 1990 Change in macrobenthic populations in the Black Sea with depth (50–200 m). *Proceedings of the Academy of Sciences of the Ukrainian SSR, Series B* 11: 68–71 (in Russian).

Zhao, W., Jiao, E., Wang, G. and Meng, X. 1992 Analysis on the variation of sediment yield in Sanchuanhe River basin in 1980s. *International Journal of Sedimentary Research* 7: 1–19.

Zhou, L., Tucker, C.J., Kaufmann, R.K., Slayback, D., Shabanov, N.V. and Myneni, R.B. 2001 Variations in northern vegetation activity inferred from satellite data of vegetation index during 1981 to 1999. *Journal of Geophysical Research – Atmospheres* 106 (D17): 20 069–83.

Zonn, I.S. 1995 Desertification in Russia: problems and solutions (an example in the republic of Kalmykia-Khalmg Tangch). *Environmental Monitoring and Assessment* 37: 347–63.

Zurayk, R.A. 1994 Rehabilitating the ancient terraced lands of Lebanon. *Journal of Soil and Water Conservation* 49: 106–12.

INDEX

A

acid rain
- acid shock, 203
- alkanisation, 202
- aquatic ecosystems
  - effect on, 203–4
  - recovery of, 211–12
- beneficial effects, 203
- buffering capacity, 202
- buildings and monuments, damage to, 207–8
- critical loads concept, 202, 209–10, 209t
- emissions reduction *see* nitrogen oxide emissions; sulphur dioxide emissions
- European Monitoring and Evaluation Programme (EMEP), 201
- experimental acidification, 204, 204t
- fertiliser use, 218–19
- fish mortality, 203–4
- forest damage (defoliation), 204–6, 205f, 206t
- 'hotspots', 200
- human health, effects on, 206–7
- lichens, loss of, 204–5
- pH scale, 198, 202
- political acceptance of, 33t
- pollutants, long-range transportation of, 200–2
- recovery from
  - chemical and biological recovery, 211
  - fresh water liming programmes, 211f, 212
  - lacustrine ecosystems, 212
  - 'memory effect', 211
  - timescales, 211–12
- soil acidification, 202
- toxic metal mobilisation, 203

aerosol sprays, 179, 179t, 183

afforestation
- acid rain, 202
- big dams, 149

Agricultural Revolution, 23, 387

agriculture
- crop yields, increasing
  - chemical farming techniques, 217
  - mechanisation, 216–17
- domestication, of plants and animals, 215–16, 215f
- expansion, 216
- human population, exponential growth of, 214, 215–16
- hunter gatherers, transition from, 215
- livestock farming, intensification of, 217
- origins of, 214
- sustainable
  - checklist for, 228–9
  - energy efficiency, 228–9, 229t
- urban civilisations, emergence of, 215
- *see also* fertilisers; irrigation; pesticides

air, undervaluation of, 25

air pollution
- geothermal energy, 307–8
- lichens, 255
- political acceptance of, 33t
- WHO, 169, 170
- *see also* motor vehicle pollution

aircraft emissions, 273

Åland Convention (1921), 347

albedo changes, 179–81

Angkor Wat civilisation, collapse of, 36

Antarctic Treaty (1959), 347, 383t

Antarctica, global convention, 383t

aquaculture (fish farming)
- global production, 229, 230
- intensive, adverse effects of, 229–30, 230f
- sewage control, 230
- techniques, 229

Aral Sea, ecological demise of, 79
- biological weapons testing, 336, 338
- hydrological parameters, 132t
- irrigation, diversion of rivers for, 130
- 'offtakes', consequences of, 130–2
- precautionary principle, rejection of, 132
- technocentric approach, dangers of, 132

arms sales, 27

Aswan High Dam, 142, 142f, 149–52, 151f, 156
- *see also* big dams

atmosphere, 1

B

Basel Convention on the Control of Transboundary Movements of Hazard Wastes and their Disposal, 290, 384t

big dams
- benefits of
  - flood protection, 142, 142f
  - foreign interests, 143, 143f
  - hydroelectricity, 142
  - irrigation, 142
  - urban areas, 143–4, 143f
- construction worldwide, 141
- environmental impacts, 22–3
  - climate changes, 149
  - cultural property, loss of, 145, 145f
  - deforestation and afforestation, 149

desiltation techniques, 148, 148f
greenhouse effect, impact on, 149
human health, 146–7
land lost/people displaced/power generated trade-off, 145, 146t
main areas of influence, 144, 144t
marine and fish populations, effects on, 152
migrants, 146
multiple impoundment, 152–3
Nile delta changes, 150–2, 151f
plant endemism, destruction of, 144
resettlement programmes, 145–6
salinity problems, 150–2, 151f, 152t
sedimentation, 147–8, 147f, 150
seismic activity, inducement of, 148–9
submerged forests, dangers of, 144–5
temporal aspect, 144
Wadi Allaqi changes, 149–50
water quality, 145
water table rise, 148
waterweed, spread of, 147
history, 141
hydroelectricity generation, 141, 142
political impacts
framework for decision-making, 155–6
international agreement and disagreement, 156, 157f
planning and consultation, need for, 155
public environmental awareness, effect of, 155
Three Gorges Project (China), development of, 154t, 155
reservoir volumes worldwide, 141
biodiversity
definition of, 248
ecology of organisms, understanding, 261–2, 261f
endemic species, 250, 264–5
extinction rates, 249
flora and fauna, threats to, 248–58
fundamental causes of loss of
agricultural, fishing and forestry products, narrowing of, 250
environment, lack of valuation of, 250
human population growth, 250
knowledge deficiency, 250
legal and institutional sanctioning of practices, 250
ownership inequities, 250
genetic diversity, 248, 249
global convention, 248, 265, 384t
habitat loss and modification
agrochemical pollution, 254
air pollution, 255
giant panda, threat to, 253–4, 253t
habitat fragmentation, 254
lowland grassland, loss of, 253
for settlement and industry, 252, 252t
tropical rainforests, human-induced extinctions in, 252–3
water pollution, 254–5
habitat protection
protected areas by continent, 258, 259t
tiger populations, 258, 260–1
World Park, proposed first, 258
hunting and trade bans, enforcement problems, 262, 263, 263f
island species
endemic species, vulnerability of, 256, 257t
exotic species, introduction of and local extinctions, 257–8
vertebrate extinctions, from hunting, 256
IUCN, Red List categories, 250t, 251, 255
K-strategists, 250–1
keystone species, 251
local involvement, 261
'mass extinction events', 249
moral and pragmatic arguments for, 258
natural extinctions, 249
off-site conservation practices
animal and plant reintroductions, 264, 264f
captive individuals, and research, 264
plant reintroductions, 264
overexploitation
hunting, and extermination of large numbers, 255–6, 256f
species loss, knock-on effect, 255
species market values, 255–6
r-strategists, 251
scientific knowledge about, 248–9
species demise, human-induced reasons for, 251–2, 252t
species extermination, 258
biogeochemical cycles, 14
biological weapons
global convention, 383t
testing, 336, 338
biomass, 1–2
as renewable energy source
dedicated energy plantations, 306
forms of, 305–6
fuel wood, 305–6, 305f
as percentage of national energy consumption, 305, 305f, 306t
biomes
classification of, 3–4
human action, impact of, 5–6
biosphere, 1
biotechnology
environmental concerns
genetic diversity, maintenance of, 227–8
genetic uniformity, crop failures attributed to, 227, 228t
GM gene escapes, risk of, 227
precautionary principle, need for, 227
'terminator technology', 227
genetic engineering
clonal propagation, 226
developing countries, disadvantaged, 226–7
'genetically modified' (GM) food, use of term, 226
recombinant DNA techniques, 226
genetic modification, as age-old process, 225–6
wheat strains, cross breeding of, 225, 226f
Black Sea *see under* ocean pollution

Blue Revolution, 230
boreal forest, 4, 4f
'Brandt line', 26f
Brundtland Commission *see* World Commission on Environment and Development
BST (growth hormone), banning of, 227

C

Canada Climate Change Development Fund (CCCDF), 195–6, 195t
Canadian International Development Agency (CIDA), 195
carbon cycle, 2, 6–7, 8f, 58
carbon dioxide emissions, 182t, 183f
  air pollutant, 273, 277, 277t
  cement manufacture, 181
  climate change, 33t, 193, 194, 194f
  and energy use, 183
  fossil fuels, 181, 183, 194, 194f
  land use changes, 183
carbon monoxide, 166, 169, 273, 277t, 281
carnivores, 8, 9f
carrying capacity
  exceeding of, 46, 46f
  fixed *versus* variable, 46–7
  intensive grazing, 75
  maximum sustainable yield, 46
  meanings of, 45
  and population, relationship between, 45–6, 45f
  tragedy of the commons, 46
Caspian Sea, variations in sea level, 106, 106f
catalytic converters, 23, 276–7, 277t
CCAMLR *see* Commission for the Conservation of Antarctic Marine Living Resources
CCCDF *see* Canada Climate Change Development Fund
CFCs *see* chlorofluorocarbons
Chernobyl disaster
  biological effects, main, 313, 313t
  ceasium accumulation, 311–12
  communities affected, downward spiral, 314f
  contaminated territories, 313
  economic costs, 313–14
  food chain, effects on, 311–12
  forest contamination, 312
  human health effects, 313
  radionuclides, dispersal of, 311–12
  soil contamination, 312–13
chlorofluorocarbons (CFCs)
  CFC control, global convention, 384t
  greenhouse effect, 181, 182t, 183–4
  stratospheric ozone, 179, 180t
CITES *see* Convention on International Trade in Endangered Species
climate
  fossil evidence, 16
  morphoclimatic regions, 3, 3f
  and vegetation, 3–5
climate change
  big dams, 149
  carbon dioxide emissions, 33t, 193, 194, 194f
  carbon sequestration, 194
  developing countries, 195–6, 195t
  direct atmospheric inputs, 179, 179t
  economic growth and energy consumption, decoupling of, 194
  global convention, 384t
  hydrological cycle, modifications to, 181
  ice ages, 177–8
  international equity issues, 192
  Kyoto Protocol, 193–4, 193t, 195
  land surface modifications, 179–81, 179t
  national climate change strategies, 195
  ozone layer-depleting substances, reduction of, 193
  policy responses
  solar radiation, periodic variations in, 177–8
  see also global warming
coastal pollution
  conservative and non-conservative types, 107, 108t
  heavy metals and organochlorine compounds, 110
  nutrient flows, increased, 94t
    algal blooms, 109
    coastal eutrophication, 108f, 109
    disease pathogens, 109
    effects of, 109
    human activities causing, 107–8
    nitrogen and phosphorous, 107, 108f
    North American seaboard, 108–9, 108f
    oxygen depletion, 109
    reduced diversity, 109
  oil spills, 110
  urban and industrial activities, 109–10
coastal zone
  characteristics of, 103
  coastal material, removal of, 104, 105f
  dam construction, 106
  erosion and sedimentation, 103–4, 104f, 105f
  historical overfishing, 106–7, 107t
  human occupation of, 103
  management and conservation efforts, 118–20, 120t
  marine floods, risk from, 104–5
  river modification, downstream impact of, 105–6
  sea level variations, 106, 106f
Commission for the Conservation of Antarctic Marine Living Resources (CCAMLR), 90
commodity exchange, unequal, 26–7, 26f, 52
common ownership, of resources
  tragedy of the commons, 25, 28, 46, 82
  unregulated emissions, 286
  valuation of resources, 41
coniferous forests, 4, 4f
continuous resources, 20t
Convention Concerning the Protection of the World Cultural and Natural Heritage, 383t
Convention on Assistance in the Case of a Nuclear Accident, 384t
Convention on Biological Diversity, 248, 265, 270
Convention on Climate Change, 384t
Convention on Early Notification of a Nuclear Accident, 384t
Convention on International Trade in Endangered Species (CITES), 262, 265

Convention on Long-Range Transboundary Air Pollution (30% club), 209
Convention on Persistent Organic Pollutants, 384t
Convention on the Conservation of Antarctic Marine Living Resources, 383t
Convention on the Conservation of Migratory Species, 383t
Convention on the International Trade in Endangered Species of Wild Fauna and Flora, 383t
Convention on the Prevention of Marine Pollution by Dumping of Wastes and Other Matters, 383t
Convention on the Prohibition of Biological Weapons, 383t
Convention on Wetlands of International Importance Especially as Wildfowl Habitats *see* Ramsar Convention
coral reefs
  biological productivity and diversity of, 110–11
  chronic overfishing, 113
  coral-bleaching phenomenon, 114
  crown of thorns starfish, population explosion of, 114
  dredging and filling, 112f, 113
  economic knock-on effects, 113
  endemism, hotspots, 265
  Great Barrier Reef, pollution increases to, 113–14, 114t
  human impacts, 111t
  mining damage, 322
  siltation, 113
  sustainable management, need for, 114
Coriolis force, 362
cornucopian approach, 31–3, 32t
critical loads concept, 133, 202, 209–10, 209t
'cultural imperialism', 27f
cyclones *see* tropical cyclones

D
dams *see* big dams
debt-for-nature swaps, 61, 382, 385–6, 385t
deep ecology approach, 31–3, 39t
deforestation *see* tropical deforestation
desert, 5
desertification
  definition of, 68–70, 69f
  desert advance, rate of, 71–2, 71f
  dryland ecology, variability of, 69–70
  Dust Bowl disaster (US Great Plains), 82
  food production
    famine, 81
    food shortages, social causes of, 81
    land productivity, reduction in, 81
  global areas affected by, 70–2, 70t
  global convention, 384t
  human and natural factors, relative roles of, 69–70
  intensive grazing
    borehole sinking, 74
    'carrying capacity', 75
    pasture use, change in, 74
    piospheres, 74–5, 75f
    rainfall variability, 75
    traditional herders, 73–4
    unpalatable shrubs, invasion by, 72–3
    vegetative cover, decrease in, 72
  invasion and settlement, 68
  land degradation, root causes of, 72, 73t
  overcultivation
    examples and root causes of, 76t
    monocultures, 75–6
    nutrient depletion, 75, 76
    soil erosion, 75–6
  'progressive dessication' theory, 68
  Sahel drought catastrophe, 82–3
  salinisation
    crop yields, impact on, 79
    irrigation schemes, poor management of, 78t, 79–80
    off-site hazards resulting from, 79
    secondary salinisation, 78–9, 78t
    water tables rise, 81
  understanding, 70–1, 81–2
  vegetation, overexploitation of
    fuelwood collection, 76–8, 77t
    vegetation clearance, long history of, 76
DNA recombinant techniques, 226
drylands, 12
  see also desertification
Dust Bowl disaster (US Great Plains), 68, 82, 242

E
Earth Summit (Rio de Janeiro 1992), 39, 41, 193, 248, 377–8, 382
earthquakes
  big dams, 148–9
  building codes, 359f, 360
  causes of, 359
  costs of, 360
  ground motion, hazard from, 359
  intensity regulation techniques, hazards of, 360
  measurement of, 359
  Mexico City earthquake, 359
  prediction of, 359
  San Francisco earthquake, 358f, 359
  soil liquefaction, and landslides, 359–60
  tsunamis, 360
  urban environments, 172
ecocentrism, 31–3
'ecocide', 340
ecofeminism, 32t
ecoimperialism, 61
ecosystems, productivity of, 1–3, 2t
EIAs *see* Environmental Impact Assessments
El Niño/Southern Oscillation, 88, 365
EMEP *see* European Monitoring and Evaluation Programme
energy *see* renewable energy sources
energy efficiency
  cogeneration of heat and power (CHP), 301–2, 301f
  commercial space, 301
  domestic energy-saving options, 300, 300t
  electricity generation, 301–2
environmental change, human forces of, 20–1, 20f
Environmental Impact Assessments (EIAs), 155, 329
environmental issues
  complexity of, 374
  global nature of, 30–1, 373
  human society elements, 374–5

multi-pollutant/multi-effect approach, 375f
not recent phenomenon, 30
political acceptance of, delayed, 33, 33t
schools of thought on, 31–3, 32t
successful tackling of, 375–6
'systemic' and 'cumulative', 373
theories explaining, 24–5, 24t
websites on, 390t
Environmental Protection Agency (EPA), 126, 301
epidemics
causes of, 364
morbidity rates, developed *versus* developing countries, 364–6
water associated, 364, 364t
cholera, 365
malaria, 365–6
EU *see* European Union
European Monitoring and Evaluation Programme (EMEP), 201
European Union (EU)
environmental policy, 41
Habitats Directive, 137–8
extrinsic resources, 20t
Exxon Valdez, oil spill, 110, 386

F

factory farming, 217
famine, 81, 343
Farraka Barrage, Bangladesh, 150–2, 151f, 152t
feedback
climate change, 180–1
global warming, 186
timescales, 12, 14f
fertilisers
fixed nitrogen, global sources of, 218, 219t
groundwater levels, rise in, 219
inadvertent impacts, 23
National Environmental Policy Plan (Netherlands), 218
nitrate pollution, 218–19
nitrogen cycle, 9
patterns of use, 218, 218t
phosphates, 218
potash, 218
river pollution, 126–8
surface water pollution, 219
fish farming *see* aquaculture
fish mortality, 203–4
fisheries
demersal species, overfishing, 85
ENSO (El Niño/Southern Oscillation), 88
Exclusive Economic Zones, 87
fish population dynamics, types of, 89
global catches, sustainable levels of, 85–6
herring, overfishing, 87
management failings
conservation, 87
fishing subsidies, 87
free and open access, 87
national and international controls, increased, 87
ownership issues, 85
population numbers, fluctuating, 88–9, 89f
shoaling pelagic species, 85, 88–9
floods
Bangladesh delta
alleviating causes, 370, 370f
beneficial *versus* harmful floods, 369–70, 369f
embankment building controversy, 369–70
causes of, 366
coastal, 366
human responses, 366–71, 367t
Mississippi flooding
channel straightening, 366–7
devastation caused by, 367, 368f, 369
prediction of, 366
river channel, 366
Thames Barrier (London), 370–1
food chain, 7–8, 9f
food production *see* agriculture; aquaculture; biotechnology; fertilisers; pesticides
food shortages, social causes of, 81
fossil evidence, 19
fossil fuels, 181, 183
fossil carbon, 6–7
non-renewable resource, 22, 300
renewable alternatives to, 194, 194f, 309
see also carbon cycle
fresh water
freshwater fish, main pressures facing, 122, 123t
global water withdrawals, 122
'hydraulic civilisations', 122
see also lakes; river management; river pollution; water pollution; wetlands

G

Gaia hypothesis, 11, 32t
GCMs (general circulation models), 185–6
GEMS *see* Global Environmental Monitoring System
General Agreement on Tariffs and Trade (GATT), 27
genetic engineering *see* biotechnology
Geneva Convention protocols, 346
geothermal energy
air pollution, 307–8
hydrothermal systems, 307
success of, 308
giant panda, threat to, 253–4, 253t
glacial-interglacial cycles *see* ice ages
Global Assessment of Human-Induced Soil Degradation (GLASOD), 235, 235t
global capitalism, 27
Global Environmental Monitoring System (GEMS), 125, 126
global warming
agricultural land use patterns, 187–8
discontinuous pattern of, 185
flooding, 369
GCMs, 185–6
geomorphological processes, 189–90, 189t
glacial and periglacial processes, 177–8, 188–9, 188f
growing seasons, increased, 187
malaria incidence, 366
permafrost line, northern movement of, 188–9, 188f

- plant and animal communities, 186–7, 187t
- sea-ice frequency, 189
- sea level rises
  - coastal regions, consequences for, 190–1
  - cyclones, increased frequency of, 192
  - flooding and inundation, increased, 191
  - low-lying countries, threat to, 191–2
  - small islands, threat to, 191, 191t
  - Venice, threat to, 190f, 191
- socio-economic systems, disruption to, 186
- soil and water conservation measures, 190
- soil moisture loss, 189
- water availability, decreased, 188
- world cereal production, decrease in, 187–8
- *see also* climate change; greenhouse gases; stratospheric ozone

glossary, 391–7
GM foods *see* biotechnology
greenhouse gases
- big dams, 149
- carbon dioxide emissions, 181, 182t, 183, 183f
- CFCs, 181, 182t, 183–4
- concentrations of, 177
- fertiliser use, nitrates pollution, 219
- global warming, 184–5
- methane emissions, 182t, 183, 184t
- Montreal Protocol, 184, 192
- natural occurrence of, 181
- political acceptance of, 33t
- *see also* carbon cycle; stratospheric ozone

Greenland National Park, proposed for first World Park, 258
Greenpeace, 92–3
Gross Domestic Product (GDP), 44
Gross National Product (GNP), 44
groundwater pollution
- major pollutants, 164, 164t
- sea-water intrusion, 165
- water table lowering, and subsidence, 164–5, 165f
- water table rise, 165–6, 166t

H

heath fritillary, decline of (UK), 261f, 262
herbivores, 8, 9f
hotspots
- acid rain, 200
- endemic species concentrations, 264–5

human health
- acid rain
  - respiratory system, effects on, 206–7
  - toxic heavy metals, effects of, 207
- big dams
  - malaria, increase in, 147
  - river blindness, decrease in, 146
  - schistosomiasis, increase in, 147
- Chernobyl disaster, 313
- coastal pollution, 109
- diseases, dispersion of, 272–3
- heavy metal pollution, 325
- itinerant miners, and spread of malaria, 327
- landfill sites, 287–8
- ozone depletion, 179
- pesticides, 223
- urban environments, 166, 169, 173, 173t
- *see also* epidemics

hunting and trade bans, 255–6, 262–3, 263f
hurricanes *see* tropical cyclones
hydrogen chloride conversion, 290
hydrological cycle, 6, 7f, 9, 181
- urban environments, 161–2, 162t
- wetlands, 134, 135t

hydropower, 142, 302–3
- *see also* big dams

hydrosphere, 1

I

IAEA *see* International Atomic Energy Agency
ice ages, 177–8, 185, 188–9, 188f
Industrial Revolution, 22, 23, 125, 317, 387
institutional reform school, 32t
International Atomic Energy Agency (IAEA), 313
International Boundary and Water Commission, 156
International Convention for the Prevention of Pollution from Ships (Protocol of 1978), 383t
International Panel on Climate Change (IPPC), 186
International Tropical Timber Organization (ITTO), 64
International Union for Conservation of Nature and Natural Resources *see* IUCN
I-PAT equation, 44–7
irrigation
- big dams, 142
- diversion of rivers for, 130
- dryland countries, 219, 220f
- global area of irrigated cropland, rise in, 219, 220f
- poor management
  - overpumping, 220–1, 221f
  - pesticides, 221
  - salinisation, 78t, 79–80
  - water tables, rising, 219
- wartime destruction, 340
- wastewater recycling, 294–5, 295t

IUCN, Red List categories, 155, 250t, 251, 255, 378
ivory trade ban, 255–6, 262

K

Keynsianism, 32t
Kyoto Protocol, 193–4, 193t, 195

L

lakes
- lacustrine degradation
  - hydrological parameters, 132t
  - irrigation, diversion of rivers for, 130
  - 'offtakes', 130–2
  - eutrophication of standing water bodies, 133
  - nutrients, 133, 133f
  - precautionary principle, rejection of, 132
  - species introductions, 132, 133
  - technocentric approach, dangers of, 132
- management
  - critical loads approach, 133–4

ecosystem approach, 133
landmines, unexploded, 342, 342t
landslides
mining, 323
soil erosion, 239
soil liquefaction, 359–60
large blue butterfly, decline of (UK), 261–2, 261f
Law of the Sea, global convention, 384t
lead pollution, 166, 169, 274–5, 275t
lichens, loss of
acid rain, 204–5
air pollution, 255
lifecycle analysis, 41
Limits to Growth school, 32t
lithosphere, 1
logging industry
concession periods, lengthening, 63
illegal logging, difficulty stopping, 60
initiatives, 64
low-impact techniques, 62–3
and wastage, 63, 63f
see also tropical deforestation

M

Malthusian ideas, of overpopulation, 24t, 47
managerialists and technical fixers approach, 32t
mangrove forest destruction
agriculture, 116, 116f
big dams, 152
cost-benefit analyses for conversion of, 118
fish farming, 230, 230f
fish ponds, 116
fuelwood, 115–16, 116t
major causes of, 115, 115t
replanting programme, 114–18
role of, 115
shrimp aquaculture, 117f, 118
sustainable resource management, 119, 120t
systematic (Viet Nam War), 340
maquis, 5
market-based approaches, 32t
Marxism, orthodox, 32t
Mayan civilisation, collapse of, 36, 37f, 46
medicinal plants, protection and use of, 64–5
Mediterranean sea, action plan for, 101
megacities
air quality in, 168–9, 168f
growth of, 159, 160f
lead content of petrol in, 275t
see also urban environments
methane emissions, 14, 14f, 182t, 183, 184t
migratory species, global convention, 383t
mining
elements and compounds mined, 317
environmental impacts, 320t
geomorphological impacts
landslides, 323
local/national scale, 323, 324f
subsidence, 323, 324–5, 324t
global economic aspects
minerals, international trade in, 318–19, 319t
reserves, global consumption of, 317–19, 318t
habitat destruction
coastal zone damage, 322–3, 323f
polymetallic nodules, (deep-sea bed mining), 322–3
smelting, and deforestation for fuel, 322
strip mining sites, 321, 322, 322f
waste dumping, 321–2
itinerant miners, and spread of malaria, 327
true costs of
environmental costs *versus* societal benefits, 331
'full-cost pricing' concept, 331
local populations, costs to, 331, 332f
sustainable practices, 331
mining damage, rehabilitation and reduction
legislative controls, 329–30
mining wastes, use of, 328, 329t
pollutants, use of, 328, 329t
precautionary principle, 329–30
waste by-products, use of, 328
waste dumps, stabilisation of, 328
mining pollution, 325
dust pollution, 326
sulphur dioxide emissions, 326
water pollution, 325–6
Montreal Protocol, 184, 192
motor vehicle pollution
alternative fuels
diesel, 277
electrically driven vehicles, 278
ethanol, 277
hydrogen, 278
LPG (liquefied petroleum gas), 277
methanol, 277
catalytic converters, 23, 276–7, 277
lead-free petrol, 274, 276, 276f
lead pollution, 274–5, 275t
megacities, lead content of petrol in, 275t

N

NAFTA *see* North American Free Trade Agreement
national strategies
environmental indicators, 378–9
National Conservation Strategies, 378
National Environmental Action Plans, 378
national strategy for achieving sustainable development (Netherlands), 378t
'command and control' approach to clean-up, 379–80
external factors, 381, 381f
National Environmental Policy Plans (NEPPs), 380–1, 380f
natural hazards
classifications of, 353–4
areal exent, 353–4
disasters, and composite hazards, 353
hazard types, 353–4, 353t
'natural' *versus* human-induced events, 353
physical geographical approaches, 352
rapid-onset *versus* slow-onset, 353

spatial, 352
hazardous areas
disaster impact, and location and preparation variables, 356
Disaster Proneness Index, of countries, 357–8, 357t
global natural catastrophes, 356, 356f
hazard risk, global increase in, 356
UN Disaster Relief Organisation (UNDRO), 357
responses to
acceptance or reduction of losses, 354, 355t
hazard events, prevention or modification of, 354, 355t
loss acceptance, 354–5
risk assessment, 355–6
timescales, 351–2
natural resources, 19–20
natural systems, sources of knowledge
fossil evidence, 17
historical archives, 17
instrumental data, 15, 16t, 17
natural archives, 17
palaeoenvironmental indicators, 17
proxy indicators, 16t, 17
negative resources, 19
Nile, River
delta changes, 150–2, 151f
flow management, 142, 142f
nitrate pollution, 107, 108f, 126–8, 218–19
nitrogen cycle, 9
nitrogen oxide emissions, 199, 200, 206, 208, 210t, 211, 273, 277t
noise pollution, 278–9, 280f
non-governmental organisations (NGOs), 33
non-renewable resources, 20t, 300
see also fossil fuels
North American Free Trade Agreement (NAFTA), 27
nuclear accidents
global conventions, 384t
see also Chernobyl disaster
nuclear power
environmental and safety concerns, 308–9
future for, 314–15
radioactivity, dangers from, 310
uranium, reuse of, 310
world electricity generation, 309–10, 310f
nuclear war, destruction caused by, 342–3, 343t
nuclear waste
dilute and disperse philosophy, 310
high-level waste, 310–11, 311t
intermediate-level waste, 310, 311t
low-level waste, 310, 311t
nutrient flows, increased, 94t, 98–9, 107–9, 108f
see also nitrate pollution

O

ocean pollution
Black Sea
Convention for the Protection of the Black Sea, 100
dam construction, 99–100, 100t
eutrophication, 98–9
fishing industry, collapse of, 100
macrobenthic species, decline of, 99, 100f
coastal waters, concentration in, 95f
contaminants, disproportionate effect of, 95–6
ocean dumping, global convention, 383t
oil spills, numbers and effects of, 86t, 96–7, 96f
open ocean, relative cleanliness of, 95
pollutants, persistance of, 95f
POPs, 98, 99f
primary causes and effects, 94t
regional seas, 100–1, 101t
TBT (tributyltin), effects of, 97
waste dumping controls
dilute-and-disperse philosophy, 98
global convention, 383t
industrial waste and sewage sludge, 97
radioactive waste, 98
waste incineration at sea, 97
oil spills
coastal pollution, 110
numbers and effects of, 86t, 96–7, 96f
ozone *see* stratospheric ozone

P

peatlands, conservation of, 137–8
permafrost line, northern movement of, 188–9, 188f
permafrost pollution, 163–4, 163f
pesticides
agricultural areas, pollution of large, 223
alternatives to
behavioural control, 224
biological control, 224–5, 225f
environmental control, 224
genetic and sterile male techniques, 224
integrated pest management, 224
resistance breeding, 224
ban on toxic and persistent, 223–4
bioaccumulation of, 222–3
DDT, 222
first uses of, 221–2
groundwater contamination, 223
human health, dangers to, 223
inadvertent impacts of, 23
in marine environment, 98
natural predators, destroyed by, 222
non-target species, effect on, 222
poisonings, 223
resistance, problem of, 222
secondary populations, problem of, 222
worldwide consumption of, 222
photovoltaic (PV) systems, 304, 304f
physical environment, human perspective on, 19–20
piospheres, 74–5, 75f
plant respiration, 7
POPs (persistent organic pollutants), 98, 99f, 384t
population, human
and environmental degradation, 21–2, 45
exponential rise in, 21, 21t
Malthusian perspective, 21
population density/relative affluence regional

classification, 376–7
and technology (I=PAT equation)
affluence, 44–5
carrying capacity, 45–7, 45f
population, 47
poverty, 45
throughput control, 44–5
precautionary principle, 41, 132, 193–4, 193t, 227, 329–30
Protocol on Substances that Deplete the Ozone Layer, 384t

R

Ramsar Convention, 134, 138–9, 139t, 265, 382, 383t
recycling *see* waste reuse
refrigerants, 183–4
refugees, 343–6, 344f, 345f
regional seas
action plans for, 101, 101t
international cooperation, 100, 101
renewable energy sources
benefits and problems of, 302
future for
deployment, barriers to, 308, 309t
environmental impacts, minimising, 308–9
fossil fuels, gradual replacement of, 309
political will, need for, 308
global energy use, 302
see also biomass; hydropower; solar power; tidal power; wind energy
renewable resources, 20t, 44t
Richter scale, 359
Rio Declaration *see* Earth Summit
river management
'catchment management plans', 130
damage, making good, 130
multiple impacts
deforestation, 125, 125f
exotic species, introduction of, 125, 126
fish extinctions, 124–5
overfishing, 125–6
regulation, conservationists against, 129
reservoirs, and wildlife conservation, 129
river modification methods, 128, 129t
salmon rehabilitation programmes, 129–30
SSSIs, 129
river pollution, 160–2, 162t
black-spots, 127
chemical fertilisers, and nitrate concentrations, 126–8, 127f
fertilisers, 126–8
pesticide contamination, 126
river quality improvements, 127–8, 127t, 128f
sewage, 126
Rossby wave propagation, 58

S

Sahel catastrophe, 76, 83, 180–1
ship pollution, global convention, 383t
Sites of Special Scientific Interest (SSSIs), 129, 269, 335
sociobiology and authoritarian ecology school, 32t
sociocultural imbalances, 20–1, 20f
cities, environmental impact of, 23
economic power, 23–4
global inequalities, 25f
commodity exchange, 26–7, 26f
consumption patterns, 26
'debt crisis', 26
economic dimension of, 26
'North'/'South' divide, 26
polluting industries, moving to less strictly regulated areas, 27, 27f
resource exploitation, 26–7
Transnational Corporations (TNCs), 27
intergenerational inequalities, 29–30, 29f
national inequalities, 28f
environmental refugees, 29
land tenure, 28
poor and disadvantaged, as victims and agents of degradation, 28–9
rural/urban divide, 28
tragedy of the commons, 28
ownership and value, 25
political power, 24
time and space scales, 30
soil conservation
agronomic measures
crop rotation, 244
mulching, 244
beneficial effects of, 245, 246t
implementation, 246–7
mechanical methods, 244, 245
soil management techniques, 244, 245
soil erosion
accelerated
agricultural techniques, inappropriate, 242–3
bush and grass burning, 240
construction activities, 240
deforestation for agriculture, 240–1, 241f
intensive arable farming, 243–4
landslides, 239
natural erosion, 238
surface destabilisation, 239
system collapse, 243, 243t
vegetation modification/removal, 239
wind erosion, 241–2
effects of
deposition, 235t
entrainment, 235t
off-site effects, 237–8, 238f
on-site effects, 236–7, 236f, 237t
transport, 235t
factors affecting, 232–3, 233f
measurement of
controlled experiments, 234
field measurements, 233–4
Global Assessment of Human-Induced Soil Degradation (GLASOD), 235, 235t
soil erosion prediction models, 234–5, 234f
soils, importance of, 232
tropical deforestation, 57–8
solar power

food chain, 7–8
government support, need for, 308
passive and active use, 304
photovoltaic (PV) systems, 304, 304f
solar thermal electric generator, 304, 305f
Somalia, herders forced into smaller ranges in, 74
Southern Oceans
global convention, 383t
species depletion in, 90
spatial scales, 13–14
see also timescales
Spitzbergen Treaty (1920), 347
SSSIs *see* Sites of Special Scientific Interest
stratospheric ozone
history of understanding and interaction with, 179, 180t
key role, 179
ozone depletion
aircraft emissions, 273
fertiliser use, nitrates pollution, 219
global convention, 384t
health, impact on, 179
political acceptance of, 33t
reduction of, 193
ozone pollution, 166, 169, 170, 190
motor vehicles, 276, 276fe
reduction of, 281–2, 282f
ultraviolet radiation, increased, 179
see also chlorofluorocarbons
sulphur dioxide emissions
acid rain, 199, 200, 204–5, 206, 208, 210, 210f, 210t
global distribution of, 199–200
motor vehicle pollution, 169–70, 169f, 277t
sources of sulphur deposits, 201–2, 202t
sulphur emission protocol, 209–10, 209t
Sumerian civilisation, collapse of, 68
suspended particulate matter (SPM), 166, 169, 273
sustainable development
definition of, 38–40
and economic growth, 43–4, 44t, 386–7
environmental resources, valuing
direct values, 42t, 43
environmental economics, 41
environmental pricing, 41
environmental value, perception of, 43
indirect values, 42t, 43
willingness to pay, 41–2
future of
human domination of nature, 387–8
institutional change, 388
socio-economic systems, changes in, 386–7, 386f
global conventions, 382, 383–4t
guiding principles of, 39–40, 39t
precautionary principle, 41
socio-economic system, and global ecosystem
environmental services, of ecosystem, 35
global biosphere, finite resources, 35–6, 36f
modes of development and environment, 3-cycle model of, 37–8, 37f
potential for collapse of, 36
urban environments, 174–5

T

taiga, 4, 4f
technocentrism, 31
technology, 22–3
temperate forests, 4
temperate grassland, 5, 5f
tidal power, 306–7, 307t
timescales
changes, typical patterns of, 13, 13f
dynamic equilibrium, 11–12, 13
events, factors influencing, 10–11
feedbacks, 12, 14f
geological, 10, 10t, 17
non-equilibrium environments, 12
resiliance, 12, 13
resistance, 12
and space scales, 14–15, 15f
thresholds, 12, 13
timelag, 12
understanding affected by, 11–12
tragedy of the commons, 25, 28, 46, 82
transport
atmospheric pollution *see* aircraft emissions; motor vehicle pollution
biospheric changes
alien species, introduction of, 270–2, 271t, 272f
disease dispersal, 272–3
species dispersal, 270
hydrological effects, 267t
land, 267t
landscape despoilation, 269
resource consumption, 269
transport routes, hazards to, 269–70, 270f
noise
major transport corridors as sources of, 279, 280f
noise levels, mitigation of, 279
sound levels, measuring acceptable, 278–9
policy
behaviour, difficulty changing, 280–2, 282f
policy approaches, 282–3, 282t
resource use by type of, 268t
tropical cyclones
conditions for, 362
damage caused by, 362
detecting and monitoring, 363
economic costs of, 362–2, 363t
impact of, 363–4
increased frequency of, 192
tropical deforestation
causes of
agriculture, 52–3, 53f, 55, 61
cattle ranching, 54
government policy and practice, 51, 52f
hydroelectric dam construction, 54
international economics, 51–2, 55, 61
mining, 54–5
by region, 51, 51t
resettlement programmes, 54
slash and burn agriculture, 54, 54f
war, 51, 52f

consequences of
climatic impacts, 58
diversity, loss of, 58–9
fires, effects of, 59–60
forest landscape, dynamic equilibrium state, 55
fragmentation, and loss of large trees, 59
hydrological cycle, 56–7, 58
indigenous peoples, effects on, 59f, 60
natural perturbations, 55
soil degradation, 57–8
deforestation, FAO definition of, 49–50
'open' and 'closed' forest, 49
political acceptance of, 33t
primary and secondary forests, distinction between, 55–6
rates, 49–50, 50f
tropical moist forest, 50
tropical forest management
clearance proposals, evaluation of, 61
debt-for-nature swaps, 61
economic policy, addressing underlying, 60–1
financial approach, changing, 61
legislation, ineffectiveness of, 60
logging industry, 60, 62–4, 63, 63f
medicinal plants, protection and use of, 64–5
natural ecosystem processes, mimicking, 62
non-timber forest products, low-intensity harvesting of, 65–6, 65t
protected areas, increasing, 62
silvicultural rules, introduction of, 62, 62f
small-scale forest dwellers, increasing control for, 66
tropical savannah, 4
tsunamis, 360
tundra, 4, 14f
typhoons *see* tropical cyclones

U

ultraviolet radiation, 212
UN Conference on Desertification (UNCOD), 68, 70, 81
UN Conference on Environment and Development (UNCED) *see* Earth Summit
UN Convention on Biological Diversity, 384t
UN Convention on Law of the Sea, 382, 384t
UN Convention to Combat Desertification, 69, 81, 384t
UN Disaster Relief Organisation (UNDRO), 357
UN Environment Programme (UNEP), 70–1, 70t, 72, 297
UN Framework Convention on Climate Change (UNFCCC), 193, 195
UNCED *see* Earth Summit
UNCOD *see* UN Conference on Desertification
UNDRO *see* UN Disaster Relief Organisation
UNDRS *see* UN Disaster Relief Organisation
UNEP *see* UN Environment Programme
UNFCCC *see* UN Framework Convention on Climate Change
United Nations International Decade for Disaster Reduction (1990s), 352
Universal Soil Loss Equation (USLE), 234–5
uranium, reuse of, 22, 310
urban environments
atmospheric pollutants, 167t
air quality monitoring programmes, 168–70, 168f, 169f, 170f
megacities, air quality in, 168–9, 168f
natural and built environments, damage to, 166, 168
suspended particulate matter (SPM), 166, 169
congestion costs, 160, 161t
developing countries
disease, 173, 173t
earthquakes, 172
industrial accidents, 172
infant mortality, 172
poor housing, 172
water supply, 173–4, 173t
ecological footprints, 159–61
garbage
hazards from uncollected, 170
recycling, 171
scavenging species, 170–1, 171f
growth rates, 159
surface water resources
'flashy' discharge regimes, 161
habitat and ecology, 162t
hydrological cycle, 161–2, 162t
morphology, 162t
permafrost pollution, 163–4, 163f
river pollution, 160–2, 162t
water quality, 161, 162t
water shortages, 163
sustainability, measures needed to improve, 174–5
measures needed to improve, 174–5, 174f
see also groundwater pollution; motor vehicle pollution
USLE *see* Universal Soil Loss Equation (USLE)

V

Vienna Convention for the Protection of the Ozone Layer, 384
volcanic eruptions
'active', 'dormant', 'extinct' volcanoes, 360–1
hazards associated with, 354, 355t, 360–2, 360t, 361f

W

war
demilitarised zones, as sanctuaries, 336, 338, 349
environmental management, neglect of, 343
environmental modification techniques, 337–8t
defoliant chemical use, 340
'ecocide', 340
rain-making programmes, 339–40
'scorched earth' policy, 339
vegetation, systematic destruction of, 339f, 340
famine, 343
infrastructure, destruction of
agricultural, 340
dam breaching, 340
industrial and municipal plants, 341
irrigation systems, 340
military vehicles, largescale movements of, 341
oilfields, damage to, 340–1
surface terrain, 341, 341f

landmines, unexploded, 342, 342t
limiting effects of, 346
Åland Convention (1921), 347
Antarctic Treaty (1959), 347
Geneva Convention protocols, 346
multilateral arms control agreements, 346, 347t
Spitzbergen Treaty (1920), 347
military preparation, costs of
biological weapons testing, 336, 338
hazardous wastes, generation and disposal of, 335–6
military spending, global scale of, 334
nuclear testing hazards, 336
nuclear weapons-producing plants, 336
resources, consumption of, 335, 335t
nuclear war, destruction caused by, 342–3, 343t
refugees, 343–6, 344f, 345f
renewable resource scarcity, 347–9, 349f
wildlife, impact on, 341–2, 346
waste
cleaner production
'closed-loop' processing cycles, 297
economic incentives, 297
economic sustainability and economic efficiency, 297
industrial examples of, 296t, 297
lifecycle analysis, 297
sources of, 285, 285f
urban environments, 170–1, 171f
waste cycles, 284
waste disposal
common ownership, and unregulated emissions, 286
hazardous waste, international movement of, 289–90
incineration, 288–9
landfill sites
cleaning up programmes, 287–8
health effects, 287–8
leakage of toxic leachates, 286–7
open to the air, 286
typical designs, 286, 287f
as sustainable option, 288
treatment, emission with or without, 285–6
waste reuse
glass recycling, 293–4
metal recycling, 293, 293f
recycling schemes, 291, 292t
scavenged resources, 291
sewage sludge, 294, 295
steel slags, recycling of, 294
tax on packaging, 292
treatment techniques, EU hierarchy of, 290, 290t
waste, as resource, 290–1
waste taxation schemes, 292–3
wastewater recycling, 294–5, 295t
water pollution
biochemical oxygen demand (BOD), 123–4
Global Environmental Monitoring System (GEMS), 125, 126
heavy metals, bioaccumulation, 124
industrial pollution, 125
micro organisms, 125
mining, 325–6
nutrient pollution, 125
sewage, 123–4
thermal pollution, 123–4
see also groundwater pollution; ocean pollution; river pollution
WCED *see* World Commission on Environment and Development
websites
international, 390t
national, 389t
WEC *see* World Energy Council (WEC)
wetlands
definition of (Ramsar Convention), 134
destruction of
conversion to cropland, 135
drainage, 135, 136f
reclamation, reasons for, 135–8
global convention, 383t
key natural functions of
habitat, 135, 135t
hydrological cycle, 134, 135t
socio-economic, 135, 135t
water quality, 134–5, 135t
protection, 138–9
Ramsar Convention 'wise use' guidelines, 138–9, 139t
whale protection
commercial whale-watching, 93–5, 93f
Convention for the Regulation of Whaling, 92
Greenpeace boycott campaigns, 92–3
International Whaling Commission, 92
overexploitation, 91–2, 256
Southern Ocean Whale Sanctuary, 93
US fisheries law, amendments to, 92
WHO *see* World Health Organisation
wind energy
wind-derived electricity, 303
wind farms, 303–4
wind turbines, 303
windmills, 303
World Bank, 60, 155, 378
World Commission on Dams, 155–6
World Commission on Environment and Development (WCED), 39
World Conservation Union *see* IUCN
World Energy Council (WEC), 300
World Health Organisation (WHO)
air pollution, 169, 170
International Drinking Water and Sanitation Decade, 173
world heritage, global convention, 383t